Advances in Experimental Medicine and Biology

Brian C.J. Moore • Roy D. Patterson
Ian M. Winter • Robert P. Carlyon
Hedwig E. Gockel
Editors

Basic Aspects of Hearing

Physiology and Perception

Springer

Editors
Brian C.J. Moore
Department of Experimental Psychology
University of Cambridge
Cambridge
United Kingdom

Roy D. Patterson
Physiology, Development and Neuroscience
University of Cambridge
Cambridge
United Kingdom

Ian M. Winter
Physiology, Development and Neuroscience
University of Cambridge
Cambridge
United Kingdom

Robert P. Carlyon
MRC-Cognition and Brain Sciences Unit
Cambridge
United Kingdom

Hedwig E. Gockel
MRC-Cognition and Brain Sciences Unit
Cambridge
United Kingdom

ISBN 978-1-4614-1589-3 ISBN 978-1-4614-1590-9 (eBook)
DOI 10.1007/978-1-4614-1590-9
Springer New York Heidelberg Dordrecht London

Library of Congress Control Number: 2013939625

Printed on acid-free paper

Springer is part of Springer Science+Business Media (www.springer.com)

Preface

This volume constitutes the Proceedings of the 16th International Symposium on Hearing (ISH), held from 23 to 27 July 2012 in St. John's College, Cambridge, UK. As is traditional for the ISH series, the emphasis was on bringing together those performing basic research on physiological and perceptual processing by the auditory system, including modelling. All chapters were submitted and subjected to preliminary editing prior to the meeting, and all chapters were available to participants before the meeting started. The timetable of ISH 2012 allowed plenty of time for discussion, and synopses of some of the discussions are included at the ends of the relevant chapters. The chapters are organised according to seven broad themes, and their order reflects the order of presentation at the meeting.

We are most grateful to those who sponsored ISH 2012 with goods or cash. These were (in alphabetical order) GNResound (Denmark), MED-EL (Austria), The Eriksholm Research Centre (part of Oticon, Denmark), Phonak (Switzerland), Starkey (USA), St. John's College Cambridge (UK) and Widex (Denmark). We are also grateful to Brian Glasberg, Cathy Schneider, Shirley Bidgood, Jackie Clark, Etienne Gaudrain, Andrew Kolarik, Arek Stasiak, Sami Alsindi, Marina Salorio-Corbetto and Sara Madsen for their help with various aspects of the running of ISH 2012 and to Eleanor Turner for a superb harp concert.

Finally, we would like to thank all authors and participants for their scientific contributions and for the lively discussions.

Cambridge, UK	Brian C.J. Moore
Cambridge, UK	Roy D. Patterson
Cambridge, UK	Ian M. Winter
Cambridge, UK	Robert P. Carlyon
Cambridge, UK	Hedwig E. Gockel

List of Previous ISH Symposia

Previous meetings in the ISH series and their respective books are:

ISH 2009 – Salamanca, Spain: The Neurophysiological Bases of Auditory Perception. Edited by E.A. Lopez-Poveda, A.R. Palmer, and R. Meddis. Springer: New York.

ISH 2006 – Cloppenburg, Germany: Hearing - From Sensory Processing to Perception. Edited by B. Kollmeier, G. Klump, V. Hohmann, U. Langemann, M Mauermann, S. Uppenkamp, and J. Verhey. Springer: New York.

ISH 2003 – Dourdan, France: Auditory Signal Processing: Physiology, Psychoacoustics, and Models. Edited by D. Pressnitzer, A. de Cheveigné, S. McAdams, and L. Collet. Springer: New York.

ISH 2000 – Mierlo, The Netherlands: Physiological and Psychophysical Bases of Auditory Function. Edited by D.J. Breebaart, A.J.M. Houtsma, A. Kohlrausch, V.F. Prijs, and R. Schoonhoven. Shaker: Maastricht.

ISH 1997 – Grantham, England: Psychophysical and Physiological Advances in Hearing. Edited by A.R. Palmer, A. Rees, A.Q. Summerfield, and R. Meddis. Whurr: London.

ISH 1994 – Irsee, Germany: Advances in Hearing Research. Edited by G. A. Manley, G. M. Klump, C. Koppl, H. Fastl, and H. Oeckinghaus. World Scientific: Singapore.

ISH 1991 – Carcans, France: Auditory Physiology and Perception. Edited by Y. Cazals, L. Demany and K. Horner. Pergamon: Oxford.

ISH 1988 – Paterswolde, Netherlands: Basic Issues in Hearing. Edited by H. Duifhuis, and J.W. Horst, H.P. Wit. Academic: London.

ISH 1986 – Cambridge, England: Auditory Frequency Selectivity. Edited by B.C.J. Moore and R.D. Patterson. Plenum: New York.

ISH 1983 – Bad Nauheim, Germany: Hearing - Physiological Bases and Psychophysics. Edited by R. Klinke, and R. Hartmann. Springer: Berlin.

ISH 1980 – Noordwijkerhout, The Netherlands: Psychophysical, Physiological and Behavioural Studies in Hearing. Edited by G. van der Brink, and F.A. Bilsen. Delft University Press: Delft.

ISH 1977 – Keele, England: Psychophysics and Physiology of Hearing. Edited by E.F. Evans, and J.P. Wilson. Academic: London.

ISH 1974 – Tutzing, Germany: Facts and Models in Hearing. Edited by E. Zwicker and E. Terhardt. Springer: Berlin.

ISH 1972 – Eindhoven, The Netherlands: Hearing Theory. Edited by B.L. Cardozo. IPO: Eindhoven.

ISH 1969 – Driebergen, The Netherlands: Frequency Analysis and Periodicity Detection in Hearing. Edited by R. Plomp and G.F. Smoorenburg. Sijthoff: Leiden.

Contents

Contributors

Kristina S. Abrams Departments of Biomedical Engineering and Neurobiology & Anatomy, University of Rochester, Rochester, NY, USA

Eisuke Adachi NTT Communication Science Laboratories, NTT Corporation, Atsugi, Kanagawa, Japan

Enzo Aguilar Instituto de Neurociencias de Castilla y León, Universidad de Salamanca, Salamanca, Spain

Instituto de Investigaciones Biomédicas de Salamanca, Universidad de Salamanca, Salamanca, Spain

Trevor R. Agus Département d'études Cognitives, Equipe Audition, Ecole Normale Supérieure, Paris, France

Laboratoire de Psychologie de la Perception (UMR CNRS 8158), Université Paris Descartes, Paris, France

Michael A. Akeroyd MRC Institute of Hearing Research (Scottish Section), Glasgow Royal Infirmary, Glasgow, UK

S. Akram Institute for Systems Research, University of Maryland, College Park, MD, USA

J. Anemüller Medizinische Physik, Universität Oldenburg, Oldenburg, Germany

Frédéric Apoux Department of Speech and Hearing Science, The Ohio State University, Columbus, OH, USA

Go Ashida, PhD Department of Biology, University of Maryland, College Park, MD, USA

Peter J. Bailey Department of Psychology, University of York, Heslington, UK

Leslie R. Bernstein, PhD Department of Neuroscience and Surgery (Otolaryngology), University of Connecticut Health Center, Farmington, CT, USA

Rainer Beutelmann Department of Animal Physiology and Behaviour Group, IBU, Carl von Ossietzky University Oldenburg, Oldenburg, Germany

Hari Bharadwaj, MS Department of Biomedical Engineering, Boston University Center for Computational Neuroscience and Neural Technology, Boston, MA, USA

Federica Bianchi Department of Electrical Engineering, Center for Applied Hearing Research, Technical University of Denmark, Lyngby, Denmark

Christopher A. Brown, PhD Department of Communication Sciences and Disorders, University of Pittsburgh, Pittsburgh, PA, USA

G.J. Brown Department of Computer Science, University of Sheffield, Sheffield, UK

Emily Buss Department of Otolaryngology/Head and Neck Surgery, University of North Carolina School of Medicine, Chapel Hill, NC, USA

Andrew J. Byrne Department of Psychology, University of Minnesota, Minneapolis, MN, USA

Samuele Carcagno, PhD Institut de Neurosciences Cognitives et Intégratives d'Aquitaine (UMR CNRS 5287), Université de Bordeaux, Bordeaux, France

Robert P. Carlyon Medical Research Council, Cognition and Brain Sciences Unit, Cambridge, UK

Laurel H. Carney Departments of Biomedical Engineering and Neurobiology & Anatomy, University of Rochester, Rochester, NY, USA

Catherine Carr, PhD Department of Biology, University of Maryland, College Park, MD, USA

Maria Chait UCL Ear Institute, London, UK

C. Chambers Equipe Audition, Ecole Normale Superieure, Paris, France

Yoojin Chung Eaton-Peabody Laboratories, Massachusetts Eye and Ear Infirmary, Boston, MA, USA

Department of Otology and Laryngology, Harvard Medical School, Boston, MA, USA

N.R. Clark Department of Psychology, University of Essex, Colchester, UK

Francisco Cervantes Constantino Department of ECE, University of Maryland, College Park, MD, USA

John F. Culling, BSc, DPhil School of Psychology, Cardiff University, Park Place, Cardiff, UK

Torsten Dau Department of Electrical Engineering, Centre for Applied Hearing Research, Technical University of Denmark, Lyngby, Denmark

S.V. David Institute for Systems Research, University of Maryland, College Park, MD, USA

Oregon Hearing Research Center, Oregon Health & Science University, Portland, OR, USA

Mitchell L. Day, PhD Eaton-Peabody Laboratories, Massachusetts Eye and Ear Infirmary, Boston, MA, USA

Department of Otology and Laryngology, Harvard Medical School, Boston, MA, USA

Jessica de Boer Institute of Hearing Research, MRC, Nottingham, UK

John M. Deeks Medical Research Council, Cognition and Brain Sciences Unit, Cambridge, UK

Bertrand Delgutte Eaton-Peabody Laboratories, Massachusetts Eye and Ear Infirmary, Boston, MA, USA

Department of Otology and Laryngology, Harvard Medical School, Boston, MA, USA

Research Laboratory of Electronics, Massachusetts Institute of Technology, Cambridge, MA, USA

Laurent Demany, PhD Institut de Neurosciences Cognitives et Intégratives d'Aquitaine (UMR CNRS 5287), Université de Bordeaux, Bordeaux, France

D. Depireux Institute for Systems Research, University of Maryland, College Park, MD, USA

Anna Dettlaff Section of Biomagnetism, Department of Neurology, University of Heidelberg, Heidelberg, Germany

Mathias Dietz Ear Institute, University College London, London, UK

Nai Ding Department of Electrical and Computer Engineering, University of Maryland, College Park, MD, USA

Vit Drga Ear Institute, University College London (UCL), London, UK

Larisa Dunai Departamento de Ingeniería Gráfica, Universitat Politècnica de València, Camino de Vera, València, Spain

Mounya Elhilali Electrical and Computer Engineering, Johns Hopkins University, Baltimore, MD, USA

Bernhard Englitz, PhD Institute for Systems Research, University of Maryland, College Park, MD, USA

Bastian Epp Department of Electrical Engineering, DTU, Lyngby, Denmark

Almudena Eustaquio-Martín, MSc Instituto de Neurociencias de Castilla y León, Universidad de Salamanca, Salamanca, Spain

Instituto de Investigaciones Biomédicas de Salamanca, Universidad de Salamanca, Salamanca, Spain

Redwan Farooq Medical Research Council, Cognition and Brain Sciences Unit, Cambridge, UK

Mark Fletcher Institute of Hearing Research, MRC, Nottingham, UK

Karl J. Friston Functional Imaging Lab, Wellcome Trust Centre for Neuroimaging, University College London (UCL), London, UK

J.B. Fritz Institute for Systems Research, University of Maryland, College Park, MD, USA

Tomofumi Fukawatase Faculty of Systems Engineering, Wakayama University, Wakayama, Japan

Shigeto Furukawa, PhD NTT Communication Science Laboratories, NTT Corporation, Atsugi, Kanagawa, Japan

Darik Gamble Laboratory of Auditory Neurophysiology, Department of Biomedical Engineering, The Johns Hopkins University, Baltimore, MD, USA

Nikodemus Gessele Division of Neurobiology, Department of Biology II, University of Munich, Martinsried, Germany

Brian R. Glasberg Department of Experimental Psychology, University of Cambridge, Cambridge, UK

Hedwig E. Gockel Medical Research Council, Cognition and Brain Sciences Unit, Cambridge, UK

David Greenberg Ear Institute, University College London, London, UK

Timothy D. Griffiths Auditory Group, Institute of Neuroscience, Medical School, Newcastle University, Newcastle upon Tyne, UK

Wellcome Trust Centre for Neuroimaging, London, UK

John H. Grose Department of Otolaryngology/Head and Neck Surgery, University of North Carolina School of Medicine, Chapel Hill, NC, USA

Ervin R. Hafter Department of Psychology, University of California, Berkeley, CA, USA

Joseph W. Hall III Department of Otolaryngology/Head and Neck Surgery, University of North Carolina School of Medicine, Chapel Hill, NC, USA

Kenneth E. Hancock Eaton-Peabody Laboratories, Massachusetts Eye and Ear Infirmary, Boston, MA, USA

Department of Otology and Laryngology, Harvard Medical School, Boston, MA, USA

William M. Hartmann Department of Physics and Astronomy, Michigan State University, East Lansing, MI, USA

Eric W. Healy Department of Speech and Hearing Science, The Ohio State University, Columbus, OH, USA

Peter Heil Department of Auditory Learning and Speech, Leibniz Institute for Neurobiology, Magdeburg, Germany

Center for Behavioral Brain Sciences, Magdeburg, Germany

Michael G. Heinz Department of Speech, Language, & Hearing Sciences and Biomedical Engineering, Purdue University, West Lafayette, IN, USA

Itai Hershenhoren Department of Neurobiology, The Silberman Institute of Life Sciences, The Interdisciplinary Center for Neural Computation, Hebrew University, Jerusalem, Israel

Haruto Hirose NTT Communication Science Laboratories, NTT Corporation, Atsugi, Kanagawa, Japan

Steven Hsiao The Solomon H. Snyder Department of Neuroscience, Zanvyl Krieger Mind/Brain Institute, The Johns Hopkins University, Baltimore, MD, USA

Juan Huang The Solomon H. Snyder Department of Neuroscience, Zanvyl Krieger Mind/Brain Institute, The Johns Hopkins University, Baltimore, MD, USA

Laboratory of Auditory Neurophysiology, Department of Biomedical Engineering, The Johns Hopkins University, Baltimore, MD, USA

Fabio Idrobo Department of Psychology, Boston University, Boston, MA, USA

Toshio Irino Faculty of Systems Engineering, Wakayama University, Wakayama, Japan

Dexter R.F. Irvine School of Psychology and Psychiatry, Monash University, Melbourne, VIC, Australia

Bionics Institute, Melbourne, VIC, Australia

D. Timothy Ives Department d'Etudes Cognitives, Ecole Normale Supérieure, Paris, France

Peter T. Johannesen, MSc Instituto de Neurociencias de Castilla y León, Universidad de Salamanca, Salamanca, Spain

Instituto de Investigaciones Biomédicas de Salamanca, Universidad de Salamanca, Salamanca, Spain

Heath G. Jones Department of Physiology and Biophysics, University of Colorado School of Medicine, Aurora, CO, USA

Søren Jørgensen Department of Electrical Engineering, Centre for Applied Hearing Research, Technical University of Denmark, Lyngby, Denmark

Philip X. Joris Laboratory of Auditory Neurophysiology, University of Leuven, Leuven, Belgium

T. Jürgens Department of Psychology, University of Essex, Colchester, UK

Sushrut Kale Department of Otolaryngology-Head & Neck Surgery, Columbia University, New York, NY, USA

Weldon School of Biomedical Engineering, Purdue University, West Lafayette, IN, USA

Sridhar Kalluri Starkey Hearing Research Center, Berkeley, CA, USA

Makio Kashino NTT Communication Science Laboratories, NTT Corporation, Atsugi, Kanagawa, Japan

Tokyo Institute of Technology, Yokohama, Kanagawa, Japan

Hideki Kawahara Faculty of Systems Engineering, Wakayama University, Wakayama, Japan

Richard Kempter Department of Biology, Institute for Theoretical Biology, Humboldt-Universität zu Berlin, Berlin, Germany

Olivia Kenyon Department of Experimental Psychology, University of Cambridge, Cambridge, UK

Angela D. Ketterer Departments of Biomedical Engineering and Neurobiology & Anatomy, University of Rochester, Rochester, NY, USA

Daniel Kish World Access for the Blind, Huntington Beach, CA, USA

Astrid Klinge-Strahl Department of Animal Physiology and Behaviour Group, IBU, Carl von Ossietzky University Oldenburg, Oldenburg, Germany

Georg M. Klump Department of Animal Physiology and Behaviour Group, IBU, Carl von Ossietzky University Oldenburg, Oldenburg, Germany

Armin Kohlrausch Philips Group Innovation, Smart Sensing & Analysis, Philips Research, Eindhoven, The Netherlands

Eindhoven University of Technology, Eindhoven, The Netherlands

Kanthaiah Koka Department of Physiology and Biophysics, University of Colorado School of Medicine, Aurora, CO, USA

B. Kollmeier Medizinische Physik, Carl von Ossietzky University, Oldenburg, Germany

Christine Köppl Institute for Biology and Environmental Sciences, and Research Center Neurosensory Science, Carl von Ossietzky University, Oldenburg, Germany

Katrin Krumbholz Institute of Hearing Research, MRC, Nottingham, UK

Sukhbinder Kumar, PhD Auditory Group, Institute of Neuroscience, Medical School, Newcastle University, Newcastle upon Tyne, UK

Neural mechanisms of human communication, Wellcome Trust Centre for Neuroimaging, London, UK

Paula T. Kuokkanen Department of Biology, Institute for Theoretical Biology, Humboldt-Universität zu Berlin, Berlin, Germany

Dave R.M. Langers Department of Otorhinolaryngology/Head and Neck Surgery, University Medical Center Groningen, University of Groningen, Groningen, NL, Netherlands

Graduate School of Medical Sciences (Research School of Behavioural and Cognitive Neurosciences), University of Groningen, Groningen, Netherlands

W. Lecluyse Department of Psychology, University of Essex, Colchester, UK

Ching-Ju Liu Auditory Behavioral Research Lab, Department of Communicative Disorders, University of Wisconsin, Madison, WI, USA

Enrique A. Lopez-Poveda, PhD Instituto de Neurociencias de Castilla y León, Universidad de Salamanca, Salamanca, Spain

Instituto de Investigaciones Biomédicas de Salamanca, Universidad de Salamanca, Salamanca, SpainDepartamento de Cirugía, Universidad de Salamanca, Salamanca, Spain

Bjoern Luebken Acoustics Group, Carl von Ossietzky University, Oldenburg, Germany

Robert A. Lutfi Auditory Behavioral Research Lab, Department of Communicative Disorders, University of Wisconsin, Madison, WI, USA

Richard F. Lyon Google Inc., Mountain View, CA, USA

Ling Ma Bioengineering Program, University of Maryland, MD, USA

Geoffrey A. Manley Cochlear and Auditory Brainstem Physiology, Department of Neuroscience, Carl von Ossietzky University Oldenburg, Oldenburg, Germany

Torsten Marquardt Ear Institute, University College London, London, UK

Toshie Matsui Department of Otorhinolaryngology, Nara Medical University, Kashihara, Japan

David McAlpine Ear Institute, University College London, London, UK

Thomas McColgan Institute for Biology II, RWTH Aachen, Aachen, Germany

Raymond Meddis Department of Psychology, University of Essex, Colchester, UK

A. Meyer Medizinische Physik, Carl von Ossietzky University, Oldenburg, Germany

B.T. Meyer Medizinische Physik, Carl von Ossietzky University, Oldenburg, Germany

Christophe Micheyl Department of Psychology, University of Minnesota – Twin Cities, Minneapolis, MN, USA

John C. Middlebrooks Departments of Otolaryngology, Neurobiology & Behavior, Cognitive Sciences, and Biomedical Engineering, University of California at Irvine, Irvine, CA, USA

Brian C.J. Moore Department of Experimental Psychology, University of Cambridge, Cambridge, UK

Louwai Muhammed Medical Research Council, Cognition and Brain Sciences Unit, Cambridge, UK

Andreas Nagy Division of Neurobiology, Department of Biology II, University of Munich, Martinsried, Germany

Sung-Il Nam Department of Otology and Laryngology, Harvard Medical School, Boston, MA, USA

Department of Otolaryngology, School of Medicine, Keimyung University, Daegu, South Korea

Israel Nelken Department of Neurobiology, Silberman Institute of Life Sciences, Hebrew University, Givat Ram, Jerusalem, Israel

Edmond and Lily Safra Center for Brain Sciences, Hebrew University, Jerusalem, Israel

Interdisciplinary Center for Neural Computation, Hebrew University, Jerusalem, Israel

Heinrich Neubauer Department of Auditory Learning and Speech, Leibniz Institute for Neurobiology, Magdeburg, Germany

Ryuichi Nisimura Faculty of Systems Engineering, Wakayama University, Wakayama, Japan

Atsushi Ochi NTT Communication Science Laboratories, NTT Corporation, Atsugi, Kanagawa, Japan

Department of Otolaryngology, Faculty of Medicine, University of Tokyo, Tokyo, Japan

Andrew J. Oxenham Department of Psychology, University of Minnesota – Twin Cities, Minneapolis, MN, USA

M.R. Panda Department of Psychology, University of Essex, Colchester, UK

Timo Parnitzke Department of Animal Physiology and Behaviour Group, IBU, Carl von Ossietzky University Oldenburg, Oldenburg, Germany

Roy D. Patterson Department of Physiology, Development and Neuroscience, Centre for the Neural Basis of Hearing, University of Cambridge, Cambridge, UK

Leyla Pinggera University Clinic for Ear, Nose and Throat Medicine, Innsbruck, Austria

Christopher J. Plack School of Psychological Sciences, The University of Manchester, Manchester, UK

Ana Polterovich Department of Neurobiology, The Silberman Institute of Life Sciences, The Interdisciplinary Center for Neural Computation, Hebrew University, Jerusalem, Israel

Daniel Pressnitzer Départements d'études cognitives, Equipe Audition, Ecole Normale Supérieure, Paris, France

Laboratoire de Psychologie de la Perception (UMR CNRS 8158), Université Paris Descartes, Paris, France

Tianshu Qu Key Laboratory on Machine Perception-Ministry of Education, Peking University Beijing, Beijing, China

Andrew P. Raimond Department of Psychology, University of Reading, Reading, UK

Lars Riecke Department of Cognitive Neuroscience, Faculty of Psychology and Neuroscience, Maastricht University, Maastricht, The Netherlands

Brian Roberts Psychology, School of Life and Health Sciences, Aston University, Birmingham, UK

Elin M. Roverud Department of Speech, Language, and Hearing Sciences, Purdue University, West Lafayette, IN, USA

Dorea R. Ruggles Department of Biomedical Engineering, Boston University Center for Computational Neuroscience and Neural Technology, Boston, MA, USA

André Rupp Section of Biomagnetism, Department of Neurology, University of Heidelberg, Heidelberg, Germany

Makoto Sakaguchi Faculty of Systems Engineering, Wakayama University, Wakayama, Japan

Kristine Sarnlertsophon The Solomon H. Snyder Department of Neuroscience, Zanvyl Krieger Mind/Brain Institute, The Johns Hopkins University, Baltimore, MD, USA

M.R. René Schädler Medizinische Physik, Carl von Ossietzky University, Oldenburg, Germany

Sven Schörnich Division of Neurobiology, Department of Biology II, University of Munich, Martinsried, Germany

Michael Schranner Division of Neurobiology, Department of Biology II, University of Munich, Martinsried, Germany

Douglas M. Schwarz Departments of Biomedical Engineering and Neurobiology & Anatomy, University of Rochester, Rochester, NY, USA

Bernhard U. Seeber Audio Information Processing, Technische Universität München, München, Germany

Catherine Semal Institut de Neurosciences Cognitives et Intégratives d'Aquitaine (UMR CNRS 5287), Université de Bordeaux, Bordeaux, France

Sahil Shah Department of Biology, University of Maryland, College Park, MD, USA

Shihab A. Shamma Department of Electrical and Computer Engineering, Institute for Systems Research, University of Maryland, College Park, MD, USA

Barbara Shinn-Cunningham Department of Biomedical Engineering, Boston University Center for Computational Neuroscience and Neural Technology, Boston, MA, USA

Jonathan Z. Simon Department of Electrical and Computer Engineering, University of Maryland, College Park, MD, USA

Department of Biology, University of Maryland, College Park, MD, USA

André Spachmann Section of Biomagnetism, Department of Neurology, University of Heidelberg, Heidelberg, Germany

Arkadiusz Stasiak Department of Physiology, Development and Neuroscience, Cambridge, UK

Mark A. Stellmack Department of Psychology, University of Minnesota, Minneapolis, MN, USA

Elizabeth A. Strickland Department of Speech, Language, and Hearing Sciences, Purdue University, West Lafayette, IN, USA

Christophe N.J. Stoelinga Auditory Behavioral Research Lab, Department of Communicative Disorders, University of Wisconsin, Madison, WI, USA

Clara Suied Département d'études cognitives, Equipe Audition, Ecole Normale Supérieure, Paris, France

Laboratoire de Psychologie de la Perception (UMR CNRS 8158), Université Paris Descartes, Paris, France

Fondation Pierre Gilles de Gennes pour la Recherche, Paris, France

Département Action et Cognition en Situation Opérationnelle, Institut de Recherche Biomédicale des Armées, Brétigny-sur-Orge, France

Robert J. Summers Psychology, School of Life and Health Sciences, Aston University, Birmingham, UK

Chihiro Takeshima College of Performing and Visual Arts, J.F. Oberlin University, Machida, Japan

C.M. Tan Department of Psychology, University of Essex, Colchester, UK

Manuel Tetschke Department of Auditory Learning and Speech, Leibniz Institute for Neurobiology, Magdeburg, Germany

Jennifer Thornton Department of Physiology and Biophysics, University of Colorado School of Medicine, Aurora, CO, USA

Simon J. Thorpe Faculté de Médecine de Rangueil, Centre de Recherche Cerveau et Cognition (UMR CNRS UPS 5549), Université Paul Sabatier, Toulouse, France

Daniel J. Tollin Department of Physiology and Biophysics, University of Colorado School of Medicine, Aurora, CO, USA

Constantine Trahiotis Department of Neuroscience and Surgery (Otolaryngology), University of Connecticut Health Center, Farmington, CT, USA

Minoru Tsuzaki Faculty of Music, Kyoto City University of Arts, Kyoto, Japan

Christian H. Uhlig Medizinische Physik, Carl von Ossietzky University, Oldenburg, Germany

Neurologische Klinik, Universität Heidelberg, Heidelberg, Germany

Stefan Uppenkamp Medizinische Physik, Carl von Ossietzky University, Oldenburg, Germany

Forschungszentrum Neurosensorik, Carl von Ossietzky University, Oldenburg, Germany

Pim van Dijk Department of Otorhinolaryngology/Head and Neck Surgery, University of Groningen, University Medical Center Groningen, Groningen, RB, Netherlands

Graduate School of Medical Sciences (Research School of Behavioural and Cognitive Neurosciences), University of Groningen, Groningen, Netherlands

Steven van de Par Acoustics Group, Carl von Ossietzky University, Oldenburg, Germany

Jesko L. Verhey Department of Experimental Audiology, Otto-von-Guericke University of Magdeburg, Magdeburg, Germany

Forschungszentrum Neurosensorik, Carl von Ossietzky University, Oldenburg, Germany

Sarah Verhulst Department of Biomedical Engineering, Center for Computational Neuroscience and Neural Technology, Boston University, Boston, MA, USA

Department of Electrical Engineering, Center for Applied Hearing Research, Technical University of Denmark, Lyngby, Denmark

Eric Verschooten Laboratory of Auditory Neurophysiology, University of Leuven, Leuven, Belgium

Neal F. Viemeister Department of Psychology, University of Minnesota, Minneapolis, MN, USA

Katharina von Kriegstein Neural mechanisms of human communication, Max Planck Institute for Human Cognitive and Brain Sciences, Leipzig, Germany

Hermann Wagner Institute for Biology II, RWTH Aachen, Aachen, Germany

Ludwig Wallmeier Division of Neurobiology, Department of Biology II, University of Munich, Martinsried, Germany

Thomas C. Walters Google Inc., Mountain View, CA, USA

Xiaoqin Wang Laboratory of Auditory Neurophysiology, Department of Biomedical Engineering, The Johns Hopkins University, Baltimore, MD, USA

Shiho Washizawa NTT Communication Science Laboratories, NTT Corporation, Atsugi, Kanagawa, Japan

Interdisciplinary Graduate School of Science and Engineering, Tokyo Institute of Technology, Yokohama, Kanagawa, Japan

Anthony J. Watkins Department of Psychology, University of Reading, Reading, UK

William M. Whitmer MRC Institute of Hearing Research (Scottish Section), Glasgow Royal Infirmary, Glasgow, UK

Lutz Wiegrebe Division of Neurobiology, Department of Biology II, University of Munich, Martinsried, Germany

Ian M. Winter Department of Physiology, Development and Neuroscience, Cambridge, UK

Jing Xia Department of Psychology, University of California, Berkeley, CA, USAStarkey Hearing Research Center, Berkeley, CA, USA

Yanbo Xu Electrical and Computer Engineering, Institute for Systems Research, University of Maryland, College Park, MD, USA

Amit Yaron Department of Neurobiology, The Silberman Institute of Life Sciences, The Interdisciplinary Center for Neural Computation, Hebrew University, Jerusalem, Israel

Ifat Yasin Ear Institute, University College London (UCL), London, UK

Pingbo Yin Electrical and Computer Engineering, Institute for Systems Research, University of Maryland, College Park, MD, USA

William A. Yost Department of Speech and Hearing Science, Arizona State University, Tempe, AZ, USA

Part I
Peripheral Processing

Chapter 1
Mosaic Evolution of the Mammalian Auditory Periphery

Geoffrey A. Manley

Abstract The classical mammalian auditory periphery, i.e., the type of middle ear and coiled cochlea seen in modern therian mammals, did not arise as one unit and did not arise in all mammals. It is also not the only kind of auditory periphery seen in modern mammals. This short review discusses the fact that the constituents of modern mammalian auditory peripheries arose at different times over an extremely long period of evolution (230 million years; Ma). It also attempts to answer questions as to the selective pressures that led to three-ossicle middle ears and the coiled cochlea. Mammalian middle ears arose de novo, without an intermediate, single-ossicle stage. This event was the result of changes in eating habits of ancestral animals, habits that were unrelated to hearing. The coiled cochlea arose only after 60 Ma of mammalian evolution, driven at least partly by a change in cochlear bone structure that improved impedance matching with the middle ear of that time. This change only occurred in the ancestors of therian mammals and not in other mammalian lineages. There is no single constellation of structural features of the auditory periphery that characterizes all mammals and not even all modern mammals.

1 Introduction

Over the past 20 years, a number of dogmata and common ways of viewing the mammalian auditory periphery have been proven to be false, due to major new fossil finds and new ways of examining fossils that allow non-destructive

G.A. Manley
Cochlear and Auditory Brainstem Physiology, Department of Neuroscience, Carl von Ossietzky University Oldenburg, Carl von Ossietzky Strasse 9–11, Oldenburg 26129, Germany
e-mail: geoffrey.manley@uni-oldenburg.de

B.C.J. Moore et al. (eds.), *Basic Aspects of Hearing*, Advances in Experimental Medicine and Biology, DOI 10.1007/978-1-4614-1590-9_1, © Springer Science+Business Media New York 2013 study of internal skull features. The following ideas have been shown to be incorrect:

(a) The middle ear of land vertebrates arose only once at the time of the water-to-land transition of the earliest tetrapods.
(b) The mammalian middle ear arose once by the addition of two extra ossicles to a pre-existing one that had arisen much earlier (as in a).
(c) All components of middle ears that are common to mammals and non-mammals (e.g., tympanic membrane, Eustachian tubes) are homologous.
(d) In mammals, the three-ossicle middle ear and the coiled cochlea are essential for each other's function and therefore arose together.

To (a) it must be noted that while there are isolated cases where some sort of tympanic middle ear can be described in very early tetrapods, there was no lineage where this structure became established and continued through the Mesozoic era (Manley 2010). The newest data suggest that, extraordinarily, all major groups of tetrapods (the ancestors of modern amphibians, of both archosaurs (including birds) and lepidosaur "reptiles" and of mammals) each independently evolved a tympanic middle ear during the Triassic period (e.g., Clack 2002). To (b) it can be observed that the middle ears of modern egg-laying monotremes (platypus and echidna) and live-bearing therian (marsupial and placental) mammals each arose separately and de novo and not through adding two extra ossicles to a pre-existing structure (Clack 2002; Manley 2010). To (c) it can be noted that mammalian middle ears arose at a different location (more ventral, over the rear part of the lower jaw) to those of non-mammals. The tympanum and the Eustachian tube arose independently of those of other vertebrates (e.g., Allin 1986). Lastly, to (d) it needs to be remembered that the three-ossicle middle ear exists or existed in a number of mammalian lineages of which only one developed a coiled cochlea. In addition, the coiled cochlea of the therian lineage developed at least 50 Ma after the three-ossicle middle ear had arisen (Wible et al. 2001).

Recent data reported using micro-CT scanning techniques that permit high-resolution studies of internal skull components of fossil mammals, such as the inner ear, now provide the basis for a new assessment of the most important evolutionary events in the history of the mammalian auditory periphery. From these data, it is possible to speculate on the selective pressures that favoured their establishment in particular mammalian lineages.

2 Middle Ears: The Why and Wherefore

As noted above, three-ossicle middle ears arose in the Triassic period in mammalian ancestors (synapsid "reptiles") that were changing their eating patterns. These changes led initially to a doubling-up of the jaw joint, bringing new supports to the joint in the form of substantial bones of the outer skull, such as the squamosum. At the same time, the complex lower jaw that had consisted of several bones was made stronger by a reduction to one single bone, the dentary. With time, the inner, older jaw joint was reduced in size and importance, and the function of jaw articulation was taken over completely by the secondary joint components. The relegated primary joint components either disappeared or became incorporated into a completely new tympanic middle ear as the incus, malleus and tympanic. This new middle ear initially lay near the rear end of the lower jaw, more ventral than an equivalent structure with a single ossicle that, during the same geological period, was arising in ancestors of different non-mammalian lineages. Thus, the selection pressures behind the origin of the mammalian middle ear lie not with hearing but with nutrition. Nonetheless, mammals gained through this development a greater sensitivity to the low-frequency sounds that they had previously only heard at higher pressures.

The basic story of this structural transition has, of course, been known since the nineteenth century (review in Maier 1990). Somewhat newer is the conviction not only that this event was unique to mammals but also that it occurred more than once, in parallel in different mammalian lineages (i.e., it is homoplastic; Fig. 1.1; Luo 2007; Martin and Luo 2005; Rich et al. 2005). The result was a middle ear that differed greatly from many modern mammalian middle ears, so much so that we would not immediately recognize it. The structures were larger, shaped differently to what is known in most modern mammals, and the ossicular chain was stiffer and not freely suspended in a large air space. This has been termed a transitional mammalian middle ear (TMME; Luo 2011; Meng et al. 2011). Although the TMME did evolve towards smaller and freer ossicles in all lineages (Fig. 1.1), only in one did it relatively quickly achieve the configuration now known from modern placentals and marsupials, viz. tiny ossicles freely suspended by ligaments in an air space. This condition has been termed the DMME, or definitive mammalian middle ear (Luo 2011; Meng et al. 2011). In the lineages to the monotremes and multituberculates, not all subgroups achieved the DMME and the result is not identical to that in therians. It is also noteworthy that in the parallel evolution of the inner ear, the cochlea of the earliest mammals was vastly different to the long, coiled cochleae of modern therians (Fig. 1.1).

3 Mammalian Inner Ears: Commonalities and Diversity

Mammalian cochleae vary absolutely in length more than those seen in any other group of vertebrates. In the immediate ancestors of mammals, among the synapsid "reptiles", the cochleae were less than 2 mm long, including the lagenar macula, meaning that the basilar membrane was probably about 1.5 mm long. The shortest known in true mammals were also about 2 mm long, seen in fossils of different lineages (e.g., Hadrocodium, Dryolestes, Henkelotherium; Luo et al. 2001, 2010; Ruf et al. 2009) and not coiled. The longest known cochleae in modern therians are those from some modern baleen whales and are more than 50 mm long, a length difference to the earliest species of more than 30-fold. All of these cochleae were supplied by three-ossicle middle ears of various stages of development (Fig. 1.1). In two mammalian lineages, the cochlea never evolved to great lengths. One of these

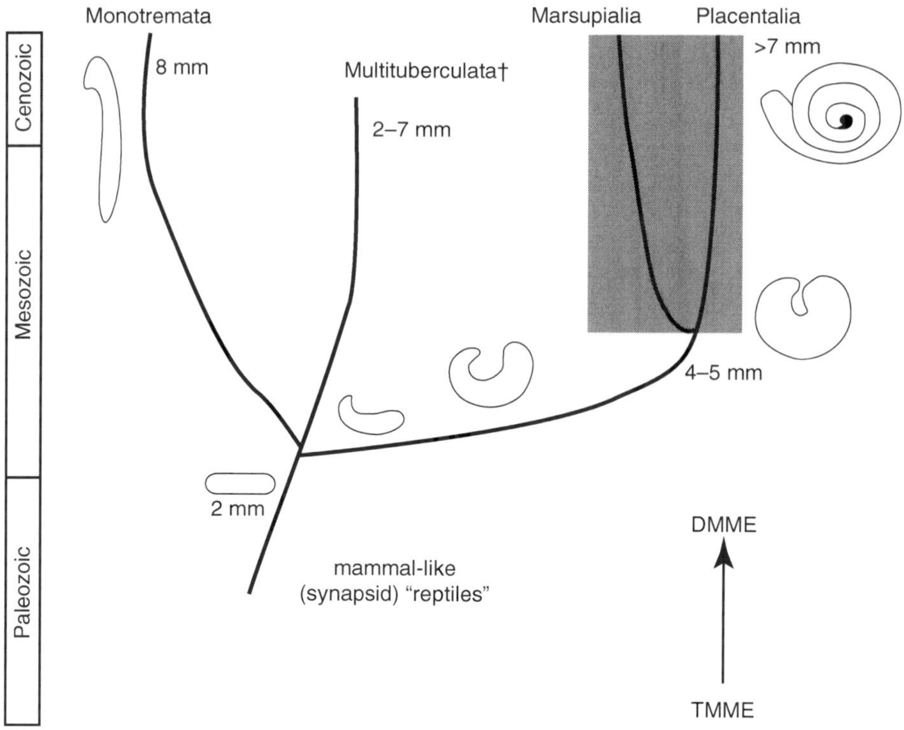

Fig. 1.1 A schematic diagram of the history of several mammalian lineages that had a common origin from synapsid "reptiles". A timescale is provided on the *left* of the diagram. At their origin about 240 Ma ago, the cochlea was only 2 mm long (lowest cochlear sketch) and the middle ear was of the transitional type (*TMME* greyscale explanation at the *lower right*). In monotreme and multituberculate lineages, the middle ear changed slowly and the cochlea only elongated somewhat during their further evolution (max 8 mm) but did not coil. At the origin of therian mammals (*middle right*), marsupials and placentals already shared a definitive mammalian middle ear (*DMME*) and a coiled cochlea with one single turn of ~5 mm length. In modern therian species, the middle ear shows significant variation, and the cochlear length varies from 7 to >50 mm

lineages led to the modern egg-laying monotremes, and the other was a long-lived and diverse lineage that died out about 30 Ma ago and is known as the multituberculates (Vater et al. 2004). Indeed, in multituberculates, after 150 Ma of the possession of a three-ossicle middle ear, the cochlea was still 2–7 mm long (Hurum 1998; Luo and Ketten 1991). This clearly demonstrates the mosaic nature of the evolution of the auditory periphery and that different components can be in quite different stages of their evolution. Even in modern monotremes, cochlear length does not exceed 8 mm (e.g., Ladhams and Pickles 1996). It is thus fair to observe that the possession of a three-ossicle middle ear does not in any way automatically lead to dramatic evolution of the cochlea. The question thus becomes more interesting: Why did the cochlea evolve in a spectacularly different way in one mammalian lineage (therian ancestors)?

A comparison of all three modern groups of mammals (monotremes, marsupials and placentals) shows that all have a hearing organ that can be termed an organ of Corti: There are two groups of hair cells that are on the inner and outer sides of a tunnel formed by pillar cells (e.g., Ladhams and Pickles 1996). This strongly suggests that this structural configuration was already present in the ancestors of mammals and that multituberculate mammals also possessed an organ of Corti. Of course, if we can assume that the monotreme cochlea represents the primitive state, during the evolution of the therian cochlea, the number of cells across the organ has been greatly reduced. Comparative evidence points to hearing being restricted to low frequencies in ancestral mammals. This indicates that the possession of an organ of Corti as such does not automatically lead to high-frequency hearing, even after more than 100 Ma of evolution; monotremes have an upper frequency limit of roughly 15 kHz, and they are relatively insensitive (Gates et al. 1974; Mills and Shepherd 2001). So what is unique about the therians that led to them evolving a long, coiled cochlea, the DMME and, later, high-frequency hearing above 20 kHz?

4 A Therian Innovation and Its Consequences

The fossil evidence indicates that after the origin of mammals, it took more than 50 Ma for the therian mammal lineage to achieve full coiling of the cochlea (Wible et al. 2001), which occurred shortly before placentals and marsupials split. At that time, the length of the hearing organ was 4–5 mm, which is shorter than the cochlea of any modern mammal. Examining all that is known about cochlear structure of the various lineages leads to the conclusion that therians developed one feature that is unique and perhaps decisive for further evolution. During the middle Jurassic era, the soft tissues of the therian cochlea became integrated with the bony canal that surrounded them. This led to the evolution of primary and secondary bony laminae that support the basilar membrane and a bony enclosure for the afferent ganglion (Luo et al. 2001, 2010; Kemp 2005; Ruf et al. 2009). This bony integration presumably improved the impedance match between the organ of Corti and the stiff middle ears of the period.

After the development of a stiffer support for the basilar membrane, the middle ear and the cochlea of therian ancestors evolved until, shortly before the marsupial and placental lineages split, both a DMME and a fully coiled cochlea (about 5 mm long) had been achieved. This became the basic structural configuration of therians. About 80 Ma ago, a marsupial cochlea existed with 1.25 turns in its coil (Meng and Fox, 1995). In modern therians, the minimum number of turns is 1.5 (e.g., in mice). Of course marsupials and especially placentals later radiated diversely into a huge number of groups that vary in the characteristics of their middle and inner ears. Some hugely elongated the cochlea, some evolved echolocation and very high-frequency hearing, and others became much larger and more low-frequency efficient. Even some very small mammals specialized for low-frequency hearing by, e.g., greatly enlarging their middle-ear bullae. Although it has been speculated that the early presence of primary and secondary laminae indicate that even the earliest therians were specialized for high-frequency hearing (Luo 2011), this is very unlikely. The distribution of secondary laminae is sporadic in modern therians; although it is always present where ultrasonic hearing is found, some placentals lack a secondary lamina, e.g., primates (some of which have quite high-frequency hearing). It can thus be surmised that the laminae were an important component of the impedance matching between early middle and inner ears, no matter what the frequency range being processed at that time. More likely is a gradual rise in the upper frequency limit. Clear fossil evidence of ultrasonic sound processing is not found in the earliest bats (Simmons et al. 2008) but probably evolved soon after in the early Cenozoic, 50–55 Ma ago.

References

Allin EF (1986) The auditory apparatus of advanced mammal-like reptiles and early mammals. In: Hotton H, MacLean PD, Roth JJ, Roth EC (eds) The ecology and biology of mammal-like reptiles. Smithsonian, Washington, D.C., pp 283–294

Clack JA (2002) Patterns and processes in the early evolution of the tetrapod ear. J Neurobiol 53: 251–264

Gates GR, Saunders JC, Bock GR, Aitkin LM, Elliot MA (1974) Peripheral auditory function in the platypus, *Ornithorhynchus anatinus*. J Acoust Soc Am 56:152–156

Hurum JH (1998) The inner ear of two Late Cretaceous multituberculate mammals, and its implications for multituberculate hearing. J Mamm Evol 5:65–93

Kemp TS (2005) The origin and evolution of mammals. Oxford University Press, Oxford

Ladhams A, Pickles JO (1996) Morphology of the monotreme organ of Corti and Macula lagena. J Comp Neurol 366:335–347

Luo Z-X (2007) Transformation and diversification in early mammal evolution. Nature 450: 1011–1019

Luo Z-X (2011) Developmental patterns in Mesozoic evolution of mammal ears. Ann Rev Ecol Evol Syst 42:355–380

Luo Z, Ketten DR (1991) CT scanning and computerized reconstructions of the inner ear of multituberculate mammals. J Vert Paleontol 11:220–228

Luo Z-X, Crompton AW, Sun AL (2001) A new mammaliaform from the early Jurassic and evolution of mammalian characteristics. Science 292:1535–1540

Luo Z-X, Rif I, Schultz JA, Martin T (2010) Fossil evidence on evolution of inner ear cochlea in Jurassic mammals. Proc Roy Soc B 278:28–34

Maier W (1990) Phylogeny and ontogeny of mammalian middle ear structures. Neth J Zool 40: 55–74

Manley GA (2010) An evolutionary perspective on middle ears. Hear Res 263:3–8

Martin T, Luo Z-X (2005) Homoplasy in the mammalian ear. Science 307:861–862

Meng J, Fox RC (1995) Therian petrosals from the Oldman and Milk River formations (Late Cretaceous), Alberta, Canada. J Vert Paleontol 15:122–130

Meng J, Wang Y, Li C (2011) Transitional mammalian middle ear from a new Cretaceous Jehol eutriconodont. Nature 472:181–185

Mills DM, Shepherd RK (2001) Distortion product otoacoustic emission and auditory brainstem response in the echidna (*Tachyglossus aculeatus*). J Assoc Res Otolaryngol 2:130–146

Rich TH, Hopson JA, Musser AM, Flannery TF, Vickers-Rich P (2005) Independent origins of middle ear bones in monotremes and therians. Science 307:910–914

Ruf I, Luo Z-X, Wible JR, Martin T (2009) Petrosal anatomy and inner ear structures of the Late Jurassic *Henkelotherium* (Mammalia, Cladotheria, Dryolestoidea): insight into the early evolution of the ear region in cladotherian mammals. J Anat 214:679–693

Simmons NB, Seymour KL, Habersetzer J, Gunnell GF (2008) Primitive Early Eocene bat from Wyoming and the evolution of flight and echolocation. Nature 451:818–821

Vater M, Meng J, Fox RC (2004) Hearing organ evolution and specialization: early and later mammals. In: Manley GA, Popper A, Fay RR (eds) Evolution of the vertebrate auditory system. Springer, New York, pp 256–288

Wible JR, Rougier GW, Novacek MJ, McKenna MC (2001) Earliest eutherian ear region: a petrosal referred to Prokennalestes from the early Cretaceous of Mongolia. Am Mus Novit 3322:1–44

Chapter 2
A Computer Model of the Auditory Periphery and Its Application to the Study of Hearing

Raymond Meddis, Wendy Lecluyse, Nicholas R. Clark, Tim Jürgens, Christine M. Tan, Manasa R. Panda, and Guy J. Brown

Abstract Computer models of the auditory periphery provide a tool for formulating theories concerning the relationship between the physiology of the auditory system and the perception of sounds both in normal and impaired hearing. However, the time-consuming nature of their construction constitutes a major impediment to their use, and it is important that transparent models be available on an 'off-the-shelf' basis to researchers. The MATLAB Auditory Periphery (MAP) model aims to meet these requirements and be freely available. The model can be used to simulate simple psychophysical tasks such as absolute threshold, pitch matching and forward masking and those used to measure compression and frequency selectivity. It can be used as a front end to automatic speech recognisers for the study of speech in quiet and in noise. The model can also simulate theories of hearing impairment and be used to make predictions about the efficacy of hearing aids. The use of the software will be described along with illustrations of its application in the study of the psychology of hearing.

1 Introduction

Auditory models come in various flavours. The model to be described aims to be a faithful simulation of physiological processes in the auditory periphery with two added layers of neurons in the auditory brainstem to make detection decisions.

R. Meddis (✉) • W. Lecluyse • N.R. Clark • T. Jürgens
C.M. Tan • M.R. Panda
Department of Psychology, University of Essex,
Wivenhoe Park, Colchester CO4 3SQ, UK
e-mail: rmeddis@essex.ac.uk

G.J. Brown
Department of Computer Science,
University of Sheffield, Sheffield, UK

B.C.J. Moore et al. (eds.), *Basic Aspects of Hearing*,
Advances in Experimental Medicine and Biology,
DOI 10.1007/978-1-4614-1590-9_2, © Springer Science+Business Media New York 2013

As such, it is an anatomical/physiological model, but the aim is to use it to help understand psychophysical phenomena such as threshold, pitch processing, speech recognition and hearing impairment.

2 Model Description

The architecture of the model is shown in Fig. 2.1. A number of features are important. The model consists of many channels each with their own best frequency (BF). This reflects the tonotopic arrangement of the auditory periphery. It also consists of a cascade of stages that reflect the sequence of successive nonlinear signal processing operations in the cochlea. It also contains feedback loops representing the acoustic reflex and medial olivocochlear (MOC) efferent suppression. Nonlinear feedback systems are difficult to approach intuitively. The model therefore acts as a visualisation tool.

Of course, such a model is only as good as its components. Fortunately, the output of individual modules can be evaluated against published physiological data. The output of each stage is expressed in terms of measurable variables such as

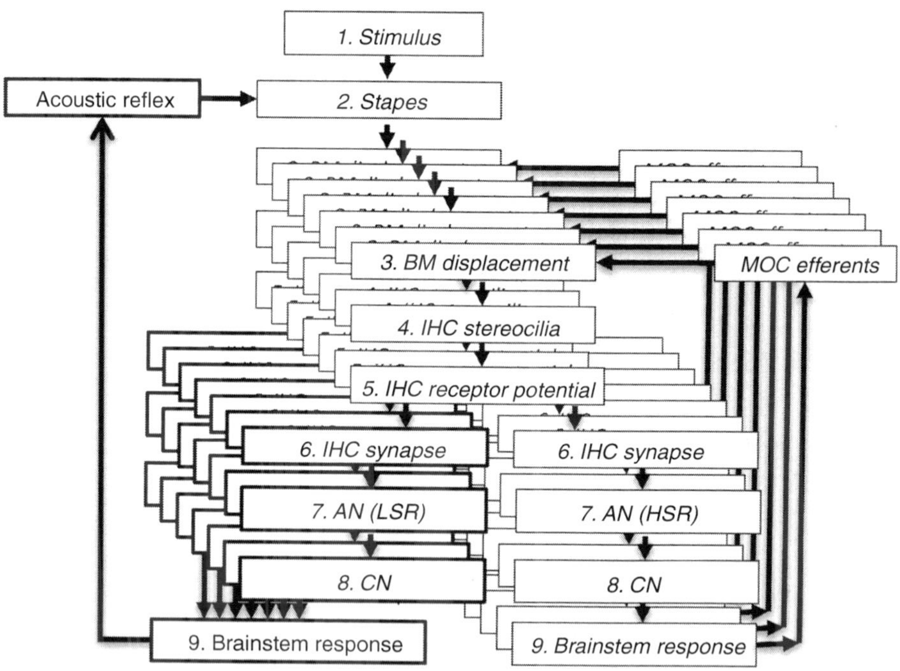

Fig. 2.1 Flow diagram of the MATLAB Auditory Periphery (MAP) model. The lower boxes on the left refer to activity driven by low spontaneous rate (LSR) fibres and forming (speculatively) part of the acoustic reflex (AR) circuit. The boxes on the right are driven by high spontaneous rate (HSR) fibres and form part of the MOC efferent circuit. *CN* cochlear nucleus

stapes displacement, basilar membrane (BM) displacement, inner hair cell (IHC) receptor potential, auditory nerve (AN) firing rate and the pattern of firing in individual brain stem neuronal units. The architecture of the model allows us to carry out pseudo physiological experiments by applying acoustic stimulation while measuring the response at the output of a particular stage and then checking against corresponding published data.

Figure 2.2 shows the output of the model at a number of stages in response to the word 'twister' presented at 50 dB SPL. Successive panels show the stimulus, the stapes response, a 21-channel BM response as well as three levels of neuronal response; the AN, cochlear nucleus (CN) chopper response and a second-level brainstem response. Figure 2.2b shows the multichannel activity in the MOC efferent. The AR is not activated at this stimulus intensity. Each panel represents an 'inspection window' for the corresponding stage.

3 Model Applications

The model is not just a computerised visual display. It has a number of applications. One is to use the AN spiking pattern as the 'front end' to another system that represents a theory of how sensory decisions are made. In the past we have used it as the input to an autocorrelation model of pitch processing and segregation of simultaneous vowels presented with different pitches. Indeed, the majority of requests from potential users of the model concern the need for a front end of this type.

One might expect that a good auditory model should make an ideal front end to an automatic speech recogniser with recognition performance close to human levels. Good performance can be achieved for speech presented in quiet but performance declines substantially in the presence of background noise. This has led us to include a simulation of the peripheral efferent system in the model because it moderates the strength of the system's response in proportion to the intensity of the background. This reduces the spread of excitation across frequency channels and produces a more stable representation. The model components representing the efferent system were first evaluated against the physiological data and then tested in studies using automatic speech recognition (ASR) techniques. The modelled efferent system includes both a MOC arrangement and a simulation of the acoustic reflex. It was possible to compare speech recognition as a function of signal-to-noise ratio (SNR) both with and without the benefit of the closed-loop multichannel efferent reflex. The unfilled squares in Fig. 2.3 show how poorly the unimproved model works as an auditory front end. A 50 % recognition rate requires 15-dB SNR. However, when the efferent pathway is enabled, performance is greatly improved. At 10-dB SNR the recognition rate rises from 30 to 90 %. The modelling exercise does not prove that the MOC is critical for perception of speech in noise, but it does illustrate how modelling can be used to explore the hypothesis. The results also show that human performance remains much better than that of the model!

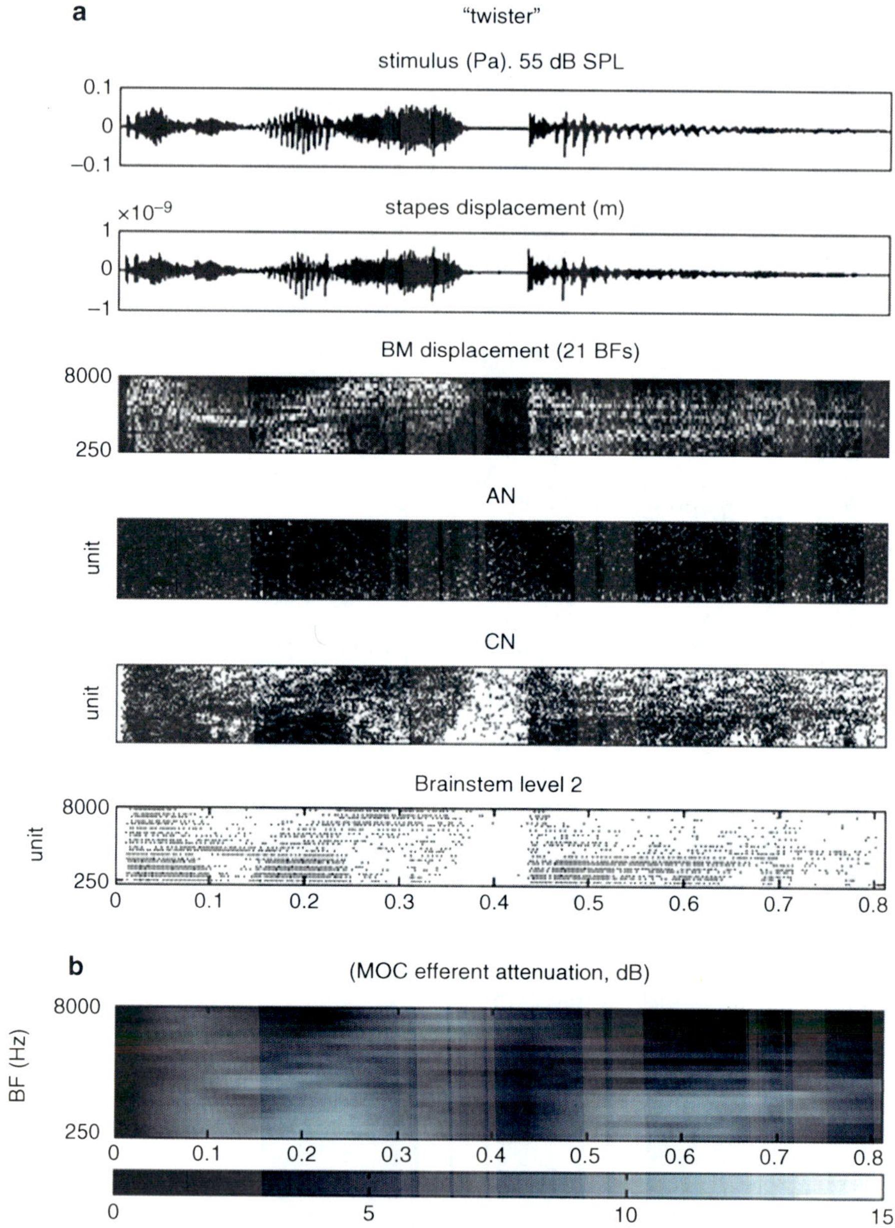

Fig. 2.2 Output from the auditory model. (**a**) Stimulus and output from five stages of the afferent part of the model (stapes, BM, AN, CN chopper, 2nd-level brainstem units). *X*-axis is time. (**b**) Activity in the efferent pathway of the model; time x channel attenuation of nonlinear DRNL input

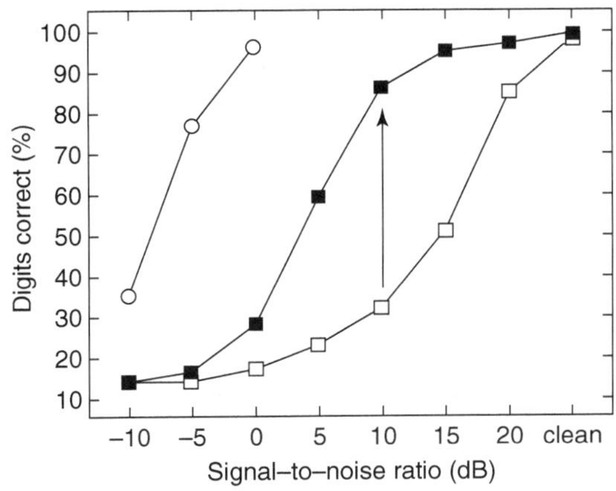

Fig. 2.3 ASR performance (% correct) as a function of SNR. The speech was connected digit triplets using both male and female talkers. The 'noise' is 20-talker babble. Representative human performance on the same test is shown as *unfilled circles*. Model performance without the efferent system is shown as *unfilled squares*. Improved performance using the efferent system is shown as *filled squares*

4 Psychophysics

Models can help understand the relationship between hearing and the underlying physiology by comparing model performance with that of human listeners in psychophysical experiments. Of course, some principle must first be established to convert the model multichannel output to a simple psychophysical response. For example, in a single-interval, adaptive tracking paradigm, the output must be converted to a 'yes' or 'no' response. Simple tasks such as detecting a tone against a silent background can be performed by creating neuronal units that never (or very rarely) spike in silence. Any response in any one of them can, therefore, be used to indicate that something has been detected.

The psychophysics of the model has been studied using three tasks: absolute thresholds, temporal masking curves (TMCs) to assess compression and psychophysical tuning curves (PTCs) to assess frequency selectivity. The latter two measurements use a forward-masking procedure where a target tone is presented *in silence* after the end of a pure-tone masker and therefore meets the basic requirement for using the model. In this way both human listeners and the MAP model can be tested using the same adaptive tracking software. All three tests were repeated in six different frequency regions. The complete set of measurements is called the hearing 'profile'. Figure 2.4 compares examples of profiles obtained using models and individual listeners. Figure 2.4a shows a profile of a young man with good hearing (dashed lines) and compares it with a profile obtained with the model. This 'good-hearing' model can then be used as a starting point for examining the consequences of different kinds of physiological pathology. Figure 2.4b shows the effect of reducing the contribution of outer hair cells (OHCs). A reduction of endocochlear potential has the effect shown in Fig. 2.4c, while Fig. 2.4d shows the effect of a reduction in the density of IHC/AN fibre synapses.

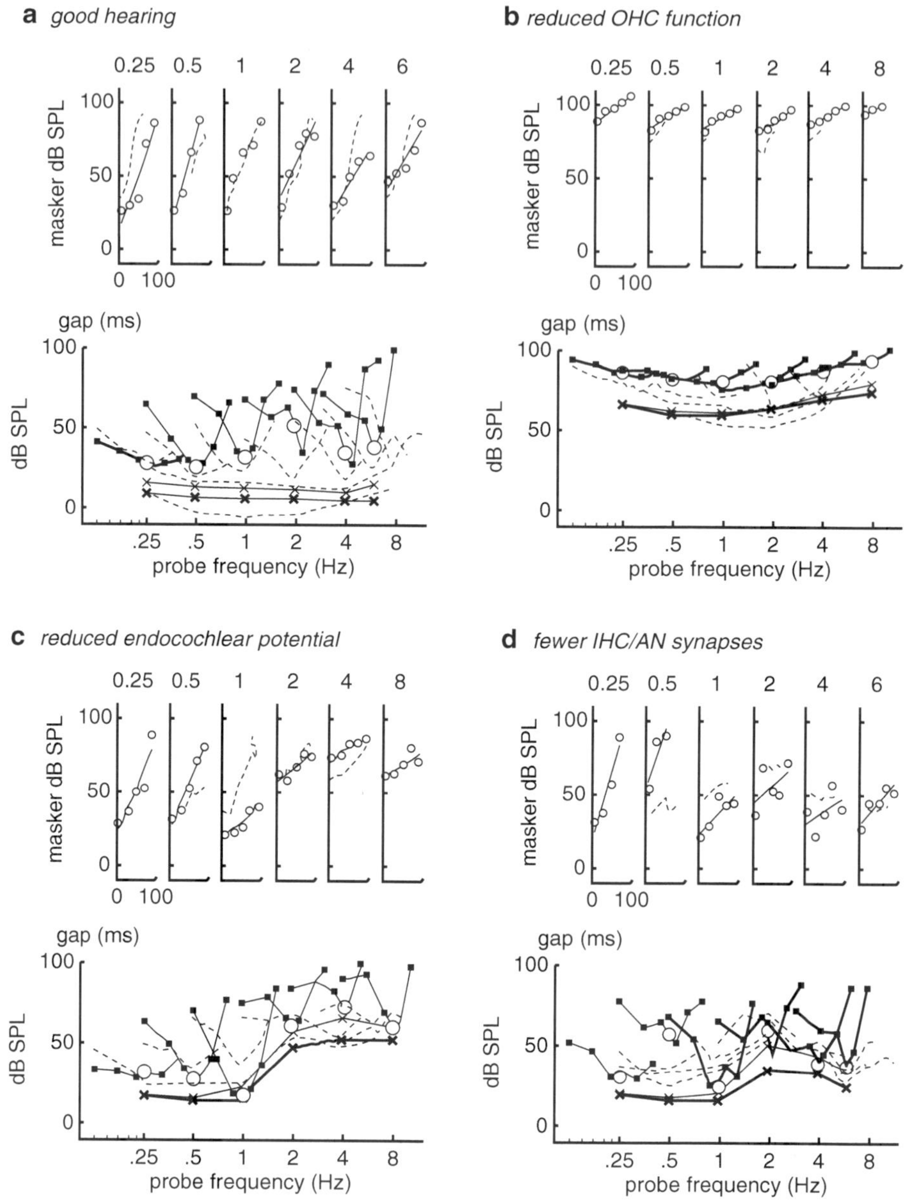

Fig. 2.4 Hearing dummies. (**a**) Good hearing. (**b**) Reduced OHC function (nonlinear path gain reduced to 6 % of original gain). (**c**) Reduced endocochlear potential (−65 mV, reduced from −100 mV). (**d**) Reduced IHC/AN synapses (reduced to 40 % of original density). In each panel, the *top row* shows TMCs at each probe frequency (in kHz). The *lower panel* shows PTCs and, below them, absolute thresholds for 16- and 250-ms tones. The continuous lines are model data. The *dashed lines* are profiles from human listeners ('NH83_R', 'IH11_R', 'IH19_R' and 'IH73_R') with profiles similar to the dummies

In each case, the pathology is simulated by changing only *one* parameter value relative to the 'good-hearing' model. In all cases, the pattern of response contains surprises that take some time to understand. This applies particularly to how the same parameter change can produce different responses at different frequencies. The profiles often have marked similarities to some of the auditory profiles measured in our hearing impaired volunteers. In each case, a real profile is presented for comparison (shown as dashed lines). A similarity between the pathological model and the individual's profile does not prove that the human subject has that particular pathology, but it is a working hypothesis supplied by the model.

5 Hearing Dummies

The original motivation for measuring patient profiles was to establish 'hearing dummy', models of the hearing of individuals with specific hearing impairments. The idea is to use these dummies for optimising the tuning of hearing aids for a given individual and to study the benefits of different hearing aid designs. The example illustrated in Fig. 2.5 shows the profile for an impaired listener (Fig. 2.5a) and the corresponding hearing dummy (Fig. 2.5b). When a new kind of hearing aid algorithm is used at the input to the dummy, the aided-model profile (Fig. 2.5d) becomes more similar to the good-hearing profile (see Fig. 2.4a). When the impaired listener is tested again (Fig. 2.5c) with the same aid settings as the model, the measured profile moves closer to the profile for good hearing.

An interesting feature of this example concerns the restoration of narrow V-shaped PTCs. The hearing aid used here was configured to restore natural instantaneous compression. The aid's algorithm is based on the architecture of the MAP model itself and represents a spin-off from the modelling exercise. However, the restoration of narrow V-shaped PTCs was not anticipated. On reflection, it could be explained by the fact that low-intensity maskers are compressed less than high-intensity maskers. There is, of course, no suggestion that the resonance characteristics of the impaired BM have been changed at all.

6 Discussion

While it is tempting to ask which one of the many auditory models now in existence is the best one, it would be a mistake to choose one and disregard the rest. Different auditory models are not just different theories; they also serve very different purposes. Each model should be judged both in terms of how well it reflects reality and how well it serves its purpose. The special function of the MAP model is to assist visualisation of what might be happening during hearing at a physiological level in the auditory periphery.

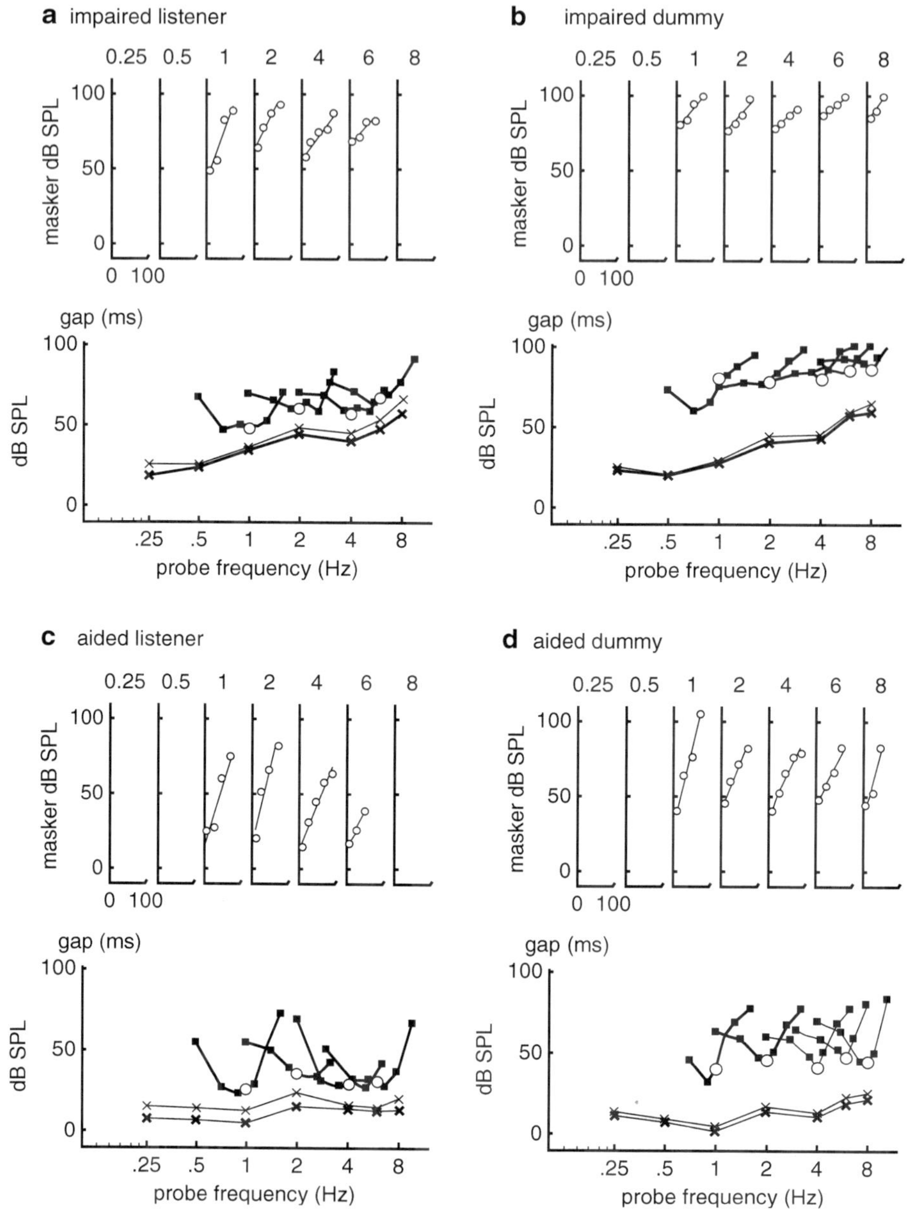

Fig. 2.5 (**a**) Auditory profile for a listener with a high-frequency hearing loss. (**b**) Profile for a corresponding hearing dummy. (**c**) Profile for the impaired listener when using the hearing aid. (**d**) Profile for the dummy when the hearing aid was used at the input to the dummy

With the MAP model much of the modelling effort is concentrated on perfecting the individual physiological modules and using realistic values for the parameters where these are known. The aim is to understand good and impaired hearing in terms of the underlying physiology and, where appropriate, its pathology. To a large extent the psychophysiological properties of the MAP model are emergent properties and sometimes come as a surprise. This was certainly the case when narrower V-shaped PTCs resulted from the application of the hearing aid algorithm (Fig. 2.5d). The selective loss of high-frequency sensitivity when the endocochlear potential was reduced (Fig. 2.4c) was also unexpected, and its explanation is subtle. The cookie-bite pattern resulting from a reduction in the number of IHC/AN synapses in Fig. 2.4d is a recent finding that remains puzzling.

Equally surprising was the finding that the MAP model had lower psychophysical thresholds for longer tones even though the model contained no component resembling an integrator.

The effect can be seen clearly in Fig. 2.4a where thresholds for a 16-ms tone (thin upper line) are consistently higher than those for a 250-ms tone (thick lower line). An integrator would be required by traditional explanations of this effect. On reflection, it was found that the reduced thresholds could be understood in terms of the probabilistic nature of the response of the decision neuron.

The software for general auditory modelling, measuring auditory profiles and running hearing dummies can be downloaded from the internet at http://www.essex.ac.uk/psychology/department/HearingLab/Welcome.html.

7 Conclusion

Computer models of the physiology of the auditory periphery can be used to explore normal and impaired hearing and sometimes spring surprises.

Comment by Fan-Gang Zeng

It is impressive that your model is able to predict a wide range of auditory behaviours including a "cookie-bite" audiogram by taking out a selective group of auditory fibres. Is it correct that your present model is largely based on average rate, without taking temporal firing patterns such as phase locking to stimulate fine structure into account?

By working with patients with auditory neuropathy, a disorder that preserves cochlear amplification function but disrupts auditory nerve functions, we found that synchronous nerve activities affect essentially all temporally based processing from microsecond-scaled interaural timing differences to minute-based loudness adaptation, particularly at low frequencies (e.g., Zeng et al. 2005). I wonder what your thoughts are on the role of temporal firing patterns in auditory behaviours and whether you plan to implement them in your model in the future.

Zeng FG, Kong YY, Michalewski HJ, Starr A (2005) Perceptual consequences of disrupted auditory nerve activity. J Neurophysiol 93:3050–3063

Reply by Meddis

The model is, in fact, sensitive to the temporal pattern of auditory nerve action potentials. Threshold decisions are not based on average AN firing rates but on coincidence-sensitive brainstem units. As such, the model should be suitable for exploring hypotheses concerning neuropathy. However, we have not yet attempted to do so.

Chapter 3
A Probabilistic Model of Absolute Auditory Thresholds and Its Possible Physiological Basis

Peter Heil, Heinrich Neubauer, Manuel Tetschke, and Dexter R.F. Irvine

Abstract Detection thresholds for auditory stimuli, specified in terms of their amplitude or level, depend on the stimulus temporal envelope and decrease with increasing stimulus duration. The neural mechanisms underlying these fundamental across-species observations are not fully understood. Here, we present a "continuous look" model, according to which the stimulus gives rise to stochastic neural detection events whose probability of occurrence is proportional to the 3rd power of the low-pass filtered, time-varying stimulus amplitude. Threshold is reached when a criterion number of events have occurred (probability summation). No long-term integration is required. We apply the model to an extensive set of thresholds measured in humans for tones of different envelopes and durations and find it to fit well. Subtle differences at long durations may be due to limited attention resources. We confirm the probabilistic nature of the detection events by analyses of simple reaction times and verify the exponent of 3 by validating model predictions for binaural thresholds from monaural thresholds. The exponent originates in the auditory periphery, possibly in the intrinsic Ca^{2+} cooperativity of the Ca^{2+} sensor involved in exocytosis from inner hair cells. It results in growth of the spike rate of auditory-nerve fibers (ANFs) with the 3rd power of the stimulus amplitude before saturating (Heil et al., J Neurosci

P. Heil (✉)
Department of Auditory Learning and Speech, Leibniz Institute for Neurobiology, Brennecke str. 6, Magdeburg 39118, Germany

Center for Behavioral Brain Sciences, Magdeburg, Germany
e-mail: peter.heil@lin-magdeburg.de

H. Neubauer • M. Tetschke
Department of Auditory Learning and Speech, Leibniz Institute for Neurobiology, Brennecke str. 6, Magdeburg 39118, Germany

D.R.F. Irvine
School of Psychology and Psychiatry, Monash University, Melbourne, VIC, Australia

Bionics Institute, Melbourne, VIC, Australia

B.C.J. Moore et al. (eds.), *Basic Aspects of Hearing*, Advances in Experimental Medicine and Biology,
DOI 10.1007/978-1-4614-1590-9_3, © Springer Science+Business Media New York 2013

31:15424–15437, 2011), rather than with its square (i.e., with stimulus intensity), as is commonly assumed. Our work therefore suggests a link between detection thresholds and a key biochemical reaction in the receptor cells.

1 Introduction

It has been known for more than a century that absolute auditory thresholds (expressed in dB) decrease with increasing stimulus duration, but the mechanisms underlying this trade-off, which is seen in every species examined, are still not fully understood. Most commonly, the threshold level–duration (TLD) functions (often termed "temporal integration functions") are interpreted as reflecting the operation of a long-time-constant temporal integrator of stimulus intensity (e.g., Plomp and Bouman 1959; Recanzone and Sutter 2008). However, this interpretation is incompatible with other experimental findings, as noted by, e.g., Gerken et al. (1990) and Viemeister and Wakefield (1991). The latter, influential, paper suggested that TLD functions can instead be explained by the assumption of the listener taking "multiple looks" at the stimulus, but this idea also raises numerous concerns (partly discussed by Viemeister and Wakefield 1991; see also Meddis and Lecluyse 2011). Here, we summarize our current thinking on this issue, based on our previous work (Heil and Neubauer 2001, 2003, 2010; Heil et al. 2006, 2008, 2011; Neubauer and Heil 2004, 2008), and we present some unpublished data to back it up. Our thinking appears to be at variance with all accounts of this issue published by others. It is most closely related to ideas recently formulated by Meddis and Lecluyse (2011) but differs in important details.

2 Methods and Results

2.1 *The LIEFTS Model Accounts for the Dependence of Detection Thresholds on Stimulus Duration and Temporal Envelope*

Inspired by the work of Gerken et al. (1990) and Solecki and Gerken (1990) and our analyses of their data (Heil and Neubauer 2003; Neubauer and Heil 2004), we measured monaural absolute thresholds of normal-hearing humans for 21 tones, with a common carrier frequency of 3.125 kHz, but differing in duration and temporal envelope (single-burst (SB), multiple-burst (MB), and onset–offset (OO) categories; Fig. 3.1a). For 7 tones, thresholds were measured twice, on different days and in different contexts. Specifically, MB tones were measured jointly with OO tones on one day and with SB tones and one OO tone on another. In this way, the reproducibility of the results could be examined, and potential day-to-day threshold

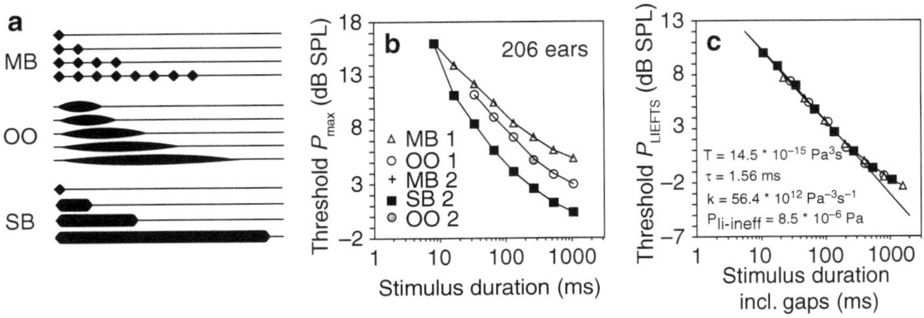

Fig. 3.1 The LIEFTS model captures absolute thresholds. (**a**) Schematic envelopes of some of the stimuli for which such thresholds were measured. They are grouped into three categories, multiple-burst (*MB*), onset–offset (*OO*), and single-burst (*SB*) stimuli. (**b**) Average monaural thresholds, in dB SPL, of 206 ears for all stimuli tested and plotted as a function of stimulus duration excluding the gaps in the *MB* series. (**c**) Thresholds after application of the model, with P_{LIEFTS} expressed in dB SPL, plotted as a function of stimulus duration including the gaps in the *MB* series. Model parameters are specified. The *thin gray line* has a slope of −1/3

fluctuations (Heil et al. 2006) could be factored out before combining the data. We employed a 3-interval-3-alternative forced-choice (3-I-3-AFC) procedure run with a 2-down-1-up rule, as described in Heil et al. (2006). Intervals were marked visually on a computer screen. Figure 3.1b plots the average thresholds for the 28 stimuli (in dB SPL, i.e., re 20 μPa) as a function of tone duration. Because of the large number of ears studied (206 in 121 participants), the associated confidence intervals are very small (not shown). The data clearly reveal that threshold SPL depends on the temporal envelope of the stimulus and, for each category, decreases progressively with increasing duration (out to durations >1 s). Although the decrease is shallower at longer durations, there is no indication that threshold asymptotes at the longer durations.

We developed a "continuous look" model, according to which the stimulus generates stochastic detection events (Heil and Neubauer 2003). Threshold is reached when a criterion number of events have occurred (probability summation). The higher the probability of detection events, the shorter the average time required to accumulate the criterion number of events, and vice versa. No long-term integration is required. Our early work (Heil and Neubauer 2003; Neubauer and Heil 2004) allowed us to conclude that the probability of occurrence of the detection events was proportional to the (effective) stimulus amplitude raised to an exponent that was most likely 3, but possibly 4 or 5. We later refined the model (Neubauer and Heil 2008) and applied it to the dependence of the latencies of first spikes (neural detection events) of ANFs on the amplitude and temporal envelope of tonal stimuli (Neubauer and Heil 2008; Heil et al. 2008). The refinement involved the addition of a leaky integrator acting on the stimulus envelope before exponentiation; the time constant was 1–2 ms, similar to the time constant of the IHC membrane. It was also assumed that the detection events are generated by an inhomogeneous Poisson process. We also confirmed the exponent to be 3.

When applying this LIEFTS model (leaky integration, event formation, temporal summation) to our psychophysical threshold data, we can ignore both spontaneous activity and a term capturing the transmission delay. The model can therefore be simplified to

$$T = \int_{-\infty}^{t_s} R(u)du = \int_{-\infty}^{t_s} k \cdot P_{li}^{\alpha}(u)du = const., \quad \text{where} \quad P_{li}(t) = 1/\tau \cdot \int_{-\infty}^{t} P(u) \cdot e^{-(t-u)/\tau}\, du \tag{3.1}$$

Here T is threshold (in $Pa^3 \cdot s$), $R(t)$ is the rate of detection events, $P(t)$ is the stimulus amplitude (in Pascals), $P_{li}(t)$ is the low-pass filtered stimulus amplitude (in Pascals), τ is the time constant of this filter (~1–2 ms), α is 3, and k is a scaling factor. The stimulus amplitude at threshold therefore varies with the duration, t_s, of the stimulus and with its temporal envelope. Figure 3.1c shows the application of this model to the thresholds shown in Fig. 3.1b. For better comparison, the ordinate in Fig. 3.1c represents the 3rd root of the integral mean of $P_{li}^{3}(t)$, P_{LIEFTS}, a quantity which can be expressed in dB SPL. If the model fully captured the data, all data points, independent of temporal envelope and duration, would fall on a single line with a slope of −1/3 in this log–log plot (or −6.67 dB per decade of duration). We find this to be the case, except for stimuli of long duration (≥0.3 s) for which thresholds exceed the model predictions slightly (by ~2 dB for the longest duration).

In the following sections we will suggest one factor that might contribute to the slight discrepancy between the data and model for long stimulus durations, provide evidence for the stochastic nature of the detection events, and confirm the value of the exponent in the model is 3.

2.2 *Deviations of Thresholds for Long-Duration Tones from the LIEFTS Model May Be Due to Lower Attention*

Here, we provide evidence for lower levels of attention during the presentation of long versus short tones, which might cause detection thresholds for long tones to be slightly higher than predicted by the LIEFTS model (Fig. 3.1c). The evidence comes from eyeblinks (EBs) measured by means of EEG and EOG in 22 subjects while they performed 3-I-3-AFC experiments as described above. EBs can be evoked by visual and (intense) auditory stimuli, but they also occur spontaneously and appear to be related to attention breaks (e.g., Nakano et al. 2009; Schleicher et al. 2008). Figure 3.2a shows the average instantaneous EB frequency during a trial. EBs are evoked by the onsets and offsets of the visual markers of the three observation intervals. EB frequency increases after the last interval and peaks sharply shortly after the button press. Figure 3.2b shows the instantaneous EB frequency over a 2-s epoch

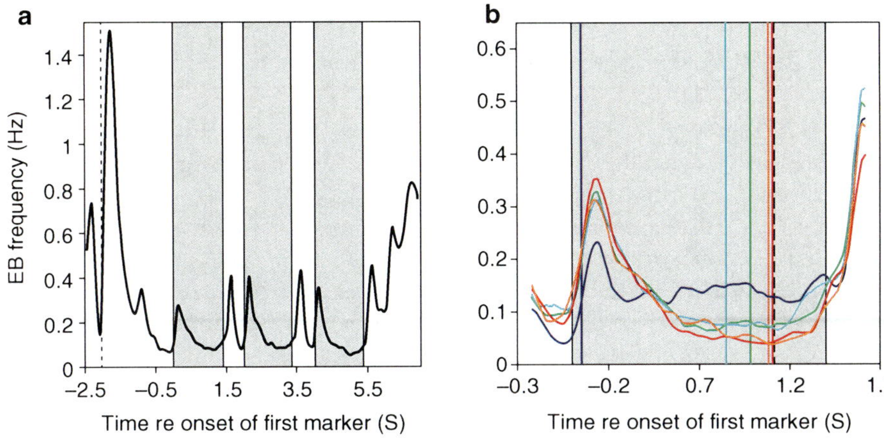

Fig. 3.2 Frequency of eyeblinks (EBs) during threshold measurements differs between long and short tones during their presentation. (**a**) EB frequency in the course of a trial of the 3-I-3-AFC threshold measurement (based on ~69.000 EBs, accumulated over all near-threshold trials, stimuli, and subjects). The visually marked observation intervals are identified by *gray bars*, the button press in response to the previous trial by the vertical *dashed line* at −2 s. (**b**) EB frequency during a 2-s epoch around the interval during which the auditory stimulus was presented. EB traces are separated according to stimulus duration. *Colored vertical lines* mark the onsets of the five different stimuli, the *dashed vertical line* their common offset

around the interval which contained the auditory stimuli. The five traces show the EB frequencies separately for the five different auditory stimuli used in these experiments (viz., 3.125-kHz tones with durations of 8.32, 30.72, 130.56, 264.32, and 1,064.32 ms), and vertical lines mark their timing within the observation intervals. The total number of EBs over the 2-s epoch is similar for the five different stimuli (~0.3 EB/epoch), but their temporal distribution is different. For the four shorter stimuli, EB frequencies drop to very low values (0.04–0.07 EB/s) around the time of stimulus presentation. For the longer stimulus, the average EB frequency during its presentation is considerably higher (~0.15 EB/s). If EBs are associated with attention breaks, so that EB frequency is inversely related to attention, then our data suggest that the average level of attention is lower during the presentation of the longest stimulus compared to the shorter stimuli. The EB patterns for those observation intervals which did not contain an auditory stimulus were very similar to that shown in Fig. 3.2b (not illustrated). This suggests that the auditory stimulus itself does not affect EB timing. Rather subjects appear to be aware when, relative to the visual markers, the auditory stimuli will occur, if they occur. This seems plausible since each track started with clearly audible stimulus levels and for each stimulus four tracks were completed by each subject.

In summary, the differential EB frequencies, and – by inference – levels of attention, during the presentation of short versus long tones may be one factor underlying the elevation of predicted thresholds for long-duration stimuli (Fig. 3.1c).

2.3 The Stochasticity of the Detection Events Can Be Shown by Means of Simple Reaction Times

The 3-I-3-AFC procedure does not allow examination of the time at which a particular stimulus is detected, but the stochastic nature of the detection events in the LIEFTS model can be tested using simple reaction times (SRTs). We have previously shown (Heil et al. 2006) that an individual SRT can be conceived of as the sum of the time required to execute the motor response and the time that a stimulus must be on before the subject detects the stimulus. We refer to this latter component as the initial effective stimulus portion. The LIEFTS model predicts that (1) the initial effective stimulus portion should most often be shorter than the stimulus duration. Hence, SRTs can be, but due to the motor component need not be, shorter than the stimulus duration. (2) For any given stimulus, which is detected on at least some of its presentations, there should be trial-to-trial variation in the duration of its initial effective portion. This causes, or contributes to, the trial-to-trial variation in SRTs. (3) When the amplitude of a test stimulus is increased, with other parameters unchanged, detection probability should increase and the average duration of the effective initial portion of the stimulus, and its variability, should decrease. Thus, mean SRT and the associated variability should decrease, so that the cumulative distributions of SRTs to test stimuli of different amplitudes diverge early and increase in steepness with increasing test stimulus amplitude. (4) When the duration of a test stimulus is increased, with other parameters unchanged, detection probability should also increase. In this case, however, the cumulative distributions of SRTs to test stimuli of different durations should have a common initial course. For any two stimuli of different durations, the common course of the SRT distributions should last as long as the duration of the shorter stimulus. (5) Since the exponent α in the LIEFTS model is associated with the (low-pass filtered) stimulus amplitude and is greater than 1, the increase of detection probability due to increasing stimulus duration by some factor is less than that due to increasing stimulus amplitude by the same factor.

We measured SRTs in a "high signal rate vigilance" procedure from 11 subjects to 3.125-kHz tones of different durations and amplitudes as described in detail in Heil et al. (2006). Figure 3.3 shows, for one representative subject, cumulative distributions of SRTs to tones of (a) fixed duration and different amplitudes and (b) fixed amplitude and different durations. All predictions listed above are verified by these data.

2.4 The Value of the Exponent $\alpha=3$ Is Verified by the Binaural Advantage at Threshold

We make the plausible assumption that at detection threshold the internal representation evoked by the stimulus has the same magnitude, independent of whether threshold is reached by stimulating the left ear (L), the right ear (R), or both ears

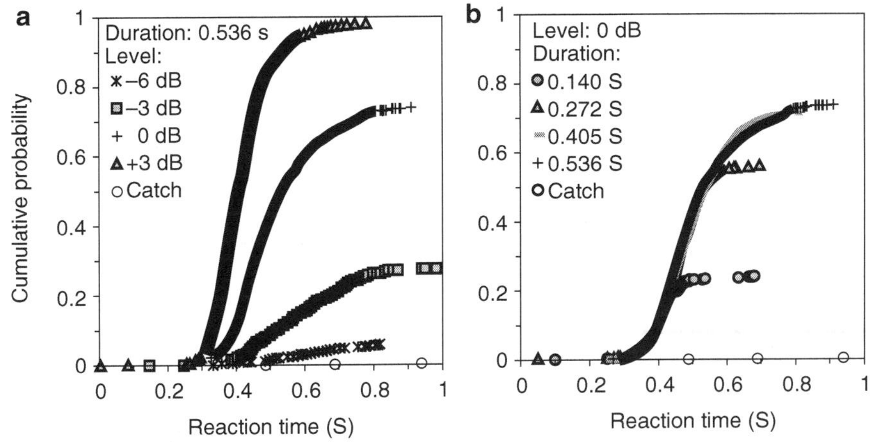

Fig. 3.3 Cumulative distributions of simple reaction times for a single subject to 3.125-kHz tones of (**a**) four different amplitudes but identical duration (0.536 s) (0 dB corresponds to 1.3 dB SPL) and (**b**) four different durations with the same amplitude. Each distribution is based on 780 stimulus presentations. The total set comprised 16 different stimuli. Stimuli were presented at intervals varying randomly between 2.5 and 4 s. The probability of false alarms, measured in 780 interspersed catch trials, was low in this subject, but can be factored out (Tiefenau et al. 2006)

(B). This assumption leads to an equation, which provides an explicit formulation of the threshold for a binaurally and diotically presented stimulus, P_B (in Pascals), as a function of the thresholds for the same stimulus presented monaurally to the left and the right ears (P_L, P_R):

$$P_B = \left(\frac{1}{P_L^{\alpha}} + \frac{1}{P_R^{\alpha}} \right)^{-\frac{1}{\alpha}} \tag{3.2}$$

The only parameter of the equation that cannot be directly measured is the exponent α. Its value, however, can be determined by comparing the observed binaural thresholds with those predicted by the equation for different α. For a fixed value of $\alpha > 0$, Eq. 3.2 predicts lower binaural than monaural thresholds. It predicts the largest binaural advantage, i.e., the difference in dB between the monaural threshold of the better ear and the binaural threshold, when the monaural thresholds are identical. It predicts the magnitude of this advantage and how it approaches zero as the threshold difference between the two ears increases. Model predictions for three integer values of α are shown in Fig. 3.4a.

We tested the model by measuring monaural (left and right) and binaural thresholds of 32 normal-hearing subjects for a set of four 3.125-kHz tones, differing in duration and temporal envelope. Neither variable affected the results, so the data from all conditions were combined. Figure 3.4a plots the binaural advantage against the absolute difference (in dB) of the monaural thresholds. Individual data points

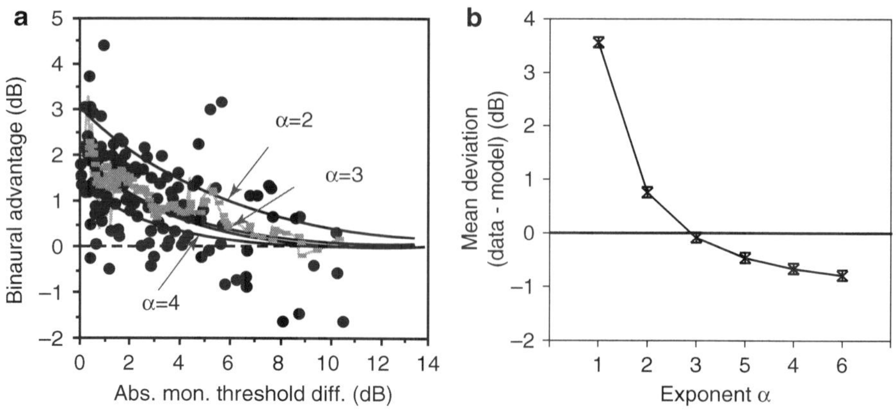

Fig. 3.4 The binaural advantage at threshold confirms a value of $\alpha=3$. (**a**) Binaural advantage, i.e., the difference between the threshold for the better ear and the binaural threshold, plotted as a function of the absolute difference between the monaural thresholds (*black circles*). The gliding average over points (*grey dashes*) reveals the decrease in the binaural advantage from about 2 dB towards zero as the monaural threshold difference increases. Predictions of Eq. 3.2 for $\alpha=2$, $\alpha=3$, and $\alpha=4$ are shown as *hyperbolic lines*. (**b**) Mean ± 1 SEM of the deviations between the 128 binaural thresholds measured and those predicted for different values of α. For $\alpha=3$, the mean deviation is close to zero

($n=128$) scatter, but a gliding average reveals the trend in the data more clearly. Figure 3.4b shows the mean deviations between the 128 measured binaural thresholds and those predicted by Eq. 3.2 for different values of α. The magnitude and sign of the mean deviation varies systematically with α. For values of $\alpha<3$, the predicted binaural thresholds are lower, and for $\alpha>3$ higher, than observed. For $\alpha=3$, the mean deviation is close to zero (−0.09 dB). For this value of the exponent, the predicted maximum binaural advantage is 2.0 dB, the advantage most commonly reported in the literature in the case of equal monaural thresholds.

3 Conclusions

Our data and analyses strongly support a probabilistic "continuous look" model of absolute detection threshold. Several lines of evidence suggest that the exponent is 3. We have previously shown that this exponent originates in the auditory periphery (Heil et al. 2008, 2011; Neubauer and Heil 2008): the mean firing rate of ANFs grows with the 3rd power of stimulus amplitude before saturating, not with its square (i.e., with stimulus intensity), as is commonly assumed (e.g., Müller et al. 1991). An intriguing possibility is that the exponent reflects the cooperative binding of Ca^{2+}-ions to the Ca^{2+}-sensor involved in fast exocytosis from the inner hair cell (Heil and Neubauer 2010; Heil et al. 2011). Ultimately, therefore, a key biochemical reaction in transmitter release might be the major determinant of the shape of the TLD functions at the perceptual level, which are so strikingly similar across different species (Heil and Neubauer 2003), suggesting a common mechanism conserved in evolution.

Acknowledgment This study was supported by the Deutsche Forschungsgemeinschaft (SFB-TRR 31, A6 to PH).

References

Gerken GM, Bhat VKH, Hutchison-Clutter M (1990) Auditory temporal integration and the power function model. J Acoust Soc Am 88:767–778

Heil P, Neubauer H (2001) Temporal integration of sound pressure determines thresholds of auditory-nerve fibers. J Neurosci 21:7404–7415

Heil P, Neubauer H (2003) A unifying basis of auditory thresholds based on temporal summation. Proc Natl Acad Sci U S A 100:6151–6156

Heil P, Neubauer H (2010) Summing across different active zones can explain the quasi-linear Ca2+-dependencies of exocytosis by receptor cells. Front Synaptic Neurosci 2:148

Heil P, Neubauer H, Tiefenau A, von Specht H (2006) Comparison of absolute thresholds derived from an adaptive forced-choice procedure and from reaction probabilities and reaction times in a simple reaction time paradigm. J Assoc Res Otolaryngol 7:279–298

Heil P, Neubauer H, Brown M, Irvine DRF (2008) Towards a unifying basis of auditory thresholds: distributions of the first-spike latencies of auditory-nerve fibers. Hear Res 238:25–38

Heil P, Neubauer H, Irvine DRF (2011) An improved model for the rate – level functions of auditory-nerve fibers. J Neurosci 31:15424–15437

Meddis R, Lecluyse W (2011) The psychophysics of absolute threshold and signal duration: a probabilistic approach. J Acoust Soc Am 129:3153–3165

Müller M, Robertson D, Yates GK (1991) Rate-versus-level functions of primary auditory nerve fibres: evidence for square-law behaviour of all fibre categories in the guinea-pig. Hear Res 55:50–56

Nakano T, Yamamoto Y, Kitajo K, Takahashi T, Kitazawa S (2009) Synchronization of spontaneous eyeblinks while viewing video stories. Proc R Soc B 276:3635–3644

Neubauer H, Heil P (2004) Towards a unifying basis of auditory thresholds: the effects of hearing loss on temporal integration reconsidered. J Assoc Res Otolaryngol 5:436–458

Neubauer H, Heil P (2008) A physiological model for the stimulus dependence of first-spike latency of auditory-nerve fibers. Brain Res 1220:208–223

Plomp R, Bouman MA (1959) Relation between hearing threshold and duration of tone pulses. J Acoust Soc Am 31:749–758

Recanzone GH, Sutter ML (2008) The biological basis of audition. Annu Rev Psychol 59: 119–142

Schleicher R, Galley S, Briest S, Galley L (2008) Blinks and saccades as indicators of fatigue in sleepiness warners: looking tired? Ergonomics 51:982–1010

Solecki JM, Gerken GM (1990) Auditory temporal integration in the normal-hearing and hearing-impaired cat. J Acoust Soc Am 88:779–785

Tiefenau A, Neubauer H, von Specht H, Heil P (2006) Correcting for false alarms in a simple reaction time task. Brain Res 1122:99–115

Viemeister NF, Wakefield GH (1991) Temporal integration and multiple looks. J Acoust Soc Am 90:858–865

Chapter 4
Cochlear Compression: Recent Insights from Behavioural Experiments

Christopher J. Plack

Abstract Although physiological measures have provided a great deal of information about the basilar membrane (BM) response of non-human mammals, it is only relatively recently that behavioural techniques have allowed researchers to measure accurately the non-linear characteristics of the human BM. These techniques are based on forward masking, in which the threshold for detecting a signal is measured in the presence of a prior masking sound. Two popular techniques, the growth of forward masking technique and the temporal masking curve technique, rely on the fact that compression in the base of the cochlea is largely restricted to frequencies close to the characteristic frequency (CF) of each place. By comparing the response to a masker with a frequency equal to that of the signal with the response to a lower-frequency masker, it is possible to infer the CF response. These measures have shown that BM compression in humans matches that of other mammals and that compression is absent in listeners with moderate-to-severe cochlear hearing loss, probably reflecting outer hair cell dysfunction. Another technique, the additivity of forward masking (AFM) technique, does not rely on a comparison between on- and off-frequency maskers, but instead measures the effect on threshold of combining two nonoverlapping maskers, an effect which is magnified by compression. The difference between thresholds in the single- and combined-masker conditions can be used to estimate compression. The AFM technique has provided evidence that strong compression extends down to low CFs in humans, a finding inconsistent with direct measures of the BM response in other mammals. Furthermore, recent AFM results suggest that there may be an additional source of compression central to the BM. This more central compression also appears to be affected by hearing loss and may reflect non-linear processes in the transduction mechanism of the inner hair cells.

C.J. Plack
School of Psychological Sciences, The University of Manchester,
Manchester M13 9PL, UK
e-mail: chris.plack@manchester.ac.uk

B.C.J. Moore et al. (eds.), *Basic Aspects of Hearing*,
Advances in Experimental Medicine and Biology
DOI 10.1007/978-1-4614-1590-9_4, © Springer Science+Business Media New York 2013

1 Introduction

Physiological studies suggest that the vibration of the basilar membrane (BM) in the mammalian cochlea is amplified in a level- and frequency-dependent manner. In the base of the cochlea, the gain is maximal for low-level components close to the characteristic frequency (CF) of each place on the BM and declines as level is increased and as frequency is moved away from CF. The result is a highly compressive input/output or "response" function for pure tones at CF, but a linear function for tones remote from CF (Ruggero et al. 1997). In the apex of the cochlea, the gain is less than in the base, but a greater range of frequencies relative to CF receives amplification and compression (Rhode and Cooper 1996). In both the apex and the base, the gain is thought to be the consequence of an active mechanism, related to electro-motility in the outer hair cells (OHCs). Damage to the OHCs results in the main characteristics of cochlear hearing loss: decreased sensitivity, reduced frequency selectivity, and an abnormally rapid growth in loudness with level.

Several different behavioural techniques have been developed for measuring cochlear gain and compression in humans. The two most popular are the growth of masking (GOM) technique (Oxenham and Plack 1997) and the temporal masking curve (TMC) technique (Nelson et al. 2001). Both of these techniques rely on the assumption that in the base of the cochlea, the response to a pure tone well below CF is linear. Hence, the effect of a forward masker well below the signal frequency ("off-frequency masker") can be used as a "linear reference" for comparison with the on-frequency forward masker in order to estimate the BM response at CF. Forward masking is used to avoid non-linear interactions between masker and signal (e.g. suppression) that may confound the results.

In the GOM technique, the masker level required to mask the signal (the masker level "at threshold") for an on- and off-frequency forward masker is measured as a function of signal level. In the TMC technique (Fig. 4.1), the signal is fixed at a low level, and the masker level at threshold for an on- or off-frequency forward masker

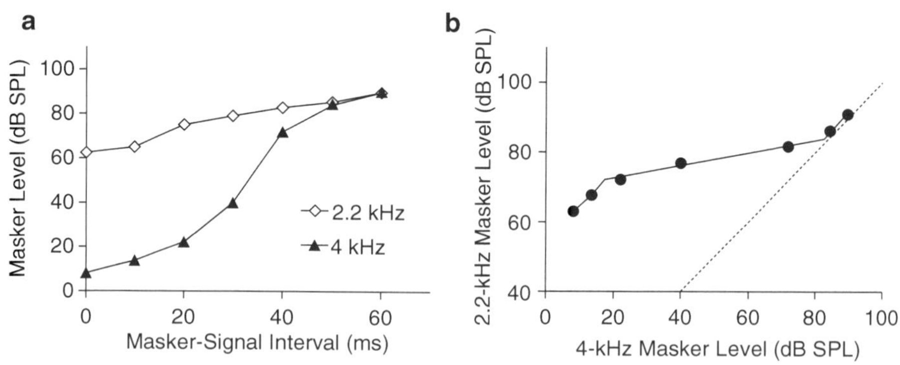

Fig. 4.1 (**a**) TMC results for a 4-kHz signal for one listener. The legend shows masker frequency. (**b**) The derived response function. The *dashed line* shows linear growth (Data are from Plack et al. (2004))

is measured as a function of masker-signal interval (to produce a TMC). It is assumed that for a given signal level or masker-signal interval, the excitation at the signal place required to mask the signal is constant. Since it is also assumed that the off-frequency masker is processed linearly at the signal place, the off-frequency masker level at threshold is effectively a measure of the BM excitation required to mask the signal (give or take an additive constant in dB). The on-frequency masker level at threshold is a measure of the input level at CF required to produce this same excitation. Hence, in both techniques, the CF response function can be derived by plotting the masker level at threshold for the off-frequency masker against the masker level at threshold for the on-frequency masker, paired by signal level (GOM) or by masker-signal interval (TMC).

2 Compression in the Base

The GOM and TMC techniques have been used to measure the BM response in the base of the cochlea (CFs of about 2 kHz and above). The results are largely consistent with direct measures in other mammals using invasive techniques such as laser interferometry (Ruggero et al. 1997). The estimated response function at CF shows a linear growth for input levels up to about 30 dB SPL, followed by a highly compressive mid-level region extending from about 30 dB SPL to about 80 dB SPL. The compression exponent (slope of the response function) in this level region is typically in the range of 0.15–0.25 dB/dB, corresponding to a compression ratio of about 5:1 (Lopez-Poveda et al. 2003; Nelson et al. 2001; Oxenham and Plack 1997). Listeners with moderate-to-severe cochlear hearing loss show a loss of gain and compression and hence a linear response function at CF (Oxenham and Plack 1997).

3 Compression in the Apex

It is known from physiological measures that compression in the apex is not frequency selective, and hence, the off-frequency masker cannot be used as a linear reference as it too will be compressed. This problem can be negotiated for the TMC technique by assuming that the rate of decay of forward masking in terms of BM excitation does not vary with frequency. The off-frequency TMC for a high signal frequency can then be used as a linear reference for *all* other frequencies (Plack and Drga 2003). For example, the 4-kHz off-frequency masker level can be plotted against the on (or off)-frequency masker level at the frequency of interest to derive a response function. Calculations such as this suggest that the basic shape of the BM response function in the apex is similar to that in the base, with a similar compression exponent (Lopez-Poveda et al. 2003; Nelson and Schroder 2004; Plack and Drga 2003). However, the range of levels that are compressed is *smaller* in the apex, and the range of frequencies relative to CF that are compressed is *greater* in the apex. Both of these findings are consistent with direct laser interferometry measures in other mammals. However, the amount of apical compression estimated in the human behavioural studies (about 5:1) is much greater than that measured directly in other mammals (about 2:1, Rhode and Cooper 1996). This could result from interspecies differences, although it is more likely a consequence of the difficulty of direct measures of the BM response in the apex, and the likelihood of damage to the active mechanism.

4 The Additivity of Forward Masking Technique

The additivity of forward masking (AFM) technique (Plack and O'Hanlon 2003) does not rely on a comparison of off- and on-frequency maskers. Instead, the technique is based on the assumption that the masking effect of two nonoverlapping forward maskers sums in a linear way. Hence, if two equally effective nonoverlapping forward maskers (M1 and M2) are combined, the internal excitation of the signal at masked threshold should increase by 3 dB (a doubling of intensity). However, if the signal is compressed prior to interacting with the maskers, the physical signal level will have to increase by more than 3 dB to produce an internal doubling in excitation. If the signal threshold in response to the individual maskers (M1 and M2) and the signal threshold in response to the combined maskers (M1 + M2) are all known, it is possible to derive the compression exponent in the level region of the signal. The AFM technique can be used without substantial modification to measure the BM response at any CF. Results with the original AFM technique have largely confirmed the GOM and TMC results and in particular have confirmed the finding that compression in the base of the cochlea is as great as that in the apex and is similarly affected by cochlear damage (Plack et al. 2008). However, recent results from a new version of the AFM technique have raised the possibility that the AFM technique may be measuring more than just BM compression.

5 Compression Central to the Basilar Membrane

The estimates of gain and compression from the GOM and TMC techniques are thought to be largely unaffected by processes central to the BM. Any subsequent processing will affect on- and off-frequency maskers equally at the signal place, and there will be no differential effect. Theoretically, however, compression estimates from the AFM technique will be affected by *any* non-linear process prior to the neural interaction of maskers and signal. The correspondence of the earlier AFM results with the GOM and TMC results suggest that any compression central to the BM is relatively insignificant. However, recent results tell a different story. Plack and Arifianto (2010) used a variant of the AFM technique in which the signal is

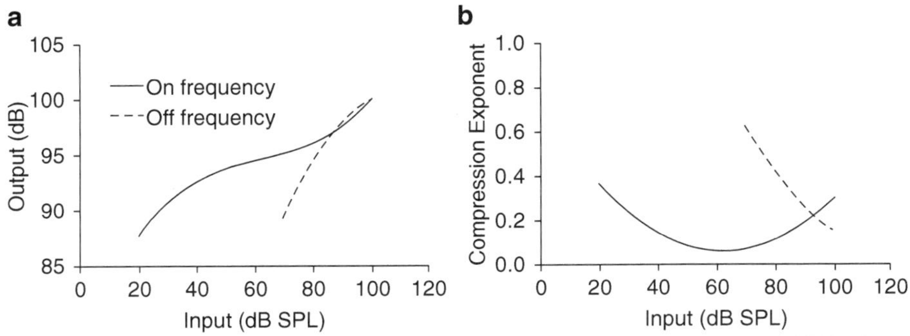

Fig. 4.2 The mean results of the masker-vary AFM experiment of Plack and Arifianto (2010) at 4 kHz. (**a**) Derived response functions for on- (4 kHz) and off- (2 kHz) frequency forward maskers. (**b**) Compression exponents derived from the response functions (i.e. the slopes of these functions). Masker durations were 20 ms (M1) and 10 ms (M2). Mean functions were derived by averaging the polynomial coefficients from the five listeners

fixed at a low level (10-dB SL) and the masker-signal gap varied. Instead of varying signal level to determine the effect on threshold of combining two maskers, the signal level is fixed and the masker levels are varied to find the threshold values for each individual masker and the two maskers combined. By comparing masker levels at threshold for each masker alone and for the combined maskers, for each masker-signal interval, it is possible to infer the BM response function. Notice that the "masker-vary" AFM technique measures compression of the *masker*. Hence, it is possible to use this technique to measure off-frequency compression at the signal place by using masker frequencies different from the signal frequency.

Plack and Arifianto found that compression estimates at 4 kHz were roughly twice as great as reported previously, both on- *and* off-frequency (Fig. 4.2). Mid-level compression was about 10:1 for the CF response, and about 2:1 for the off-frequency response. Plack and Arifianto included conditions in which the maskers and signal covered a total duration of 40 ms: too short for substantial olivocochlear efferent activation at the time of presentation of the signal, which may affect compression estimates (Jennings et al. 2009). So how do we reconcile the earlier "signal-vary" AFM results with the more recent results? First, the earlier experiments used higher-level signals, rather than a fixed low-level signal. The maximum excitation on the BM produced by the signal moves to places with higher CFs as level is increased, places for which the signal frequency is below CF and hence is processed more linearly. In addition, previous studies used a longer-duration M1, which may produce efferent effects, reducing compression estimates (Plack and Arifianto 2010).

These new AFM results are in a sense consistent with the GOM and TMC results, since the ratio between on- and off-frequency compressions is about 5:1 in all cases. So it could be that the off-frequency BM response is not really linear in the base. However, this would be inconsistent with the known physiology of the cochlea. The simplest way of reconciling the masker-vary AFM results with the earlier results is to assume that there is a source of compression, with a ratio of about 2:1, *central to the BM*. A possible location of this compression is the *IHCs* (Lopez-Poveda and Eustaquio-Martin 2006; Plack and Arifianto 2010). It is known that the receptor potential of the IHCs grows compressively with BM velocity at medium to high levels (Cheatham and Dallos 2001; Patuzzi and Sellick 1983), due to saturation of the transducer currents at the highest levels, and the basolateral potassium currents (Kros and Crawford 1990). If temporal processing is affected by IHC (or more central) compression, then a second question arises: Is this post-BM compression also affected by cochlear hearing loss?

6 Experiment: Compression at Low Levels in Normal and Impaired Ears

6.1 Method

Eight normal-hearing (NH) listeners and four ears from three listeners with cochlear hearing loss (HI) were tested. The "signal-vary" AFM technique (Plack and O'Hanlon 2003) was used. The pure-tone signal had a duration of 44 ms including 2-ms, raised-cosine, onset and offset ramps. The pure-tone maskers had durations of 20 ms (M1) and 10 ms (M2), including 5-ms ramps. The maskers and signal were presented contiguously in the order M1, M2, and signal. The signal frequency was between 3.6 and 4 kHz, positioned at peaks and valleys in the threshold microstructure for each individual. The maskers had the same frequency as the signal. In the first phase, the signal was fixed at 8 dB SL and presented with either M1 or M2. The levels of M1 and M2 were adaptively varied separately to determine threshold. In the second phase, the maskers were presented at these equally effective levels. Signal threshold was determined in the presence of M1, M2, and M1 and M2 combined. These values were used to calculate the compression exponent.

6.2 Results and Discussion

Figure 4.3 shows the mean compression ratios, converted back into compression exponents (see Plack and O'Hanlon 2003). The exponents for the NH listeners are about 0.5 (2:1 compression), but the HI ears show exponents greater than one, indicating a slight expansive non-linearity. T-tests revealed that differences between NH and HI are significant at both peak and valley locations ($p<0.05$ in each case). The results suggest that NH listeners have a compressive non-linearity at levels close to absolute threshold (8 dB SL corresponded to a mean of 17 dB SPL for the peak location and 13 dB SPL for the valley location), levels at which the BM response is

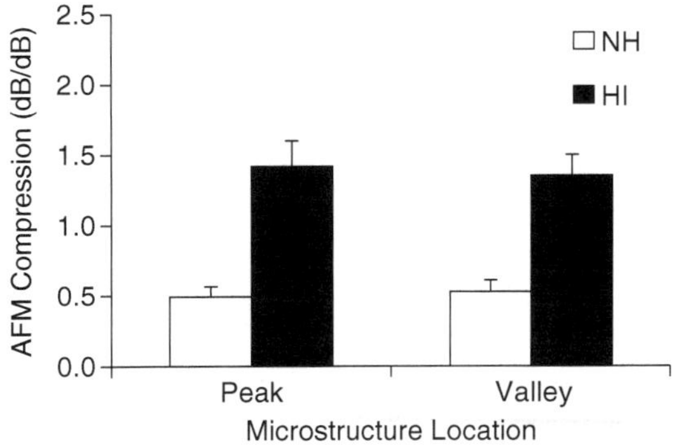

Fig. 4.3 Compression exponents at low levels derived from the signal-vary AFM results for NH and HI listeners as a function of location in the threshold microstructure. *Error bars* are standard errors

usually assumed to be linear. These results are consistent with those of Plack et al. (2008) at slightly higher levels (over 20 dB SPL). The compression appears to be absent in ears with cochlear hearing loss, although the sample is admittedly quite small. The compression could arise in the IHCs, as it is known that IHC dysfunction is associated with cochlear hearing loss. However, it is not clear that IHC receptor potentials show compressive growth near absolute threshold: most compression is observed at higher levels (Cheatham and Dallos 2001; Patuzzi and Sellick 1983).

7 Conclusions

1. Behavioural experiments suggest strong compression (about 5:1) in the basal region of the human BM at medium levels for tones close to CF. Compression in the apex is similar to that in the base except that apical compression occurs over a smaller range of levels and a wider range of frequencies relative to CF.
2. Recent AFM results suggest an additional source of compression (about 2:1) central to the BM that affects the temporal processing of sounds and that is reduced by cochlear hearing loss. It is possible that the location of this compression is the IHCs.

References

Cheatham MA, Dallos P (2001) Inner hair cell response patterns: implications for low-frequency hearing. J Acoust Soc Am 110:2034–2044

Jennings SG, Strickland EA, Heinz MG (2009) Precursor effects on behavioral estimates of frequency selectivity and gain in forward masking. J Acoust Soc Am 125:2172–2181

Kros CJ, Crawford AC (1990) Potassium currents in inner hair cells isolated from the guinea-pig cochlea. J Physiol 421:263–291

Lopez-Poveda EA, Eustaquio-Martin A (2006) A biophysical model of the inner hair cell: the contribution of potassium currents to peripheral auditory compression. J Ass Res Otolaryngol 7:218–235

Lopez-Poveda EA, Plack CJ, Meddis R (2003) Cochlear nonlinearity between 500 and 8000 Hz in listeners with normal hearing. J Acoust Soc Am 113:951–960
Nelson DA, Schroder AC (2004) Peripheral compression as a function of stimulus level and frequency region in normal-hearing listeners. J Acoust Soc Am 115:2221–2233
Nelson DA, Schroder AC, Wojtczak M (2001) A new procedure for measuring peripheral compression in normal-hearing and hearing-impaired listeners. J Acoust Soc Am 110:2045–2064
Oxenham AJ, Plack CJ (1997) A behavioral measure of basilar-membrane nonlinearity in listeners with normal and impaired hearing. J Acoust Soc Am 101:3666–3675
Patuzzi R, Sellick PM (1983) A comparison between basilar membrane and inner hair cell receptor potential input–output functions in the guinea pig cochlea. J Acoust Soc Am 74:1734–1741
Plack CJ, Arifianto D (2010) On- and off-frequency compression estimated using a new version of the additivity of forward masking technique. J Acoust Soc Am 128:771–786
Plack CJ, Drga V (2003) Psychophysical evidence for auditory compression at low characteristic frequencies. J Acoust Soc Am 113:1574–1586
Plack CJ, O'Hanlon CG (2003) Forward masking additivity and auditory compression at low and high frequencies. J Ass Res Otolaryngol 4:405–415
Plack CJ, Drga V, Lopez-Poveda EA (2004) Inferred basilar-membrane response functions for listeners with mild to moderate sensorineural hearing loss. J Acoust Soc Am 115:1684–1695
Plack CJ, Oxenham AJ, Simonson A, O'Hanlon CG, Drga V, Arifianto D (2008) Estimates of compression at low and high frequencies using masking additivity in normal and impaired ears. J Acoust Soc Am 123:4321–4330
Rhode WS, Cooper NP (1996) Nonlinear mechanics in the apical turn of the chinchilla cochlea in vivo. Aud Neurosci 3:101–121
Ruggero MA, Rich NC, Recio A, Narayan SS, Robles L (1997) Basilar-membrane responses to tones at the base of the chinchilla cochlea. J Acoust Soc Am 101:2151–2163

Chapter 5
Improved Psychophysical Methods to Estimate Peripheral Gain and Compression

Ifat Yasin, Vit Drga, and Christopher J. Plack

Abstract It is possible that previous psychophysical estimates of basilar membrane gain and compression using temporal masking curve (TMC) and additivity of forward masking (AFM) methods using long-duration maskers (>30 ms) could have been affected by activation of the medial olivocochlear reflex (MOCR) (Jennings et al. 2009; Plack and Arifianto 2010). In experiment 1, AFM and TMC methods were compared to a new fixed-duration masking curve (FDMC) method in which the combined masker and signal stimulus duration is fixed at 25 ms. Estimates of compression were found to be not significantly different for TMC, FDMC and AFM methods. Estimates of gain were similar for TMC and FDMC methods. Maximum compression was associated with a significantly lower input masker level using the FDMC compared to the TMC method. In experiment 2, the FDMC method was used to investigate the effect of efferent activation on gain and compression estimates by presenting a precursor sound prior to the combined masker-signal stimulus. Estimated gain decreased as precursor level increased, and increased as the silent interval between the precursor and combined masker-signal stimulus increased, consistent with a decay of the efferent response.

1 Introduction

Several psychophysical masking paradigms have been used to estimate cochlear gain and compression in humans. In the temporal masking curve (TMC) method (Nelson et al. 2001), masker levels at threshold are obtained for a low-level signal

I. Yasin (✉) • V. Drga
Ear Institute, University College London (UCL), 332 Grays Inn Road, London WC1X 8EE, UK
e-mail: i.yasin@ucl.ac.uk

C.J. Plack
School of Psychological Sciences,
The University of Manchester, Manchester M13 9PL, UK

B.C.J. Moore et al. (eds.), *Basic Aspects of Hearing*,
Advances in Experimental Medicine and Biology,
DOI 10.1007/978-1-4614-1590-9_5,

in the presence of off- and on-frequency forward maskers as a function of the temporal interval between signal and masker. A plot of the off- vs. on-frequency masker level paired by masker-signal silent interval provides an estimate of the basilar membrane (BM) response function. Compression exponents can be estimated from slopes of the response functions. In the additivity of forward masking (AFM) method, signal threshold is obtained in the presence of two equally effective and temporally nonoverlapping forward maskers (e.g. Plack and O'Hanlon 2003). An energy-summation model predicts that two equally effective maskers, when combined, should result in a 3-dB increase in signal threshold, compared to that obtained with individual maskers. When the signal level is compressed, the physical signal level will need to be increased by more than 3 dB (excess masking) in order to maintain signal threshold. The compression exponent can be estimated from the amount of excess masking.

An efferent neural pathway from the superior olivary complex via the olivocochlear bundle can reduce the cochlear gain applied over time to the BM response to a sound (Liberman et al. 1996). Reported time delays for the onset of the efferent response from non-invasive human studies are about 31–43 ms (James et al. 2002). However, most TMC and AFM methods have used combined masker-signal durations of greater than 50 ms; hence, results may have been affected by the efferent response (Jennings et al. 2009).

Experiment 1 compared estimates of gain and compression using AFM (with short- and long-duration maskers), TMC (with long-duration maskers) and a novel method, the fixed-duration masking curve (FDMC). In the FDMC method, masker level at threshold for a signal is obtained in the presence of off- and on-frequency forward maskers for different durations of the signal and masker, with a combined total masker and signal duration of 25 ms, thereby minimizing efferent effects. If longer masker durations activate the MOCR, then estimates of compression and gain should be similar for the TMC and AFM methods using long-duration maskers, but differ from the estimates obtained using FDMC and AFM methods using short-duration maskers. Experiment 2 used the FDMC method to investigate the timing- and level-dependent characteristics of efferent activity using precursor sounds of varying levels (20–80 dB SPL) presented from 0 to 200 ms prior to the combined FDMC masker-signal stimulus.

2 Experiment 1: Comparison of TMC, AFM and FDMC Methods

2.1 Stimuli and Conditions

Seven normal-hearing listeners participated. For both TMC and FDMC methods, the signal was a 4-kHz sinusoid presented at 10 dB SL. Masker levels at threshold were obtained for an on-frequency (4 kHz) or off-frequency (1.8 kHz) sinusoidal forward masker. For the TMC method, masker levels at threshold were obtained for a 6-ms signal and 104-ms masker, at masker-signal silent intervals of 10, 20, 30, 40, 50, 60, 70, 80 and 90 ms (see schematic in Fig. 5.1a). For the FDMC method the masker-plus-signal total duration was always 25 ms, and the masker-signal silent interval was 0 ms. Across trial blocks, signal steady-state portions varied from 0 to 15 ms in 2.5 ms increments, while the complementary masker steady-state portions varied from 15 down to 0 ms in 2.5 ms decrements (see Fig. 5.1b).

For the AFM method (as described in Plack and O'Hanlon 2003), the signal was a 6-ms, 4-kHz sinusoid and the forward maskers, M1 and M2, were samples of white noise strictly bandlimited between 3,500 and 4,500 Hz. The duration of M1 was either short (10 ms) or long (200 ms). The duration of M2 was always 10 ms. The maskers and signal (when present) were temporally nonoverlapping as shown in Fig. 5.1c. The AFM method had two phases. Phase 1 measured masker levels at threshold for which M1 and M2 were equally effective in masking a 10- or 30-dB SL signal. Phase 2 measured signal level at threshold in the presence of either M1 alone, M2 alone or M1 and M2 combined (M1 + M2), where maskers were fixed at the levels obtained in phase 1 to mask a 10- or 30-dB SL signal.

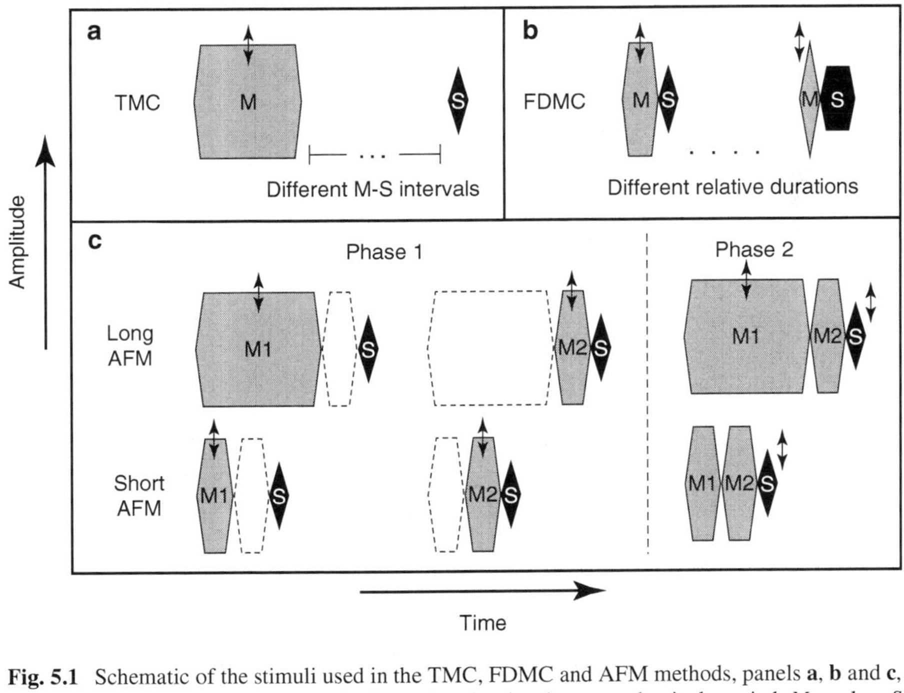

Fig. 5.1 Schematic of the stimuli used in the TMC, FDMC and AFM methods, panels **a**, **b** and **c**, respectively. *Double-headed arrows* indicate the stimulus that was adaptively varied. *M* masker, *S* signal

2.2 Results

The mean TMC and FDMC results are presented in Fig. 5.2 (left and right panels, respectively). Linear regression provided good fits (average $R^2 = 0.995$) to both the mean off-frequency TMC and FDMC data, with slopes of 0.32 and 2.13 dB/ms, respectively. The overall shape of both on- and off-frequency TMCs is similar to that reported previously (e.g. Plack and Drga 2003; Lopez-Poveda et al. 2003), and the overall shape of the FDMC function is similar to TMC functions reported here and in earlier studies.

The mean AFM results from phases 1 and 2 of the AFM experiment are presented in Fig. 5.3. The left panel presents data from phase 1, and the middle and right panels present mean data from phase 2 with a short or long M1 masker, respectively. Thresholds obtained for the M1 + M2 masker are greater than thresholds for short M1, long M1 or M2 masker alone, and excess masking (>3 dB difference) is observed at the higher level.

Mean response functions were obtained from the AFM data for each listener by applying the fitting procedure described by Plack and Arifianto (2010). Inferred BM response (input–output (I/O)) functions for the TMC and FDMC data were derived by plotting off-frequency vs. on-frequency masker levels, paired by masker-signal silent interval (TMC) or by signal duration (FDMC). The derived I/O functions are shown in Fig. 5.4 (left panel). The TMC and FDMC data points are connected by faint dashed lines representing the 3rd-order polynomial fit to the I/O functions. The response functions show typical non-linear I/O characteristics; i.e. a steep portion (for low input levels) followed by a shallow portion (input levels around 40–60 dB SPL). Compression exponents were obtained by differentiating the 3rd-order polynomial fits to the I/O functions and are shown in Fig. 5.4 (right panel). Two measures of compression were obtained: an average value of compression exponent for inputs between 40 and 60 dB SPL (CE_{40-60}) and a minimum value (CE_{Min}). The

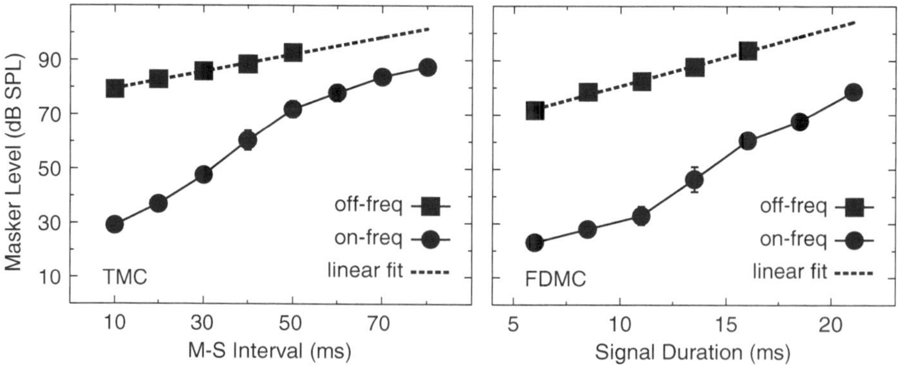

Fig. 5.2 Mean TMC (*left panel*) and FDMC (*right panel*) results, showing the masker level at threshold as a function of masker-signal silent interval (TMC) or signal duration (FDMC) for an on-frequency (*circles*) and an off-frequency masker (*squares*). Linear fits to the off-frequency mean data are shown by the overlaid *dashed line*. Standard *error bars* are also shown

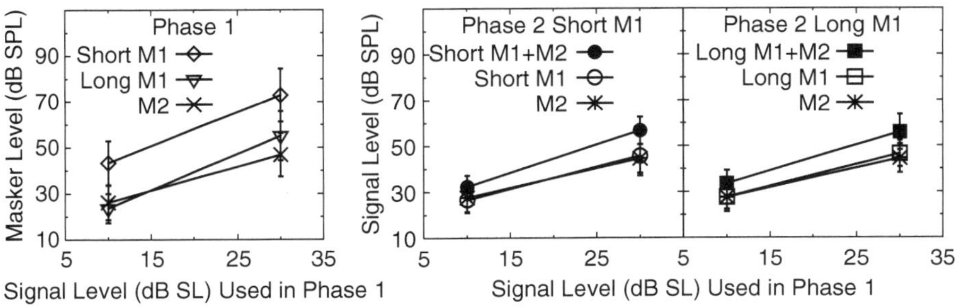

Fig. 5.3 Mean results of the AFM experiment. *Left panel* shows the results of phase 1. *Middle* and *right panels* show the results of phase 2. Phase 1 and 2 masker-signal levels at threshold are shown for a 10- or 30-dB SL signal

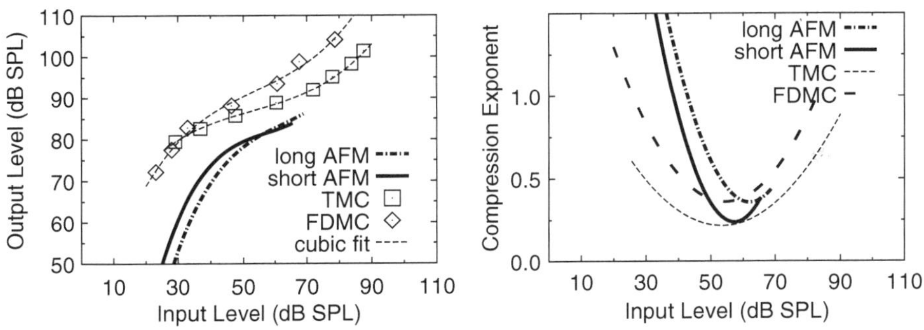

Fig. 5.4 Mean derived I/O functions (*left panel*) and compression exponents (*right panel*) for TMC and FDMC methods (*open squares* and *diamonds*, respectively), overlaid by a 3rd-order polynomial fit to the data shown by a *dashed line*. Also shown are mean derived I/O functions and compression exponents for AFM with a long masker (*dashed-plus-dotted line*) and AFM with short masker (*solid bold line*). Note that the vertical locations of the AFM I/O functions are arbitrary

maximum level of gain ($Gain_{Max}$) was estimated as the difference between off- and on-frequency maskers for the smallest masker-signal silent interval of 10 ms (TMC) or the shortest signal duration (FDMC).

Prior to the analysis, data outliers were eliminated by use of Tukey's method, which defines outliers as greater than 1.5 interquartile ranges below the 25th percentile or above the 75th percentile (Tukey 1977). One-way ANOVAs with main factor of method (4 levels) did not show a significant effect for values of either CE_{40-60} or CE_{Min}, but revealed a significant effect for the input masker level associated with CE_{Min}, $F_{(3,15)}=3.30$, $p<0.05$, with effect size, $\eta^2=0.40$. Post hoc paired t-tests (Bonferroni corrected) revealed that the FDMC method resulted in CE_{Min} at lower input masker levels than the TMC method [mean difference between input masker levels = 3.40, SD = 1.86, $t(5)=4.61$, p(two-tailed) <0.01]. Estimates of $Gain_{Max}$ from TMC and FDMC methods did not differ significantly. Pearson's r revealed a significant negative correlation between absolute threshold and values of $Gain_{Max}$ estimated from TMC [$r(5)=-0.82$, p(one-tailed) <0.05] and FDMC

[$r(5) = -0.74$, p(one-tailed) < 0.05] methods. Absolute thresholds were also found to be highly correlated with input levels associated with CE_{Min} for TMC [$r(5) = 0.93$, p(two-tailed) < 0.05], FDMC [$r(5) = 0.97$, p(two-tailed) < 0.01] and long AFM [$r(5) = 0.87$, p(two-tailed) < 0.05].

3 Experiment 2: Effect of Precursor Level and Temporal Delay

3.1 Stimuli and Conditions

Five listeners participated. The masker and signal characteristics were the same as for Experiment 1 for FDMCs except that the combined masker-signal stimulus was preceded by a 500-ms precursor noise (1-kHz wide noise band centred at 4 kHz), presented at 20, 40, 60 and 80 dB SPL, each at precursor-to-masker (P-M) silent intervals of 0, 50, 100 and 200 ms. A condition was also run with no precursor using the same within-trial timing as for a P-M interval of 0 ms. Since the precursor acted as a forward masker on the signal, precursor-only masked thresholds were first measured. The signal level was then set at 10 dB above this level to measure the FDMCs.

3.2 Results

The mean results are presented in Fig. 5.5. The left panel presents gain as a function of precursor level; the right panel presents gain as a function of P-M interval. For a given P-M interval, increasing precursor level appears to reduce the gain (estimated as the difference in masker levels for on- and off-frequency maskers for the shortest

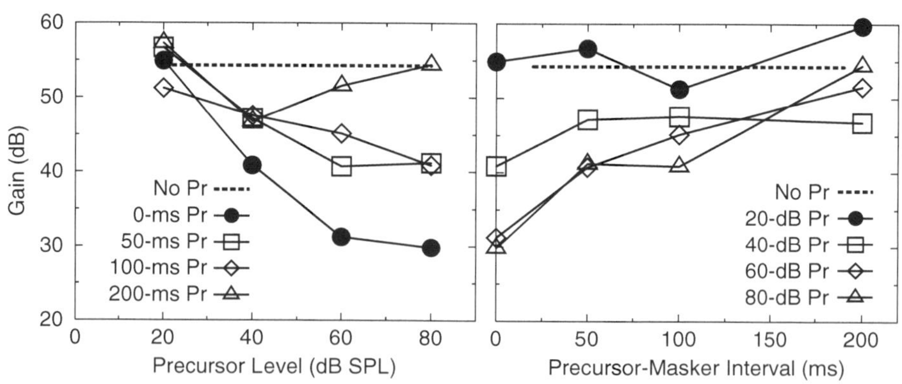

Fig. 5.5 Mean results for the precursor experiment. Estimated gain is shown as a function of precursor level for the parameter of P-M silent interval (*left panel*) and as a function of P-M silent interval for the parameter of precursor level (*right panel*). The *horizontal dashed line* represents the estimated gain in the absence of a precursor

signal duration). For a given precursor level, increasing the P-M silent interval results in an increase in estimated gain. A two-way ANOVA with main factors of precursor level and P-M silent interval showed a significant effect of precursor level [$F_{(3,12)}=14.09$, $p<0.01$, with effect size, $\eta^2=0.78$], P-M interval [$F_{(3,12)}=5.28$, $p<0.05$, with effect size, $\eta^2=0.57$] and a significant interaction between precursor level and P-M interval [$F_{(9,36)}=2.94$, $p<0.05$, with effect size, $\eta^2=0.42$]. It appears that effect of P-M interval on recovery of gain increases with precursor level.

4 Discussion

Estimates of compression, CE_{40-60} and CE_{Min}, are not significantly different between TMC and FDMC methods and are on average similar to those reported by previous TMC studies (e.g. Nelson et al. 2001; Lopez-Poveda et al. 2003; Plack and Drga 2003; Yasin and Plack 2003; Rosengard et al. 2005). Estimates of CE_{40-60} and CE_{Min} are also not significantly different for AFM with short or long maskers, although they indicate less compression than previous studies using the signal-vary AFM method with a long- or short-duration M1 masker (Plack and O'Hanlon 2003; Plack and Arifianto 2010; Plack et al. 2008). The value of CE_{Min} is associated with a lower input masker level using the FDMC method compared to the TMC method, consistent with the suggestion that activation of the efferent response could displace the operating point of OHC reverse transduction affecting feedback to the BM (e.g. Murugasu and Russell 1996). However, such a shift in operating point may be expected to be associated with a reduction in gain, which was not observed in the present study.

FDMCs were also obtained with and without a precursor. As P-M silent interval was increased, the gain increased indicating a recovery from the efferent effect at longer P-M silent intervals around 200 ms, corresponding to the estimated MOCR offset and decay (Backus and Guinan 2006) and the decrease in the temporal effect seen for 100-ms precursor durations (Roverud and Strickland 2010). As the precursor level was increased, gain was reduced, by about 5.9 dB per 10-dB increase in precursor level (for a 0-ms P-M silent interval), near the upper end of the estimate reported by Strickland (2008). As temporal delay between precursor and masker was increased to 100 ms, the rate of gain reduction fell to about 1.3 dB per 10-dB increase in precursor level, indicative of the interaction between activation of efferent gain reduction (due to increasing precursor level) and the decay of the efferent response (due to increasing P-M delay).

5 Conclusions

1. Estimates of the compression exponent were found to be similar between TMC, FDMC and AFM methods.
2. The new FDMC technique can produce reliable estimates of cochlear gain and compression, in the absence of a confound from efferent activation by the masker.

3. The FDMC technique may also provide an unbiased measure of efferent gain reduction by a precursor.
4. Preliminary results are broadly consistent with the expected effects on gain of precursor level and precursor-signal time interval.

Acknowledgements The research was supported by EPSRC grant EP/H022732/1.

References

Backus BC, Guinan JJ Jr (2006) Time-course of the human medial olivocochlear reflex. J Acoust Soc Am 119:2889–2904

James AL, Mount RJ, Harrison RV (2002) Contralateral suppression of DPOAE measured in real time. Clin Otolaryngol 27:106–112

Jennings SG, Strickland EA, Heinz MG (2009) Precursor effects on behavioural estimates of frequency selectivity and gain in forward masking. J Acoust Soc Am 125:2172–2181

Liberman MC, Puria S, Guinan JJ Jr (1996) The ipsilaterally evoked olivocochlear reflex causes rapid adaptation of the $2f_1$-f_2 distortion product otoacoustic emission. J Acoust Soc Am 99:3572–3584

Lopez-Poveda EA, Plack CJ, Meddis R (2003) Cochlear nonlinearity between 500 and 8000 Hz in listeners with normal hearing. J Acoust Soc Am 113:951–960

Murugasu E, Russell IJ (1996) The effect of efferent stimulation on basilar membrane displacement in the basal turn of the guinea pig cochlea. J Neurosci 16:325–332

Nelson DA, Schroder AC, Wojtczak M (2001) A new procedure for measuring peripheral compression in normal-hearing and hearing-impaired listeners. J Acoust Soc Am 110:2045–2064

Plack CJ, Arifianto D (2010) On- and off-frequency compression estimated using a new version of the additivity of forward masking technique. J Acoust Soc Am 128:771–785

Plack CJ, O'Hanlon CG (2003) Forward masking additivity and auditory compression at low and high frequencies. J Assoc Res Otolaryngol 4:405–415

Plack CJ, Oxenham AJ, Simonson A, O'Hanlon CG, Drga V, Arifianto D (2008) Estimates of compression at low and high frequencies using masking additivity in normal and impaired ears. J Acoust Soc Am 123:4321–4330

Plack CJ, Drga V (2003) Psychophysical evidence for auditory compression at low characteristic frequencies. J Acoust Soc Am 113:1574–1586

Rosengard PS, Oxenham AJ, Braida LD (2005) Comparing different estimates of cochlear compression in listeners with normal and impaired hearing. J Acoust Soc Am 117:3028–3041

Roverud E, Strickland EA (2010) The time course of cochlear gain reduction measured using a more efficient psychophysical technique. J Acoust Soc Am 128:1203–1213

Strickland EA (2008) The relationship between precursor level and the temporal effect. J Acoust Soc Am 123:946–954

Tukey J (1977) Exploratory data analysis. Addison-Wesley Publishing Co., Reading

Yasin I, Plack CJ (2003) The effects of a high-frequency suppressor on tuning curves and derived basilar membrane response estimates. J Acoust Soc Am 114:322–332

Chapter 6
Contralateral Efferent Regulation of Human Cochlear Tuning: Behavioural Observations and Computer Model Simulations

Enrique A. Lopez-Poveda, Enzo Aguilar, Peter T. Johannesen, and Almudena Eustaquio-Martín

Abstract In binaural listening, the two cochleae do not act as independent sound receptors; their functioning is linked via the contralateral medial olivo-cochlear reflex (MOCR), which can be activated by contralateral sounds. The present study aimed at characterizing the effect of a contralateral white noise (CWN) on psychophysical tuning curves (PTCs). PTCs were measured in forward masking for probe frequencies of 500 Hz and 4 kHz, with and without CWN. The sound pressure level of the probe was fixed across conditions. PTCs for different response criteria were measured by using various masker-probe time gaps. The CWN had no significant effects on PTCs at 4 kHz. At 500 Hz, by contrast, PTCs measured with CWN appeared broader, particularly for short gaps, and they showed a decrease in the masker level. This decrease was greater the longer the masker-probe time gap. A computer model of forward masking with efferent control of cochlear gain was used to explain the data. The model accounted for the data based on the assumption that the sole effect of the CWN was to reduce the cochlear gain by ~6.5 dB at 500 Hz for low and moderate levels. It also suggested that the pattern of data at 500 Hz is the result of combined broad bandwidth of compression and off-frequency listening. Results are discussed in relation with other physiological and psychoacoustical studies on the effect of activation of MOCR on cochlear function.

E.A. Lopez-Poveda, PhD (✉)
Instituto de Neurociencias de Castilla y León, Universidad de Salamanca,
Calle Pintor Fernando Gallego 1, 37007 Salamanca, Spain

Instituto de Investigaciones Biomédicas de Salamanca, Universidad de Salamanca,
Salamanca, Spain

Departamento de Cirugía, Facultad de Medicina, Universidad de Salamanca,
Salamanca, Spain
e-mail: ealopezpoveda@usal.es

E. Aguilar • P.T. Johannesen, MSc • A. Eustaquio-Martín, MSc
Instituto de Neurociencias de Castilla y León, Universidad de Salamanca,
Calle Pintor Fernando Gallego 1, 37007 Salamanca, Spain

Instituto de Investigaciones Biomédicas de Salamanca, Universidad de Salamanca, Salamanca, Spain

B.C.J. Moore et al. (eds.), *Basic Aspects of Hearing*,
Advances in Experimental Medicine and Biology,
DOI 10.1007/978-1-4614-1590-9_6,

1 Introduction

Human auditory perception depends on the frequency- and level-dependent gain and tuning characteristics of the human cochlea. It is not yet possible to directly measure these characteristics in living subjects for obvious reasons. There exist, however, innocuous psychoacoustical techniques that, with reasonable assumptions, allow us to infer such characteristics in an approximate manner. A wealth of data pertaining to human nonlinear cochlear frequency selectivity has been collected over the years using these techniques. These data have been used to develop theories of human auditory perception as well as computer models of human cochlear processing. Some of these theories and models have influenced the design of auditory prostheses and automatic speech recognition systems.

In developing these models and applications, it has been commonly assumed that the response characteristics of the human cochlea are fixed. That is, that the characteristics inferred in laboratory conditions using simple monaural sounds are representative of cochlear responses in natural, binaural listening to time-varying complex sounds, such as speech or music. This assumption is probably wrong.

Indeed, the central auditory system has the capacity to modulate cochlear responses. Medial olivo-cochlear (MOC) efferent fibres project to the outer hair cells, changing their motility and thus the operation characteristics of the cochlear amplifier. It is this amplifier that determines the nonlinear characteristics of the cochlear response. Ipsilateral and/or contralateral sounds can cause MOC efferent fibres to initiate a reflex in a few tens of milliseconds. This means that the MOC reflex (MOCR) is almost certainly activated during typical binaural listening conditions and that it modulates the characteristics of human cochlear responses dynamically.

Our long-term goal is to use behavioural data and methods to characterize and model human cochlear nonlinearity and tuning in natural binaural listening conditions. Here, we summarize baseline results obtained using a broadband contralateral noise (CWN). We also introduce here a new time-domain computer model of forward masking with efferent control. In the model, it is assumed that the CWN elicits the MOCR and this reduces the cochlear gain to low- and moderate-level sounds. The model is crucial for explaining some seemingly paradoxical aspects of the behavioural data. A comprehensive account of the work summarized here can be found in Aguilar et al. (2013).

2 Experimental Studies

2.1 Methods

Psychophysical tuning curves (PTCs) were obtained for probe frequencies of 500 Hz and 4 kHz in the presence and in the absence of a CWN. The durations of the probes and the maskers were 10 and 200 ms, respectively. The level of the CWN was fixed at 60 dB SPL. Physiological experiments show that this level is enough to activate the MOCR without activating the middle-ear acoustic reflex (Lilaonitkul and Guinan 2009). The CWN had a duration of 1,210 ms. It started 500 ms before and ended well after the stimuli used to measure the PTCs. PTCs were measured in forward masking using a method similar to that of Lopez-Poveda et al. (2007). PTCs were measured for different masker-probe time intervals (2, 10, and 50 ms) to assess tuning over a range of levels. The effect of the CWN on the internal (i.e., post cochlear compression) rate of recovery from forward masking was assessed by measuring a linear reference temporal masking curve (TMC) for a probe frequency of 4 kHz and a masker frequency of 1,600 Hz (Lopez-Poveda et al. 2003). It is important to note that the sound pressure level (SPL) of the probes was identical with and without the CWN. It was held constant at 10 dB above individual absolute threshold for the probe measured without the CWN.

2.2 *Results and Discussion*

2.2.1 The Effect of the CWN on the Rate of Recovery from Forward Masking

The slope of a linear reference TMCs is thought to reflect the rate of recovery from forward masking without the influence of cochlear compression (Nelson et al. 2001). The linear reference TMCs measured with and without the CWN had virtually identical slopes (0.16 dB/ms) (Fig. 6.1). Assuming that the effect of the CWN was to elicit the MOCR, this suggests that the MOCR had no effect on the recovery from forward masking per se.

Figure 6.1 also suggests that the CWN shifted the mean linear reference TMC downwards by ~1.14 dB. This would be consistent with the expected effect of the CWN activating the MOCR; it reduces cochlear gain and thereby the response to the fixed-SPL probe. The observed shift, however, was neither statistically significant nor consistent across subjects. Indeed, it was smaller than the masker-level step size (2 dB) used in our adaptive procedure. This suggests that either the CWN did not activate the MOCR at 4 kHz or that the effect of MOCR activation on the fixed-SPL 4-kHz probe was so small that it became undetectable by our measuring method. Physiological

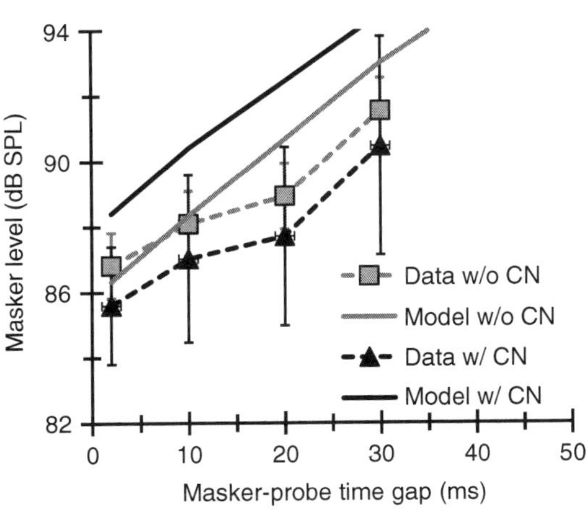

Fig. 6.1 Linear reference TMCs obtained with and without the CWN. *Symbols* and *dotted lines*: mean experimental data; *error bars* illustrate ± one standard deviation. *Continuous lines*: computer model simulations

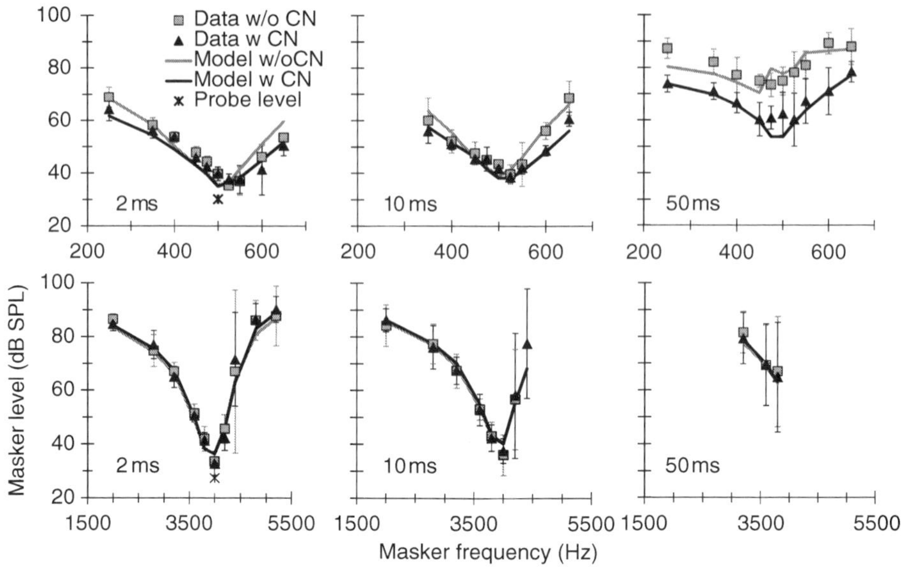

Fig. 6.2 Mean behavioural PTCs (*symbols*) and model PTCs (*lines*) at 500 Hz (*top*) and 4 kHz (*bottom*), with and without the CWN. Each panel is for a different masker-probe time interval, as indicated in the *bottom-left corner* of the panel. *Error bars* are associated to the *symbols* and illustrate ± one standard deviation. *Asterisks* in the left-most panels depict the probe levels

studies demonstrate that a 60-dB SPL CWN is sufficient to evoke the MOCR (Lilaonitkul and Guinan 2009). Therefore, the latter explanation seems more likely.

2.2.2 The Effects of the CWN on the PTCs

The CWN had no significant effect on the PTCs at 4 kHz (Fig. 6.2, bottom). At 500 Hz, by contrast, the CWN led to lower masker levels at masked threshold in some conditions (Fig. 6.2, top). This effect occurred for all masker frequencies at the longest masker-probe time interval (50 ms). For the shorter intervals (2 and 10 ms), it only occurred for masker frequencies remote from the probe frequency, thus broadening the near-threshold PTCs.

The present results appear inconsistent with the study of Vinay and Moore (2008) on the effect of a CWN on near-threshold PTCs. They showed that the CWN had a different effect on the PTCs depending on the probe frequency. At 500 Hz, the CWN typically shifted the low-frequency side of the PTCs *upwards* and increased the tuning significantly. The difference with the present results may reflect methodological differences between studies. Indeed, Vinay and Moore (2008) measured PTCs using simultaneous rather than forward masking, and their probe and CWN levels differed from those used here.

At first sight, the pattern of the present results appears inconsistent with the effects of electrical activation of the MOCR on high-frequency basilar membrane (BM) responses, which shifts only the tip of the tuning curves upwards (Cooper and Guinan 2006). The computer model simulations described below will show that this inconsistency is more apparent than real.

3 Computer Model Simulations

A computer model of forward masking with efferent control was developed and used to test the assumption that the pattern of experimental data was consistent with the hypothesis that the CWN reduces cochlear gain by activating the MOCR. The model was inspired by the BM-temporal window model (Plack et al. 2002). In the current version of the model, the dual-resonance nonlinear (DRNL) filter (Meddis et al. 2001) was replaced by a version with efferent attenuation (Ferry and Meddis 2007). The latter simulates physiological observations at the level of the BM and auditory nerve by means of a single parameter that attenuates the input signal to the nonlinear path of the standard DRNL filter. Another novelty of the current model is that it accounts for the off-frequency listening effects at 500 Hz. This was deemed necessary because no precaution was taken experimentally to minimize off-frequency listening effects on PTCs that may have occurred due to the brief duration of the probe (10 ms).

The model was implemented in the time domain and evaluated for identical stimuli and conditions as used in the experiments. Its parameters were optimized automatically as follows. First, the parameters of the standard DRNL filter (without efferent attenuation) *and* the temporal window were optimized simultaneously to mimic the 4-kHz PTCs and the linear reference TMCs measured without the CWN. The resulting temporal window parameters were held constant for the other test frequencies and conditions. It was assumed that the rate of recovery from forward masking is constant across probe frequencies and that it is unaffected by CWN (see Fig. 6.1). Second, the parameters of the standard DRNL filter (without efferent attenuation) were optimized at 500 Hz using the experimental PTCs measured without the CWN. Lastly, the parameter controlling the amount of efferent attenuation was optimized to maximize the fit between simulated and experimental PTCs measured with CWN. The value of this parameter was allowed to vary across probe frequencies.

Importantly, rather than fitting the model to the mean data, a different set of model parameters was obtained for each individual data set, and the individual model responses were averaged and compared to the mean experimental PTCs.

3.1 Results and Discussion

The grey continuous line in Fig. 6.1 illustrates the mean linear reference TMC resulting from optimizing the model to simultaneously fit the experimental linear reference TMCs and the 4-kHz PTCs. Its slope was slightly steeper than that of the corresponding experimental curve (0.22 vs. 0.16 dB/ms). Although the difference in slope was not statistically significant, the goodness of fit to the experimental PTCs became poorer when the temporal window parameters were independently optimized to match the slope of the linear reference TMCs. This suggests that the actual recovery from forward masking could be slightly steeper than suggested by the measured linear reference TMC. This is not unreasonable considering that the experimental curve is based on measured values and it disregards runs that called for values higher than the maximum system output (105 dB SPL). The experimental linear reference TMC might have been steeper if the masker level had been allowed to go higher.

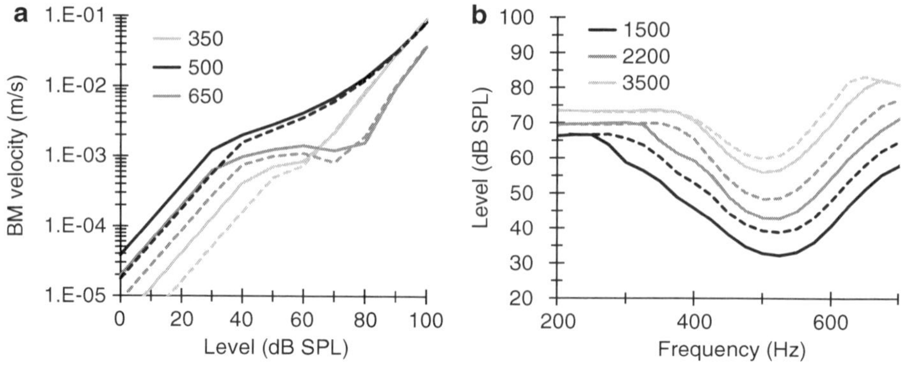

Fig. 6.3 BM responses in the 500-Hz model for a specific subject without (*continuous lines*) and with (*dashed lines*) efferent attenuation of −6.9 dB. (**a**) Input/output curves for three stimulus frequencies, as indicated by the legend (in Hz). (**b**) Tuning curves for three different response criteria, as indicated by the legend (in μm/s). Note that this pattern of responses is broadly consistent with the known physiological effects of MOCR activation on BM responses

The model reproduced the PTCs measured with and without CWN very well, both qualitatively and quantitatively (Fig. 6.2). This is rather remarkable considering that the effect of the CWN was accounted for by a single model parameter, namely, the attenuator to the input signal to the nonlinear path of the DNRL filter (Ferry and Meddis 2007). In the model, the mean value of this attenuator was −6.76 and +1.12 dB at 500 Hz and 4 kHz, respectively. The difference across test frequencies was significant ($p \sim 0.01$, $N=3$, two-tailed paired Student's t-test).

The efferent control of the BM stage in the current model is physiologically plausible (see Fig. 6.3 and Ferry and Meddis (2007)). Note also that the simulated BM tuning curves (Fig. 6.3b) are broader than the corresponding behavioural PTCs (Fig. 6.2). Still, the model mimics the behavioural PTCs rather well, including the paradoxical effects of the CWN described above. In the 500-Hz model, those effects are the result of using an identical probe SPL with and without the CWN combined with a broad bandwidth of compression and off-frequency listening. What is important here is that the model supports the hypothesis that the observed effect of the CWN on the PTC is due to MOCR activation and that the magnitude of the effect is significantly greater at 500 Hz than at 4 kHz. It is also important to stress that the behavioural data are likely affected by MOCR effects, but the form and magnitude of these effects may *not* be inferred directly from the data. Instead, a nonlinear computer model is necessary to properly infer MOCR effects from the data.

4 Overall Discussion

The present results suggest that the effects of MOCR activation by CWN are stronger at 500 Hz than at 4 kHz. This is consistent with recent human otoacoustic emission studies that have reported that CWNs have stronger effects on compression threshold (Bhagat and Carter 2010) and on cochlear response latency (Francis and Guinan 2010) at lower than at higher test frequencies. They are inconsistent, however, with other behavioural studies that report that ipsilateral precursors designed to activate the MOCR change cochlear gain and tuning at 4 kHz (Jennings et al. 2009). The reason for this inconsistency is uncertain, but there is evidence that the ratio of ipsilateral to contralateral MOCR may vary with test frequency, as reviewed by Guinan (2006). The effect of ipsilateral precursors at 500 Hz is uncertain.

5 Conclusions

Contralateral broadband noises have a measurable effect on behavioural PTCs at 500 Hz but not at 4 kHz. The pattern of the effect is explained by a computer model of forward masking based on the sole assumption that the contralateral noise reduces BM sensitivity to low- and moderate-level sounds by MOCR activation.

The results show that the response in the human cochlea depends on the presence and level of any contralateral sounds. This should be taken into account when developing auditory filterbanks and related applications. Further research is necessary to fully characterize the size of these changes, their dependence on the characteristics of the contralateral sounds, and their significance in perception.

Comment by Glennis Long

How sure are you that the middle ear muscle reflex (MEMR) did not contaminate your estimates of the effects of efferent activation? The contralateral stimulation (CAS) level used has been shown to evoke an MEMR is some individuals (Guinan et al. 2003). Furthermore a MEMR can be facilitated when additional stimuli are used (Kawase et al. 1997) so that even if the individuals do not have a MEMR to the contralateral stimulus alone, the MEMR may well be evoked when the maskers are combined with the CAS.

Guinan JJ Jr, Backus BC, Lilaonitkul W, Aharonson V (2003) Medial olivocochlear efferent reflex in humans: otoacoustic emission (OAE) measurement issues and the advantages of stimulus frequency OAEs. J Assoc Res Otolaryngol 4:521–540

Kawase T, Hidaka H, Takasaka T (1997) Frequency summation observed in the human acoustic reflex. Hear Res 108:37–45

Reply by Lopez-Poveda

This comment is very useful. There are three possibilities: (a) that our 60 dB SPL contralateral noise by itself activated the MEMR; (b) that the high level tonal maskers by themselves activated the MEMR; and (c) that the high level maskers *combined* with the contralateral noise activated the MEMR. Case (b) would not pose a problem because we are interested in the differential effect of the contralateral noise. As for case (a), we have experimentally confirmed that our 60 dB SPL contralateral white noise did not activate the MEMR over the range of masker frequencies used in our experiments. This control experiment was done using the method of Lilaonitkul and Guinan (2009) and involved the use of low level (40 dB SPL) test tones, so we cannot rule out case (c). We may control for possibility (c) by repeating the control experiment using higher level test tones. Unfortunately, I do not think we will be able to use test tone levels as high as 95–100 dB SPL (the highest masker levels used in our experiments) because our probe microphone is unreliable at these levels. That said, our model accounts for the data based solely on the assumption that the effect of the contralateral noise is to reduce cochlear gain. This supports the idea that the measured effects are due to MOCR activation by the contralateral noise; or, at least, that MOCR and MEMR effects are indistinguishable.

Lilaonitkul W, Guinan JJ Jr (2009) Reflex control of the human inner ear: a half-octave offset in medial efferent feedback that is consistent with an efferent role in the control of masking. J Neurophysiol 101:1394–1406

Acknowledgements Work supported by a grant from the Spanish Ministry of Economy and Competitiveness (ref. BFU2009-07909) to EALP and by a doctoral studentship of the Chilean CONICYT to EA.

References

Aguilar E, Eustaquio-Martin A, Lopez-Poveda EA (2013) Contralateral efferent reflex effects on threshold and supra-threshold psychoacoustical tuning curves at low and high frequencies. J. Assoc. Res. Otolaryngol. DOI: 10.1007/s10162-013-0373-4

Bhagat SP, Carter PH (2010) Efferent-induced change in human cochlear compression and its influence on masking of tones. Neurosci Lett 485:94–97

Cooper NP, Guinan JJ Jr (2006) Efferent-mediated control of basilar membrane motion. J Physiol 576:49–54

Ferry RT, Meddis R (2007) A computer model of medial efferent suppression in the mammalian auditory system. J Acoust Soc Am 122:3519–3526

Francis NA, Guinan JJ Jr (2010) Acoustic stimulation of human medial olivocochlear efferents reduces stimulus-frequency and click-evoked otoacoustic emission delays: Implications for cochlear filter bandwidths. Hear Res 267:36–45

Guinan JJ (2006) Olivocochlear efferents: anatomy, physiology, function, and the measurement of efferent effects in humans. Ear Hear 27:589–607

Jennings SG, Strickland EA, Heinz MG (2009) Precursor effects on behavioral estimates of frequency selectivity and gain in forward masking. J Acoust Soc Am 125:2172–2181

Lilaonitkul W, Guinan JJ Jr (2009) Human medial olivocochlear reflex: effects as functions of contralateral, ipsilateral, and bilateral elicitor bandwidths. J Assoc Res Otolaryngol 10:459–470

Lopez-Poveda EA, Plack CJ, Meddis R (2003) Cochlear nonlinearity between 500 and 8000 Hz in listeners with normal hearing. J Acoust Soc Am 113:951–960

Lopez-Poveda EA, Barrios LF, ves-Pinto A (2007) Psychophysical estimates of level-dependent best-frequency shifts in the apical region of the human basilar membrane. J Acoust Soc Am 121:3646–3654

Meddis R, O'Mard L, Lopez-Poveda EA (2001) A computational algorithm for computing nonlinear auditory frequency selectivity. J Acoust Soc Am 109:2852–2861

Nelson DA, Schroder AC, Wojtczak M (2001) A new procedure for measuring peripheral compression in normal-hearing and hearing-impaired listeners. J Acoust Soc Am 110:2045–2064

Plack CJ, Oxenham AJ, Drga V (2002) Linear and nonlinear processes in temporal masking. Acta Acust/Acustica 88:348–358

Vinay, Moore BCJ (2008) Effects of activation of the efferent system on psychophysical tuning curves as a function of signal frequency. Hear Res 240:93–101

Chapter 7
Modeling Effects of Precursor Duration on Behavioral Estimates of Cochlear Gain

Elin M. Roverud and Elizabeth A. Strickland

Abstract Physiological data show that preceding sound can reduce cochlear amplifier gain via the medial olivocochlear reflex (MOCR). Our laboratory has used forward masking as a tool to look for evidence of this effect psychophysically, which has led us to reassess mechanisms of forward masking. The traditional temporal window model (TWM) proposes that masking occurs via an excitatory mechanism that integrates within a temporal window. We propose that gain reduction may also contribute to forward masking. In our stimulus paradigm, growth of masking is measured with an off-frequency forward masker to estimate the basilar membrane input/output (I/O) function. The gain of the I/O function is reduced when an on-frequency precursor is introduced, consistent with a gain reduction hypothesis. Recently we explored the time course of this estimated gain reduction by examining the effect of precursor duration (Roverud and Strickland 2010). In that study, thresholds initially increased with increasing precursor duration, then decreased for longer durations. This result is not consistent with solely excitatory masking, but may reflect gain reduction by the MOCR. If the precursor is long enough, it could be influenced by the gain reduction it elicited. In the present study, we examine the effect of precursor duration with an on-frequency precursor and an off-frequency precursor. If intense enough, an off-frequency precursor may reduce gain at the signal frequency place. However, assuming it has no gain at the signal place, it would not be influenced by the reduction in gain, regardless of its duration. We developed a modified TWM that includes time-varying gain reduction by the precursor, resulting in an adapting I/O function. Results are modeled with the standard TWM and the TWM with gain reduction.

E.M. Roverud (✉) • E.A. Strickland
Department of Speech, Language, and Hearing Sciences,
Purdue University, Heavilon Hall, 500 Oval Drive,
West Lafayette, IN 47907, USA
e-mail: eroverud@purdue.edu

B.C.J. Moore et al. (eds.), *Basic Aspects of Hearing*,
Advances in Experimental Medicine and Biology,
DOI 10.1007/978-1-4614-1590-9_7,

1 Introduction

Threshold for a signal may be elevated by a preceding sound even with no temporal overlap, a phenomenon known as forward masking. A well-known theory proposes that forward masking is due to the integration of excitation from the masker and signal within a temporal window. This has been suggested to reflect a limit in the temporal acuity of the auditory system (Oxenham and Moore 1994). This temporal window model (TWM) does not describe physiological mechanisms of forward masking, but may represent a number of effects including neural adaptation. Another possible mechanism of forward masking is the medial olivocochlear reflex (MOCR), which reduces the gain of the cochlear amplifier in response to sound (Guinan 2006). The MOCR is sluggish, with a delay from elicitor onset of approximately 25 ms and gradual buildup; there is a similar offset delay and decay from elicitor offset (Backus and Guinan 2006). Thus, a sound following an elicitor may be processed by a cochlea with reduced gain.

We have explored this gain reduction effect psychophysically using a forward masking paradigm. Growth of masking is measured with an off-frequency masker to estimate the basilar membrane input/output (I/O) function. The estimated gain of the I/O function is reduced when a long on-frequency precursor, intended to elicit the MOCR, is presented prior to the masker. One potential avenue for distinguishing the mechanisms of forward masking may lie in the expected timing of the effects. The sluggishness of the MOCR may mean that gain reduction begins later than integration of excitation and persists longer.

In Roverud and Strickland (2010), we examined the time course of forward masking gain reduction by manipulating precursor duration and delay. In that study, for some subjects, a shorter (50-ms) precursor resulted in a greater threshold shift (gain reduction) than a longer (100-ms) one. This pattern, which will be called rollover, is inconsistent with integration of excitation. The amount of rollover in Roverud and Strickland (2010) was only a few dB and could possibly be attributed to random variability. However, in Oxenham and Plack (2000), signal threshold tended to roll over with increasing masker duration (from 30 to 200 ms) when there was a 20-ms delay between masker and signal but not in the no-delay condition, which argues against a random variability explanation. In both studies, rollover was seen when there was a 20-ms delay (filled either with silence or with a masker) between precursor offset and signal onset.

Our proposed explanation of rollover is that, if long enough, the gain of an on-frequency precursor will be affected by the gain reduction it elicited. This could render later-occurring portions of the precursor less effective at excitatory masking and at reducing gain for the following signal. To test this explanation, we compared the effect of precursor duration with on- and off-frequency precursors. The off-frequency precursor, if intense enough, could reduce gain at the signal frequency place. However, assuming this precursor has no gain at the signal place, it would not be influenced by gain reduction. We predicted that the effect of precursor duration would be different for on- and off-frequency precursors, with no rollover for off-frequency precursors. We compared predictions of the data with the TWM and a TWM that incorporates gain reduction.

2 Methods

Data for three normal-hearing participants (ages 22–26) will be discussed. All conditions, with the exception of quiet threshold, consisted of a precursor, masker, and signal presented sequentially. The signal was a 4-kHz, 6-ms sinusoid with 3-ms $\cos^2$ ramps. Ramps for all other stimuli were 5 ms. The masker was a 2.4-kHz, 20-ms sinusoid. Following this short masker, the signal falls within the MOCR onset delay and should not be influenced by gain reduction by the masker. The precursor was 0.8, 2.4, or 4 kHz. The 2.4- and 4-kHz precursors, with durations ranging from 10 to 150 ms, were intended to elicit gain reduction at the signal frequency. Even with the shortest precursor, 10 ms, the delay between precursor onset and signal onset exceeds the MOCR onset delay and should be within the buildup of gain reduction (Roverud and Strickland 2010). High-pass noise with a lower cutoff frequency of 4.8 kHz was present in all conditions to limit off-frequency listening. The stimuli were generated digitally and presented to the listener's right ear through an ER-2 insert earphone. Thresholds are the average of at least two runs with standard deviations less than 5 dB. If learning was observed, only later thresholds were included in the average.

3 Procedures

In Exp. I, once quiet threshold for the signal was determined, off-frequency growth of masking (GOM) data were collected for a range of fixed signal and masker levels to estimate the cochlear I/O function. A 0.8-kHz, 100-ms, 40-dB SPL precursor was present for the GOM conditions to maintain similar temporal characteristics across experiments. This was the control condition with no gain reduction (Jennings et al. 2009).

In Exp. II, for each subject, a masker level that produced a signal threshold on the linear portion of the GOM function was selected for the remainder of the study to provide a fixed amount of excitatory masking. The shift in this masked signal threshold by a 4-kHz (on-frequency) and a 2.4-kHz (off-frequency) precursor was determined. On-frequency precursor levels were 40, 50, and 60 dB SPL. Off-frequency precursor levels were set at 85, 90, and 95 dB SPL. These levels were selected to produce the same signal threshold as the on-frequency precursors for at least one duration, so that any deviations at other durations between the two frequencies would be evident. The difference in masked signal threshold between the 2.4- or 4-kHz precursor and the control condition (the 0.8-kHz precursor) will be taken to be the gain reduction produced by the precursor (Roverud and Strickland 2010).

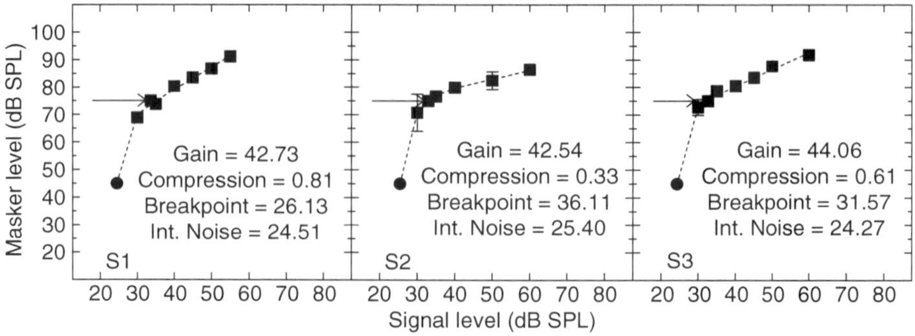

Fig. 7.1 GOM data (*filled squares*) with I/O function fits (*dashed lines*). *Filled circles* are quiet threshold. *Arrows* indicate fixed masker levels for Exp. II

4 Results: Experiment I

GOM functions are shown in Fig. 7.1. The masker level fixed in Exp. II is indicated by an arrow. Each GOM function was fitted with a piecewise linear function described by Yasin and Plack (2003) and modified by Jennings and Strickland (2010) to include an internal noise parameter. The Yasin and Plack (2003) equations are

$$L_{\text{out}} = L_{\text{in}} + G, \quad L_{\text{in}} \leq BP1, \tag{7.1}$$

$$L_{\text{out}} = c \times L_{\text{in}} + BP1 \times (1 - c) + G, \quad BP1 < L_{\text{in}} \leq 100, \tag{7.2}$$

where L_{in} is input signal level, L_{out} is masker level, G is gain, $BP1$ is breakpoint, and c is the slope between $BP1$ and 100 dB. The modification by Jennings and Strickland (2010) included parameter α to represent constant low-level internal noise. The intensity of α is subtracted from the intensity levels of L_{in} and $BP1$ in Eqs. 7.1 and 7.2. The parameter estimates of the fits are presented in Fig. 7.1.

5 Results: Experiment II

On-frequency precursor data are shown as filled symbols in Fig. 7.2. Open symbols are off-frequency precursor data. The dashed horizontal line is threshold for the control condition. The difference between each threshold and this line is assumed to reflect the amount of gain reduction by the precursor. In some panels, on-frequency precursor thresholds increase with increasing precursor duration up to 50 ms and then roll over, forming an oscillating pattern. In other panels, thresholds increase with duration up to about 50 ms then plateau for longer durations. Off-frequency

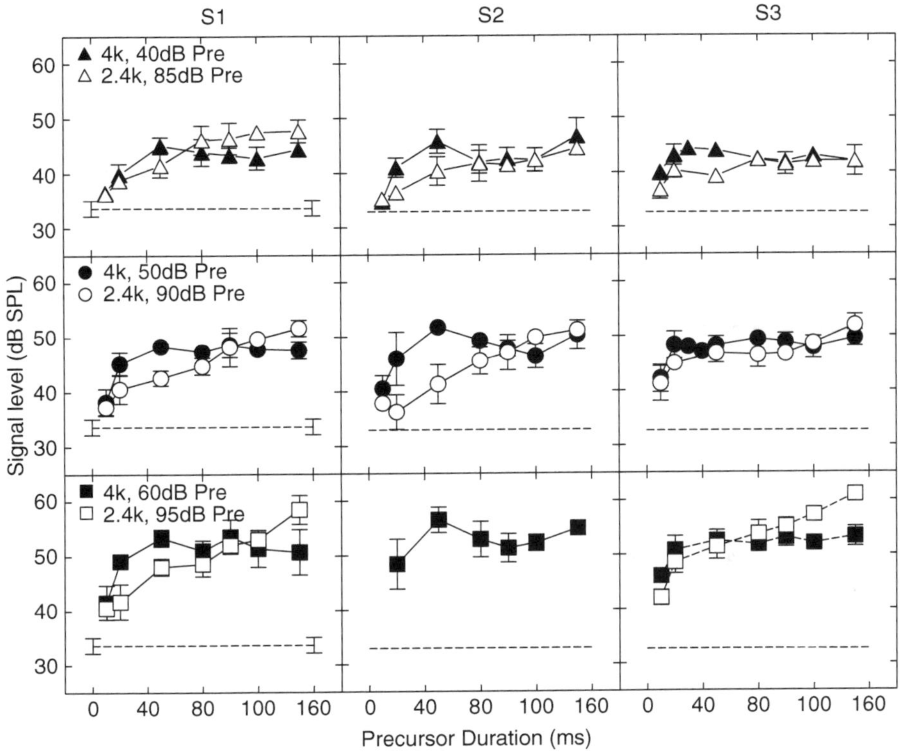

Fig. 7.2 Precursor duration data for on-frequency (*filled symbols*) and off-frequency (*open symbols*) precursors. *Dashed lines* are thresholds with a 0.8-kHz precursor

precursor thresholds generally increased with increasing duration. A repeated-measures mixed model ANOVA was performed to evaluate the effects of precursor frequency, level, and duration on threshold. We also evaluated the effect of subject. All main effects were significant ($p<0.01$). All two-way interactions were also significant ($p<0.01$), including a significant interaction of frequency × duration, $F(6,396)=27.19$ ($p<0.0001$), which supports our hypothesis that the effect of duration differs for the two precursor frequencies. A three-way interaction of subject × frequency × duration was not significant.

6 Modeling

We used MATLAB (MathWorks, Natick, MA) to model our results. We compared the standard TWM to a modified TWM with history-dependent gain variation simulating the MOCR. The TWM in its standard form includes four main modules (Oxenham and Moore 1994). The first is peripheral filtering. Second is a compressive nonlinearity

Table 7.1 Parameter estimates and RMS errors for TWM and TWM-GR fits

TWM	T_1	T_2		w	β		SNR	RMS error
S1	99.05	0		0	−5.90		0.05	3.07
S2	1	73.13		0.13	10.66		0.19	3.77
S3	69.57	0		0	1.45		0.07	3.83
TWM-GR	T_1	β	*max ΔG*	**del**	L_{win}	t_{win}	**SNR**	**RMS error**
S1	2.87	0.48	32.89	6	164	33.21	0.75	1.76
S2	73.13	8.21	23.40	16	152	21.01	0.45	2.05
S3	69.57	4.85	18.50	26	79	6.50	0.06	1.86

representing the basilar membrane I/O function, followed by half-wave rectification. Third is a sliding integrating window. The window for forward masking is

$$W(t) = (1 - w)exp(t / T_1) + w\,exp(t / T_2) \quad (7.3)$$

where W describes window shape, t is time relative to the window center, T_1 and T_2 are time constants, and w is the weighting of the relative contribution of T_1 and T_2. The fourth module is a decision mechanism where the maximum ratio of masker + signal to masker-alone (SNR) is determined. It is assumed that this SNR is constant across conditions.

In our version of the TWM, instead of a filtering module, stimuli were classified as on- or off-frequency relative to the signal and were represented as half-wave rectified envelopes. Output levels were determined on a point-by-point basis with a sampling rate of 10 kHz. Off-frequency stimuli had a linear representation at the signal frequency; only the intercept β was free to vary (Output = Input − β). On-frequency stimuli were subjected to compressive nonlinearity (Eqs. 7.1 and 7.2 with no internal noise). G, c, and $BP1$ remained fixed at the original parameter estimates (Fig. 7.1). Internal noise limited the output minimum of all stimuli. The on- and off-frequency outputs were combined and convolved with the forward masking window (Eq. 7.3). This procedure was repeated for each condition over a range of signal levels. Threshold predictions were determined by minimizing RMS error on β, T_1, T_2, w, and SNR. For each subject, all data were fit at once. Parameter estimates and RMS errors are presented in Table 7.1.

In the gain reduction version of the TWM, TWM-GR, there was a feedback loop between the I/O function and an integrating window that determined the amount of gain change with an imposed MOCR delay (Fig. 7.3). Off-frequency precursors have no gain to adjust, but are capable of eliciting gain reduction with high enough output. The equations for calculating the on-frequency I/O function at each sample, modified from Jennings and Strickland (2010), are

$$Gadapt(t) = G, \quad t \leq del / 1{,}000 \times SR \quad (7.4)$$

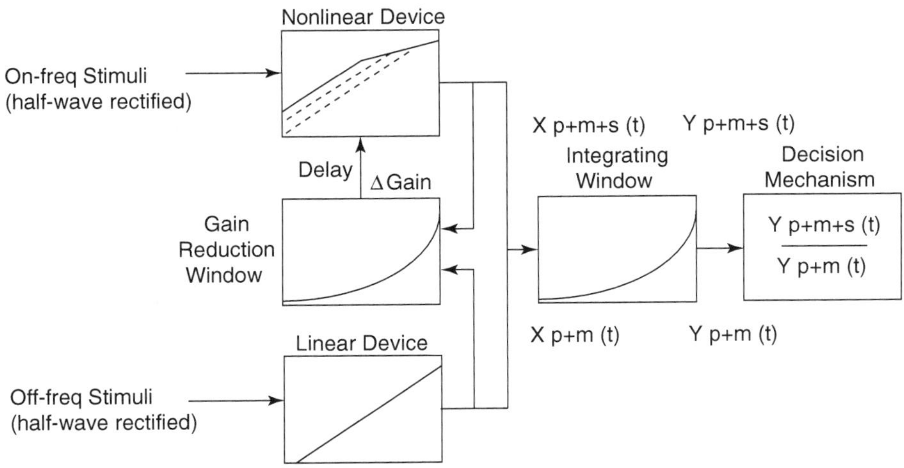

Fig. 7.3 Schematic of the TWM-GR

$$Gadapt(t) = G - \Delta G\big(t - ((del/1{,}000) \times \mathrm{SR})\big), \quad t > del/1{,}000 \times SR \tag{7.5}$$

$$BP1adapt(t) = \big(G + BP1 - c \times BP1 - Gadapt(t)\big)/1 - c \tag{7.6}$$

where G, $BP1$, and c are the original parameter estimates (Fig. 7.1), t is the sampling point, del is delay in ms, SR is sampling rate, and ΔG is the output of the gain reduction window.

The gain reduction window is intended to simulate the MOCR. Threshold and maximum elicitor levels for this window were set as the output levels for inputs of 30 and 60 dB SPL, respectively, to contain the precursor levels in this study. Once scaled to the window, outputs of on- and off-frequency stimuli from sample 1 to t were multiplied by

$$T_{\mathrm{Win}} = exp(L_{\mathrm{win}}/\tau_{\mathrm{win}}) \tag{7.7}$$

where L_{win} is length and τ_{win} is the time constant of the window. Output of this window corresponds to amount of gain change, ΔG, which is some proportion of parameter $max\ \Delta G$. On- and off-frequency outputs were combined and convolved with the forward masking window (Eq. 7.3) with parameter w fixed at 0 (one time constant), because the second time constant occurs in the gain reduction window (Eq. 7.7). Parameters for the TWM-GR are T_1, β, $max\ \Delta G$, del, L_{win}, τ_{win}, and SNR. T_l maximum was limited to the largest time constant from the TWM fit. The fits for both TWM and TWM-GR are shown for one subject (S2) in Fig. 7.4 (also see Table 7.1).

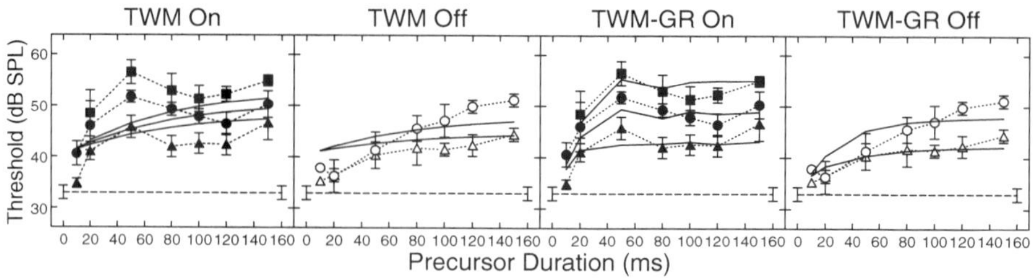

Fig. 7.4 TWM and TWM-GR fits to on- and off-frequency precursor duration data for S2 (*solid lines*). On-frequency data for all three precursor levels are plotted in the *first panel* and again in the *third panel*. Off-frequency precursor data are repeated in the *second* and *fourth panels*

7 Discussion and Conclusion

The change in signal threshold with precursor duration differed for on- and off-frequency precursors. This difference was better predicted by a model incorporating temporal integration and gain reduction (TWM-GR) than by one including only temporal integration (TWM). Previous studies have shown little to no difference between models based on temporal integration and models based on adaptation in predicting forward masking data (Oxenham 2001; Ewert et al. 2007). In these models, the adaptation is placed beyond the level of the cochlea. In the TWM-GR, the adaptation is based on the MOCR and thus is assumed to be at the level of the cochlea and to follow a sluggish time course. Because gain reduction applies to the on-frequency precursor but not the off-frequency one, the TWM-GR model is able to predict the differences between the two precursor frequencies with duration. Specifically, the TWM-GR can predict rollover with on-frequency precursors, while the TWM cannot (Fig. 7.4).

In conclusion, we suggest that there is an additional time-varying mechanism contributing to forward masking, gain reduction, which operates over a longer time course. This idea is supported by known auditory physiology (the MOCR). In contrast to other mechanisms of forward masking which may reflect a limitation in the system's temporal acuity, gain reduction may reflect a time-sensitive process which adapts the system to the acoustic environment.

Acknowledgment This work was supported by a grant to the second author: NIH (NIDCD) R01 DC008327.

References

Backus BC, Guinan JJ Jr (2006) Time-course of the human medial olivocochlear reflex. J Acoust Soc Am 119:2889–2904

Ewert SD, Hau O, Dau T (2007) Forward masking: temporal integration or adaptation? In: Kollmeier B, Klump G, Hohmann V, Langemann U, Mauermann M, Uppenkamp S, Verhey J (eds) Hearing – from sensory processing to perception. Springer, Berlin, pp 165–174

Guinan JJ Jr (2006) Olivocochlear efferents: anatomy, physiology, function, and the measurement of efferent effects in humans. Ear Hear 27:589–607

Jennings SG, Strickland EA (2010) The frequency selectivity of gain reduction masking: analysis using two equally-effective maskers. In: Lopez-Poveda A, Palmer AR, Meddis R (eds) The neurophysiological bases of auditory perception. Springer, New York, pp 46–58

Jennings SG, Strickland EA, Heinz MJ (2009) Precursor effects on behavioral estimates of frequency selectivity and gain in forward masking. J Acoust Soc Am 125:2172–2181

Oxenham AJ (2001) Forward masking: adaptation or integration? J Acoust Soc Am 109:732–741

Oxenham AJ, Moore BCJ (1994) Modeling the additivity of nonsimultaneous masking. Hear Res 80:105–118

Oxenham AJ, Plack CJ (2000) Effects of masker frequency and duration in forward masking: further evidence of the influence of peripheral nonlinearity. Hear Res 150:258–266

Roverud E, Strickland EA (2010) The time course of cochlear gain reduction measured using a more efficient psychophysical technique. J Acoust Soc Am 128:1203–1214

Yasin I, Plack CJ (2003) The effects of a high-frequency suppressor on tuning curves and derived basilar-membrane response functions. J Acoust Soc Am 114:322–332

Chapter 8
Is Overshoot Caused by an Efferent Reduction in Cochlear Gain?

Mark Fletcher, Jessica de Boer, and Katrin Krumbholz

Abstract Under certain conditions, detection of a masked tone is improved by a preceding sound ("precursor"). This phenomenon is referred to as the "temporal effect" or "overshoot". A prevalent model of overshoot, referred to as the "gain reduction model", posits that overshoot is caused by efferent reduction in cochlear gain mediated by the medial olivocochlear (MOC) bundle. The model predicts that reduction in cochlear gain will reduce masking when masking is suppressive or when masking is excitatory and the signal-to-masker ratio is high. This study was aimed at testing the validity of these predictions. It consisted of two experiments. The first experiment investigated the relative contributions of suppressive versus excitatory masking to overshoot. The signal was a short 4-kHz tone pip, and the masker and precursor were limited to contain energy either only within (on-frequency) or only outside (off-frequency) the cochlear filter around the signal frequency. The on-frequency masker would be expected to cause mainly excitatory masking, whereas masking by the off-frequency masker would be expected to be mainly suppressive. Only the off-frequency masker and precursor yielded significant overshoot. This suggests that measurable overshoot requires suppressive masking. The second experiment sought to quantify the effect of a precursor on cochlear suppression more directly by measuring the amount of suppression caused by a 4.75-kHz suppressor on a lower-frequency (4-kHz) suppressee with and without a precursor present. Suppression was measured using a forward-masking paradigm. While we found large suppression and large overshoot, we found no reduction in suppression by the precursor. This is contrary to the gain reduction model. Taken together, our results indicate that measurable overshoot requires off-frequency masking and that off-frequency overshoot must be caused by a mechanism other than MOC-mediated reduction in cochlear suppression.

M. Fletcher (✉) • J. de Boer • K. Krumbholz
Institute of Hearing Research, MRC,
University Park, NG7 2RD, Nottingham, UK
e-mail: markf@ihr.mrc.ac.uk

B.C.J. Moore et al. (eds.), *Basic Aspects of Hearing*,
Advances in Experimental Medicine and Biology,
DOI 10.1007/978-1-4614-1590-9_8,

1 Introduction

The medial olivocochlear (MOC) bundle is that part of the auditory efferent pathway that has direct influence on cochlear function. It is thought to play an important role in protecting the auditory system from overexposure and has also been suggested to facilitate the perception of sounds such as speech in noisy environments. However, the exact perceptual consequences of MOC activation remain poorly understood. One perceptual phenomenon that has been linked to MOC function is the so-called overshoot or temporal effect (von Klitzing and Kohlrausch 1994). Overshoot refers to the improvement in the audibility of a masked tone by a preceding sound, or "precursor". The precursor is assumed to activate the MOC system and thereby reduce the gain of the cochlear amplifier. Reduction in cochlear gain would be expected to reduce masking when masking is suppressive (Strickland 2004) or when masking is excitatory and the signal-to-masker ratio is high (Strickland 2001). The most excitatory masking occurs when the masker is at the signal frequency (on-frequency), and suppressive masking occurs when the masker contains energy at frequencies remote from the signal frequency (off-frequency).

Previous studies suggest that the most overshoot is caused when the masker and precursor contain off-frequency energy. In fact, most studies have failed to find any overshoot when the masker and precursor were limited to contain only on-frequency energy (e.g. Fastl 1977). This study consists of two experiments. The aim of Experiment 1 was to reassess the relative contributions of on- versus off-frequency energy to overshoot. This experiment revealed what conditions yield measurable overshoot. The aim of Experiment 2 was then to investigate whether this overshoot was caused by MOC-mediated reduction in cochlear gain.

2 Experiment 1

Three conditions were included to quantify the relative contributions to overshoot from on- versus off-frequency energy within the masker and precursor. The signal was a short (15-ms) 4-kHz tone pip, which was presented just (5 ms) after the onset of a masker. The masker had a duration of 105 ms. In some conditions, the masker was preceded by a precursor, which was 305 ms long. We also measured the signal detection threshold in the presence of the precursor alone to ascertain how much masking was caused by the precursor. In the "broad/broad" condition, the masker and precursor were broadband [ten equivalent rectangular bandwidths (ERBs; Glasberg and Moore 1990) wide] noises and thus contained energy both within and outside the cochlear filter around the signal. In the "notch/notch" condition, the masker and precursor were broadband noises (ten ERBs as in broad/broad) but with a spectral notch (three ERBs wide) centred at the signal frequency, so they contained only off-frequency energy. In the "narrow/narrow" condition, the masker and precursor were narrowband noises, filtered to contain only on-frequency energy (one ERB around the signal frequency). All stimuli were gated on and off with 5-ms quarter-cosine ramps. The precursor and masker were cross-faded to create a constant intensity envelope. The masker and precursor were presented at the same level. The level was varied from 30 to 75 dB SPL per ERB in 15-dB steps, because some of the previous studies have found the amount of overshoot to depend on level (e.g. Bacon and Smith 1991). All thresholds were measured using a three-interval, three-alternative, forced-choice task, with a two-down one-up adaptive tracking procedure. Five normal-hearing participants (three male, 18–25 years) took part in this experiment. The experiment was conducted in a sound-attenuated booth (IAC). The stimuli were generated digitally using MATLAB (25-kHz sampling rate), D-A converted using TDT System 3 (24-bit amplitude resolution), and presented monaurally (left ear) via Sennheiser HD600 headphones.

In all three conditions, the signal detection threshold depended strongly on the masker and precursor level, as would be expected [main effect of level: broad/broad, $F(1.7,6.8)=461.64$, $p=0.000$; notch/notch, $F(1.3,5.0)=80.18$, $p=0.000$; narrow/narrow, $F(3.0,12.0)=703.48$, $p=0.000$; Fig. 8.1a]. In the broad/broad condition, we found a small, but statistically significant, overshoot of, on average, 2.9 dB [main effect of precursor: $F(1.0,4.0)=10.78$; $p=0.029$; left panel in Fig. 8.1b]. The notch/notch condition also yielded significant overshoot [4.7 dB on average; $F(1,4)=11.16$; $p=0.029$; Fig. 8.1b, middle panel]. The dependence of overshoot on level was nonsignificant for both conditions [level-by-precursor interaction; broad/

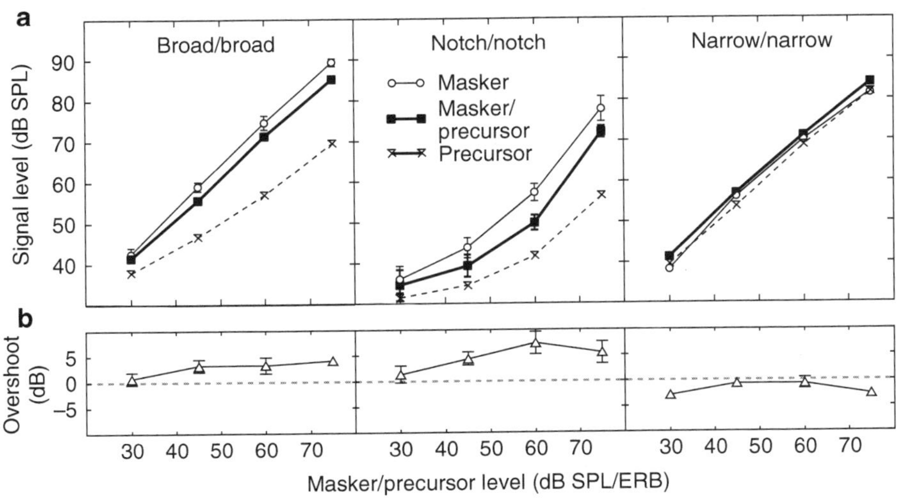

Fig. 8.1 Average signal detection thresholds for five participants as a function of the masker/precursor level (**a**). The *open circles* show the thresholds for the masker alone, the *filled squares* show the thresholds for the masker plus precursor and the *crosses* and *dashed lines* show the thresholds for the precursor alone (see legend in rightmost panel). Different stimulus conditions are shown in different panels (see panel titles). The amount of overshoot (i.e. the threshold difference between the masker-alone and masker-plus-precursor conditions) for each condition is shown in the lower panels (**b**). *Error bars* show the standard error of the mean (SE)

broad: $F(3.0,12.0)=2.40$, $p=0.119$; notch/notch: $F(2.2,8.7)=3.68$, $p=0.067$], and there was large interindividual variability in overshoot (0.7–5 dB for broad/broad, 0.7–8.9 dB for notch/notch). No overshoot was found in the narrow/narrow condition (Fig 8.1b, right panel). In fact, the precursor caused an average increase of the signal detection threshold in the narrow/narrow condition ("negative" overshoot). This negative overshoot was small (1.8 dB) but significant [$F(1.0,4.0)=8.95$, $p=0.040$].

These results are consistent with previous findings. The amount of overshoot found in the broad/broad and notch/notch conditions was slightly smaller, but generally comparable to the amount of overshoot found in previous studies under similar stimulus conditions (von Klitzing and Kohlrausch 1994; Strickland 2001, 2004). Our results indicate that off-frequency energy in the masker and precursor is required for overshoot to be measurable. This is consistent with the results of Bacon and Smith (1991). The negative overshoot in the narrow/narrow condition can be interpreted in several ways. On the one hand, it might mean that no overshoot occurred in this condition. The signal-to-masker ratio was slightly smaller in the narrow/narrow than broad/broad condition and may have been too small for overshoot to be effective. On the other hand, it may be that overshoot occurred, but the effect was countered by other factors. In particular, the signal may have been detected in frequency channels remote from the signal frequency ("off-frequency listening"; Patterson and Nimmo-Smith 1980) to which overshoot would not be expected to apply. Alternatively, any overshoot effect in the narrow/narrow condition may have been overridden by additional masking produced by the precursor; the precursor produced relatively more forward masking in the narrow/narrow condition than in either of the other two conditions. In order to test these possibilities, we measured overshoot in two further conditions.

In the first condition, referred to as "narrow/broad", the masker was broadband as in the broad/broad condition, but the precursor was narrowband as in the narrow/narrow condition. This condition was expected to yield only on-frequency overshoot (due to the precursor being narrowband) while removing the possibility of off-frequency listening and also increasing the signal-to-masker ratio. In the second condition, referred to as "broad/narrow", the masker was narrowband and the precursor was broadband. This condition was intended to reduce forward masking by the precursor and thereby prevent the masking from overriding any overshoot effect. The broadband precursor might also elicit stronger MOC activation (Lilaonitkul and Guinan 2009) and thus cause more overshoot.

Figure 8.2 shows that both conditions yielded significant negative overshoot [narrow/broad:−3.3 dB; $F(1.0,4.0)=63.94$, $p=0.001$; broad/narrow: −6.1 dB; $F(1.0,4.0)=73.42$, $p=0.001$].

The absence of positive overshoot in the narrow/broad condition (Fig. 8.2, left panels) precludes the possibility that the absence of positive overshoot in the narrow/narrow condition was due to off-frequency listening or too low a signal-to-masker ratio. As in the narrow/narrow condition, the negative overshoot in the narrow/broad condition may have been due to the additional masking caused by the precursor. However, additional masking by the precursor cannot explain the absence of positive overshoot in the broad/narrow condition (Fig. 8.2, right panels).

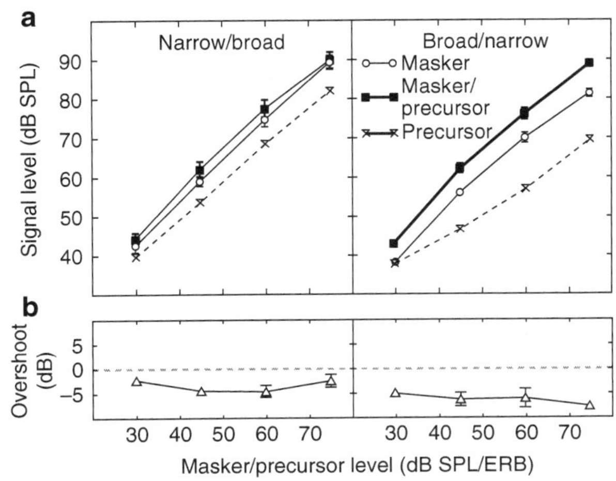

Fig. 8.2 Average signal detection thresholds (**a**) and overshoot (**b**) for the narrow/broad (*left panels*) and broad/narrow (*right panels*) conditions, plotted as in Fig. 8.1

The broadband precursor in the broad/narrow condition produced less masking than the narrowband precursor in both the narrow/narrow and narrow/broad conditions, yet the amount of negative overshoot was greater (−6.1 dB versus −1.8 and −3.3 dB; $p=0.002$ and 0.059). It is possible that the negative overshoot observed in the broad/narrow condition was due to informational ("transient") masking (Bacon and Moore 1987), which would be assumed to arise at central rather than peripheral processing levels.

Taken together, the results from Experiment 1 indicate that both the masker and precursor must contain off-frequency energy for measurable positive overshoot to occur; conditions in which the masker and/or precursor are limited to on-frequency channels yield either no or negative overshoot. The aim of Experiment 2 was to test whether overshoot in off-frequency conditions is caused by MOC-mediated reduction in cochlear gain.

3 Experiment 2

According to the gain reduction model, overshoot in off-frequency conditions is due to reduction in suppressive masking. Suppressive masking occurs when the masker and signal are presented simultaneously and is maximal when the masker is above the signal in frequency. In contrast to excitatory masking, suppressive masking occurs, not through "swamping" of the signal response by the masker response, but by a masker-induced reduction in the size of the signal response. Houtgast (1972) showed that the degree of suppression of the signal response can be measured by measuring the amount of masking that the signal exerts on a subsequent probe stimulus. In this experiment, we used Houtgast's method to estimate the reduction in suppressive masking brought about by the precursor in an overshoot paradigm where the masker and precursor are above the signal in frequency.

Previous work has shown that such paradigms tend to yield large overshoot (Bacon and Viemeister 1985).

The signal was a 12.5-ms tone pip at 4 kHz, presented at a fixed level of 25 dB SL. The masker was also a tone pip, gated simultaneously with the signal and above the signal in frequency (4.75 kHz). First, we determined the masker level required to fully mask the signal. Then, we measured the masking exerted by the signal on a subsequent probe stimulus both with and without the masker present. The probe was a 2.5-ms tone pip at the same frequency as the signal (4 kHz) presented straight after the signal offset (0 ms). By suppressing the signal response, the masker would be expected to reduce the ability of the signal to mask the probe and thus reduce the probe detection threshold. This improvement in the probe threshold would be assumed to represent an estimate of how much suppression has occurred. Suppression was estimated both with and without a precursor present. The precursor was a pure tone at the masker frequency (4.75 kHz), created by extending the masker by 250 ms (Fig. 8.3, upper panel). According to the gain reduction model of overshoot, the precursor should reduce suppression by the masker and thereby restore the ability of the signal to mask the probe. A control experiment showed that the precursor caused no direct masking of the probe. To ensure that the precursor was able to produce overshoot, overshoot was measured using the same precursor and masker. As in the suppression measurements, the signal was at 4 kHz and gated on together with the masker, but had a shorter duration of 2.5 ms (Fig. 8.3, lower panel). All stimuli were gated on and off with 2.5-ms quarter-cosine ramps. A different set of five normal-hearing participants (four male, 18–25 years) took part in this experiment.

The suppression measurement indicated that the 4.75-kHz masker caused considerable suppression of the 4-kHz signal: the masker reduced the masking exerted by the signal on the probe by, on average, 8.7(±1.2) dB (compare white and grey bars in Fig. 8.4a). In the overshoot measurement, the precursor caused a sizeable overshoot effect, reducing the signal detection threshold by 10.7(±2.2) dB, on average (compare white and black bars in Fig. 8.4b). In contrast, the precursor failed to restore the masking effect of the signal on the probe in the suppression measurement (compare grey and black bars in Fig. 8.4a). In fact, it caused a small but significant reduction in the probe detection threshold [1.5(±0.2) dB]. This indicates

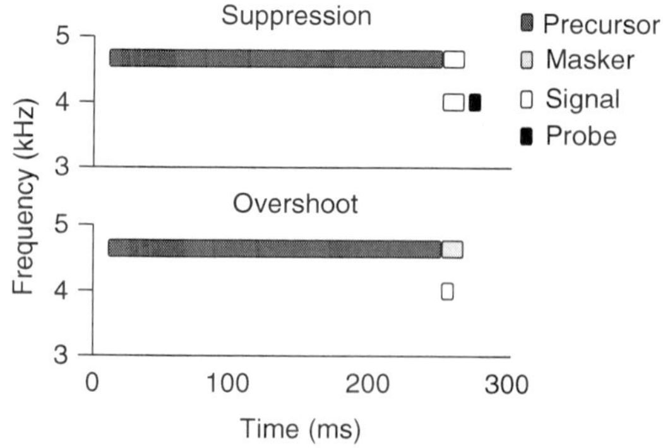

Fig. 8.3 Schematic representation of the suppression (*upper panel*) and overshoot (*lower panel*) measurements in Experiment 2. The precursor is shown in *dark grey*, the masker is shown in *light grey* and the signal is shown in *white* and the probe in *black* (see legend)

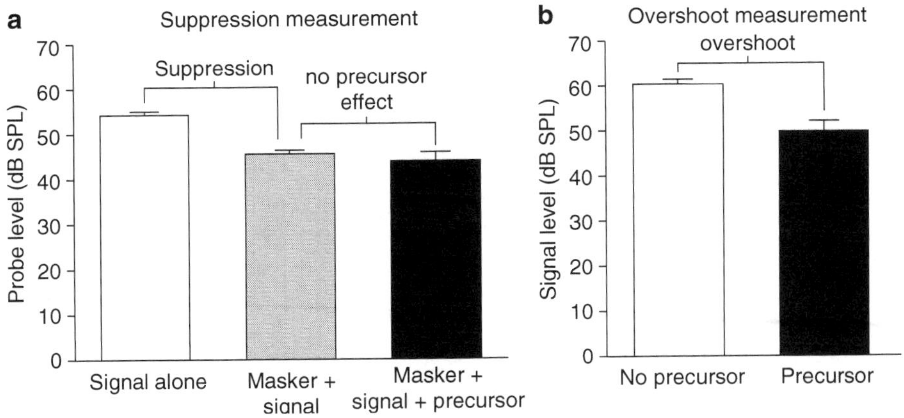

Fig. 8.4 (**a**) Average probe detection thresholds for the suppression measurement. The *white bar* shows the probe threshold in the presence of the signal alone, the *grey bar* shows the threshold for the masker and signal and the *black bar* shows the threshold when the masker and signal are preceded by the precursor. (**b**) Average signal detection thresholds for the overshoot measurements. The *white bar* shows the signal threshold in the presence of the masker alone, and the *black bar* shows the threshold when the masker is preceded by the precursor. *Error bars* show the SE

that the observed overshoot effect (i.e. the reduction in the signal detection threshold by the precursor) cannot have been due to the precursor reducing suppressive masking of the signal. This is contrary to the gain reduction model of overshoot.

4 Discussion

In this study, we found that only conditions where both the masker and precursor contain off-frequency energy yield measurable overshoot (Experiment 1). The results suggested that, when the precursor and/or masker are limited to contain only on-frequency energy, overshoot is either absent or is overridden by other factors, such as masking caused by the precursor itself or informational (transient) masking by the transition from the precursor to the masker. According to the gain reduction model of overshoot, overshoot in off-frequency conditions is caused by MOC-induced reduction in suppressive masking. However, Experiment 2 suggests that this hypothesis does not apply. It showed that an off-frequency precursor that was shown to produce a large overshoot effect did not cause any measurable reduction in suppressive masking.

Overall, our results suggest that overshoot is caused by mechanisms other than cochlear gain reduction by reflexive activation of the MOC bundle. Several alternative explanations have been proposed. For instance, it is possible that off-frequency overshoot is caused not by MOC-induced reduction of cochlear suppression but by adaptation of lateral inhibition at the neural level (see, e.g. Nelson and Young 2010). The effect of lateral inhibition would be expected to be similar to the effect of suppression, but would persist beyond the end of the eliciting stimulus. Thus, in Experiment 2, lateral inhibition should have affected the signal and probe stimuli similarly, so any change in the amount of inhibition brought about by the precursor would not have been expected to cause a measurable change in the probe threshold. Alternatively, overshoot may be caused by transient masking, whereby the response to the transient signal is confused with the response to the masker onset (Bacon and Moore 1987). Finally, Scharf et al. (2008) suggested that overshoot may involve selective attention: the precursor might reduce the ability of the masker to capture attention away from the signal and thereby make the signal easier to detect. There is some evidence that, in the auditory system, selective attention may have a direct modulatory effect on cochlear gain and that this effect is mediated by the MOC system. In particular, it has been shown that MOC suppression is greater for unattended than for attended sounds (Maison et al. 2001). Thus, while overshoot may not be explainable by reflexive MOC activation, the MOC system may, because of its possible role in selective attention, still play an important part in generating overshoot.

References

Bacon SP, Moore BCJ (1987) Transient masking and the temporal course of simultaneous tone-on-tone masking. J Acoust Soc Am 81:1073–1077

Bacon SP, Smith MA (1991) Spectral, intensive, and temporal factors influencing overshoot. Q J Exp Psychol 43:373–399

Bacon SP, Viemeister NF (1985) Simultaneous masking by gated and continuous sinusoidal maskers. J Acoust Soc Am 78:1220–1230

Fastl H (1977) Temporal masking effects: II. Critical band noise masker. Acustica 36:317–330

Glasberg BR, Moore BCJ (1990) Derivation of auditory filter shapes from notched-noise data. Hear Res 47:103–138

Houtgast T (1972) Psychophysical evidence for lateral inhibition in hearing. J Acoust Soc Am 51:1885–1894

Lilaonitkul W, Guinan JJ (2009) Human medial olivocochlear reflex: effects as functions of contralateral, ipsilateral, and bilateral elicitor bandwidths. J Assoc Res Otolaryngol 10: 459–470

Maison S, Micheyl C, Collet L (2001) Influence of focused auditory attention on cochlear activity in humans. Psychophysiology 38:35–40

Nelson PC, Young ED (2010) Neural correlates of context-dependent perceptual enhancement in the inferior colliculus. J Neurosci 30:6577–6587

Patterson RD, Nimmo-Smith I (1980) Off-frequency listening and auditory-filter asymmetry. J Acoust Soc Am 67:229–245

Scharf B, Reeves A, Giovanetti H (2008) Role of attention in overshoot: frequency certainty versus uncertainty. J Acoust Soc Am 123:1555–1561

Strickland EA (2001) The relationship between frequency selectivity and overshoot. J Acoust Soc Am 109:2062–2073

Strickland EA (2004) The temporal effect with notched-noise maskers: analysis in terms of input-output functions. J Acoust Soc Am 115:2234–2245

von Klitzing R, Kohlrausch A (1994) Effect of masker level on overshoot in running- and frozen-noise maskers. J Acoust Soc Am 95:2192–2201

Chapter 9
Accurate Estimation of Compression in Simultaneous Masking Enables the Simulation of Hearing Impairment for Normal-Hearing Listeners

Toshio Irino, Tomofumi Fukawatase, Makoto Sakaguchi, Ryuichi Nisimura, Hideki Kawahara, and Roy D. Patterson

Abstract This chapter presents a unified gammachirp framework for estimating cochlear compression and synthesizing sounds with inverse compression that cancels the compression of a normal-hearing (NH) listener to simulate the experience of a hearing-impaired (HI) listener. The compressive gammachirp (cGC) filter was fitted to notched-noise masking data to derive level-dependent filter shapes and the cochlear compression function (e.g., Patterson et al., J Acoust Soc Am 114:1529–1542, 2003). The procedure is based on the analysis/synthesis technique of Irino and Patterson (IEEE Trans Audio Speech Lang Process 14:2222–2232, 2006) using a dynamic cGC filterbank (dcGC-FB). The level dependency of the dcGC-FB can be reversed to produce inverse compression and resynthesize sounds in a form that cancels the compression applied by the auditory system of the NH listener. The chapter shows that the estimation of compression in simultaneous masking is improved if the notched-noise procedure for the derivation of auditory filter shape includes noise bands with different levels. Since both the estimation and resynthesis are performed within the gammachirp framework, it is possible for a specific NH listener to experience the loss of a specific HI listener.

T. Irino (✉) • T. Fukawatase • M. Sakaguchi • R. Nisimura • H. Kawahara
Faculty of Systems Engineering, Wakayama university,
930 Sakaedani, Wakayama 640-8510, Japan
e-mail: irino@sys.wakayama-u.ac.jp

R.D. Patterson
Department of Physiology, Development and Neuroscience,
Centre for the Neural Basis of Hearing, University of Cambridge,
Downing Site, Cambridge, Cambridgeshire CB2 3EG, UK

B.C.J. Moore et al. (eds.), *Basic Aspects of Hearing*,
Advances in Experimental Medicine and Biology,
DOI 10.1007/978-1-4614-1590-9_9,

1 Introduction

There have previously been attempts to simulate hearing impairment in a form that allows normal-hearing (NH) listeners to experience the difficulty that hearing-impaired (HI) listeners have in everyday life. For example, conductive hearing loss can be simulated simply by attenuating sound with a graphic equalizer. It has been difficult, however, to simulate the impairment associated with the loss of cochlear compression because the signal processing scheme requires reliable measures of both the compression and the frequency selectivity of the cochlea. Compression has been measured using forward masking with growth-of-masking curves (GOMs; Oxenham and Plack 1997) and temporal-masking curves (TMCs; Nelson et al. 2001). It can also be measured with simultaneous, notched-noise masking. The simultaneous technique has the advantage of measuring level-dependent filter shapes simultaneously with the associated compression (Baker and Rosen 2002; Patterson et al. 2003). This avoids the problems involved in combining data from different experiments and listeners.

In this chapter, we propose a unified gammachirp framework for estimating compression and frequency selectivity and then synthesizing sounds with inverse compression that cancels the compression of a normal-hearing (NH) listener to simulate the perception of a hearing-impaired (HI) listener.

2 Simulation of Hearing Impairment

2.1 Compression of NH vs HI and Inverse Compression

The procedure for simulating a hearing impairment is illustrated with the input-output (IO) functions for a NH (circles) and a HI (asterisks) listener in Fig. 9.1. Hearing level (HL) is assumed to occur at the intersection of the individual's IO function and the threshold level of the internal representation. What is required is a sound processor having "inverse compression" (squares) that will convert the IO function of the NH person into that of the HI person.

2.2 Compressive Gammachirp Filter

The block diagram of the compressive gammachirp filter (cGC) is illustrated in Fig. 9.2. The absolute Fourier spectrum, $|G_{CC}(f)|$, is defined as

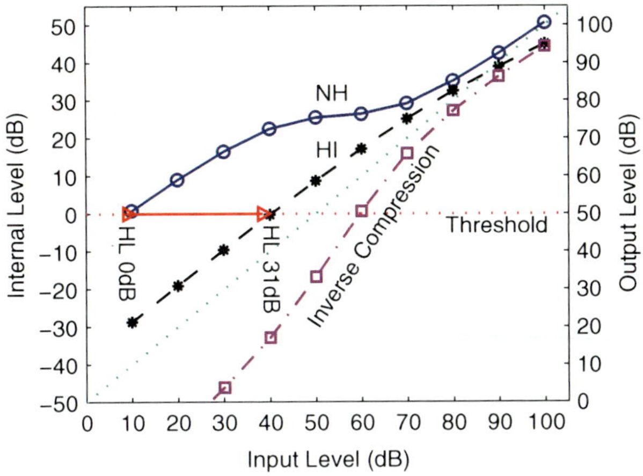

Fig. 9.1 Schematic input-output functions of NH (*circles*, HL = 0 dB) and HI (*asterisks*, HL = 31 dB) listeners, left ordinate. IO function for a signal processor with the desired inverse compression (*squares*), right ordinate

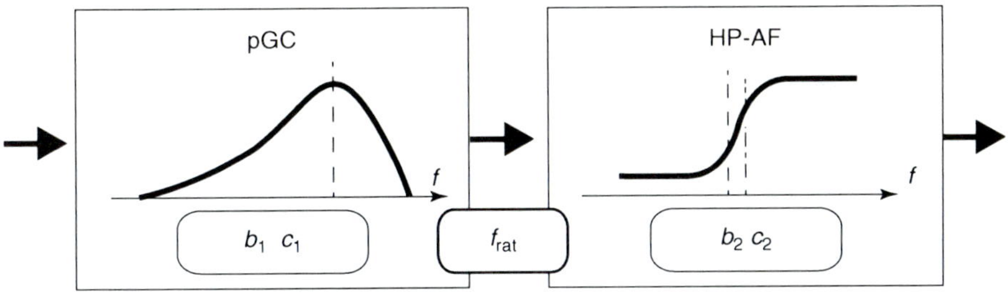

Fig. 9.2 Structure of cGC. The relative position of pGC and HP-AF is defined by a level-dependent parameter, f_{rat}

$$
\begin{aligned}
|G_{CC}(f)| &= \left[a_\Gamma |G_\Gamma(f)| \cdot \exp(c_1\theta_1(f))\right] \cdot \exp(c_2\theta_2(f)) \\
&= |G_{CP}(f)| \cdot \exp(c_2\theta_2(f)) \\
\theta_1(f) &= \arctan\{(f - f_{r1}) / b_1 \mathrm{ERB_N}(f_{r1})\} \\
\theta_2(f) &= \arctan\{(f - f_{r2}) / b_2 \mathrm{ERB_N}(f_{r2})\}
\end{aligned}
\tag{9.1}
$$

where $\mathrm{ERB_N}(f)$ is the equivalent rectangular bandwidth of average NH subjects. Conceptually, this compressive gammachirp is composed of a level-independent "passive" gammachirp filter (pGC), $|G_{CP}(f)|$, which represents the passive basilar membrane, and a level-dependent high-pass asymmetric function (HP-AF), $\exp(c_2\theta_2(f))$, that simulates the active mechanism in the cochlea. To realize the compressive characteristics, we defined the ratio f_{rat} between the peak frequency of pGC and the center frequency of the HP-AF as

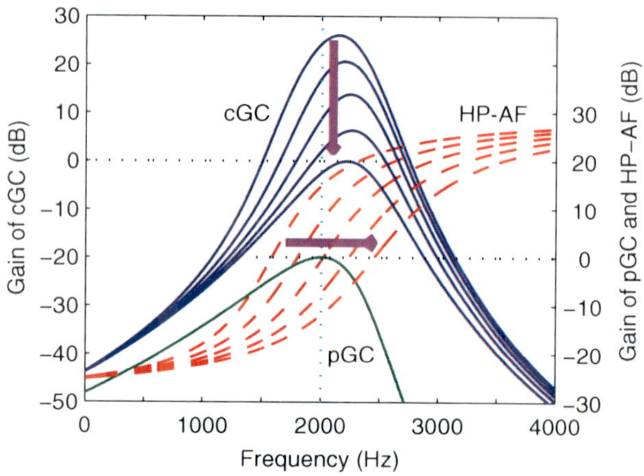

Fig. 9.3 The implementation of compression in the cGC. The *arrows* show the shift of the HP-AF and the gain of the cGC when the input sound level increases

$$f_{\mathrm{rat}} = f_{\mathrm{rat}}^{(0)} + f_{\mathrm{rat}}^{(1)} \cdot P_{\mathrm{gcp}} \tag{9.2}$$

where P_{gcp} is the total level at the output of pGC in dB and $f_{rat}^{(1)}$ is positive for compression. Figure 9.3 illustrates that the HP-AF shifts up in frequency (rightward arrow) and the peak gain of the cGC decreases (downward arrow) as the input sound level increases.

2.3 Implementation of Inverse Compression

Inverse compression is produced by reversing the arrows in Fig. 9.3, thereby inverting the sign of $f_{rat}^{(1)}$. First, we define the center level for inversion, $P_c^{(r)}$, as

$$f_{rat} = f_{rat}^{(0)} + f_{rat}^{(1)} \cdot P_c^{(r)} + f_{rat}^{(1)} \cdot \left(P_{gcp} - P_c^{(r)}\right).$$

By substituting $f_{rat}^{(1r)} (< 0)$ for $f_{rat}^{(1)} (> 0)$, the relative position of HP-AF moves leftward as level increases:

$$\begin{aligned} f_{rat}^{(rev)} &= f_{rat}^{(0)} + f_{rat}^{(1)} \cdot P_c^{(r)} + f_{rat}^{(1r)} \cdot \left(P_{gcp} - P_c^{(r)}\right) \\ &= \left\{ f_{rat}^{(0)} + \left(f_{rat}^{(1)} - f_{rat}^{(1r)}\right) \cdot P_c^{(r)} \right\} + f_{rat}^{(1r)} \cdot P_{gcp} \\ &= f_{rat}^{(0r)} + f_{rat}^{(1r)} \cdot P_{gcp}. \end{aligned} \tag{9.3}$$

Equation 9.3 is the same as Eq. 9.2 but with different coefficients.

For the signal processor, we used the dynamic cGC filterbank (dcGC-FB) of Irino and Patterson (2006) for both the analysis and resynthesis of sounds. The dcGC-FB consists of banks of pGC filters, HP-AFs, and feed-*forward* level controllers which together produce NH compression using Eq. 9.2. For inverse compression, the input sound is analyzed with a dcGC-FB in the form of Eq. 9.3, and then, the output sound is resynthesized linearly with the overlap-and-add method after compensation for the phase lag in each filter.

We performed a simulation to confirm the inverse compression concept. The resynthesized sound with inverse compression was analyzed by the original dcGC-FB having NH compression. The sample sounds were sinusoids with frequencies at octaves of 250 Hz and a word spoken by a male. It was confirmed that the IO function of the complete system was virtually linear for these sounds. The parameter values were $f_{rat}^{(1r)} = -0.016$ and $P_c^{(r)} = 65$. The fact that it is not a simple inversion of the default value of $f_{rat}^{(1)} (= 0.0109)$ is probably due to the overlap of adjacent auditory filters in frequency and a small change in the level-estimation circuit of the dcGC-FB.

3 Estimation of Compression Using Asymmetric-Level Masking

3.1 Stability of Threshold Measurement

The accuracy of inverse compression requires accurate estimation of frequency selectivity *and* compression for both the NH and HI listeners. Irino et al. (2010) used forward masking to estimate compression (Nelson et al. 2001) and constrain the auditory filter shapes derived from simultaneous masking. But listeners had difficulty detecting the brief, faint probe tone following closely after a loud masker, and there were relatively large differences between consecutive measurements of the same condition. This stands in contrast to the stable threshold data obtained with the notched-noise method recommended by Moore and Glasberg (1981) for estimating auditory filter shape (standard deviations of 2 dB or less).

3.2 Asymmetric Noise-Band Levels

Traditionally, level-dependent frequency selectivity is measured with a notched-noise masker in which the levels of the lower and upper bands are the same (as shown in the left-hand panel of Fig. 9.4). However, Nelson et al. (2001) noted that in forward-masking experiments, the most sensitive estimates of compression were obtained when there was a substantial difference between the levels of the on-frequency and off-frequency maskers. Accordingly, we used *asymmetric-level* masking (ALM) with the conditions illustrated in the right-hand panel of Fig. 9.4. The fitting procedure described in Patterson et al. (2003) was used for the data with

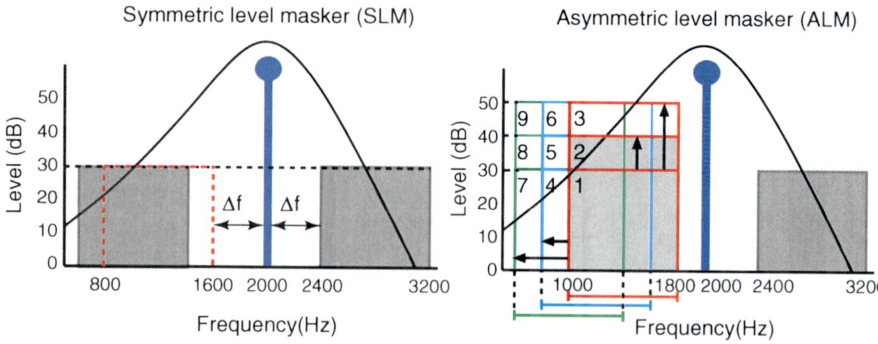

Fig. 9.4 Configuration of the noise bands in the masking experiment. The traditional symmetric-level masker (*SLM*) is on the left; the asymmetric-level masker (*ALM*) is on the right; it has nine lower-band conditions (3 levels × 3 shifts) while the upper band is fixed

ALM. The level dependency in Eq. 9.2 is a function of the output level of the pGC filter, P_{gcp}, which can be calculated from any notched-noise masker with arbitrary spectrum level. It is not like the conventional fitting procedure with a specified input SPL (Moore 2012).

3.3 *The ALM Experiment*

Experiments were performed to evaluate the effectiveness of ALM. The signal, a 2.0-kHz tone, was masked by either a symmetric-level masker (SLM) or an asymmetric-level masker (ALM). In the SLM part of the experiment, the noise level was 30, 40, or 50 dB, and there were 14 different notch configurations (42 conditions in total). In the ALM part of the experiment, there were 9 additional conditions as illustrated in Fig. 9.4. Threshold in each condition was measured twice. The procedure was basically the same as in the standard notched-noise experiment. Six young NH listeners participated in the experiment after giving informed consent. For each listener, thresholds that differed by 2 dB or less were used to estimate frequency selectivity and compression as suggested by Moore and Glasberg (1981). The set of thresholds was fitted with the cGC using the procedure described in Patterson et al. (2003). Good fits were obtained for five listeners; the fitting procedure did not converge for one listener due to insufficient data.

Figure 9.5 shows the results for one listener: those in the left-hand panel are based on both SLM and ALM conditions; those in the right-hand panel are based solely on the SLM conditions.

The slopes of the IO functions are 0.39 and 0.63 dB/dB; the rms errors between the measured and predicted thresholds were 2.40 and 2.28 dB, respectively. In general, inclusion of the ALM data reduced a listener's IO slope and brought it into the range of slopes derived from forward-masking experiments (0.2–0.4 dB/dB) (e.g., Nelson, et al. 2001; Plack et al. 2004).

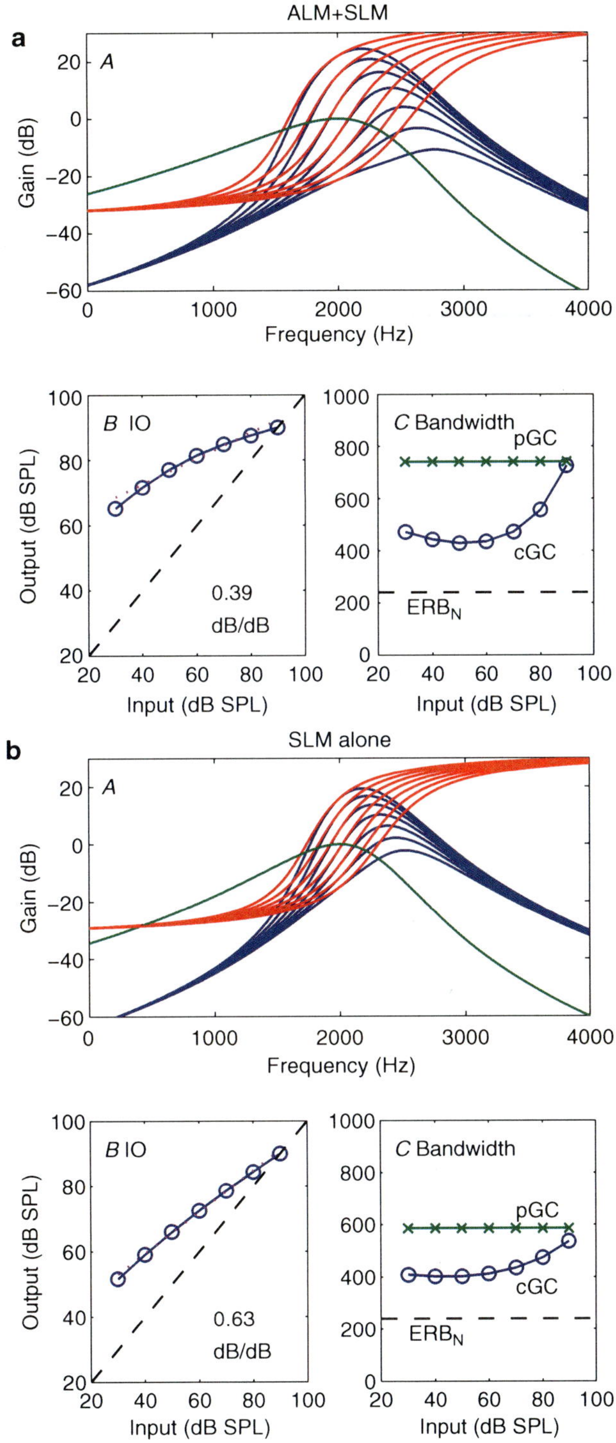

Fig. 9.5 Filter shapes (*A*), IO functions (*B*), and bandwidth functions (*C*) for a NH listener using ALM and SLM data (**a**) or SLM data alone (**b**)

The average IO slopes for the five listeners and their standard deviations were 0.40 (±0.14) dB/dB for SLM data with ALM data and 0.53 (±0.18) dB/dB for SLM data alone. The slopes are shallower when ALM data is included. The rms errors were 2.57 (±0.70) dB for SLM with ALM and 2.49 (±0.69) dB for SLM alone. The difference is only 0.08 dB on average even though the number of points is about 20 % greater when the ALM data is included. The results suggest that the inclusion of ALM conditions improves the estimation of compression and alters masking levels near the center frequency and in the tail of the filter. The changes are due to the inclusion of the lower-band conditions where level changes without a change in frequency.

4 Conclusion

A unified gammachirp framework can be used to simulate the experience of one HI listener for a given NH listener. The procedure depends on the accurate measurement of both compression and frequency selectivity in the cochlea which is achieved by including asymmetric levels in the notched-noise masker. The sounds resynthesized with inverse compression accurately cancel the compression of the NH listener. Currently, the procedure requires manual tuning to locate the coefficients for accurate inverse compression, but it should be possible to automate convergence of the fitting process in the future.

Acknowledgement This work was supported by a JSPS Grant-in-Aid (B21300069).

References

Baker RJ, Rosen S (2002) Auditory filter nonlinearity in mild/moderate hearing impairment. J Acoust Soc Am 111:1330–1339

Irino T, Patterson RD (2006) A dynamic compressive gammachirp auditory filterbank. IEEE Trans Audio Speech Lang Process 14:2222–2232

Irino T, Takahashi H, Kawahara H, Patterson RD (2010) Constraining the derivation of auditory filter shape with temporal masking curves. ARO 33th Midwinter meeting, #329, Anaheim, 2010

Nelson DA, Schroder AC, Wojtczak M (2001) A new procedure for measuring peripheral compression in normal-hearing and hearing impaired listeners. J Acoust Soc Am 110: 2045–2064

Moore BCJ (2012) An introduction to the psychology of hearing, 6th edn. Emerald Group Pub, Bingley

Moore BCJ, Glasberg BR (1981) Auditory filter shapes derived in simultaneous and forward masking. J Acoust Soc Am 70:1003–1014

Oxenham AJ, Plack CJ (1997) A behavioral measure of basilar membrane nonlinearity in listeners with normal and impaired hearing. J Acoust Soc Am 101:3666–3675

Patterson RD, Unoki M, Irino T (2003) Extending the domain of center frequencies for the compressive gammachirp auditory filter. J Acoust Soc Am 114:1529–1542

Plack CJ, Drga V, Lopez-Poveda EA (2004) Inferred basilar-membrane response functions for listeners with mild to moderate sensorineural hearing loss. J Acoust Soc Am 115:1684–1695

Chapter 10
Modelling the Distortion Produced by Cochlear Compression

Roy D. Patterson, D. Timothy Ives, Thomas C. Walters, and Richard F. Lyon

Abstract Lyon (J Acoust Soc Am 130:3893–3904, 2011) has described how a cascade of simple asymmetric resonators (CAR) can be used to simulate the filtering of the basilar membrane and how the gain of the resonators can be manipulated by a feedback network to simulate the fast-acting compression (FAC) characteristic of cochlear processing. When the compression is applied to complex tones, each pair of primary components produces both quadratic and cubic distortion tones (DTs), and the cascade architecture of the CAR-FAC system propagates them down to their appropriate place along the basilar membrane, where they combine additively with each other and any primary components at that frequency. This suggests that CAR-FAC systems might be used to study the role of compressive distortion in the perception of complex sounds *and* that behavioural measurements of cochlear distortion data might be useful when tuning the parameters of CAR-FAC systems.

1 Introduction

In a classic paper, Goldstein (1967) used a cancellation-of-beats technique to measure the magnitude of the cubic DTs produced by a pair of primary sinusoids in the cochlea. More recently, Pressnitzer and Patterson (2001) used the same technique to

R.D. Patterson (✉)
Department of Physiology, Development and Neuroscience,
Centre for the Neural Basis of Hearing, University of Cambridge,
Downing Site, Cambridge CB2 3EG, UK
e-mail: rdp1@cam.ac.uk

D.T. Ives
Department d'Etudes Cognitives, Ecole Normale Supérieure,
29 Rue d'Ulm, 75005 Paris, France

T.C. Walters • R.F. Lyon
Google Inc., 1600 Amphitheatre Parkway, Mountain View, CA 94043, USA

B.C.J. Moore et al. (eds.), *Basic Aspects of Hearing*,
Advances in Experimental Medicine and Biology,
DOI 10.1007/978-1-4614-1590-9_10, © Springer Science+Business Media New York 2013

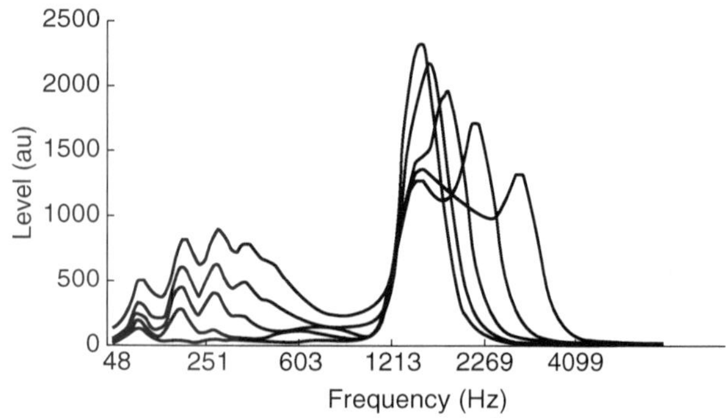

Fig. 10.1 Auditory spectra for five complex tones with the same lowest component (15). The fundamental was 100 Hz and the tones had 2, 3, 5, 9 or 17 consecutive harmonics, all with the same level, 54 dB SPL

measure the spectrum of low-frequency quadratic distortion tones (qDTs) produced by multi-harmonic complexes with consecutive harmonics of a 100-Hz fundamental (F0), starting with the 15th harmonic. In their first experiment, the tones had 11 primaries in cosine phase. The sound level was 54 dB SPL per component; the overall level was 65 dB SPL. They measured the level of the DTs at the first four harmonics of F0 and showed that qDT magnitude decreased as harmonic number increased. In their third experiment, they demonstrated that qDT magnitude was strongly dependent on the relative phases of the primaries; when the phases of successive primaries were alternated between 0° and 90°, the magnitude of the first and third qDTs decreased, whereas the magnitude of the second and fourth qDTs was unchanged or increased slightly. The CAR-FAC system of Lyon (2011) was found to produce distortion tones with similar properties insofar as it produced DTs at the first four harmonics when the primaries were in cosine phase and the magnitudes of the DTs at the first and third harmonics decreased when the phase of adjacent primaries was alternated between 0° and 90°.

In their main experiment (the second), Pressnitzer and Patterson (2001) measured the magnitude of the qDT at 100 Hz as they increased the number of primaries (NP) in a cosine phase tone. Tones with 2, 3, 5, 9 or 17 harmonics were used to evoke, respectively, 1, 2, 4, 8 or 16 first-order difference tones between adjacent pairs of primaries in the complex tone. The magnitude of the qDT at 100 Hz was observed to increase by close to 3 dB per doubling of the number of first-order difference tones, indicating an orderly summation of distortion components in the cochlea, as would be expected. Figure 10.1 shows the auditory spectra (AS) produced by the CAR-FAC system for the five complex tones. Each of the AS contains a band of activity in the region of the primaries above 1,500 Hz and a band of activity in the region of the first five harmonics of 100 Hz. For convenience, these two regions of the AS will be referred to as the primary spectrum (PS) and the quadratic distortion spectrum (qDS).

When there are only two primaries, there is only one component in the qDS and it is at 100 Hz, as expected. As NP doubles, the magnitude of the qDT at 100 Hz increases and the range of components in the qDS increases; that is, primaries that are farther apart in frequency contribute qDTs at higher difference frequencies. In the CAR-FAC system, the magnitude of the qDT increases with qDT frequency from 100 to 300 Hz; above this frequency, qDT magnitude decreases. In contrast, in the data of Pressnitzer and Patterson (2001), the qDT at 100 Hz has the greatest magnitude, and magnitude decreases monotonically as qDT frequency increases.

In this example, where the level of the primaries is fixed, the magnitude of the qDS grows with the number of primaries in absolute terms, and it grows even more with respect to the magnitude of the PS. This is because, locally, cochlear compression is driven by the overall level of the input. Thus, as NP increases and the PS broadens above 1,500 Hz, the response at 1,500 Hz decreases as the compression increases. Suppression grows within the PS as NP increases, and as a result, the PS develops edge tones when NC is 8 or more.

These initial simulations indicate that the compression applied by CAR-FAC systems produces distortion spectra that are similar in many respects to those observed by Pressnitzer and Patterson (2001), although the magnitudes of the qDTs in the CAR-FAC system do not decrease monotonically with increasing frequency as they do in the behavioural data.

Quadratic distortion products are known to have a pronounced effect on the lower limit of melodic pitch (Pressnitzer et al. 2001; Pressnitzer and Patterson 2001, 2001) and the detection of low-frequency tones in the presence of high-frequency bands of amplitude-modulated noise (Wiegrebe and Patterson 1999). Accordingly, we examined the degree to which Lyon's (2011) CAR-FAC system could explain the DS produced by complex tones.

2 The Effects of Compressive Distortion in CAR-FAC Systems

The parameter values for the CAR-FAC system were those referred to as "fit 507" of the "pole-zero filter cascade" (Lyon 2011, Fig. 6, PZFC). This version of the system provides a good fit to a wide range of notched-noise masking data, and it produces pronounced quadratic distortion. The CAR-FAC system was used to generate sets of auditory spectra for a complex tone presented at levels varying from 40 to 80 dB SPL in 10 dB steps. The fundamental of the tone was 200 Hz and it had 11 adjacent harmonics. The lowest component, LC, of the tones was varied from 1 to 16 in doublings to survey the parameter space and locate typical distortion patterns. Broadly speaking, tones with LC equal 1, 2 or 4 produced similar patterns of AS, and high-frequency tones with LC equal 8 or 16 produced similar patterns of AS, so the results will be described for two tones: one with LC = 2 and the other with LC = 8.

2.1 Auditory Spectra for High-Frequency Complex Tones (LC 8)

Three sets of AS were generated for LC 8 (right-hand column of Fig. 10.2). The AS for the standard CAR-FAC system are shown in the middle-right panel. The PS show that the primaries are not resolved; the DS reveal resolved peaks at the first

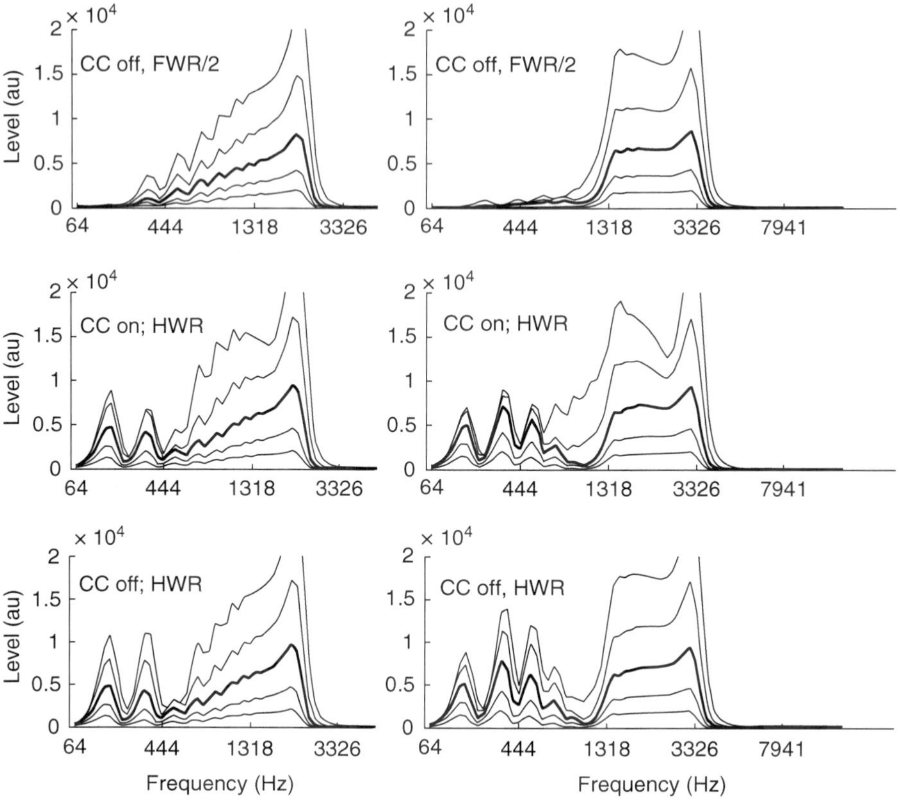

Fig. 10.2 Auditory spectra for a 200-Hz complex tone with 11 harmonics at levels from 40 to 80 dB SPL in 10-dB steps; LC is 2 (*left column*) or 8 (*right column*). The *middle row* shows the AS produced with the CAR-FAC system. The *lower row* shows the AS when the cubic compressor is turned off; the *upper row* shows the AS when the cubic compressor is turned off *and* HWR is replaced with FWR/2

four harmonics of the fundamental. The sequence of PS shows that the system is compressive; in the range 40–60 dB SPL, each 10-dB step in stimulus level produces a roughly equal change in the level of the PS. As the level of the tone increases to 70 and then 80 dB SPL, the magnitude of the increase in the PS increases. Nevertheless, the system remains compressive; the increases in AS level are far less than an order of magnitude in both cases. Over the same range of levels (60–80 dB SPL), the magnitude of the DS increases at a slower rate than it does at lower levels. Moreover, the pattern of PS activity changes above 60 dB SPL: there is an increase in suppression in the central region of the PS, leading to more pronounced edge tones; there is an increase in the magnitude of the fourth harmonic of F0; and there is an increase in the response in the region between the PS and the qDS. These are the characteristics of cubic distortion as they appear in the AS of the CAR-FAC system.

The *cubic compressor* in the CAR-FAC system reduces the amplitude of the output of each section, by a small proportion of the cube of what would otherwise be the output, before the output passes to the next stage. In the model, it is associated with the operation of the outer hair cell. The input/output function for the compressor is radially symmetric, so it would be expected to introduce cubic distortion tones (cDTs) of the form $2f_1–f_2$ rather than qDTs of the form $f_2–f_1$ (where f_1 and f_2 are the frequencies of two primaries). For a complex tone, cubic distortion appears as a spectrum of harmonics cascading down in frequency and magnitude from the low-frequency side of the PS. It will be referred to as the cubic portion of the distortion spectrum (cDS), and this is what appears in the AS at the higher tone levels (70 and 80 dB SPL). For comparison, the bottom-right panel of Fig. 10.2 shows the set of AS generated when the cubic compressor is turned off (CC off). The cDS that appears in the middle-right panel at 80 dB SPL disappears, and the suppression in the PS is greatly reduced, confirming that these are aspects of cubic compression. Note also that the qDS persists at low frequencies in the bottom-right panel when the cubic compressor is turned off. Indeed, the magnitude of the qDS increases at the higher tone levels (70 and 80 dB SPL) indicating that in this model, the cubic distortion suppresses the qDS at high stimulus levels.

The CAR-FAC system has an automatic-gain-control (AGC) network (Lyon 2011, Fig. 3); it computes a continuous estimate of the magnitude at the output of each filter stage and, as the magnitude increases, the network reduces the gain of the filter. This is why the degree of PS compression is similar in the middle and bottom panels of the right-hand column.

The output of each filter stage is half-wave rectified to simulate the dominant aspect of inner hair-cell processing before it is passed to the AGC network, which consists of a cascade of four low-pass filters with time constants of about 3, 12, 48 and 200 ms. Their contributions are combined with increasing weights (1.0, 1.4, 2.0 and 2.8, respectively) to produce the current estimate of tone magnitude in a given channel. The top-right panel shows the AS generated for the same tones when the rectification is switched from half wave to full wave. This largely eliminates the rapid oscillation of gain associated with HWR, and it largely eliminates the quadratic distortion. The FWR signal has about double the magnitude of the HWR signal, and so the FWR output was divided by 2 to ensure that the estimate going into the AGC network was about the same for the two forms of rectification. The compression range is about the same in the top and bottom panels, and the range is also similar to that in the middle-right panel.

In summary, the quadratic distortion produced by the default CAR-FAC system arises in the AGC network and is closely associated with the HWR that represents the operation of the inner hair cell in the model, rather than the cubic compressor that represents the operation of the outer hair cell in the model.

2.2 *Auditory Spectra for Low-Frequency Complex Tones (LC 2)*

When LC is 8 or more, the quadratic and cubic distortion products appear in separate frequency regions and their interaction is minimal. Pressnitzer and Patterson (2001) chose a relatively high LC value (15) to minimize the interaction of quadratic and cubic distortion and so provide uncontaminated measurements of the quadratic distortion. However, the complex tones of speech and music typically have prominent low-order harmonics where quadratic and cubic effects of compression might be expected to interact. CAR-FAC systems have the advantage that all of the effects of compression appear in the AS for any sound, and thus, they make it possible to investigate the complicated interaction of primaries and distortion tones in everyday sounds. The left-hand column of Fig. 10.2 shows sets of AS like those in the right-hand column but for tones in which LC is 2 rather than 8. As noted above, similar AS are generated by tones with lowest components in the range 1–4.

The AS in the top-left panel were produced with the cubic compressor turned off (CC off) and the HWR replaced by FWR/2, so these AS show the effects of compression with the minimum of distortion. For the lower levels, 40–60 dB SPL, the AS largely reflect the energy in the stimulus at the output of the filterbank. The lowest harmonics appear as separate peaks and there is minimal activity at the fundamental. Then, as frequency increases and resolution decreases, multiple components pass through the filters and the magnitude of the AS rises. The upper edge tone does not appear until about 60 dB SPL, indicating that suppression does not play a large role in determining the shape of these AS at low to moderate levels. At higher levels, 70 and 80 dB SPL, the slope of the spectrum increases because the bandwidths of the filters increase at the higher levels. At the same time, the relative magnitude of the upper edge tone increases indicating that suppression may affect the slope of the PS at higher levels.

The middle-left panel shows the AS with cubic compression (CC on) and HWR. At the lower levels, 40–60 dB SPL, the high-frequency end of the PS exhibits slightly less compression (the displacement of the AS increases slightly relative to that in the top-left panel), but the AS is still largely determined by the energy in the primaries. At higher levels, the rate of increase in PS level increases, and the relative strength of the upper edge tone increases. However, the main difference, relative to the AS with CC off and FWR/2, appears in the region of harmonics 1–3: a prominent distortion tone appears at the fundamental (200 Hz) where the stimulus has no energy, and there is an increase in the level of activity at the second harmonic (400 Hz). In contrast, the level of the third harmonic (600 Hz) in the middle-left panel *de*creases with respect to that in the top-left panel where the distortion was minimized. This means that in this CAR-FAC system, primaries in cosine phase can generate distortion tones whose aggregate magnitude at a specific frequency is greater than that of the primary at that frequency and the phase of the distortion tone is such that it partially cancels the primary. There are no perceptual measurements that might confirm or refute the prediction that the internal magnitude of the third harmonic is less than that of the adjacent harmonics for multi-harmonic tones of this form.

Finally, compare the AS in the middle-left panel with those in the bottom-left panel where the cubic compressor is turned off. The rectification mode (HWR) is the same in these two versions of the system. At the lower levels, 40–60 dB SPL, the PS and DS are essentially the same, which confirms that it is the AGC network that determines the compression in the system. At the higher levels, there is no evidence of cubic distortion on the lower flank of the PS in the bottom-left panel, as expected. At the same time, the levels of the fundamental and second harmonic *in*crease, indicating that cubic distortion interferes with quadratic distortion in the full system (middle-left panel) when the stimulus contains low-order harmonics.

3 Conclusion

CAR-FAC systems provide a means of investigating the wide range of distortion components produced by cochlear compression when the stimulus is a multi-harmonic tone and the distortion components interact on the basilar membrane. Perceptual measures of cochlear distortion are sufficiently accurate to assist in determining which of the many forms of CAR-FAC system might provide the most promising models of cochlear processing.

Comment by Diek Duifhuis

First, I would like to make a general comment that relates to the peripheral modelling that has been presented in this chapter. I appreciate that at least one of the models extends beyond the Helmholtz type (cf Duifhuis 2012, Fig. 3.5) in which longitudinal coupling simply does not exist, and DPs and OAEs are not propagated. In terms of the cochlear biophysics, any model that takes longitudinal coupling into account is a step forward (see Wegel and Lane 1924, Fig. 7b; Duifhuis 2012, Fig. 3.6).

But now coming to your analysis, I have two problems:

1. Apparently we agree that the cochlea nonlinearity (NL) is compressive. Therefore, a Taylor expansion gives an inefficient approximation because all terms in the series expand: all terms beyond power 1 are expanding. An alternative terminology can be built on the use of even- or odd-order NL elements [The even-order family is defined by $f(y)=f(-y)$, the odd-order family by $f(y)=-f(-y)$]. Any time-invariant NL can be decomposed into an odd- and an even term. The latter is responsible for the difference tones and even harmonics, while the former is responsible for the odd-order intermodulation products such as $2f1-f2$, and for odd harmonics. Your example of the half-wave rectifier (of any shape) can also be decomposed into an odd and an even part, which in this case will be of similar amplitudes. So a half-wave rectifier should generate significant even-order as well as odd-order distortion.
2. Lyon's model considers propagation through the cochlear partition. It neglects coupling through the fluid. Moreover, it is simplified to a one-way block diagram structure, solved in the frequency domain (s-plane) for which the accuracy in dealing with NL phenomena remains unknown.

A time-domain analysis provides a valid alternative, and real-time modeling or hardware implementation is not too far off. Currently, the time domain analysis does allow a reliable estimation of OAEs. Examples are given in Duifhuis (2012), Chapter 10.

References

Duifhuis H (2012) Cochlear mechanics. Introduction to a time domain analysis of the nonlinear cochlea. Springer, New York

Wegel RL, Lane CE (1924) The auditory masking of one pure tone by another and its probable relation to the dynamics of the inner ear. Phys Rev 23:266–285

Response by Patterson

We agree that the Volterra expansion and the Taylor expansion are very misleading when it comes to predicting the levels of compressive distortion products (DPs). We used the terms "quadratic" and "cubic" simply to make a well known distinction between the DPs in the region of the fundamental and those just below the primaries. We did not have space for a discussion of terminology in this short paper. While we agree that mathematical terms "odd-order" and "even-order" are more accurate, the even-order function is not very useful when trying to explain the low-frequency DPs produced by complex tones.

We agree that a real-time model of cochlear processing that produces the correct spectrum of DPs in response to complex tones would be a great advance, especially if it could be made into an integrated circuit. In the meantime, we think Lyon's CAR-FAC system remains a useful tool for investigating the origin of the DPs produced by complex tones and their interaction in the cochlea.

References

Goldstein JL (1967) Auditory nonlinearity. J Acoust Soc Am 41:676–689

Lyon RF (2011) Cascades of two-pole–two-zero asymmetric resonators are good models of peripheral auditory function. J Acoust Soc Am 130:3893–3904

Pressnitzer D, Patterson RD (2001) Distortion products and the perceived pitch of harmonic complex tones. In: Breebaart D, Houtsma A, Kohlrausch A, Prijs V, Schoonhoven R (eds) Physiological and psychophysical bases of auditory function. Shaker BV, Maastrict, pp 97–104

Pressnitzer D, Patterson RD, Krumbholz K (2001) The lower limit of melodic pitch. J Acoust Soc Am 109:2074–2084

Wiegrebe L, Patterson RD (1999) Quantifying the distortion products generated by amplitude-modulated noise. J Acoust Soc Am 106:2709–2718

Part II
Temporal Fine Structure and Pitch

Chapter 11
How Independent Are the Pitch and Interaural-Time-Difference Mechanisms That Rely on Temporal Fine Structure Information?

Shigeto Furukawa, Shiho Washizawa, Atsushi Ochi, and Makio Kashino

Abstract The temporal fine structure (TFS) of acoustical signals, represented as the phase-locking pattern of the auditory nerve, is the major information for listeners performing a variety of auditory tasks, e.g., judging pitch and detecting interaural time differences (ITDs). Two experiments tested the hypothesis that processes for TFS-based pitch and ITD involve a common mechanism that processes TFS information and the efficiency of the common mechanism determines the performance of the two tasks. The first experiment measured the thresholds for detecting TFS-based pitch shifts (Moore and Moore, J Acoust Soc Am 113:977–985, 2003) and for detecting ITD for a group of normal-hearing listeners. The detection thresholds for level increments and for interaural level differences were also measured. The stimulus was a harmonic complex (F0 = 100 Hz) that was spectrally shaped for the frequency region around the 11th harmonic. We expected a *positive* correlation between the pitch and ITD thresholds, based on the hypothesis that a common TFS mechanism plays a determinant role. We failed to find evidence for a positive

S. Furukawa, PhD (✉)
NTT Communication Science Laboratories
NTT Corporation, 3-1 Morinosato-Wakamiya, 243-1098 Atsugi, Kanagawa, Japan
e-mail: furukawa.shigeto@lab.ntt.co.jp

S. Washizawa • M. Kashino
NTT Communication Science Laboratories
NTT Corporation, 3-1 Morinosato-Wakamiya, 243-1098 Atsugi, Kanagawa, Japan

Interdisciplinary Graduate School of Science and Engineering
Tokyo Institute of Technology, 4259 Nagatsuta-cho, Midori-ku, Yokohama, 226-8503 Kanagawa, Japan

A. Ochi
NTT Communication Science Laboratories
NTT Corporation, 3-1 Morinosato-Wakamiya, 243-1098 Atsugi, Kanagawa, Japan

Department of Otolaryngology, Faculty of Medicine, University of Tokyo, 7-3-1 Hongo Bunkyo-ku, 113-8655 Tokyo, Japan

B.C.J. Moore et al. (eds.), *Basic Aspects of Hearing*, Advances in Experimental Medicine and Biology, DOI 10.1007/978-1-4614-1590-9_11,

correlation, hence no support for the above hypothesis. The second experiment examined whether perceptual learning with respect to detecting TFS-based pitch shifts via training would transfer to performance in other untrained tasks. The stimuli and tasks were the same as those used in the first experiment. Generally, training in the pitch task improved performance in the (trained) pitch task, but *degraded* the performance in the (untrained) ITD task, which was unexpected on the basis of the hypothesis. No training effect was observed in the other untrained tasks. The results imply that the pitch and ITD processes compete with each other for limited neural resources.

1 Introduction

In the auditory periphery, the temporal fine structure (TFS) of acoustical signals is encoded as the pattern of phase locking of the auditory nerve firing. This information plays essential roles in relation to various aspects of auditory perception, including pitch perception and lateralization based on interaural time difference (ITD). It can be considered that pitch perception (note that this study focuses on the TFS, although it is not the only factor that determines pitch perception) and ITD-based lateralization are the result of some kinds of autocorrelation and interaural cross-correlation, respectively, of the TFS information. A natural hypothesis is that the processes for pitch and ITD involve a common mechanism that processes TFS information and the efficiency of the common mechanism is the major factor determining the performance of the two tasks. This hypothesis has important implications as a basis for both a fundamental understanding of the auditory system and for developing psychophysical methods to diagnose auditory dysfunctions (cf., Hopkins and Moore 2010). The present study was conducted to test this hypothesis, employing two different experimental paradigms.

2 Experiment 1: Correlation of Individual Performances Across Tasks

If processes for TFS-based pitch and ITD involve a common mechanism whose efficiency determines a listener's sensitivity to the two properties, sensitivity measures of individual listeners could reveal a correlation between the two properties. This experiment measured detection thresholds for TFS-based pitch and ITD. Additional threshold measurements of the interaural level difference (ILD) and level increment were made as control measures, representing binaural and monaural sensitivity to acoustical properties that would not require TFS information.

2.1 Methods

Twenty-two listeners with normal audiometric thresholds (age, 19–43 years old; 10 males and 12 females) participated in the experiment. The stimulus was a spectrally shaped multicomponent complex (SSMC), which was a harmonic complex with a fundamental frequency (F0) of 100 Hz, consisting of the 7th to 14th harmonics (added in the sine phase). The spectral envelope of the stimulus had a flat passband (5×F0 centered at 1,100 Hz) and sloping edges (Moore and Moore 2003). The overall level of the complex was 54 dB SPL. Threshold-equalizing noise (TEN; Moore et al. 2000), extending from 125 to 15,000 Hz, was added to mask combination tones and to help ensure that the audible parts of the excitation patterns evoked by the harmonic and frequency-shifted tones were the same in the pitch task (described later). The TEN level at 1 kHz was set at 30 dB/ERB$_N$, which was 15 dB below the level of the 1,100 Hz component.

The listeners performed the following four tasks: (1) *Pitch task*: The listeners were required to detect a common frequency shift imposed on the individual components of the SSMC with the spectral envelope remaining unchanged. The listeners were expected to use the pitch change as the major cue for the task, and the pitch change was largely the result of changes in the TFS (de Boer 1956; Moore and Moore 2003; Schouten et al. 1962). A detection threshold for the frequency shift was determined with a method that was essentially the same as the "TFS1" test developed by Moore and Sek (2009). A two-interval two-alternative forced choice (2I-2AFC) procedure was used in the 2-up/1-down transformed up-down method (Levitt 1970). The "signal" and "non-signal" intervals in the 2I-2AFC method contained HSHS and HHHH sequences, respectively, where H indicates a harmonic complex (i.e., original SSMC) and S indicates a frequency-shifted SSMC. The listener was required to indicate the signal interval (HSHS). An H or S tone had a duration of 100 ms including 20-ms raised cosine ramps, and there were 100-ms silent intervals between the tones in a sequence. The maximum frequency shift was limited to 0.5 F0. When adaptive tracking failed to converge within this limit, trials with a shift of 0.5 F0 were repeated 30 times. In that case, the proportion of correct trials was converted to d', and then the "threshold" was derived on the assumption that d' is proportional to the frequency shift (Hopkins and Moore 2007). (2) *ITD task*: In a 2I-2AFC trial, SSMCs (400-ms duration including 100-ms raised cosine ramps) in the two intervals had ITDs of $+\Delta$ITD and $-\Delta$ITD μs, respectively. The raised cosine ramps at the onset and offset of the stimulus were synchronized between the two ears. The listeners were required to indicate the direction of the ITD change between the two intervals based on the laterality of sound images. The ΔITD value at the threshold was determined using the 2-up/1-down transformed up-down method. (3) *ILD task*: In a 2I-2AFC trial, the listeners were required to indicate the direction of the ILD change based on the laterality of sound images. The procedure was generally the same as that of the ITD task, except that ILDs of $+\Delta$ILD and $-\Delta$ILD dB, instead of ITDs, were imposed on the stimuli and the raised cosine ramps were 20 ms. (4) *Level task*: In a 2I-2AFC trial, the listeners were required to indicate an interval containing 400-ms long SSMC whose central 200-ms portion was incremented in level by ΔL dB, while the other non-signal interval contained an original SSMC. The stimulus duration included 20-ms raised cosine ramps. The ΔL value at the threshold was determined using the 2-up/1-down transformed up-down method.

In all the tasks, the adaptive tracking for estimating the threshold was repeated at least six times, and average across the thresholds was computed. Correct-answer feedback was provided for each trial. Geometric averages were obtained for the pitch and ITD tasks. The stimulus was delivered monaurally (pitch and level tasks) or binaurally (ITD and ILD tasks) through headphones.

2.2 Results

Figure 11.1 compares thresholds obtained from the 22 listeners for pairs of the four tasks. For the pitch and ITD tasks, the thresholds were converted to a logarithmic scale when plotting the data and computing the correlation coefficients. A statistically significant positive correlation was found for pairs of ITD and ILD tasks ($r=0.55$; $p=0.008$) and of ILD and level tasks ($r=0.62$; $p=0.002$), implying that those pairs of tasks rely (partially) on common mechanisms: The performance of the two tasks was determined to some extent by the efficiency (or the amount of "noise") of the common mechanism. The pair of pitch and ITD tasks showed a weak negative correlation ($r=-0.26$), which was, however, not statistically significant ($p=0.247$).

3 Experiment 2: Effects of Training on TFS-Based Pitch Task

Experiment 2 was conducted as another approach to studying the interaction of the two processes. A perceptual learning paradigm was employed. Listeners underwent training on the pitch task for a certain period, and the thresholds for the pitch and ITD tasks (plus other control tasks) were compared between sessions before and after the training. If the thresholds for an untrained ITD task have changed after training on the pitch task, that could constitute evidence indicating interactions between the two processes.

3.1 Methods

Twenty listeners with normal audiometric thresholds (age, 18–29 years old; 6 males and 14 females) participated in the experiment. None of the listeners had prior experience of participating in psychophysical experiments. The listeners were divided

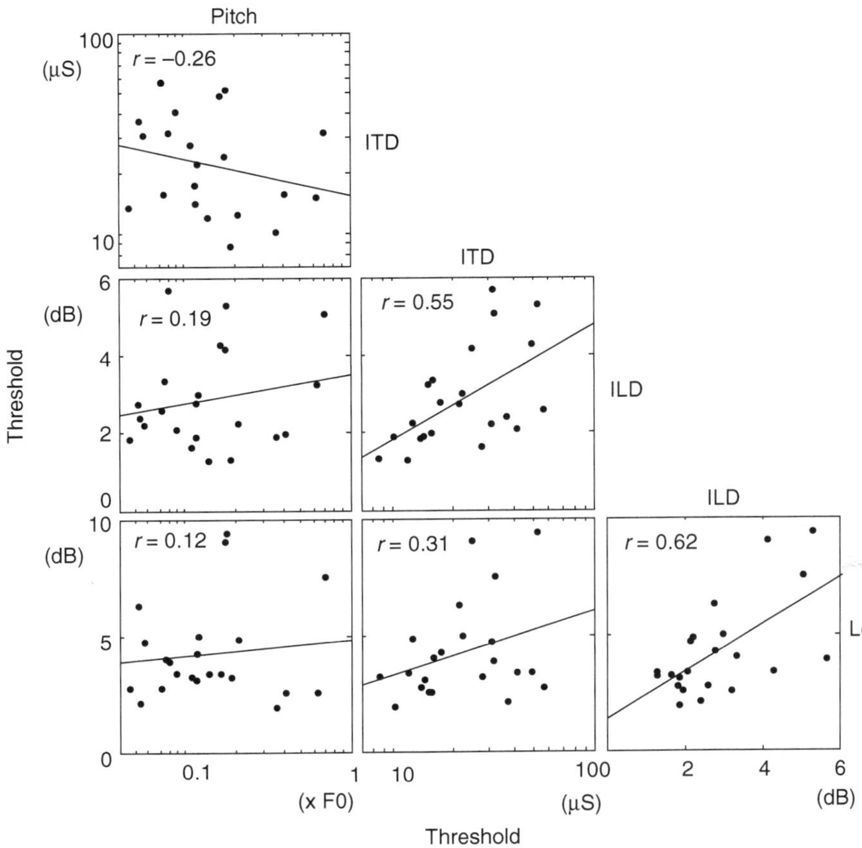

Fig. 11.1 Comparisons of thresholds between tasks. Each panel represents one combination of tasks as labeled. Each *dot* represents one listener

into two groups, namely, *trained* and *control* groups. Each group consisted of 10 listeners.

A total of five tasks were performed. Four of these were identical to the tasks used in Experiment 1, and the fifth one was a newly employed *ITD-high task*. The ITD-high task was the same as the ITD task, except that a transposed stimulus (a 4-kHz tone, amplitude-modulated with a half-wave rectified 125-Hz tone; Bernstein and Trahiotis 2002) was used instead of the SSMC.

The listeners in both groups participated in 4 days of test sessions for all the five tasks. The test sessions on the first day (S0) were intended to familiarize the listeners with the general procedures and the stimuli used in the experiment. The second day (S1) was in the same week as S0, but the days were not consecutive. For the trained group, a 2-week training period (described later) started in the week after S1. The control group experienced the same number of days as the training sessions, but no psychophysical measurements or training was performed. The third and

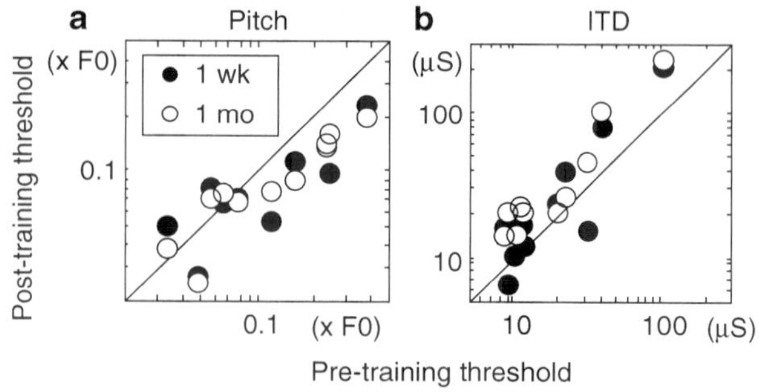

Fig. 11.2 Comparisons of (**a**) post-training and (**b**) pre-training thresholds. *Filled* and *open symbols* indicate thresholds obtained 1 week and 1 month after the pitch training, respectively. Each symbol represents one listener

fourth days of the test sessions (S2 and S3, respectively) were 1 week and 1 month, respectively, after the last training session day. On each test session day, the five tasks were conducted in a random order. For each day and task, three threshold values were obtained by repeating the adaptive tracking, and the average of the three values was taken as that day's threshold for the task. The thresholds on S1, S2, and S3 were regarded as pre-training, 1-week post-training, and 1-month post-training thresholds and were subject to data analysis.

For the trained group, the listeners underwent training in the pitch task, in which adaptive tracking was repeated 12 times per day. The 2-week training period included weekends and so were 12-day training sessions.

3.2 Results

Figure 11.2a compares post-training thresholds with pre-training values for the pitch task (trained task), for the trained group of listeners. The symbols in the plot are generally below the diagonal line, indicating that the thresholds became lower after the training (i.e., performance improved). There was no such trend for the control group (data not shown). A repeated measures analysis of variance (ANOVA) on the thresholds (in a logarithmic scale) was conducted with factors of the test day (within-subject factor) and the listener group (between-subject factor). The main effects of the two factors were not significant (test day $F(2,36)=1.38$, $p=0.2651$; group $F(1,36)=2.32$, $p=0.1454$), but there was a marginal interaction of the test day and the listener group ($F(2,36)=3.25$, $p=0.0505$). When focusing on the trained group, an ANOVA indicated a significant effect of the test day ($F(2,18)=4.25$, $p=0.0309$). The difference between post- and pre-training thresholds was marginally significant for the 1-week post-training data ($p=0.073$, paired t-test) and was significant for the 1-month post-training data ($p=0.032$). A closer examination of the plot revealed that the degree of improvement tended to be smaller in listeners with lower pre-training thresholds. This can be explained as a floor effect, which was why the statistical test showed only a marginal significance for the 1-week post-training data. An ANOVA on the control group showed no effect of the test day ($F(2,18)=0.77$, $p=0.4763$).

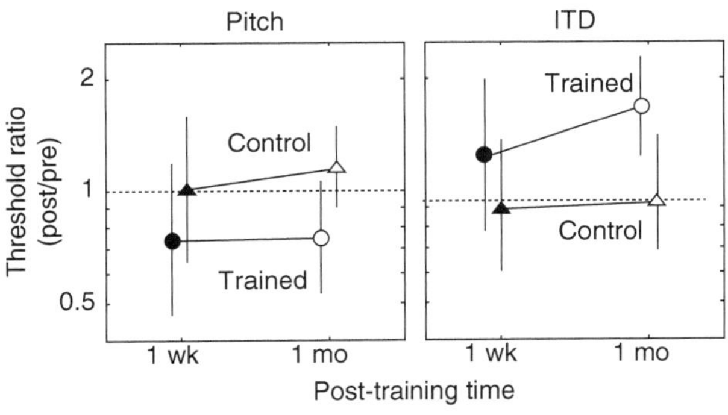

Fig. 11.3 Post- versus pre-training ratio of thresholds. Ratios lower and greater than 1 indicate deteriorated and improved performance, respectively. The abscissa indicates the time after the training period. The *symbols* and *error bars* indicate the means and standard deviations across ten listeners

The thresholds for the ITD task (untrained task) are shown in Fig. 11.2b. The 1-week post-training thresholds were not significantly different from the pre-training thresholds ($p=0.174$), although there was a slight tendency for listeners with higher pre-training thresholds to exhibit elevated post-training thresholds. Thresholds obtained 1 month after the training session exhibited significant *increases* relative to the pre-training data ($p=0.0005$). A repeated measure ANOVA with factors of the test day and the listener group indicated a significant main effect of the test day ($F(2,36)=3.58, p=0.0382$), but not for the listener group ($F(1,36)=0.34, p=0.5697$). There was a significant interaction of the test day and the listener group ($F(2,36)=3.95$, $p=0.0280$). Separate ANOVAs indicated a significant effect of the test day for the trained group ($F(2,18)=7.26$, $p=0.0049$), but not for the control group ($F(2,18)=0.03$, $p=0.9715$).

For the other control tasks, i.e., the ITD-high, ILD, and level tasks, the pre- and post-training thresholds were not significantly different (criterion $p=0.1$). This indicates that the aversive effect of the pitch training on the ITD task was not a simple procedural effect. The ITD-high and ILD tasks adopted the same lateralization procedure as the ITD task, but did not exhibit the training effect.

The effect of training in the pitch task is quantified as the ratio of the post- and pre-training thresholds, and the distribution of this ratio is shown in Fig. 11.3. This representation of the data confirms the above observations. In the pitch task (left panel), the ratios for the trained group tended to be below 1 (improved performance), while those for the control group were distributed around 1 (no change). In the ITD task (right panel), the ratios for the trained group tended to be greater than 1 (deteriorated performance), while those for the control group were distributed around 1.

There was no apparent relationship between the sizes of the training effects in the pitch and ITD tasks. The threshold ratio showed no correlation between the tasks ($r=0.21$ (1 week), −0.11 (1 month)).

4 Discussion

An initial hypothesis of the present study was that processes for TFS-based pitch and for ITD involve a common mechanism that processes TFS information and the efficiency of the common mechanism determines the performance in the two tasks.

Under this hypothesis, it was expected that the performance of individual listeners in the pitch and ITD tasks should correlate positively with each other (Experiment 1) and that the improvement in performance in the pitch task resulting from training in that task could be generalized to the untrained ITD task (Experiment 2). Experiment 1, however, provided no evidence for supporting the expectation. The result of Experiment 2 was even opposite to the expectation (i.e., the performance in the ITD task degraded).

The puzzling results of Experiment 2 can be explained in terms of the listener's weighting on the acoustical cues available when conducting the tasks. In the pitch task, the temporal envelope of the stimulus was always the same and thus was not a useful cue. Instead, the TFS information can be the major cue, although other types of information, such as distortion products by cochlear nonlinearity and the excitation pattern, are also arguably potential cues (Micheyl et al. 2010; Oxenham et al. 2009). In the ITD task, both the (ongoing) envelope and TFS information can be used (since ITDs were imposed on both of those properties), but the envelope information may be less "noisy" in the auditory system. In Experiment 2, training in the pitch task could discourage the listeners from using the envelope information. Thus, after the training, listeners in the trained group might have tended to rely more on the TFS cue than before, which resulted in improved performance in the pitch task but in poorer performance in the ITD task.

Another explanation is that both the pitch and ITD tasks relied on a common module, in which the two tasks competed with each other for limited neural resources: The proportions of the resources assigned for the two tasks might be determined through relatively short-term (a few weeks) training. The medial superior olive is a candidate for such a "common module," which is thought to incorporate neuronal circuits capable of processing monaural and binaural TFS information and to have a certain degree of plasticity (Grothe 2000; Siveke et al. 2012).

It is noteworthy that the pitch-training-induced elevation of the ITD thresholds (Experiment 2) became statistically significant only on the day 1 month after the training but not 1 week after training. This implies that the process involved in the observed across-task transfer of the training effect has a relatively long time constant. This is in general agreement with the notion that neuronal changes underlying generalization are distinct from those underlying learning in the trained task and have a longer time course (Wright et al. 2010).

References

Bernstein LR, Trahiotis C (2002) Enhancing sensitivity to interaural delays at high frequencies by using "transposed stimuli". J Acoust Soc Am 112:1026–1036

de Boer E (1956) Pitch of inharmonic signals. Nature 178:535–536

Grothe B (2000) The evolution of temporal processing in the medial superior olive, an auditory brainstem structure. Prog Neurobiol 61:581–610

Hopkins K, Moore BC (2007) Moderate cochlear hearing loss leads to a reduced ability to use temporal fine structure information. J Acoust Soc Am 122:1055–1068

Hopkins K, Moore BC (2010) Development of a fast method for measuring sensitivity to temporal fine structure information at low frequencies. Int J Audiol 49:940–946

Levitt H (1970) Transformed up-down methods in psychoacoustics. J Acoust Soc Am 49: 467–477

Micheyl C, Dai H, Oxenham AJ (2010) On the possible influence of spectral- and temporal-envelope cues in tests of sensitivity to temporal fine structure. J Acoust Soc Am 127:1809–1810

Moore GA, Moore BC (2003) Perception of the low pitch of frequency-shifted complexes. J Acoust Soc Am 113:977–985

Moore BC, Sek A (2009) Development of a fast method for determining sensitivity to temporal fine structure. Int J Audiol 48:161–171

Moore BC, Huss M, Vickers DA, Glasberg BR, Alcantara JI (2000) A test for the diagnosis of dead regions in the cochlea. Br J Audiol 34:205–224

Oxenham AJ, Micheyl C, Keebler MV (2009) Can temporal fine structure represent the fundamental frequency of unresolved harmonics? J Acoust Soc Am 125:2189

Schouten J, Ritsma RJ, Cardozo B (1962) Pitch of the residue. J Acoust Soc Am 34:1418–1424

Siveke I, Leibold C, Schiller E, Grothe B (2012) Adaptation of binaural processing in the adult brainstem induced by ambient noise. J Neurosci 32:462–473

Wright BA, Wilson RM, Sabin AT (2010) Generalization lags behind learning on an auditory perceptual task. J Neurosci 30:11635–11639

Chapter 12
On the Limit of Neural Phase Locking to Fine Structure in Humans

Philip X. Joris and Eric Verschooten

Abstract The frequency extent over which temporal fine structure is available in the human auditory system has recently become a topic of much discussion. It is common, in both the physiological and psychophysical literature, to encounter the assumption that fine structure is available to humans up to about 5 kHz or even higher. We argue from existing physiological, anatomical, and behavioral data in animals, combined with behavioral and anatomical data in humans, that it is unlikely that the human central nervous system has access to fine structure above a few kHz.

1 Introduction

The issue of temporal versus place coding has occupied hearing researchers for over a century. It is not just an academic issue as it is relevant for the remediation of hearing impairment, e.g., via cochlear implants. Unfortunately, it is at present not possible to apply the most incisive behavioral and physiological techniques within the same species. Thus, physiological findings in nonhuman species are typically invoked to interpret human hearing, which raises questions regarding the physiological similarity between laboratory and human species. We argue that coding of fine structure in human is limited to lower frequencies than in cat.

P.X. Joris (✉) • E. Verschooten
Laboratory of Auditory Neurophysiology, University of Leuven,
Herestraat 49 bus 1021, Leuven B-3000, Belgium
e-mail: philip.joris@med.kuleuven.be

B.C.J. Moore et al. (eds.), *Basic Aspects of Hearing*,
Advances in Experimental Medicine and Biology,
DOI 10.1007/978-1-4614-1590-9_12, © Springer Science+Business Media New York 2013

2 Neural Phase Locking to Fine Structure

Broadly defined, phase locking of sensory neurons refers to their ability to represent some temporal aspect of a stimulus waveform. By "representation" is usually meant an increased and/or decreased probability that the neuron fires action potentials (spikes) at specific time points of the stimulus waveform. In the context of auditory physiology, it is customary to distinguish phase locking to fine structure, envelope, and transients, which differ in their underlying physiological mechanisms. Here, we use the term phase locking in the restricted sense of a modulation of firing rate locked to stimulus fine structure, i.e., to instantaneous, cycle by cycle, pressure fluctuations. Phase locking to fine structure can in principle be studied with any sound, but in practice single tones have been the stimulus of choice. They indeed deserve their status of purity in this context because of their perfectly flat envelope. Ignoring onset or offset transients, the sustained part of the neuronal response allows a straightforward assessment of phase locking to fine structure. For reasons of space, we base our arguments mainly on pure tone responses in the cat – the animal which has been characterized most extensively – but the general trends we describe are, at least partly, observed in guinea pig, rabbit, chinchilla, gerbil, monkey, and barn owl.

3 Human Behavioral Limits Reflecting Phase Locking to Fine Structure

Many attributes of auditory perception have been proposed to involve or require neural phase locking to fine structure. Because changes in stimulus waveform are typically accompanied by changes in response dimensions other than fine structure (spatial pattern of excitation, temporal envelope), assessment of its role requires indirect arguments. The exception is binaural hearing, for which it is beyond doubt that neural coding and processing of fine structure are essential. Many studies have explored sensitivity to interaural differences in phase and time. Again, pure tones are purest: for any other stimulus, interactions between stimulus components may create envelope fluctuations underlying behavioral sensitivity.

Humans show exquisite sensitivity to interaural phase differences up to 1.4 kHz (Zwislocki and Feldman 1956). A similar limit on the use of fine structure is suggested by nontonal stimuli (Klumpp and Eady 1956; Schiano et al. 1986). Besides ITD jnds, other perceptual attributes that are necessarily based on a binaural comparison of fine structure show a similar upper frequency limit (Licklider et al. 1950; Perrott and Musicant 1977) of ~1.5 kHz. Dunai and Hartmann (2011) recently traced the upper frequency limit for tonal ITD discrimination in fine steps: in all four subjects, the highest frequency at which ITD sensitivity could be measured was 1,400 or 1,450 Hz.

4 Physiological Limits of Phase Locking to Fine Structure

Early studies indicated that the existence range of pure-tone neural phase locking extends up to several kHz (Rose et al. 1967; Johnson 1980). The most used metric is the vector strength (Goldberg and Brown 1969). A plot across nerve fibers of maximal vector strength as a function of characteristic frequency (CF), in response to CF tones, suggests a low-pass filter (Johnson 1980). Likewise, a within-fiber plot of maximal vector strength as a function of frequency has a monotonically decreasing shape (Rose et al. 1968; Joris et al. 1994a). When other metrics are used, such as the temporal width of the period histogram (Joris et al. 1994b) or width and height of autocorrelograms (Louage et al. 2004), the low-pass filter characterization appears less adequate. For simplicity we here take the upper limit of phase locking as the highest frequency at which significant phase locking is found, but the exact definition is not critical to our argument. The upper limit of phase locking differs between species (Weiss and Rose 1988). In cat, vector strength starts to decrease at about 1 kHz and becomes insignificant at ~5 kHz.

The auditory nerve projects to the cochlear nucleus. Here, the phase-locking limit clearly differs between subnuclei and cell types. In the anteroventral cochlear nucleus (AVCN), bushy cells show exquisite phase locking, but their upper cutoff of significant phase locking is somewhat lower than in the AN (~4 kHz, Bourk 1976; Rhode and Smith 1986; Blackburn and Sachs 1989; Joris et al. 1994a). AVCN stellate cells can show excellent phase locking at very low frequencies (Joris et al. 1994b), but phase locking is clearly degraded in terms of upper frequency limit (Bourk 1976; Rhode and Smith 1986; Blackburn and Sachs 1989), and the cells are biased towards high CFs (Melcher 1993). The posteroventral cochlear nucleus contains neurons with exquisite phase locking to low-frequency tones (Godfrey et al. 1975; Rhode and Smith 1986; Smith et al. 2005). The nucleus is biased towards high CFs, and its output neurons receive highly convergent input from the auditory nerve and show wide tuning. Octopus cells show sustained phase-locked responses at low frequencies, but for tones above ~2 kHz, it is rare to get more than a single pure onset response (Godfrey et al. 1975; Rhode and Smith 1986; Joris and Smith 2011). Multipolar cells do show sustained responses at all frequencies but have a lower upper limit of phase locking than the auditory nerve, as do octopus cells (Rhode and Smith 1986). Finally, the dorsal cochlear nucleus shows poor phase locking to fine structure (Goldberg and Brownell 1973) and is also biased towards high CFs (Spirou et al. 1993). In summary, it appears that in terms of upper limit, bushy cells best preserve the phase locking present in the AN. This is of little surprise since these cells receive a small number of powerful axosomatic terminals (the end bulbs and modified end bulbs of Held) from the auditory nerve (Ryugo and Rouiller 1988; Sento and Ryugo 1989) and have short membrane time constants by virtue of their intrinsic properties (reviewed in Young and Oertel 2004).

The main projection targets of bushy cells are the binaural circuits in the superior olivary complex. Spherical bushy cells give excitatory projections to the medial and lateral superior olive (LSO and MSO). Globular bushy cells give excitatory projections to glycinergic, inhibitory nuclei (the medial and lateral nuclei of the trapezoid body, MNTB, and LNTB) which in turn project to LSO and MSO. Both kinds of bushy cells show exquisite phase locking (reviewed by Joris and Smith 2008) and have a variety of specializations in their innervation, biophysical properties, and anatomy, including one of the largest synaptic terminals in the brain (the calyx of Held on the MNTB).

The use by the central nervous system of phase locking to fine structure in the peripheral auditory system is dramatically illustrated in the sensitivity, arising in the superior olivary complex, to interaural time differences (ITDs) to pure tones (equivalently referred to as sensitivity to interaural phase differences, IPDs). ITD sensitivity is generated both in the MSO and the LSO. Recordings in MSO show prominent ITD sensitivity to pure tones (Goldberg and Brown 1969; Yin and Chan 1990), but the upper limit has, unfortunately, not been characterized in cat. The LSO has only a small low-CF representation, and its ITD sensitivity to tones is weaker, in a number of respects, than in MSO (Joris and Yin 1995; Tollin and Yin 2005). Importantly, ITD sensitivity signifies a recoding from a temporal to a rate code. Put simplistically, phase locking to fine structure is "used" in MSO and LSO and is translated to a code of average rate. Thus, even though data at the level of the MSO are scarce, limits of ITD sensitivity in neurons at higher levels give insight in the highest frequency at which phase locking to fine structure affects the output rate of neurons.

The best studied projection target of the binaural nuclei in the superior olivary complex is the inferior colliculus (IC). In cat, the highest frequencies at which ITD sensitivity has been reported in IC are just below 3 kHz (2,860 Hz in Rose et al. 1966; 2.8 kHz in Yin and Kuwada 1983). This ITD sensitivity is thought to be largely inherited from the superior olivary complex. It is unclear whether this sensitivity is further modified by monaural phase-locked inputs at levels above MSO and LSO, e.g., through an interaction between monaural phase-locked input from the AVCN and binaural ITD-sensitive input from superior olivary complex. Phase locking in IC neurons is clearly limited to much lower frequencies than in auditory nerve or cochlear nucleus (Kuwada et al. 1984). Cochlear nucleus neurons that directly project to IC do not phase lock to the high frequencies to which bushy cells phase lock, so even if phase locking to fine structure would directly alter properties of ITD sensitivity in the IC, this could only be the case at low frequencies and would not affect the upper limit of ITD sensitivity.

Sensitivity to ITDs of fine structure is evident up to 3 kHz in cat thalamus (Ivarsson et al. 1988). Interestingly, significant IPD sensitivity in cat primary auditory cortex was found up to 2.5 kHz (Reale and Brugge 1990), though it is not clear from that report whether higher frequencies were tested. Thus, while the upper limit of phase locking to fine structure degrades with ascending anatomical levels, this does not appear to be the case for the upper frequency limit of ITD sensitivity, which is similar at different anatomical levels. This is not surprising. Once the temporal code is converted to a rate code, the sensitivity can simply be passed on to higher structures, and specializations such as large terminals and fast membranes are no longer required.

ITD sensitivity to fine structure is easiest to test with pure tones but is of course also evident in response to other sounds. Importantly, in response to broadband stimuli, the ITD sensitivity of neurons with CF between 1 and 3 kHz can be dominated by envelope even when these same neurons show excellent ITD sensitivity to pure tones (Joris 2003). A psychophysical corollary is the transition of binaural performance based on fine structure to that based on envelope, which starts at a much lower frequency (by more than an octave, Bernstein and Trahiotis 1996) than the behavioral upper limit of ITD sensitivity to fine structure of 1.4 kHz, mentioned in the previous section.

5 Behavioral Limits of Phase Locking in Cat

The upper limit of ~3 kHz reported in the cat for physiological ITD sensitivity to fine structure fits rather well with binaural behavioral measurements. In a dichotic lateralization task with tones (detection of left-right reversals), the upper limit was put between 2 and 3 kHz (Wakeford and Robinson 1974). A later study using finer steps found an upper limit of 2.8 kHz (Jackson et al. 1996; Heffner HE and Heffner RS, personal communicated, 2012).

6 Comparative Anatomy of the Brainstem

At a coarse level of description, the human auditory brainstem conforms to a basic mammalian plan and can be said to be similar to that of laboratory animals. A closer look shows obvious differences in size, connectivity, and even existence of brainstem nuclei between different species. Such differences are most obvious with species in particular acoustic habitats (marine and burrowing mammals) or with particular auditory specializations (bats). Even leaving such species out of consideration, comparative anatomical studies show systematic trends in size of brainstem auditory nuclei, which are correlated with body size and/or hearing range. For example, within the primate order, the MSO increases in size from lower to higher primates, while the opposite is true for the LSO (Moore and Moore 1971). The inferior quality of human material available and the lack of connectional data severely limit neuroanatomical knowledge on the human auditory brainstem. Nevertheless, cytoarchitectural and cytological features of different cell types, observed in laboratory animals, are recognizable in human tissue and indicate the presence of a well-developed ITD circuit centered on MSO, while the LSO circuit (including globular bushy cells, MNTB neurons, LSO) is more disputed but anyhow less prominent (Moore and Osen 1979; Richter et al. 1983; Adams 1996; Bazwinsky et al. 2003; Kulesza 2007, 2008; Grothe et al. 2010). These anatomical data are consistent with the characterization of humans as "low-frequency animals" but are in themselves neutral regarding the upper frequency limit of phase locking to fine structure relative to experimental animals. For comparison, toothed whales, which are also exceedingly difficult to study experimentally, show striking anatomical differences with terrestrial mammals regarding axon number and diameter in the auditory nerve and central fiber tracts and in volume of auditory brainstem nuclei (reviewed in Ridgway 2000). Such differences can be interpreted as specializations subserving temporal processing. There are no indications along those lines in the human auditory brainstem.

In fact, there are two anatomical observations that would hamper rather than favor a high temporal limit in humans. The first is the innervation of multiple hair cells by single nerve fibers (Nadol 1983). Second, the auditory nerve is substantially longer (~2.5 cm) in humans than in cat, but axon diameters are similar (Moore 1987). Dispersion between input fibers is one possible explanation for the somewhat lower phase-locking limit in bushy cells than in the auditory nerve.

7 Summary

To conclude, we first emphasize what we are *not* saying. Our point is not that there is no neural fine structure in humans above 1.4 kHz. There likely is, just as in cat, where the upper limit for phase locking in the auditory nerve is higher than the binaural behavioral limit. Neither is our point that neural coding of fine structure is only behaviorally relevant in binaural hearing. Our point is that the binaural system is neurally optimized to make use of fine structure, that there is no physiological evidence in animals that the monaural neural system would make use of fine structure at higher frequencies than the binaural system, and that there are no indications of anatomical specializations in humans towards phase locking at higher frequencies than in experimental animals. We therefore consider it parsimonious to regard the binaural behavioral limit of 1.4 kHz as a general upper bound for the use of fine structure by the human auditory system.

Acknowledgments The authors are supported by the Fund for Scientific Research – Flanders (G.0714.09 and G.0961.11) and Research Fund K. U. Leuven (OT/09/50).

References

Adams JC (1996) Neural circuits in the human auditory brainstem. In: Ainsworth WA, Greenberg S (eds) Auditory basis of speech perception. Keele University, Keele, pp 39–44

Bazwinsky I, Hilbig H, Bidmon HJ, Rubsamen R (2003) Characterization of the human superior olivary complex by calcium binding proteins and neurofilament H (SMI-32). J Comp Neurol 456:292–303

Bernstein LR, Trahiotis C (1996) The normalized correlation: accounting for binaural detection across center frequency. J Acoust Soc Am 100:3774–3784

Blackburn CC, Sachs MB (1989) Classification of unit types in the anteroventral cochlear nucleus: PST histograms and regularity analysis. J Neurophysiol 62:1303–1329

Bourk TR (1976) Electrical responses of neural units in the anteroventral cochlear nucleus of the cat. PhD thesis, MIT, Cambridge

Dunai L, Hartmann WM (2011) Frequency dependence of the interaural time difference thresholds in human listeners. J Acoust Soc Am 129:2485

Godfrey DA, Kiang NYS, Norris BE (1975) Single unit activity in the posteroventral cochlear nucleus of the cat. J Comp Neurol 162:247–268

Goldberg JM, Brown PB (1969) Response of binaural neurons of dog superior olivary complex to dichotic tonal stimuli: some physiological mechanisms of sound localization. J Neurophysiol 22:613–636

Goldberg JM, Brownell WE (1973) Discharge characteristics of neurons in anteroventral and dorsal cochlear nuclei of cat. Brain Res 64:35–54

Grothe B, Pecka M, McAlpine D (2010) Mechanisms of sound localization in mammals. Physiol Rev 90:983–1012

Ivarsson C, De Ribaupierre Y, De Ribaupierre F (1988) Influence of auditory localization cues on neuronal activity in the auditory thalamus of the cat. J Neurophysiol 59:586–606

Jackson LL, Heffner HE, Heffner RS (1996) Species differences in the upper limit of binaural phase discrimination. Assoc Res Otolaryngol Abs 19:63

Johnson DH (1980) The relationship between spike rate and synchrony in responses of auditory-nerve fibers to single tones. J Acoust Soc Am 68:1115–1122

Joris PX (2003) Interaural time sensitivity dominated by cochlea-induced envelope patterns. J Neurosci 23:6345–6350

Joris PX, Smith PH (2008) The volley theory and the spherical cell puzzle. Neuroscience 154: 65–76

Joris P, Smith P (2011) Octopus cells: the temporally most precise neurons in the brain? Assoc Res Otolaryngol Abs 34:227

Joris PX, Yin TC (1995) Envelope coding in the lateral superior olive. I. Sensitivity to interaural time differences. J Neurophysiol 73:1043–1062

Joris PX, Carney LH, Smith PH, Yin TC (1994a) Enhancement of neural synchronization in the anteroventral cochlear nucleus. I. Responses to tones at the characteristic frequency. J Neurophysiol 71:1022–1036

Joris PX, Smith PH, Yin TC (1994b) Enhancement of neural synchronization in the anteroventral cochlear nucleus. II. Responses in the tuning curve tail. J Neurophysiol 71:1037–1051

Klumpp RG, Eady HR (1956) Some measurements of interaural time difference thresholds. J Acoust Soc Am 28:859–860

Kulesza RJ Jr (2007) Cytoarchitecture of the human superior olivary complex: medial and lateral superior olive. Hear Res 225:80–90

Kulesza RJ Jr (2008) Cytoarchitecture of the human superior olivary complex: nuclei of the trapezoid body and posterior tier. Hear Res 241:52–63

Kuwada S, Yin TCT, Syka J, Buunen TJF, Wickesberg RE (1984) Binaural interaction in low-frequency neurons in inferior colliculus of the cat. IV. Comparison of monaural and binaural response properties. J Neurophysiol 51:1306–1325

Licklider JCR, Webster JC, Hedlun JM (1950) On the frequency limits of binaural beats. J Acoust Soc Am 22:468–473

Louage DH, van der Heijden M, Joris PX (2004) Temporal properties of responses to broadband noise in the auditory nerve. J Neurophysiol 91:2051–2065

Melcher JR (1993) The cellular generators of the brainstem auditory evoked potential. PhD thesis, MIT, Cambridge

Moore JK (1987) The human auditory brain stem as a generator of auditory evoked potentials. Hear Res 29:33–43

Moore JK, Moore RY (1971) A comparative study of the superior olivary complex in the primate brain. Folia Primatol 16:35–51

Moore JK, Osen KK (1979) The cochlear nuclei in man. Am J Anat 154:393–418

Nadol JB Jr (1983) Serial section reconstruction of the neural poles of hair cells in the human organ of Corti. I. Inner hair cells. Laryngoscope 93:599–614

Perrott DR, Musicant AD (1977) Rotating tones and binaural beats. J Acoust Soc Am 61: 1288–1292

Reale RA, Brugge JF (1990) Auditory cortical neurons are sensitive to static and continuously changing interaural phase cues. J Neurophysiol 64:1247–1260

Rhode WS, Smith PH (1986) Encoding timing and intensity in the ventral cochlear nucleus of the cat. J Neurophysiol 56:261–286

Richter EA, Norris BE, Fullerton BC, Levine RA, Kiang NYS (1983) Is there a medial nucleus of the trapezoid body in humans. Am J Anat 168:157–166

Ridgway SH (2000) The auditory central nervous system of dolphins. In: Au WWL, Popper AN, Fay RR (eds) Hearing by whales and dolphins, Springer Handbook of Auditory Research. Springer, New York, pp 273–293

Rose JE, Gross NB, Geisler CD, Hind JE (1966) Some neural mechanisms in the inferior colliculus of the cat which may be relevant to localization of a sound source. J Neurophysiol 29: 288–314

Rose JE, Brugge JF, Anderson DJ, Hind JE (1967) Phase-locked response to low-frequency tones in single auditory nerve fibers of the squirrel monkey. J Neurophysiol 30:769–793

Rose JE, Brugge JF, Anderson DJ, Hind JE (1968) Patterns of activity in single auditory nerve fibres of the squirrel monkey. In: de Reuck AVS, Knight J (eds) Ciba foundation symposium on hearing mechanisms in vertebrates. J&A Churchill, London, pp 144–157

Ryugo DK, Rouiller EM (1988) Central projections of intracellularly labeled auditory nerve fibers in cats: morphometric correlations with physiological properties. J Comp Neurol 271: 130–142

Schiano JL, Trahiotis C, Bernstein LR (1986) Lateralization of low-frequency tones and narrow bands of noise. J Acoust Soc Am 79:1563–1570

Sento S, Ryugo DK (1989) Endbulbs of held and spherical bushy cells in cats: morphological correlates with physiological properties. J Comp Neurol 280:553–562

Smith PH, Massie A, Joris PX (2005) Acoustic stria: anatomy of physiologically characterized cells and their axonal projection patterns. J Comp Neurol 482:349–371

Spirou GA, May BJ, Wright DD, Ryugo DK (1993) Frequency organization of the dorsal cochlear nucleus in cats. J Comp Neurol 329:36–52

Tollin DJ, Yin TC (2005) Interaural phase and level difference sensitivity in low-frequency neurons in the lateral superior olive. J Neurosci 25:10648–10657

Wakeford OS, Robinson DE (1974) Lateralization of tonal stimuli by the cat. J Acoust Soc Am 55:649–652

Weiss TF, Rose C (1988) A comparison of synchronization filters in different auditory receptor organs. Hear Res 33:175–180

Yin TCT, Chan JK (1990) Interaural time sensitivity in medial superior olive of cat. J Neurophysiol 64:465–488

Yin TCT, Kuwada S (1983) Binaural interaction in low-frequency neurons in inferior colliculus of the cat. II. Effects of changing rate and direction of interaural phase. J Neurophysiol 50:1000–1018

Young ED, Oertel D (2004) Cochlear nucleus. In: Shepherd GM (ed) The synaptic organization of the brain. Oxford University Press, Oxford, pp 125–163

Zwislocki J, Feldman RS (1956) Just noticeable differences in dichotic phase. J Acoust Soc Am 28:860–864

Chapter 13
Effects of Sensorineural Hearing Loss on Temporal Coding of Harmonic and Inharmonic Tone Complexes in the Auditory Nerve

Sushrut Kale, Christophe Micheyl, and Michael G. Heinz

Abstract Listeners with sensorineural hearing loss (SNHL) often show poorer thresholds for fundamental-frequency (F0) discrimination and poorer discrimination between harmonic and frequency-shifted (inharmonic) complex tones, than normal-hearing (NH) listeners—especially when these tones contain resolved or partially resolved components. It has been suggested that these perceptual deficits reflect reduced access to temporal-fine-structure (TFS) information and could be due to degraded phase locking in the auditory nerve (AN) with SNHL. In the present study, TFS and temporal-envelope (ENV) cues in single AN-fiber responses to band-pass-filtered harmonic and inharmonic complex tones were measured in chinchillas with either normal-hearing or noise-induced SNHL. The stimuli were comparable to those used in recent psychophysical studies of F0 and harmonic/inharmonic discrimination. As in those studies, the rank of the center component was manipulated to produce different resolvability

S. Kale
Department of Otolaryngology-Head & Neck Surgery, Columbia University, 630 W. 168th St., P&S 11-452, New York, NY 10032, USA

Weldon School of Biomedical Engineering, Purdue University, 500 Oval Drive, West Lafayette, IN 47907, USA
e-mail: sk3646@columbia.edu

C. Micheyl
Department of Psychology, University of Minnesota, N628 Elliott Hall, 75 East River Road, Minneapolis, MN 55455, USA
e-mail: cmicheyl@umn.edu

M.G. Heinz (✉)
Department of Speech, Language, & Hearing Sciences and Biomedical Engineering, Purdue University, 500 Oval Drive, West Lafayette, IN 47907, USA
e-mail: mheinz@purdue.edu

B.C.J. Moore et al. (eds.), *Basic Aspects of Hearing*, Advances in Experimental Medicine and Biology,
DOI 10.1007/978-1-4614-1590-9_13,

conditions, different phase relationships (cosine and random phase) were tested, and background noise was present. Neural TFS and ENV cues were quantified using cross-correlation coefficients computed using shuffled cross correlograms between neural responses to REF (harmonic) and TEST (F0- or frequency-shifted) stimuli. In animals with SNHL, AN-fiber tuning curves showed elevated thresholds, broadened tuning, best-frequency shifts, and downward shifts in the dominant TFS response component; however, no significant degradation in the ability of AN fibers to encode TFS or ENV cues was found. Consistent with optimal-observer analyses, the results indicate that TFS and ENV cues depended only on the relevant frequency shift in Hz and thus were not degraded because phase locking remained intact. These results suggest that perceptual "TFS-processing" deficits do not simply reflect degraded phase locking at the level of the AN. To the extent that performance in F0- and harmonic/inharmonic discrimination tasks depend on TFS cues, it is likely through a more complicated (suboptimal) decoding mechanism, which may involve "spatiotemporal" (place-time) neural representations.

1 Introduction

In normal-hearing (NH) listeners, fundamental-frequency (F0) discrimination thresholds are usually lower (better) when low-rank harmonics are present than when only high-rank harmonics are available (e.g., Houtsma and Smurzynski 1990). Moreover, in hearing-impaired (HI) listeners, F0-discrimination thresholds for complex tones containing low-rank harmonics are often elevated compared to those measured using comparable stimuli in NH listeners (e.g., Moore et al. 2006). A traditional explanation for these findings is that accurate F0 discrimination requires having access to "resolved" harmonics, which in turn depends on cochlear frequency selectivity—which is often adversely affected by sensorineural hearing loss (SNHL). An alternative explanation is that accurate F0 discrimination depends on temporal-fine-structure (TFS) cues, specifically, intervals between TFS peaks under temporal-envelope (ENV) maxima, and that these cues are degraded by SNHL (e.g., Hopkins and Moore 2007). However, the effects of SNHL on the neural representation of TFS and ENV information at the level of the auditory nerve (AN) have not been thoroughly studied. In this study, we used a combination of computational modeling and neurophysiological recordings in NH and HI (noise-exposed) chinchillas in order to (1) assess the impact of SNHL on the neural representation of TFS and ENV cues for the discrimination of F0, or coherent frequency shifts, for complex tones similar to those used in recent psychophysical studies (Moore et al. 2009), as a function of the lowest harmonic number present in the stimulus, and (2) predict the thresholds that could in theory be obtained by using either all of the information (including TFS, ENV, and place information) contained in AN-fiber responses or solely place and average firing-rate information.

2 Methods

2.1 Experimental Procedures

Single-fiber AN recordings were obtained in nine NH and six HI chinchillas using standard procedures (Kale and Heinz 2010). Noise-induced hearing loss was produced by presenting an octave-wide band of noise centered at 500 Hz continuously for 2 h at ~117 dB SPL. This acoustic overexposure resulted in mild to moderate hearing loss over the CF range of 0.3–6 kHz, consistent with previous studies in chinchillas showing mixed hair-cell losses well beyond the octave-wide exposure band (Harding and Bohne 2007). The animals were allowed to recover for 4–6 weeks prior to the recordings. Except for the impaired-fiber CFs, which were determined manually (as in Liberman 1984), CF, threshold, and Q_{10} were determined using an automated algorithm. Spike times were measured with 10-μs resolution. All procedures were approved by the Animal Care and Use Committee of Purdue University.

2.2 Stimuli

The stimuli were as similar as possible to those used in a recent psychophysical study of F0 and harmonic versus inharmonic (H-I) discrimination (Moore et al. 2009). The "reference" (REF) stimulus was a harmonic complex tone. The "test" (TEST) stimulus was either another harmonic complex tone with a different F0 (as in the F0-discrimination task) or an inharmonic complex tone produced by shifting the frequencies of all components in the REF complex upward by a constant amount in Hz (as in the H–I discrimination task). F0 shifts of 0.04, 0.1, and 0.5 % were tested in most fibers; larger shifts were also tested in some fibers, whenever possible. Constant frequency shifts of 0.04, 0.1, or 0.5 % of the F0 were tested. The complexes were band-pass filtered with a fifth-order Butterworth filter; the filter passband contained approximately five components. The center frequency of the REF complex matched the CF of the fiber. The rank of the lowest component contained within the 3-dB passband of the stimulus (hereafter referred to as the "harmonic rank") was manipulated by varying the F0 of the stimulus. Four harmonic ranks (2, 4, 6, and 20) were tested. The component starting phases were either cosine or randomized. All stimuli were presented in background noise. The level of the noise was set 10 dB below the masked threshold defined as the noise level required to just mask the response to the suprathreshold tone complex.

2.3 Analysis Metrics

Shuffled correlograms were used to quantify within-fiber TFS and ENV coding (e.g., Louage et al. 2004). Shuffled autocorrelograms (SACs) were computed by computing inter-spike intervals (ISIs) across spike trains obtained in response to a single polarity of the stimulus. Shuffled cross-polarity correlograms (SCCs) were obtained by computing ISIs across spike trains obtained in response to positive and negative polarities of the stimulus. Stimulus polarity inversion inverts the fine structure while keeping the envelope unchanged. Therefore, differences between SAC and SCC (*difcor*) functions reflect TFS information, whereas averages of SAC and SCC (*sumcor*) functions represent ENV information. Neural cross-correlation coefficients for *difcor* and *sumcor* functions were used to quantify the degree of similarity in neural responses to the TFS (ρTFS), or to the ENV (ρENV), of the REF and TEST stimuli (Heinz and Swaminathan 2009). ρTFS (or ρENV) values close to 1 indicate a high degree of similarity in neural responses to TFS (or ENV)—and, therefore, poor discriminability—of the REF and TEST stimuli.

2.4 Optimal-Observer Model

To determine the just noticeable differences (JNDs) for F0 discrimination and H–I discrimination that could be achieved using the information contained in neural responses to the REF and TEST stimuli used in the present study, we used an approach similar to that used by Heinz et al. (2001). Simulated neural responses obtained using a physiologically realistic AN model (Zilany and Bruce 2006) were analyzed using an optimal-observer model (Siebert 1970). Two versions of the model were considered: rate-place (RP) and all-information (AI) (Heinz et al. 2001); the RP observer uses only spike-count information, whereas the AI observer uses the actual spike times (see Heinz et al. (2001) for details). Both models rest on the assumption that AN spiking is well described as a nonhomogeneous Poisson process. Responses from 150 AN-model fibers, with CFs ranging from 150 to 2,500 Hz, were simulated.

3 Results

Figure 13.1 shows examples of power spectral densities of *difcor* and *sumcor* functions computed using the responses of a noise-exposed fiber (CF = 1.54 kHz) to REF (harmonic) stimuli, for different values of the harmonic rank (N). As N increased from 2 to 20, the number of peaks in the *difcor* function increased (Fig. 13.1a–d), reflecting the increase in the number of TFS components falling in the passband of the fiber's tuning curve. Note that for $N = 2$, the most dominant *difcor* (TFS) component corresponded to a frequency nearly 1 octave below the estimated CF of the fiber (Fig. 13.1a). In contrast, for $N = 20$, the frequency of the most dominant *difcor*

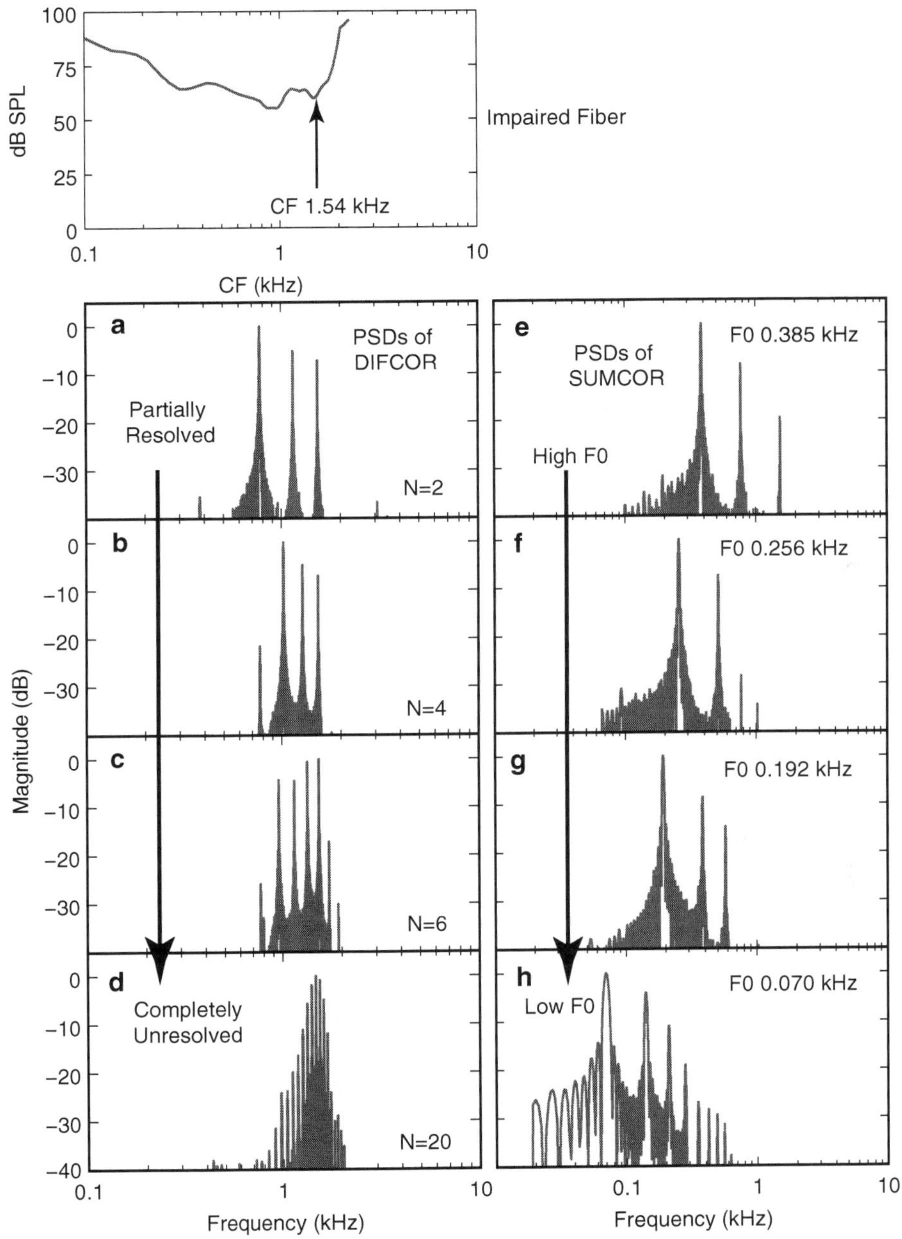

Fig. 13.1 Following SNHL, the most dominant TFS response component was below the estimated cochlear CF (*arrow in top left panel*, Liberman 1984). (**a–d**) Normalized power spectra of difcor functions. (**e**, **f**) Same for sumcor functions. Each row represents a rank and F0 condition

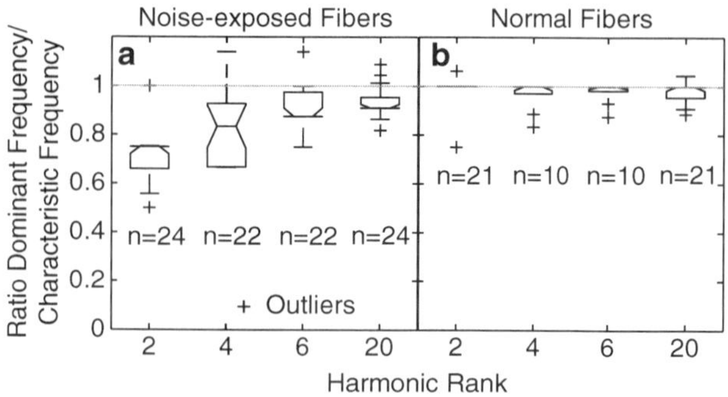

Fig. 13.2 Ratio of the most dominant TFS response component to cochlear CF for a population of normal and noise-exposed fibers. (**a**) Noise-exposed fibers. (**b**) Normal fibers. Number of fibers (*n*) that went into the computation of median and interquartile ranges is shown below each box-whisker plot. Outliers are indicated by "+"

peak corresponded to the estimated CF (compare Fig. 13.1a with 13.1d). The *sumcor* functions reflected F0-related periodicities and distortion components in the modulation spectrum (Fig. 13.1e, f). Qualitatively similar results were obtained across the entire population of impaired fibers. These results show that even when the stimulus contains only unresolved harmonics, the temporal responses of AN fibers convey both TFS and ENV information, and this is the case even for impaired AN fibers.

To quantify the mismatch between the frequency of the most dominant TFS component and the estimated cochlear CF, the ratio of these two quantities was computed for every fiber. For noise-exposed fibers, the ratio was lower than 1 on average for low harmonic ranks ($N=2$ or 4), and it slowly approached 1 as N increased toward 20 (Fig. 13.2). These results provide further evidence that following SNHL, there is a mismatch between the CF of a fiber and the frequency of the TFS component that is most strongly represented in the ISIs of this fiber; this mismatch is most marked for fibers responding to partially resolved components. In contrast, for normal fibers, the ratio was always around one, indicating a good match between cochlear CF and the most dominant TFS response component for all harmonic-rank conditions.

Figure 13.3a, b shows the correlation coefficients, ρTFS and ρENV, as a function of the F0 difference (ΔF0) between REF and TEST stimuli. Although both coefficients decreased as ΔF0 increased, for ρENV, the decrease became less and less steep as N increased. The latter effect can be explained by considering that for a given CF, the F0 was lower for $N=20$ than for $N=2$ and hence the ΔF0 in Hz was lower for $N=20$ than for $N=2$. Consistently across the population of normal and impaired fibers, ρENV corresponding to a frequency shift of 0.5 % (of F0) was higher for $N=20$ than for $N=2$ (Fig. 13.3c, d). Further analysis of the data showed that a given ΔF in Hz produced equivalent changes in ρTFS and ρENV. These results suggest that based on within-fiber spike-timing information, TFS and ENV cues for the discrimination of frequency shifts in Hz are equally strong. Figure 13.3e, f shows that for both fiber populations, the ρTFS value corresponding to a ΔF0 of 0.5 % was similar for $N=2$ and $N=20$.

For H–I discrimination, ρENV was saturated near 1 (Fig. 13.3g, h). The latter result is consistent with the fact that the temporal envelope of a sound is unaffected by a coherent frequency shift of all the components in the sound, and it demonstrates that this was also the case after cochlear filtering—at least for small frequency

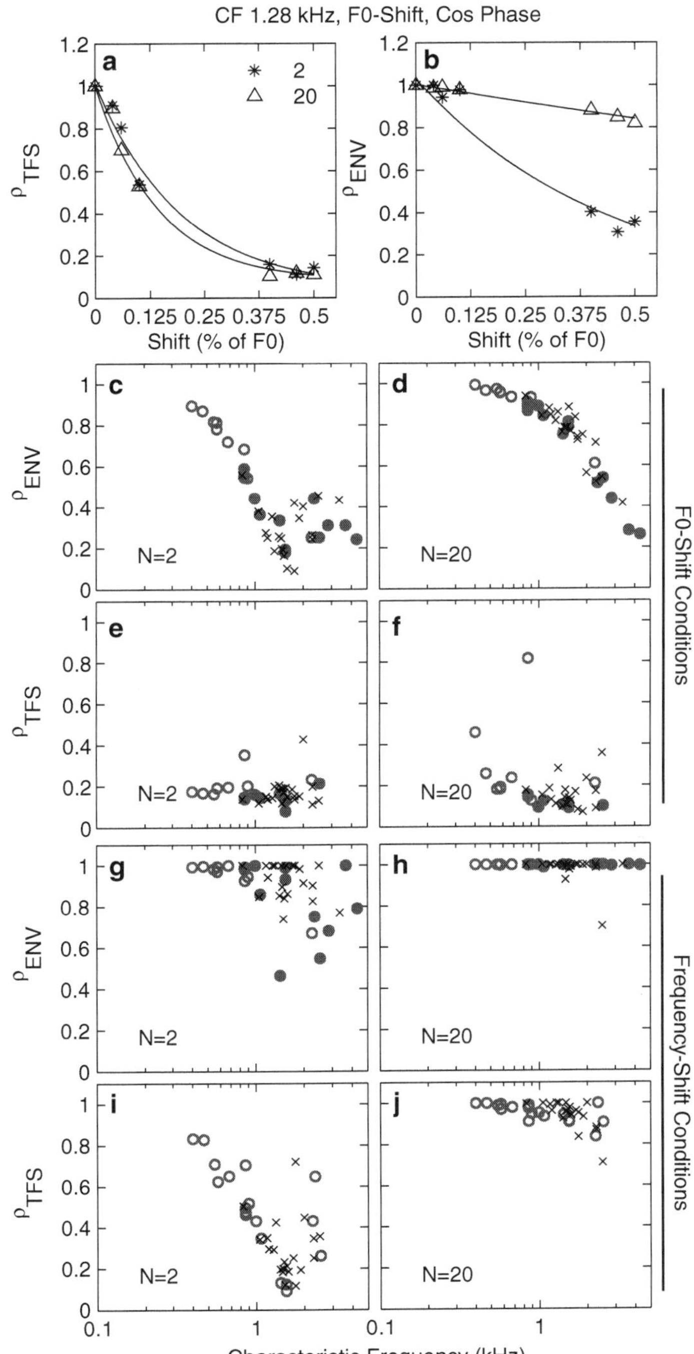

Fig. 13.3 ρ_{TFS} (**a**) and ρ_{ENV} (**b**) as a function of ΔF0 (in % of F0) are shown for a normal fiber. ρ_{TFS} and ρ_{ENV} corresponding to ΔF0 of 0.5 % are shown for the populations of normal (*x*) and impaired (*o*) fibers for F0 discrimination (**c–f**) and H–I discrimination (**g–j**). Filled circles are fibers with significantly broadened tuning

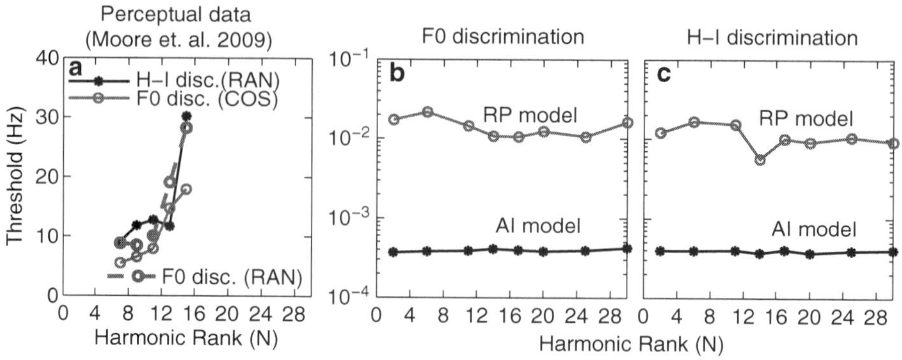

Fig. 13.4 (**a**) F0- and H–I discrimination thresholds in humans (Moore et al. 2009), converted to frequency shifts of the center component in Hz. Thresholds are averaged across three F0 conditions (35, 50, and 100 Hz) and correspond to $d'=1$. (**b**) F0-discrimination thresholds predicted using the AI (*asterisks*) and RP (*circles*) models. (**c**) Same as (**b**) but for H–I discrimination

shifts, for the cosine phase condition, and for the stimuli and fibers considered here. Overall, the ρ metrics across the NH and HI fiber populations were quite similar, suggesting that the ability of AN fibers to encode F0 or frequency shifts in the timing of their discharges (phase locking to the TFS or the ENV) is not affected by SNHL.

3.1 Comparison with Psychophysical Data

Figure 13.4a shows JNDs for F0 and JNDs for frequency shifts (H–I discrimination) measured in human listeners (Moore et al. 2009) using complex tones similar to those used in the current study, replotted here in terms of the frequency shift (in Hz) of the center component. Our decision to transform these thresholds into frequency shifts of the center component in Hz, rather than as percentages, was motivated by our finding that ρTFS and ρENV values were equal for equal changes in either the F0 or the TFS components near CF, in Hz. Using this metric, the thresholds measured in the two tasks (F0 and H–I discrimination) were comparable suggesting that performance in these two tasks may have been based on the same cue, the usability of which depends on the magnitude of the shift (in Hz) of the center component.

Figure 13.4b, c shows predicted JNDs (in Hz) corresponding to $d'=1$ for F0 discrimination and H–I discrimination, respectively, based on AI or RP information only. The shift in the TFS component near CF was the stimulus parameter to be discriminated in both tasks. Note that the predicted thresholds do not depend on N. This result is inconsistent with the psychophysical results illustrated in Fig. 13.4a but is consistent with our neural data showing that the decrease in ρTFS as a function of $\Delta F0$ (or ΔF) did not depend on N (Fig. 13.3a).

4 Discussion

The results of this study indicate that the main effect of SNHL on TFS encoding at the level of single AN fibers is a mismatch between the fiber's cochlear CF (estimated based on the tuning curve) and the frequency of the most dominant TFS component in the neural response. No significant effect of SNHL on the ability of AN fibers to phase lock to the TFS of complex tones was found. This suggests that poorer F0- and frequency-shift discrimination performances (or thresholds) in listeners with SNHL than in NH listeners are due to factors other than degraded phase locking to TFS at the level of AN. It is possible that these perceptual deficits originate beyond the AN and/or are due to factors such as mismatches between TFS and place (tonotopic) information. Based on analyses of actual and simulated neural responses to harmonic and inharmonic complex tones, neither F0-discrimination performance nor frequency-shift detection performance (i.e., H–I discrimination) should be expected to degrade as the rank (*N*) of the lowest component in the stimulus increases. This stands in sharp contrast with psychophysical data, which show marked increases in F0- and H–I discrimination thresholds as *N* increases (Moore et al. 2009). Therefore, it appears that performance in these perceptual tasks either does not depend on TFS cues or does depend on TFS cues but via a complex and suboptimal decoding mechanism that has yet to be identified and that may involve a combination of place and temporal information.

Acknowledgments This work was supported by NIH R01-DC009838 (SK and MGH) and R01-DC05216 (CM). Some of the results presented in this chapter are described in greater detail elsewhere (Kale et al. submitted).

References

Harding GW, Bohne BA (2007) Distribution of focal lesions in the chinchilla organ of Corti following exposure to a 4-kHz or a 0.5-kHz octave band of noise. Hear Res 225:50–59

Heinz MG, Swaminathan J (2009) Quantifying envelope and fine-structure coding in auditory-nerve responses to chimaeric speech. J Assoc Res Otolaryngol 10:407–423

Heinz MG, Colburn HS, Carney LH (2001) Evaluating auditory performance limits: I. One-parameter discrimination using a computational model for the auditory nerve. Neural Comput 13:2273–2316

Hopkins K, Moore BC (2007) Moderate cochlear hearing loss leads to a reduced ability to use temporal fine structure information. J Acoust Soc Am 122:1055–1068

Houtsma AJM, Smurzynski J (1990) Pitch identification and discrimination for complex tones with many harmonics. J Acoust Soc Am 87:304–310

Kale S, Heinz MG (2010) Envelope coding in auditory nerve fibers following noise-induced hearing loss. J Assoc Res Otolaryngol 11:657–673

Kale S, Micheyl C, Heinz MG. Implications of within-fiber temporal coding for perceptual studies of F0-discrimination and discrimination of harmonic and inharmonic tone complexes. (submitted)

Liberman MC (1984) Single-neuron labeling and chronic cochlear pathology. I. Threshold shift and characteristic-frequency shift. Hear Res 16:33–41

Louage DH, Van Der Heijden M, Joris PX (2004) Temporal properties of responses to broadband noise in the auditory nerve. J Neurophysiol 91:2051–2065
Moore BC, Glasberg BR, Hopkins K (2006) Frequency discrimination of complex tones by hearing-impaired subjects: evidence for loss of ability to use temporal fine structure. Hear Res 222:16–27
Moore BC, Hopkins K, Cuthbertson S (2009) Discrimination of complex tones with unresolved components using temporal fine structure information. J Acoust Soc Am 125:3214–3222
Siebert WM (1970) Frequency discrimination in auditory system – place or periodicity mechanisms? Proc IEEE 58:723–750
Zilany MSA, Bruce IC (2006) Modeling auditory-nerve responses for high sound pressure levels in the normal and impaired auditory periphery. J Acoust Soc Am 120:1446–1466

Chapter 14
A Glimpsing Account of the Role of Temporal Fine Structure Information in Speech Recognition

Frédéric Apoux and Eric W. Healy

Abstract Many behavioral studies have reported a significant decrease in intelligibility when the temporal fine structure (TFS) of a sound mixture is replaced with noise or tones (i.e., vocoder processing). This finding has led to the conclusion that TFS information is critical for speech recognition in noise. How the normal auditory system takes advantage of the original TFS, however, remains unclear. Three experiments on the role of TFS in noise are described. All three experiments measured speech recognition in various backgrounds while manipulating the envelope, TFS, or both. One experiment tested the hypothesis that vocoder processing may artificially increase the apparent importance of TFS cues. Another experiment evaluated the relative contribution of the target and masker TFS by disturbing only the TFS of the target or that of the masker. Finally, a last experiment evaluated the relative contribution of envelope and TFS information. In contrast to previous studies, however, the original envelope and TFS were both preserved – to some extent – in all conditions. Overall, the experiments indicate a limited influence of TFS and suggest that little speech information is extracted from the TFS. Concomitantly, these experiments confirm that most speech information is carried by the temporal envelope in real-world conditions. When interpreted within the framework of the glimpsing model, the results of these experiments suggest that TFS is primarily used as a grouping cue to select the time-frequency regions corresponding to the target speech signal.

F. Apoux (✉) • E.W. Healy
Department of Speech and Hearing Science,
The Ohio State University,
1070 Carmack Rd., Columbus, OH 43210, USA
e-mail: fred.apoux@gmail.com

B.C.J. Moore et al. (eds.), *Basic Aspects of Hearing*,
Advances in Experimental Medicine and Biology,
DOI 10.1007/978-1-4614-1590-9_14, © Springer Science+Business Media New York 2013

1 Introduction

Many behavioral studies have reported a significant decrease in intelligibility when the temporal fine structure (TFS) of a sound mixture is replaced with noise or tones using vocoder processing (see Füllgrabe et al. 2006, for a review). Accordingly, it has been suggested that the information conveyed by the TFS of a sound mixture plays an important role in speech recognition in noise. While the deleterious effect of vocoder processing on speech recognition in noise is not debatable, it is unclear whether the observed drop in performance can be exclusively attributed to the loss of TFS information. Indeed, the effect of vocoder processing is twofold. First, the information potentially conveyed by the TFS is entirely removed from the sound mixture. A second effect, however, is that the vocoded sound mixture is composed of "sounds" that share the same TFS or carrier. More specifically, the sound mixture consists of a sum of envelopes imposed on a single carrier that is completely independent from the original individual sounds. In this situation, is it still appropriate to consider that the sound mixture is composed of several distinct sounds? Alternatively, is it appropriate to conclude from these vocoder studies that TFS is important for speech recognition in noise? On one hand, it may be argued that listeners are asked to process one sound as if it was composed of two or more individual sounds. On the other hand, it may be argued that the single vs. double carrier comparison is a valid assessment of the role of TFS.

2 The Single Carrier Limitation

A first experiment was designed to provide some clarification on the consequences of having a single carrier. The stimuli were processed according to the scheme described in Smith et al. (2002). The envelope was obtained from one sound mixture, while the TFS was obtained from a "different" sound mixture. The two sound mixtures, however, were composed of the same target and masker signals. The only difference between the two mixtures was the signal-to-noise ratio (SNR) at which the individual sounds were added. To obtain the single carrier conditions, the SNR of the sound mixture from which the TFS was extracted was set to −1,000 or 1,000 dB.

2.1 Methods

Five normal-hearing listeners were presented with a series of two concurrent sentences. One sentence, the target sentence, was randomly selected from the speech perception in noise (SPIN) test. The other sentence, the masker sentence, was randomly selected from the AzBio test (Spahr et al. 2012). Target and masker sentences were added at different SNRs. One set of sentences was added at −12 or 0 dB, and the other set was added at −1,000, 0, or 1,000 dB. After addition, the sound mixtures were filtered into 30 contiguous frequency bands ranging from 80 to 7,563 Hz. The filtering roughly simulated the frequency selectivity of the normal auditory system. The envelope and TFS were extracted from each band using Hilbert decomposition as described in Apoux et al. (2011). The envelope extracted from a given band was then imposed on the TFS of the corresponding band obtained from the same stimulus mixed at a different SNR. The resulting amplitude-modulated TFS was then band-pass filtered using the initial analysis filter. The 30 amplitude-modulated TFSs were finally summed to create the final stimulus. It should be noted that the two sound mixtures from which the envelope and TFS were independently obtained differed only in their initial SNR (i.e., the same target and masker sentences were used). In other words, the final stimuli consisted of the envelope of a sound mixture at x dB SNR and the TFS of the *same* sound mixture at y dB SNR. The SNR of the sound mixture from which the envelope was obtained and that of the sound mixture from which the TFS was obtained will be referred to as SNR_{env} and SNR_{tfs}, respectively. The overall level of the 30 summed bands was normalized and calibrated to produce 65 dBA. For comparison, a traditional vocoder condition (Voc) was also tested. In this condition, the original TFS in each band was replaced with a speech-shaped noise (constant spectrum level below 800 Hz and 6 dB/oct roll-off above 800 Hz). All six combinations of SNR_{env} (−12 and 0 dB) and SNR_{tfs} (−1,000, 0, and 1,000 dB) were tested. Two combinations of SNR_{env} and Voc were also tested, resulting in a total of eight conditions.

2.2 Results

The results for the eight conditions tested in Exp. 1 are shown in Fig. 14.1. Average sentence recognition scores are plotted as a function of SNR_{env} with SNR_{tfs}/Voc as parameter. As can be seen, performance increased with increasing SNR_{env}. This increase was roughly similar for all SNR_{tfs}/Voc conditions. In other words, the pattern of results did not differ across the two SNR_{env} conditions. Sentence recognition scores were always best in the 0 dB SNR_{tfs} condition. They decreased noticeably in the −1,000 and 1,000 dB SNR_{tfs} conditions. Scores in the Voc condition were the poorest. This pattern of results is remarkable in at least two ways. First, sentence recognition at 1,000 dB SNR_{tfs} was always poorer than at 0 dB SNR_{tfs}. This is surprising because intelligibility typically increases with increasing SNR. Second, sentence recognition at 1,000 dB SNR_{tfs} was no different from that at −1,000 dB. Again, this pattern does not follow the typical relationship between intelligibility and SNR. Taken together, these data support our initial assertion that replacing the original TFS of a sound mixture may affect speech intelligibility in more than one way and that it cannot be simply reduced to a loss of TFS information.

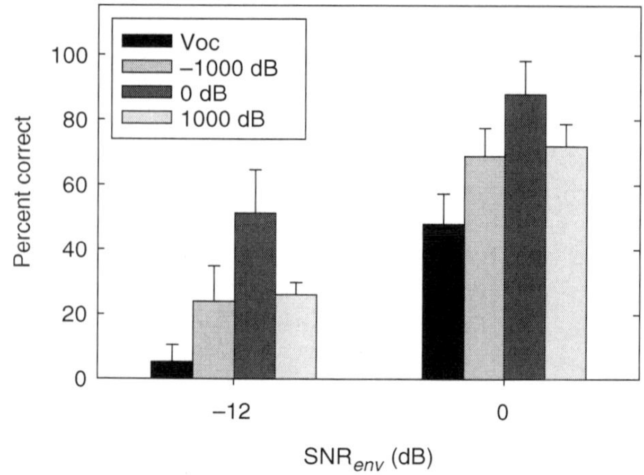

Fig. 14.1 Average sentence recognition scores as a function of SNR$_{env}$ with SNR$_{tfs}$ as parameter. The vocoder condition (Voc) is also plotted as a parameter. *Error bars* indicate one standard deviation

3 Role of the Target and Masker Fine Structure

Experiment 1 showed that stimuli with only the TFS of the target (i.e., 1,000 dB SNR$_{tfs}$) are not more intelligible than those with only the TFS of the masker (i.e., −1,000 dB SNR$_{tfs}$). The results of this experiment, however, should not be taken to imply that the target and the masker TFS make equal contribution to speech recognition in noise. The fact that they were obtained with vocoder-like stimuli (i.e., stimuli having a single carrier) strongly limits their applicability to situations in which both carriers are preserved. Moreover, it may be assumed according to the results of previous TFS studies that the TFS of the target is more important than that of the masker. This assumption is based on the finding that vocoder processing is more detrimental to speech recognition in fluctuating than in steady backgrounds, suggesting that it is when the representation of the masker is poorest (i.e., in the masker dips) that the benefit from preserved target TFS is the largest.

A study by Apoux and Healy (2011) recently assessed this assumption. The authors manipulated independently the TFS of the target and that of the masker to evaluate their individual contributions to speech recognition in noise. This evaluation included four masker types: a speech-shaped noise (SSN), a speech-modulated noise (SMN), a time-reversed sentence (TRS), and a sentence (SPE). All four maskers were added to the target /a/-consonant-/a/ stimuli at −6 or 0 dB SNR. For each combination of masker type and SNR, four processing conditions were implemented. A first condition, referred to as UNP, consisted of the unprocessed stimuli. The remaining conditions involved vocoder processing. In one condition, only the target was vocoded (VOC$_t$). In another condition, only the masker was vocoded (VOC$_m$). In the last condition, the entire sound mixture was vocoded (VOC$_{tm}$). This last condition is analogous to the traditional vocoder condition.

Consonant identification scores averaged across 20 normal-hearing listeners are shown in Fig. 14.2. Because the patterns were very similar for the two SNRs, only the data from the −6 dB condition are presented. As pointed out by the authors, two patterns emerged from these data. One pattern was only observed in the SSN

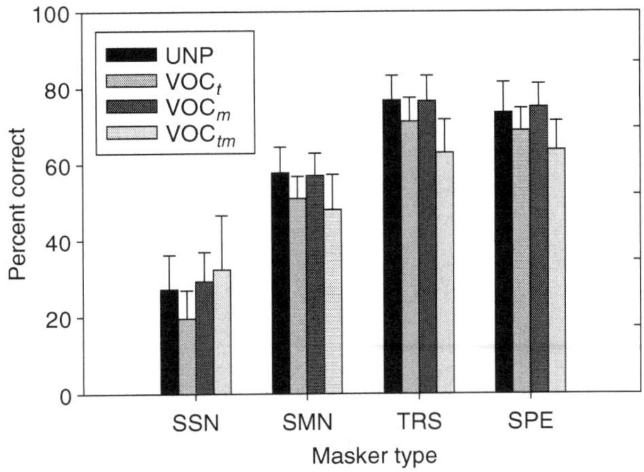

Fig. 14.2 Average percent correct scores for consonant identification as a function of masker type with VOC processing as parameter (unprocessed (UNP), target only (VOC_t), masker only (VOC_m), and entire stimulus (VOC_{tm})). *Error bars* indicate one standard deviation

condition. It involved no effect (i.e., scores equivalent to UNP) of vocoding only the masker (VOC_m) or the entire sound mixture (VOC_{tm}). The fact that scores were not reduced as a result of VOC_{tm} in steady noise is consistent with previous TFS work. More surprisingly, this pattern also involved an effect of vocoding only the target (VOC_t). The other pattern involved a drop in intelligibility when vocoding only the target (VOC_t) or the entire sound mixture (VOC_{tm}). The fact that scores were reduced as a result of VOC_{tm} in modulated noise is also consistent with previous TFS work. Again, no effect of vocoding only the masker was observed. The statistical significance of all the above effects was confirmed by a multiple pairwise comparison (corrected paired *t*-tests).

One interpretation of the above data is that the normal auditory system does not rely heavily on the nature of the masker TFS to extract speech from noise. Most of the segregation cues seem to be provided by the target signal. In other words, listeners would tend to focus on the TFS of the target signal to uncover the time-frequency regions containing a relatively undistorted view of local signal properties, the so-called glimpses (Cooke 2006; Apoux and Healy 2009). The glimpses would be subsequently used to form a representation of this target signal. Logically, this strategy is no longer effective when the TFS of the entire sound mixture is vocoded (VOC_{tm}). More surprisingly, the strategy also seems to fall apart when only the target is vocoded. While this last result may be interpreted as evidence that speech information is conveyed by the TFS, it may simply reflect a better "extractability" of the original speech TFS.

4 Relative Contribution of Envelope and Fine Structure

Apoux and Healy (2011) demonstrated that even when two carriers are somehow preserved in the sound mixture, speech recognition may be affected by the introduction of distortions in the TFS. This result is consistent with our findings in Exp. 1 showing that the nature of the single carrier plays a role as vocoder processing was more detrimental than preserving only the target or the masker carrier. More importantly, this last result suggests that the effect of vocoder processing cannot be attributed exclusively to the single carrier artifact discussed previously. Vocoder processing, however, may still not be appropriate to investigate the contribution of TFS cues to speech recognition in noise. A different approach was therefore needed. The approach used in Exp. 1 appeared to have a number of advantages to evaluate the role of TFS. First, it is possible to preserve the TFS of the target and that of the masker. Second, one may independently vary the SNR in the TFS and that in the envelope. If TFS cues are critical for speech recognition, a strong relationship should be observed between performance and SNR in the TFS.

4.1 Methods

Twenty normal-hearing listeners participated in Exp. 2. They were presented with two concurrent sentences. One sentence, the target sentence, was randomly selected from the SPIN test. The other sentence, the masker sentence, was randomly selected from the AzBio test. As in Exp. 1, the final stimuli consisted of the envelope of a sound mixture at x dB SNR and the TFS of the *same* sound mixture at y dB SNR. However, the SNRs ranged from −18 to 0 dB in 6-dB steps. All combinations of SNR_{env} and SNR_{tfs} were tested, resulting in a total of 16 conditions. Other methodological and procedural details were identical to those in Exp. 1.

4.2 Results

In Fig. 14.3, average sentence recognition scores are plotted as a function of SNR_{tfs} with SNR_{env} as parameter. The bold line connects the four data points obtained in the conditions in which SNR_{env} and SNR_{tfs} were identical and, therefore, corresponds to the baseline performance (i.e., reference function, REF). As expected, baseline performance increased with increasing SNR. Because the influence of SNR_{env} and SNR_{tfs} is combined in this REF function, it cannot provide any indication about which cues drove performance. The experimental data, however, are relatively straightforward. As can be seen, all four functions are regularly spaced. Since SNR_{env} is the parameter, this suggests that the level of distortion in the envelope played a strong role in the present experiment. More importantly, all four functions are nearly horizontal. Since SNR_{tfs} is on the abscissa, it may be concluded that this factor did not play a critical role in the present experiment. Indeed, as SNR_{tfs} increased, intelligibility remained essentially the same. This pattern was observed irrespective of SNR_{env}.

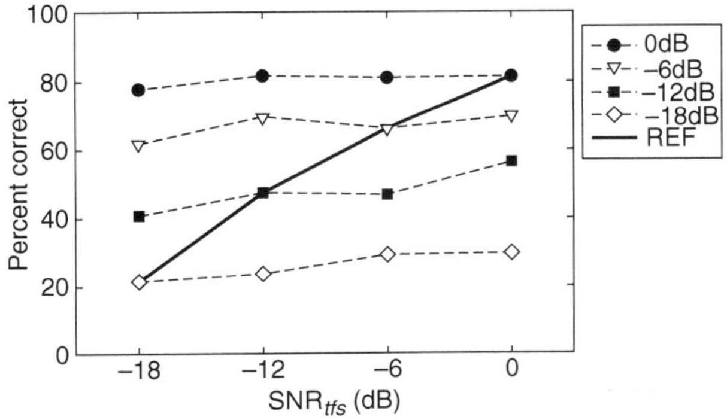

Fig. 14.3 Average sentence recognition scores as a function of SNR$_{tfs}$ with SNR$_{env}$ as parameter. The *bold line* (REF) connects the data points for which SNR$_{env}$ and SNR$_{tfs}$ were equal

5 Concluding Remarks

Experiment 1 suggested that the detrimental effect of vocoder processing observed in previous TFS studies may be attributed to at least two factors. A first factor is the substitution of the two (or more) original carriers with only one. This raises the question of the nature of the sound mixture after vocoder processing, as only one carrier is used to convey the envelopes. It was argued here that such stimuli should not be considered a sound mixture. In any case, the negative effect of having only one carrier in the sound mixture was evident in Exp. 1 as performance dropped substantially even when the target TFS was used as the sole carrier. A second factor is the relationship between the remaining carrier and the complex envelope. Experiment 1 showed that when no relationship exists (i.e., Voc), performance drops farthest. Whether it is the TFS of the target or that of the masker that is preserved, however, does not seem to be a significant factor.

The second factor mentioned above may seem in contradiction with the results of Apoux and Healy (2011). Indeed, these authors showed a differential role of the target and masker TFS in that the nature of the masker TFS did not affect performance, while that of the target did. The conditions, however, are not comparable as the sound mixtures in Apoux and Healy (2011) had two carriers. Accordingly, the apparent discrepancy between Exp. 1 and Apoux and Healy (2011) was most likely attributable to the single carrier artifact.

Thus far, all of the above experiments somehow indicated a contribution of TFS to speech recognition in noise. In Exp. 2, this contribution was limited, at best. One way to reconcile all of the results reported here is to assume that the TFS does not convey any speech information and is only used to detect the glimpses. In addition, it may be argued that the lack of effect of SNR$_{tfs}$ revealed that listeners were actually able to use TFS cues even at the least favorable SNRs. Such a possibility existed because the masker was another sentence and, therefore, contained a number of silent gaps. Decreasing the SNR to −18 dB reduced the target level in these gaps, but it was presumably not sufficient to prevent the listeners from detecting the glimpses. In conclusion, it is suggested that a dichotomy exists between the envelope and TFS in that the envelope provides the actual speech information, while the role of TFS is more or less limited to carrying this information.

References

Apoux F, Healy EW (2009) On the number of auditory filter outputs needed to understand speech: further evidence for auditory channel independence. Hear Res 255:99–108

Apoux F, Healy EW (2011) Relative contribution of target and masker temporal fine structure to the unmasking of consonants in noise. J Acoust Soc Am 130:4044–4052

Apoux F, Millman RE, Viemeister NF, Brown CA, Bacon SP (2011) On the mechanisms involved in the recovery of envelope information from temporal fine structure. J Acoust Soc Am 130: 273–282

Cooke MP (2006) A glimpsing model of speech perception in noise. J Acoust Soc Am 119: 1562–1573

Füllgrabe C, Berthommier F, Lorenzi C (2006) Masking release for consonant features in temporally fluctuating background noise. Hear Res 211:74–84

Smith ZM, Delgutte B, Oxenham AJ (2002) Chimaeric sounds reveal dichotomies in auditory perception. Nature 416:87–90

Spahr AJ, Dorman MF, Litvak LM, Van Wie S, Gifford RH, Loizou PC, Loiselle LM, Oakes T, Cook S (2012) Development and validation of the AzBio sentence lists. Ear Hear 33:112–117

Chapter 15
Assessing the Possible Role of Frequency-Shift Detectors in the Ability to Hear Out Partials in Complex Tones

Brian C.J. Moore, Olivia Kenyon, Brian R. Glasberg, and Laurent Demany

Abstract The possible role of frequency-shift detectors (FSDs) was assessed for a task measuring the ability to hear out individual "inner" partials in a chord with seven partials uniformly spaced on the ERB_N-number (Cam) scale. In each of the two intervals in a trial, a pure-tone probe was followed by a chord. In one randomly selected interval, the frequency of the probe was the same as that of a partial in the chord. In the other interval, the probe was mistuned upwards or downwards from the "target" partial. The task was to indicate the interval in which the probe coincided with the target. In the "symmetric" condition, the frequency of the mistuned probe was midway in Cams between that of two partials in the chord. This should have led to approximately symmetric activation of the up-FSDs and down-FSDs, such that differential activation provided a minimal cue. In the "asymmetric" condition, the mistuned probe was much closer in frequency to one partial in the chord than to the next closest partial. This should have led to differential activation of the up-FSDs and down-FSDs, providing a strong discrimination cue. Performance was predicted to be better in the asymmetric than in the symmetric condition. The results were consistent with this prediction except when the probe was mistuned above the sixth (second highest) partial in the chord. To explain this, it is argued that activation of FSDs depends both on the size of the frequency shift between successive components and on the pitch strength of each component.

B.C.J. Moore (✉) • O. Kenyon • B.R. Glasberg
Department of Experimental Psychology, University of Cambridge,
Downing Street, Cambridge, CB2 3EB, UK
e-mail: bcjm@cam.ac.uk

L. Demany, PhD
Institut de Neurosciences Cognitives et Intégratives d'Aquitaine (UMR CNRS 5287),
Université de Bordeaux, 146 rue Leo Saignat, Bordeaux F-33076, France

B.C.J. Moore et al. (eds.), *Basic Aspects of Hearing*,
Advances in Experimental Medicine and Biology,
DOI 10.1007/978-1-4614-1590-9_15,

1 Introduction

The ability to hear out partials in complex tones provides a basic measure of the frequency selectivity of the auditory system (Plomp 1964) and is important for theories of pitch perception for complex tones (Moore and Gockel 2011). A widely used task for measuring this ability is based on that described by Roberts and Bregman (1991) and by Moore and Ohgushi (1993). On each trial, a sinusoidal tone (the "probe") is followed by a multi-partial complex (the "chord"). The probe is slightly higher or lower in frequency than one of the partials in the chord, called the "target". The target is selected randomly on each trial. Subjects are asked to indicate whether the probe is higher or lower in pitch than the target. Good performance on this task requires practice and appropriate training (Moore et al. 2006), the task is affected by perceptual confusion effects (Moore et al. 2009, 2012), and the task is very difficult for hearing-impaired subjects to perform (Bernstein 2006; Moore and Glasberg 2011).

Moore and Glasberg (2011) measured the ability to hear out partials using a task that removed the need to judge the direction of a pitch change, which some subjects find difficult (Semal and Demany 2006). In each of the two observation intervals, a probe was followed by a chord. In one randomly selected interval, the frequency of the probe was the same as that of one of the partials (the target) in the chord. In the other interval, the probe and the chord were the same, except that the target partial was omitted. The task was to indicate the interval in which the target partial was present. A problem with this task is that it might be performed by learning the difference in timbre between the "complete" chord (with all partials) and the "incomplete" chord (with one partial missing), rather than by hearing out the target partial.

The task used in the present study was intended to avoid the problems discussed above: it did not require discrimination of the direction of a frequency change, it was expected to minimize perceptual confusion effects, and it could not be performed by comparing the chord across the two intervals of a trial. The task is illustrated by the schematic spectrograms in Fig. 15.1. In each of the two observation intervals, a probe was followed by a chord. In one randomly selected interval, the frequency of the probe was the same as that of one of the partials (the target) in the chord. In the other interval, the probe was mistuned upwards or downwards from the target. The task of the subject was to indicate the interval in which the probe coincided with the target in frequency.

A major motivation for this study was to provide experimental evidence relating to the possible existence of frequency-shift detectors (FSDs), which are neural systems sensitive to frequency changes in a specific direction. Evidence for the existence of FSDs was provided by Demany and Ramos (2005). They presented an inharmonic chord with components spaced by at least 0.5 octave followed by a single pure tone that was either identical to a partial in the chord or was halfway in frequency between two partials. These two types of sequence could not be reliably discriminated. However, if the single tone was slightly (e.g. one semitone) lower or higher in frequency than one of the partials in the chord, a pitch shift was perceived

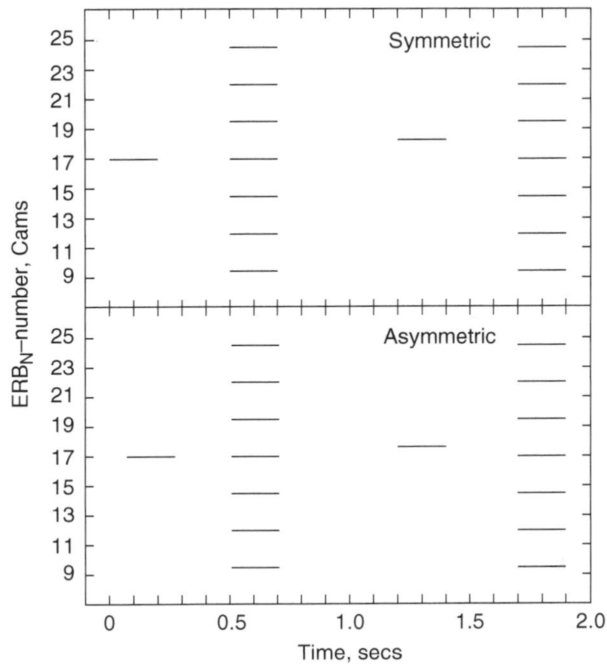

Fig. 15.1 Schematic spectrograms of the stimuli used for the symmetric condition (*top*) and asymmetric condition (*bottom*). See text for details

and subjects could distinguish the two types of sequence. Demany and Ramos explained this effect in terms of FSDs. They argued that, like the motion-detection system in vision, subjects are sensitive to the balance of responses between FSDs sensitive to upward and downward frequency shifts. They also argued that the sensitivity of the FSDs varied with the magnitude of the frequency shift. When the probe was midway in frequency between two partials in the chord or coincided with a partial in the chord, the up-FSDs and down-FSDs would have been equally activated, so the FSDs did not provide a useful discrimination cue. When the probe was slightly mistuned from one of the partials in the chord, the up-FSDs and down-FSDs would have been differentially activated, providing a discrimination cue. Subsequent work has suggested that the FSDs are optimally activated by a shift of about 7 % (Demany et al. 2009, 2010, 2011).

The possible influence of FSDs in the present task was assessed by comparing two conditions. The chord contained partials that were equally spaced on the ERB$_N$-number scale (Glasberg and Moore 1990), which has units called Cams (Moore 2012). This means that all partials were roughly equally well resolved in the auditory periphery (Moore et al. 2006). For medium frequencies, the "optimal" frequency shift for activating the FSDs corresponds to a shift on the ERB$_N$-number scale of about 0.5–0.6 Cams. In the "symmetric" condition (Fig. 15.1, top panel), the frequency of the mistuned probe was midway in Cams between two partials in the chord. This should have led to reasonably symmetric activation of the up-FSDs and down-FSDs, such that differential activation provided a minimal cue. In the "asymmetric" condition (Fig. 15.1, bottom panel), the mistuned probe was much closer in frequency to one partial in the chord than to the next closest partial. This should have led to differential activation of the up-FSDs and down-FSDs, providing a strong discrimination cue (in comparison to the interval in which the probe coincided in frequency with a partial in the chord). If the FSDs do not influence performance in this task, then performance should be better for the symmetric than for the asymmetric condition, since in the interval in which the probe is mistuned, the mistuning from the target is greater for the former than for the latter. In contrast, if the FSDs do play a role, performance might better in the asymmetric than in the symmetric condition.

2 Method

2.1 *Stimuli and Procedure*

The partials in the chord were separated by 2.5 Cams, so the chord was inharmonic. The relationship between Cam value and frequency, f (Hz), was assumed to be as suggested by Glasberg and Moore (1990):

$$\text{Cam} = 21.4\log_{10}(0.00437f + 1) \quad (15.1)$$

Uniform spacing on the ERB_N-number scale was chosen because the salience of frequency changes is roughly constant across centre frequencies when the extent of the change is expressed in Cams (Hermes and van Gestel 1991). The spacing of 2.5 Cams was chosen since it led to a substantial difference between the symmetric and asymmetric conditions.

In each observation interval, a 200-ms sinusoidal probe was followed after a 300-ms silent interval by a 200-ms chord. All stimuli had 20-ms raised-cosine ramps and durations are specified between half-amplitude points. The intervals were separated by 500 ms of silence. In one randomly chosen interval, the probe coincided in frequency with a partial (the target) in the chord. In the other interval, the probe differed in frequency from the target. The task was to indicate the interval in which the probe coincided with the target. The target was selected randomly from one trial to the next. Only the inner partials (2–6) were used as targets. Feedback as to the correct answer was provided after each trial. The duration of the stimuli was chosen to avoid ceiling effects that might have occurred if a longer duration had been used (Moore and Ohgushi 1993; Moore et al. 2006).

The frequencies of the partials in the chord corresponded to 9.5, 12, 14.5, 17, 19.5, 22, and 24.5 Cams (corresponding to 408.5, 606, 864, 1,202, 1,645, 2,224, and 2,983 Hz). For the symmetric condition, the probe was separated by 1.25 Cams from the target and hence fell midway between two partials, which was expected to lead to roughly equal activation of the up-FSDs and down-FSDs. For the asymmetric condition, the probe was separated by 0.625 Cams from the target and by 1.875 Cams from the next nearest partial. The shift of 0.625 Cams was expected to lead to near-optimal activation of the FSDs, while the shift of 1.875 Cams was expected to lead to reduced activation of the FSDs, giving differential activation of the up-FSDs and down-FSDs. This differential activation was predicted to lead to better performance for the asymmetric condition.

For each observation interval, the frequencies of all components (both probe and chord) were multiplied by a common factor randomly chosen from a uniform distribution between 0.9 and 1.1. This was done to prevent subjects from using the frequency of the probe as a cue. In a given block of trials, each of the inner partials in the chord was selected as the target ten times: five with the probe mistuned downwards and five with it mistuned upwards. Each block was repeated at least five times, so each partial in the chord was the target at least 50 times. Stimuli were generated digitally and presented via one earpiece of a Sennheiser HD580 headphone. The level of the probe and of each partial in the chord was 60 dB SPL.

2.2 *Subjects*

Six subjects with normal hearing were tested. Their ages ranged from 21 to 67 years. Most subjects had previously taken part in a similar experiment using 1,000-ms stimuli rather than 200-ms stimuli. Hence, they had several hours of experience in a similar task.

3 Results

For each target and each subject, scores were averaged for the case when the mistuned probe was lower and higher in frequency than the target. These average scores are plotted for each subject in Fig. 15.2. For statistical analysis, the scores were transformed to RAU (Studebaker 1985). The scores in RAU were averaged across subjects and transformed back to percent correct. The dashed lines show the means obtained in this way.

A within-subjects analysis of variance was conducted, with factors condition (symmetric and asymmetric) and partial number for the target (2–6). There was no significant main effect of condition, but there was a significant effect of partial number: $F(4, 20)=2.99$, $p=0.044$. There was a highly significant interaction between condition and partial number: $F(4, 20)=7.06$, $p<0.001$. This reflects the fact that scores were higher for the asymmetric than for the symmetric condition for partials 2, 3, 4, and 5, but the reverse was true for partial 6. The mean scores in RAU are given in Table 15.1. Overall, the results are consistent with the prediction based on FSDs for partials 2–5, but not for partial 6.

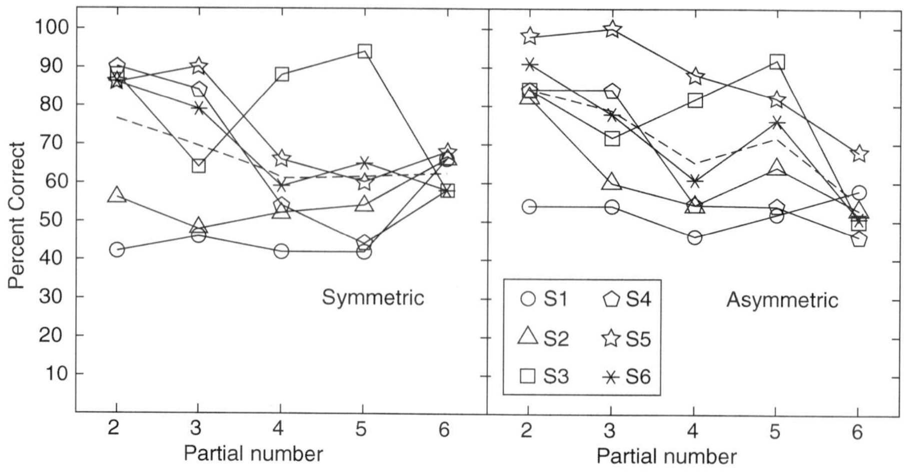

Fig. 15.2 The score for each partial for the symmetric (*left*) and asymmetric (*right*) conditions. Each *symbol* represents one subject, and the *dashed lines* show the means

Table 15.1 Mean score in RAU for each condition and each partial number

	Target partial				
Condition	2	3	4	5	6
Symmetric	76.0	69.0	60.3	60.7	61.6
Asymmetric	85.0	78.3	64.4	70.5	54.1

The poor scores for the asymmetric condition for partial 6 were mainly due to the case when the probe was mistuned upwards from the target partial, i.e. when the nominal frequency of the probe was 2,395 Hz. That case is considered in more detail below.

4 Discussion

The pattern of results found when the sixth partial was the target may be related to the fact that the pitch of the upper "edge" partial in a chord (the highest partial) is more salient than that of the inner partials (Moore and Ohgushi 1993; Moore et al. 2006). The strength of activation, A, of FSDs may depend both on the amount of frequency shift, Δf, between two successive sinusoidal components, C1 and C2, and on the strength of the pitches evoked by those components. This might be modelled by

$$A = \Phi(\Delta f) \times S(\text{C1}) \times S(\text{C2}), \tag{15.2}$$

where $\Phi(\Delta f)$ is a function representing activation as a function of frequency shift, S(C1) represents the pitch strength of the first component (the probe), and S(C2) represents the pitch strength of the second component.

Table 15.2 Calculated amount of activation, *A*, of FSDs for various conditions

Condition	Probe freq. (Hz)	Partial number	Freq. of partial (Hz)	Shift Cams	$\Phi(\Delta f)$	*S*(C2)	*A*
Symmetric	2,587 (mis up)	6	2,222	−1.25	0.531	0.2	0.106
Symmetric	2,587 (mis up)	7	2,980	+1.25	0.531	0.6	0.319
Asymmetric	2,395 (mis up)	6	2,222	−0.625	0.990	0.2	0.198
Asymmetric	2,395 (mis up)	7	2,980	+1.875	0.190	0.6	0.114
Both	2,222 (tuned)	7	2,980	+2.5	0.058	0.6	0.035
Both	2,222 (tuned)	5	1,643	−2.5	0.058	0.2	0.012

To illustrate how this equation might work for our stimuli, we assume that *S*(C1) is 1 (maximum pitch strength) and that *S*(C2) is 0.2 for an inner partial (e.g. the sixth) and 0.6 for the top partial (the seventh). We assume further that $\Phi(\Delta f)$ has the form of the function relating d' to frequency shift suggested by Demany et al. (2009), except that the frequency shift is expressed in Cams rather than cents:

$$\Phi(\Delta f, \text{Cams}) = 9.41(\Delta f)^{1.4} \times \exp(-2.55\Delta f) \tag{15.3}$$

The rightmost column of Table 15.2 shows the strength of activation of the FSDs calculated from Eqs. 15.2 and 15.3. Consider first the symmetric condition. In the observation interval where the probe was mistuned upwards ("mis up") from the sixth partial, the most important frequency shifts were a downward shift to the sixth partial and an upward shift to the seventh partial. The difference between *A* for the up-FSDs and down-FSDs was 0.319 − 0.106 = 0.213. In the observation interval where the probe coincided with the sixth partial (bottom two rows), there was an upward shift to the seventh partial and a downward shift to the fifth partial. In this case, the FSDs were only weakly activated (because of the small values of $\Phi(\Delta f)$), and the difference between *A* for the up-FSDs and down-FSDs was 0.023. Thus the balance of activation of the up- and down-FSDs differed markedly across the two intervals, providing a potential cue.

Consider now the asymmetric condition. In the observation interval where the probe was mistuned upwards from the sixth partial, the most important frequency shifts were again a downward shift to the sixth partial and an upward shift to the seventh partial. The difference between *A* for the up-FSDs and down-FSDs was 0.198 − 0.114 = 0.084. In the observation interval where the probe coincided with the sixth partial, the difference between *A* for the up-FSDs and down-FSDs was again 0.023. Thus the balance of activation of the up- and down-FSDs differed only slightly across the two intervals, providing a minimal detection cue. This could explain why performance was worse for the asymmetric than for the symmetric condition for the sixth partial.

The values for *S*(C2) used here were arbitrarily chosen, and the exact shape of the function $\Phi(\Delta f)$ is not known. Furthermore, both the values of *S*(C2) and the function $\Phi(\Delta f)$ may vary across subjects. The main point of this example is to illustrate how, in principle, greater salience of the upper edge partial might lead to better performance for the symmetric than for the asymmetric condition when the target was the sixth partial.

For the symmetric condition, asymmetric activation of the FSDs would have been minimal for the third, fourth, and fifth partials, but performance was still well above chance. Thus, for the 2.5-Cam spacing used, subjects probably could sometimes hear out the pitches of individual partials in the chord. In the asymmetric condition, the additional cues presumably provided by the FSDs could potentially be used without hearing out the target partial consciously. Nevertheless, for the FSDs to be activated, the partials would have to be resolved to some extent in the auditory periphery. Hence, the results for the asymmetric condition may give a better indication of the extent to which partials are resolved in the auditory periphery.

In summary, the results support the idea that subjects are sensitive to the difference in activation of up-FSDs and down-FSDs. They also support the idea that the amount of activation of FSDs depends on the pitch strength of successive frequency components.

Acknowledgements The work of Moore and Glasberg was supported by the MRC (UK) (Grant number G0701870). Thanks to Hedwig Gockel and Bob Carlyon for helpful comments.

References

Bernstein JG (2006) Pitch perception and harmonic resolvability in normal-hearing and hearing-impaired listeners. PhD thesis, MIT, Cambridge

Demany L, Ramos C (2005) On the binding of successive sounds: perceiving shifts in nonperceived pitches. J Acoust Soc Am 117:833–841

Demany L, Pressnitzer D, Semal C (2009) Tuning properties of the auditory frequency-shift detectors. J Acoust Soc Am 126:1342–1348

Demany L, Semal C, Cazalets JR, Pressnitzer D (2010) Fundamental differences in change detection between vision and audition. Exp Brain Res 203:261–270

Demany L, Semal C, Pressnitzer D (2011) Implicit versus explicit frequency comparisons: two mechanisms of auditory change detection. J Exp Psychol Hum Percept Perform 37:597–605

Glasberg BR, Moore BCJ (1990) Derivation of auditory filter shapes from notched-noise data. Hear Res 47:103–138

Hermes DJ, van Gestel JC (1991) The frequency scale of speech intonation. J Acoust Soc Am 90:97–102

Moore BCJ (2012) An introduction to the psychology of hearing, 6th edn. Brill, Leiden, The Netherlands

Moore BCJ, Glasberg BR (2011) The effect of hearing loss on the resolution of partials and fundamental frequency discrimination. J Acoust Soc Am 130:2891–2901

Moore BCJ, Gockel H (2011) Resolvability of components in complex tones and implications for theories of pitch perception. Hear Res 276:88–97

Moore BCJ, Ohgushi K (1993) Audibility of partials in inharmonic complex tones. J Acoust Soc Am 93:452–461

Moore BCJ, Glasberg BR, Low K-E, Cope T, Cope W (2006) Effects of level and frequency on the audibility of partials in inharmonic complex tones. J Acoust Soc Am 120:934–944

Moore BCJ, Glasberg BR, Jepsen ML (2009) Effects of pulsing of the target tone on the audibility of partials in inharmonic complex tones. J Acoust Soc Am 125:3194–3204

Moore BCJ, Glasberg BR, Oxenham AJ (2012) Effects of pulsing of a target tone on the ability to hear it out in different types of complex sounds. J Acoust Soc Am 131:2927–2937

Plomp R (1964) The ear as a frequency analyzer. J Acoust Soc Am 36:1628–1636

Roberts B, Bregman AS (1991) Effects of the pattern of spectral spacing on the perceptual fusion of harmonics. J Acoust Soc Am 90:3050–3060

Semal C, Demany L (2006) Individual differences in the sensitivity to pitch direction. J Acoust Soc Am 120:3907–3915

Studebaker GA (1985) A "rationalized" arcsine transform. J Speech Hear Res 28:455–462

Chapter 16
Pitch Perception: Dissociating Frequency from Fundamental-Frequency Discrimination

Andrew J. Oxenham and Christophe Micheyl

Abstract High-frequency pure tones (>6 kHz), which alone do not produce salient melodic pitch information, provide melodic pitch information when they form part of a harmonic complex tone with a lower fundamental frequency (F0). We explored this phenomenon in normal-hearing listeners by measuring F0 difference limens (F0DLs) for harmonic complex tones and pure-tone frequency difference limens (FDLs) for each of the tones within the harmonic complexes. Two spectral regions were tested. The low- and high-frequency band-pass regions comprised harmonics 6–11 of a 280- or 1,400-Hz F0, respectively; thus, for the high-frequency region, audible frequencies present were all above 7 kHz. Frequency discrimination of inharmonic log-spaced tone complexes was also tested in control conditions. All tones were presented in a background of noise to limit the detection of distortion products. As found in previous studies, F0DLs in the low region were typically no better than the FDL for each of the constituent pure tones. In contrast, F0DLs for the high-region complex were considerably better than the FDLs found for most of the constituent (high-frequency) pure tones. The data were compared with models of optimal spectral integration of information, to assess the relative influence of peripheral and more central noise in limiting performance. The results demonstrate a dissociation in the way pitch information is integrated at low and high frequencies and provide new challenges and constraints in the search for the underlying neural mechanisms of pitch.

A.J. Oxenham (✉) • C. Micheyl
Department of Psychology, University of Minnesota – Twin Cities,
75 East River Pkwy, Minneapolis, MN 55455, USA
e-mail: oxenham@umn.edu

B.C.J. Moore et al. (eds.), *Basic Aspects of Hearing*,
Advances in Experimental Medicine and Biology,
DOI 10.1007/978-1-4614-1590-9_16,

1 Introduction

Pitch – the perceptual correlate of acoustic waveform periodicity – plays an important role in music, speech, and animal vocalizations. Periodicity is represented in the time waveform and can be extracted via the autocorrelation of either the waveform or a representation of the stimulus after peripheral processing (e.g., cochlear filtering and hair-cell transduction), with a peak in the autocorrelation function occurring at (and at multiples of) the waveform period (e.g., Licklider 1954; Meddis and O'Mard 1997). Alternatively, periodicity can be represented via the power spectrum before or after auditory transformations, with spectral peaks occurring at harmonics of the fundamental frequency (F0) (Cohen et al. 1995; Schroeder 1968; Wightman 1973).

Whether pitch perception depends on peripheral timing (e.g., autocorrelation) or rate-place (e.g., spectral) information (or both) remains a central question in auditory research. The fact that the autocorrelation is the Fourier transform of the power spectrum limits the extent to which this question can be answered via psychoacoustics or simple stimulus manipulations. Instead, it is necessary to rely on putative limitations of the auditory system in order to infer how information is extracted. For instance, physiological data on cochlear tuning, along with psychophysical estimates of frequency selectivity, have suggested that harmonics beyond about the sixth to tenth are not spectrally resolved and so do not provide spectral information. Thus, the fact that pitch can be heard with harmonics all above the tenth has been used as evidence for a pitch code that does not depend on a spectral representation in the auditory periphery (e.g., Houtsma and Smurzynski 1990; Kaernbach and Bering 2001).

Similarly, based on physiological data from other species (Palmer and Russell 1986; Rose et al. 1967), it has often been assumed that phase locking to pure tones in the human auditory nerve is degraded above 1–2 kHz and is not available beyond about 4 kHz, thereby potentially ruling out temporal coding of pure tones above about 4 kHz (Sek and Moore 1995). Consistent with a temporal code for pitch, humans' ability to discriminate the frequency of pure tones degrades rapidly between 4 and 8 kHz, as would be expected if temporal information were necessary for accurate discrimination (Moore 1973). In addition, studies have found that melody perception becomes difficult or impossible when all the tones are above 4–5 kHz, even when the melodies are highly familiar (e.g., Attneave and Olson 1971; Licklider 1954), but see Burns and Feth (1983). Thus, the upper limit of melodic pitch, and perhaps even the upper limit of the musical instruments, may be due to the physiological phase-locking limits of the auditory nerve.

Here we review some data suggesting that pure tones well above 6 kHz, which alone produce very poor melodic pitch perception, can, when combined within a harmonic complex, produce a pitch that is salient enough to carry a melody. We explore this phenomenon further, estimating the accuracy of the representation of the individual tones, along with the complex tones, by measuring frequency (or F0) difference limens. We find an interesting dissociation between the relationship between pure-tone and complex-tone pitch discrimination at low (<4 kHz) and high (>6 kHz) frequencies. Finally, some possible explanations of these findings are offered.

2 Melody Perception with All Components Above 6 kHz

2.1 Methods

Details of the methods can be found in Oxenham et al. (2011). Briefly, listeners were presented with two four-note melodies and were asked to judge whether the two melodies were the same or different. The notes from each melody were selected randomly from the diatonic (major) scale within a single octave. On half the trials, the second melody was the same as the first; on the other half of the trials, the second or third note of the second melody was changed up or down by one scale step. The tones were presented at a level of 55 dB SPL per component (65 dB SPL per component for pure tones below 3 kHz) and were embedded in a background threshold-equalizing noise (TEN) with a level of 45 dB SPL per equivalent rectangular auditory-filter bandwidth (ERB_N) (Moore et al. 2000), with an additional TEN, presented at 55 dB SPL per ERB_N, and low-pass filtered at 6 kHz. Each note was 300 ms, including 10-ms onset and offset ramps, and was separated from the next by 200 ms.

Three conditions were tested in the main experiment. In the first (pure-tone low), the first melody was constructed from pure tones on a diatonic scale between 500 and 1,000 Hz, and the second melody was constructed from pure tones between 2,000 and 4,000 Hz, i.e., two octaves higher than the first melody. In the second condition (pure-tone high), the first melody was constructed from pure tones between 1,500 and 3,000 Hz, and the second melody was constructed from pure tones between 6,000 and 12,000 Hz. In the third condition (complex-tone high), the first melody was constructed from pure tones between 1,000 and 2,000 Hz, and the second melody was constructed from harmonic complex tones with F0s between 1,000 and 2,000 Hz, band-pass filtered with corner frequencies of 7,500 and 16 kHz and 30-dB/octave slopes, so that all components below 6 kHz fell below their masked threshold in the background noise.

Six young listeners with audiologically normal hearing passed the screening to ensure that their 16-kHz pure-tone thresholds in quiet were no higher than 50 dB SPL. Listeners were presented with the melodies in blocks of 60 trials, each with 20 trials per condition (10 same and 10 different). After practice consisting of 2 blocks, a total of 10 blocks were run per condition, yielding 200 trials per subject and condition. No feedback was provided.

2.2 Results

The results are shown in Fig. 16.1. As expected, performance in the pure-tone low condition was good, with an average score of about 85 % correct (based on the unbiased maximum percent correct for the obtained value of d'), and performance in the pure-tone high condition was poor, with an average score of about 59 % correct. Performance in the complex-tone high condition was good (94 %)

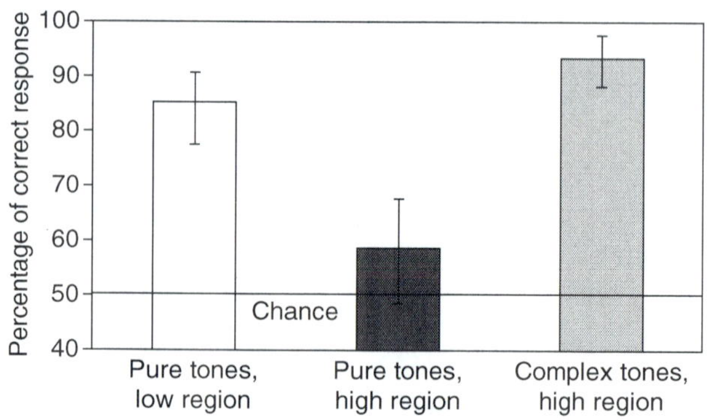

Fig. 16.1 Percentage of correct responses in the melody-discrimination task. *Error bars* show 95 % confidence intervals

and was not significantly different from that for the pure-tone low condition, despite the fact that none of the audible components in the complex tone was below 6 kHz.

2.3 *Discussion*

The results suggest that tones above 6 kHz can elicit a salient pitch, sufficient for melody recognition, when they combine to form a harmonic complex tone with a lower F0 (in this case between 1 and 2 kHz). A less interesting interpretation would be that the components are at least partly unresolved and that listeners are sensitivity to the waveform (or temporal envelope) repetition rate of between 1 and 2 kHz, rather than to the individual frequencies. This explanation was unlikely because the tones were presented in random phase, which weakens envelope cues, and because the repetition rates were so high (1–2 kHz) that sensitivity to envelope pitch was expected to be very poor (Carlyon and Deeks 2002). Indeed, two control conditions (run with new groups of six subjects), involving shifted harmonics and dichotic presentation, produced results consistent with predictions based on the processing of individual components, rather than the temporal envelope: when the harmonics were shifted to produce inharmonic complexes but with an unchanged temporal envelope rate, performance dropped to near-chance levels of about 55 %. On the other hand, when the harmonics were presented to opposite ears, performance remained high and not significantly different from that for the original (dichotic) condition. Overall, the results are not consistent with the previously defined "existence region" of pitch (Ritsma 1962). The results also highlight an interesting dissociation, whereby high-frequency tones, which alone do not induce a salient pitch, combine within a complex tone to elicit a salient pitch.

3 Comparing Frequency and F0 Difference Limens

There are different possible explanations for why a dissociation in pitch salience is observed between high-frequency pure tones and complex tones. One possible explanation is that the upper limit for perceiving melodic pitch is determined at a level higher than the auditory nerve, perhaps due to lack of exposure to high (>4 kHz) F0s in normal acoustic environments. Thus, when individual high-frequency tones are presented, they elicit a pitch beyond the "existence region" of melodic pitch. However, when the high-frequency tones are presented in combination with other harmonically related tones, they elicit a pitch corresponding to the F0, which falls within the "existence region."

Another explanation is that the limits of melodic pitch perception are determined peripherally and that multiple components elicit a more salient pitch simply due to a combination of multiple independent information sources. Based on the results from lower frequencies, this explanation seems less likely: little or no improvement in pitch discrimination is found when comparing the results from individual pure tones with the results from a complex tone comprised of those same pure tones (Faulkner 1985; Goldstein 1973). However, similar measurements have not been made at high frequencies, so it is unclear whether a similar pattern of results would be observed.

3.1 Methods

Seven young normal-hearing listeners participated in this experiment, screened as before for detection thresholds in quiet at 16 kHz no higher than 50 dB SPL. Difference limens for complex-tone F0 (F0DLs) and difference limens for pure-tone frequency (FDLs) were measured. Two nominal F0s were tested: 280 Hz and 1,400 Hz. FDLs were measured for two sets of eight nominal frequencies, corresponding to harmonics 5–12. The complex tones were generated by band-pass-filtering broadband harmonic complexes, such that only harmonics 5–11 of the complex tones at the nominal F0s were within the filter passband. All tones were presented at a level of 55 dB SPL per component within the filter passband. The filter slopes were 30 dB/octave. As in the previous experiment, random component starting phases were used on each presentation. The tones were 300-ms long each, including 10-ms onset and offset ramps, and both pure and complex tones were presented in the same combination of broadband noise and low-pass TEN that was used in the previous experiment. The background noise started 200 ms before the onset of the first tone and ended 200 ms after the offset of the last tone, for the current trial. The interstimulus interval was 500 ms.

The F0DLs and FDLs were measured using both a 2I-2AFC task and a 3I-3AFC task. The first requires a labeling (up vs. down), whereas the second requires only identification of the interval that was different. For both tasks, thresholds were measured using an adaptive, two-down one-up procedure (Levitt 1971). The stimuli were generated digitally and played out via a soundcard (Lynx Studio L22) with 24-bit resolution and a sampling frequency of 48 kHz. They were presented monaurally to the listener via Sennheiser HD 580 headphones.

3.2 Data Analysis

The individual FDLs within each of the two spectral regions were used to compute predicted F0DLs for the respective spectral region using the following equation:

$$\hat{\theta} = \left(\sum_{n=5}^{12} \theta_n^{-2} \right)^{-\frac{1}{2}}, \tag{16.1}$$

where $\hat{\theta}$ is the predicted F0DL and θ_n denotes the FDL measured using a nominal test frequency corresponding to the nth harmonic of the considered nominal F0. The equation stems from the general equation for predicting sensitivity based on multiple, statistically independent observations, assuming an optimal (maximum-likelihood ratio) observer (Goldstein 1973; Green and Swets 1966). The measured and predicted thresholds were log-transformed before statistical analyses using repeated measure analyses of variance (ANOVAs).

3.3 Results

Figure 16.2 shows the mean F0DLs and FDLs for the two F0s and two tasks (2- or 3-AFC). Considering first the complex-tone F0DLs, a two-way RMANOVA was performed on the log-transformed data, with the task (2I-2AFC, 3I-3AFC) and F0 (280 Hz, 1,400 Hz) as within-subject factors. No significant main effect of task was observed [$F(1, 6)=0.08$, $p=0.786$]. The effect of nominal F0 just failed to reach significance [$F(1, 6)=5.57$, $p=0.056$]. No significant interaction between the two factors was observed either [$F(1, 6)=0.015$, $p=0.906$].

Considering next the FDLs, a three-way RMANOVA on the log-transformed FDLs yielded a significant difference between the low- and high-frequency regions [$F(1, 6)=99.31$, $p<0.0005$], reflecting the fact that FDLs were generally larger in the high-frequency than low-frequency region. A significant main effect of harmonic number was observed [$F(7, 42)=6.29$, $p<0.0005$], as was a significant interaction between the frequency region and harmonic number [$F(7, 42)=7.08$, $p<0.0005$], consistent

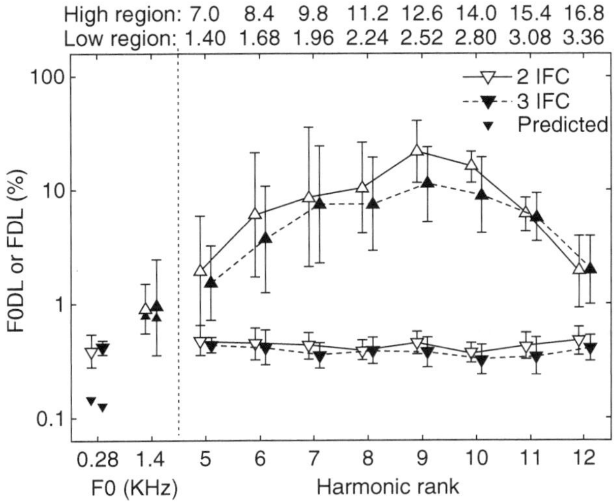

Fig. 16.2 Mean F0DLs (*left*) and FDLs (*right*). The numbers at the top are the pure-tone frequencies (kHz) for the high-frequency (*upward-pointing triangles*) and low-frequency (*downward-pointing triangles*) regions. *Error bars* show 95 % confidence intervals

with the observation that FDLs varied more with harmonic number in the high- than in the low-frequency region. Finally, a significant main effect of task [$F(1, 6)=8.28$, $p=0.028$] was observed: on average, FDLs measured with the 3I-2AFC task were slightly, but significantly, smaller than the FDLs measured with the 2I-2AFC task. The seemingly better FDLs at the highest frequencies may be due to loudness or audibility cues, as the higher-frequency tones become less audible due to the band-pass filter, and to the steeply rising absolute-threshold curve, at and above 16 kHz.

The small filled symbols represent the predicted F0DLs based on optimal integration of the information from the individual components. In the low-frequency region, the predicted F0DL was lower (better) than the measured F0DL. In contrast, in the high-frequency region, the predicted F0DL was on par with the measured F0DL.

3.4 Discussion

The finding of poorer-than-predicted F0DLs in the low-frequency region is in line with earlier studies (Gockel et al. 2005; Goldstein 1973; Moore et al. 1984b) and suggests either that the human auditory system is unable to optimally combine information across frequency when estimating F0 or that mutual interference between simultaneous harmonics degrades the peripheral information on which the optimal processor operates (e.g., Moore et al. 1984a). However, the fact that a different pattern of results was observed in the high-frequency region complicates the interpretation, as it is not immediately clear why mutual interference should occur at low but not at high frequencies or why information integration should be optimal at high, but not at low, frequencies. Again, multiple interpretations are possible.

One is that FDLs and F0DLs are not limited by peripheral-coding limitations at low frequencies, but by a more central "noise" source at the level of pitch representations. That may explain the limited benefit of combining several low-frequency harmonics. However, at very high frequencies, a more peripheral "noise" may begin to dominate performance, perhaps due to poorer phase locking, and "swamp" the influence of more central noise sources. Once the peripheral noise (which may be uncorrelated across channels) becomes dominant, then more optimal integration of information may occur.

In summary, the comparison of FDLs and F0DLs has revealed an interesting difference between the pattern of results at low and high frequencies. Further exploration of this difference may help in the search for the basic coding principles of pitch perception.

Acknowledgments Data from Experiment 1 were published in Oxenham et al. (2011). We thank Adam Loper for help with data collection. The work was supported by NIH grant R01 DC 05216.

References

Attneave F, Olson RK (1971) Pitch as a medium: a new approach to psychophysical scaling. Am J Psychol 84:147–166

Burns EM, Feth LL (1983) Pitch of sinusoids and complex tones above 10 kHz. In: Klinke R, Hartmann R (eds) Hearing - physiological bases and psychophysics. Springer, Berlin, pp 327–333

Carlyon RP, Deeks JM (2002) Limitations on rate discrimination. J Acoust Soc Am 112: 1009–1025

Cohen MA, Grossberg S, Wyse LL (1995) A spectral network model of pitch perception. J Acoust Soc Am 98:862–879

Faulkner A (1985) Pitch discrimination of harmonic complex signals: residue pitch or multiple component discriminations. J Acoust Soc Am 78:1993–2004

Gockel H, Carlyon RP, Plack CJ (2005) Dominance region for pitch: effects of duration and dichotic presentation. J Acoust Soc Am 117:1326–1336

Goldstein JL (1973) An optimum processor theory for the central formation of the pitch of complex tones. J Acoust Soc Am 54:1496–1516

Green DM, Swets JA (1966) Signal detection theory and psychophysics. Krieger, New York

Houtsma AJM, Smurzynski J (1990) Pitch identification and discrimination for complex tones with many harmonics. J Acoust Soc Am 87:304–310

Kaernbach C, Bering C (2001) Exploring the temporal mechanism involved in the pitch of unresolved harmonics. J Acoust Soc Am 110:1039–1048

Levitt H (1971) Transformed up-down methods in psychoacoustics. J Acoust Soc Am 49: 467–477

Licklider JCR (1954) "Periodicity" pitch and "place" pitch. J Acoust Soc Am 26:945

Meddis R, O'Mard L (1997) A unitary model of pitch perception. J Acoust Soc Am 102: 1811–1820

Moore BCJ (1973) Frequency difference limens for short-duration tones. J Acoust Soc Am 54: 610–619

Moore BCJ, Glasberg BR, Shailer MJ (1984) Frequency and intensity difference limens for harmonics within complex tones. J Acoust Soc Am 75:550–561

Moore BCJ, Huss M, Vickers DA, Glasberg BR, Alcantara JI (2000) A test for the diagnosis of dead regions in the cochlea. Br J Audiol 34:205–224

Oxenham AJ, Micheyl C, Keebler MV, Loper A, Santurette S (2011) Pitch perception beyond the traditional existence region of pitch. Proc Natl Acad Sci U S A 108:7629–7634

Palmer AR, Russell IJ (1986) Phase-locking in the cochlear nerve of the guinea-pig and its relation to the receptor potential of inner hair-cells. Hear Res 24:1–15

Ritsma RJ (1962) Existence region of the tonal residue. I. J Acoust Soc Am 34:1224–1229

Rose JE, Brugge JF, Anderson DJ, Hind JE (1967) Phase-locked response to low-frequency tones in single auditory nerve fibers of the squirrel monkey. J Neurophysiol 30:769–793

Schroeder MR (1968) Period histogram and product spectrum: new methods for fundamental-frequency measurement. J Acoust Soc Am 43:829–834

Sek A, Moore BCJ (1995) Frequency discrimination as a function of frequency, measured in several ways. J Acoust Soc Am 97:2479–2486

Wightman FL (1973) The pattern-transformation model of pitch. J Acoust Soc Am 54:407–416

Chapter 17
Pitch Perception for Sequences of Impulse Responses Whose Scaling Alternates at Every Cycle

Minoru Tsuzaki, Chihiro Takeshima, and Toshie Matsui

Abstract The purpose of this study is to investigate the sufficient "similarity" between consecutive auditory events for the auditory system to define the fundamental period for pitch perception. It is possible to contaminate the periodicity of harmonic complex tones by scaling the impulse response in the time domain at every other cycle. Scale-alternating wavelet sequences (SAWS) in which two impulse responses with different scaling factors alternated were generated based on impulse responses obtained from Japanese vowels spoken by a male speaker. Preliminary listening to such signals indicated that the perceived pitch went down an octave relative to the original when the scaling factor exceeded a certain degree. In the first experiment, pitch matching was measured as a function of the scaling factor by the method of adjustment where the comparison stimuli were completely periodic with adjustable base periods. The pitch shift was discontinuous against the base period, chromatic continuum. In the second experiment, pitch matching was investigated with comparison stimuli whose odd harmonics were attenuated. This procedure provides a stimulus continuum where the pitch moved up an octave without changing its pitch chroma. The attenuation of the odd harmonics needed to match the SAWS varied systematically as a function of the degree of scaling. The relation between pitch matching and the peak height along the time interval axis of the stabilized auditory image is discussed.

M. Tsuzaki (✉)
Faculty of Music, Kyoto City University of Arts, Kyoto, Japan
e-mail: minoru.tsuzaki@kcua.ac.jp

C. Takeshima
College of Performing and Visual Arts, J.F. Oberlin University, Machida, Japan

T. Matsui
Department of Otorhinolaryngology, Nara Medical University, Kashihara, Japan

B.C.J. Moore et al. (eds.), *Basic Aspects of Hearing*,
Advances in Experimental Medicine and Biology,
DOI 10.1007/978-1-4614-1590-9_17,

1 Introduction

1.1 Resonance and Periodicity as a Signature of Sound Source

When a certain body is hit, it generates a sound that conveys the resonance characteristics of the body. If the body has a cavity that contains a common medium, such as the atmosphere, this resonance systematically reflects the size of the body. Thus, the body resonance could serve as a cue for sound source identification. On the other hand, it seems a general strategy for mammals to use the quasiperiodic excitation as a way of communication. Since the body size also constrains this periodicity to some degree, the fundamental frequency with its harmonics can provide another cue for sound source identification. It has been demonstrated that these two features function as effective cues for auditory grouping.

In terms of signal processing, this process of sound generation can be modeled in the framework of source-filter theory. The resonance characteristics are represented as the frequency transfer function of the certain filter, which has a continuous spectrum. The periodic excitation of the source is represented as a harmonic structure which is a discrete spectrum. Acoustic signals are given by the multiplication of these two aspects in the frequency domain or by overlapping and adding each impulse response with every cycle of the fundamental period in the time domain. In the first approximation, we can understand that the pitch and timbre perceptions are the psychological correlates of the filter characteristics and the periodicity of the source, respectively.

In natural stimuli such as vowels, waveforms are seldom periodic in the strict sense. It has been argued that a certain nonlinear coupling should be considered to model the real situation (Titze 2008). Nevertheless, the auditory system manages to extract a stable pitch as well as a stable timbre. A question may be asked how tolerant the auditory system is towards the perturbation in impulse responses delivered at each cycle to perceive the original periodicity (or pitch).

1.2 Scale Alternating Wavelet Sequence

In this chapter, the perturbation was provided by the resonance scaling because it is one of the significant characteristics for source identity. The discrimination threshold of the resonant scaling was about 5 %. Its alternation between consecutive segments could break up the vowel sequences into two streams sorted by each scaling factor.

The test stimuli in this chapter will be called "scale-alternating wavelet sequences (SAWSs)," where an original wavelet of the impulse response of a certain vowel and its scaled version alternate with each other at each periodic, "glottal" cycle. The general purpose was to find a certain degree of scaling factor that causes some perceptual change. The scaling of wavelets was performed by dilating or contracting the original wavelet along the linear time axis. This corresponds to shifting the original frequency transfer function downwards or upwards along the log frequency axis. Each wavelet was overlapped and added periodically along the time axis. Accordingly, the original wavelet was overlapped and added at the odd cycles, while the scaled version was added at the even cycles. The outcome of this manipulation is equivalent to what is obtained by mixing two vowel sounds whose vocal tract features are original or scaled, respectively, with one half of the original fundamental frequency and with a temporal disparity equal to the original fundamental period.

Tsuzaki et al. (2012) demonstrated that the resonance difference could be an effective cue for source segregation. If the resonance scale is an absolute and compelling cue for source segregation, we might hear two persons with different body sizes speaking the same vowel with the same pitch that is an octave below to that of the "original" when the factor of the resonance scaling exceeds a certain limit. Another possible percept might be that of a person with an intermediate body size speaking the vowel with the original pitch. Preliminary listening to a couple of versions of the SAWSs disclosed that neither of these was the case. Perception of two persons was not convincing, while it was also difficult to imagine any single sound source to produce such a sound. The most clear and dominant impression was that the pitch was shifted an octave down compared to the original. Therefore, pitch matching experiments were carried out.

2 Experiment 1 (F0 Pitch Matching)

2.1 *Purpose*

The purpose of the first experiment was to investigate the amount of pitch shift as a function of the factor of the resonance scaling by adjusting the fundamental frequency (F0) of the comparison stimulus.

2.2 *Stimulus*

The method of generating the SAWS was outlined in Sect. 1.2. Each original wavelet was obtained from Japanese vowels "a," "i," and "u" uttered by an adult, male speaker. For each vowel, a series of vocal tract transfer functions was calculated using the F0-adaptive spectral smoothing method in STRAIGHT (Kawahara et al. 2008). A sample transfer function was picked up from an arbitrary frame for a steady portion. The original wavelet was generated by applying the inverse FFT to this transfer function with a phase characteristics containing a random phase dispersion in the high-frequency region. The scaled wavelets were generated by dilating or contracting the original wavelet along the time axis. The scaling factor was 0.50, 0.80, 0.83, 0.87, 0.91, 1.00, 1.04, 1.09, 1.14, 1.19, 1.25, or 2.00, and there were 13 values in total. A scaling factor less than 1.00 means contraction, which corresponds to shrinkage of the vocal tract; a factor greater than 1.00 means dilation, which corresponds to enlargement of the vocal tract.

The SAWS was synthesized by overlapping and adding the original and the scaled wavelets alternatively at a certain periodic time grid. The period was 8.0 or 8.3 ms. Subsequently, it will be referred as the overlap-and-add period, and its inverse will be referred as the overlap-and-add frequency. The pitch interval between these two overlap-and-add frequencies was about 64 cents. The comparison stimuli for the pitch matching task were synthesized by overlapping and adding just the original wavelet for the same vowel class as the test stimulus with various fundamental periods. It ranged from −1,400 to +1,400 cents relative to 125 Hz using a 35.4 cents step. All the manipulations for the stimulus synthesis were digitally performed with 44.1 kHz sampling rate and 16 bit quantization using MATLAB (MathWorks).

2.3 Procedure

Each listener was required to match the pitch of the comparison stimulus to that of the test stimulus by adjusting the F0 of the comparison stimulus. The experiment was controlled by a computer (Apple iMac G5) with a DSP (Symbolic Sounds Capybara 320 + Kyma). On the GUI, four buttons were prepared, i.e., "Listen," "Up," "Down," and "OK." The "Listen" button triggered the stimulus presentation, and each listener could listen to the succession of the test and comparison stimuli. The "Up" and "Down" buttons changed the F0 of the comparison stimulus by one step upwards or downwards, respectively. Listeners were allowed to push these two buttons several times without listening to the stimuli. When they felt satisfied with the adjustment, they were required to push the "OK" button, which triggered the next condition of the test stimulus.

An experimental session was comprised of 156 matches, which were two repetitions of 78 combinations of the vowel categories (3), the overlap-and-add frequencies (2), and the scaling factor (13). It took 1 or 2 h to complete one session. The stimuli were presented diotically via a headphone (Sennheiser HD 600) through a headphone amplifier (Luxman P-1). The presentation level was 65 dB SPL with the A weighting. A 6 dB roving of presentation level was introduced to minimize the possibility of listeners relying on the loudness difference when performing the matches.

2.4 Listeners

Eight undergraduates and one graduate of Faculty of Music, Kyoto City University of Arts, participated in the experiment. All the listeners claimed to possess absolute pitch. None had any severe listening problem. All were paid for the participation.

2.5 Results

The difference between the overlap-and-add frequency and the F0 of the matched comparison was converted into cent value for each of the test stimuli and each listener. The occurrences were counted for each scaling factor by pooling the listeners, vowel categories, and overlap-and-add frequencies. Their percentages relative to the total number of the matches per scaling factor are depicted by the size and gray scale of each disk in Fig. 17.1.

The response distributions were discrete: most of matches were concentrated near 0 cent or −1,200 cents (an octave below). The frequencies of the octave below matching tended to increase as the scaling factor moved away from the original, i.e., 1.00. An asymmetry was observed between the contraction (<1.00) and dilation

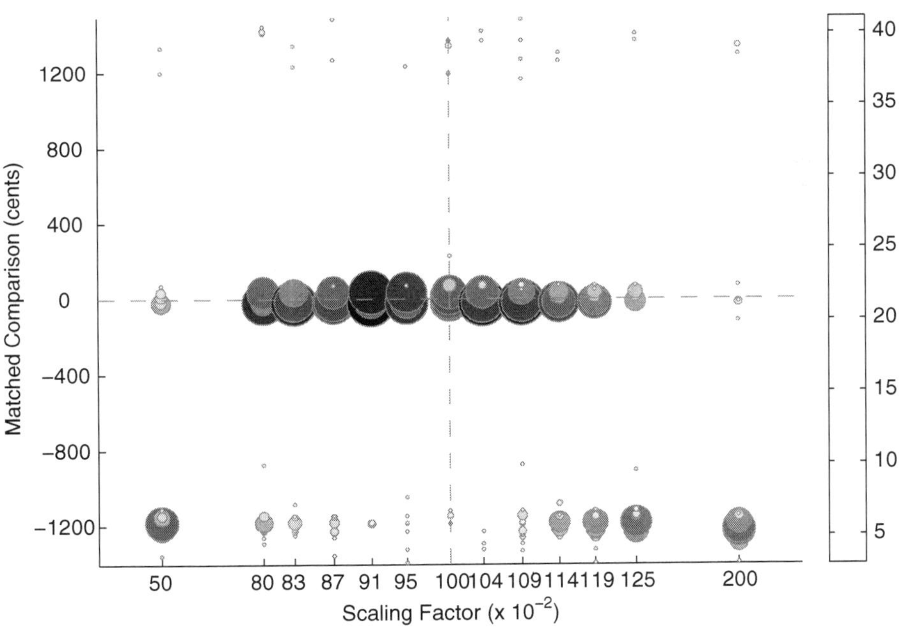

Fig. 17.1 Percentages of pitch-matched F0 relative to the overlap-and-add frequency

(>1.00) sides. The tendency of the octave downshift was more prominent for the dilation side than for the contraction side.

For each scaling factor, a test of the homogeneity of the distribution compared to the reference condition was performed by counting the number of responses in three bins; (a) −100 to +100 cents, (b) −1,300 to −1,100 cents, and (c) others. The null hypothesis of homogeneity was rejected for all scaling factors except for 0.95 and 1.04.

2.6 Discussion

The results demonstrated that an octave down pitch shift occurred when the scaling factor exceeded a certain amount. The distribution of pitch matching was discontinuous along the F0 axis, i.e., it was quite rare to match an F0 that would correspond to the perception of a different chroma. This observation suggests that the pitch shift caused by alternating the two wavelets is one with no change in chroma. The familiar change in pitch used for musical melodies is accompanied with chroma changes and accomplished by changing the F0. Contrastingly, a pitch shift with no chroma change can be achieved by attenuating the odd harmonics (Patterson et al. 1993). In the second experiment, the pitch shift of the SAWSs was investigated by matching to a harmonic complex tone whose odd-numbered harmonics were attenuated.

3 Experiment 2 (Odd-Harmonic Reduction Pitch Matching)

3.1 Purpose

The purpose of the second experiment was to measure the pitch shift of the scaling alternating wavelet sequences in terms of the attenuation level of the odd harmonics of the harmonic complex tones.

3.2 Stimulus

The test stimuli were the SAWSs generated as in Experiment 1. In Experiment 2, they were generated for five Japanese vowels, "a," "e," "i," "o," and "u." The overlap-and-add period was 8 ms. The other details were the same as in Experiment 1. However, the scaling factor 1.09 was not tested because of a bug in the experimental configuration.

The important difference from Experiment 1 was the way of making the comparison stimulus. One can shift the pitch of the original harmonic complex an octave "upwards" by attenuating the odd harmonics. On the other hand, an octave "downward" shift was expected by alternating the scaled and original wavelets. To generate appropriate comparison stimuli, vowel sounds, i.e., harmonic complex tones, with 16 ms base period were generated as a first step, and a series of comparison stimuli were generated with the attenuation factor up to −40 dB in 1-dB steps. The odd-harmonic attenuation was accomplished by adding the delayed version of the original waveform with the required amount. The delay necessary to attenuate the odd harmonics was 8 ms for this experiment.

3.3 *Procedure*

The experimental procedure was the same as in Experiment 1 except for the number of the test stimuli. The total number of pitch matches required for a listener was 120. The experimental session took 1 or 2 h per listener.

3.4 *Listeners*

Six undergraduates of Faculty of Music, Kyoto City University of Arts, participated in the experiment. All the listeners claimed to possess absolute pitch. None had any severe listening problem.

3.5 *Results*

The distribution of matching was unimodal along the axis of odd-harmonic attenuation in contrast to Experiment 1 where it was bimodal along the F0 axis. The boxplots are depicted as a function of the scaling factor in Fig. 17.2.

The degree of scaling could be a good indicator of the attenuation of the odd harmonics. The reference stimulus whose scaling factor was 1.0 matched to the comparison with about 22 dB attenuation in median. This amount of attenuation corresponds to what has been reported to induce an upward pitch shift by an octave in previous studies (Patterson et al. 1993). For both sides of scaling, i.e., contraction and dilation, the matched attenuation level gradually decreased as the degree of scaling increased. An asymmetry between the contraction and dilation sides was observed as in Experiment 1.

3.6 *Discussion*

The observed results indicate that the pitch shift caused by alternating two different scaled wavelets was an octave shift with no chroma change. Compared to the pitch matching by F0 in Experiment 1, the pitch matching by attenuating the odd

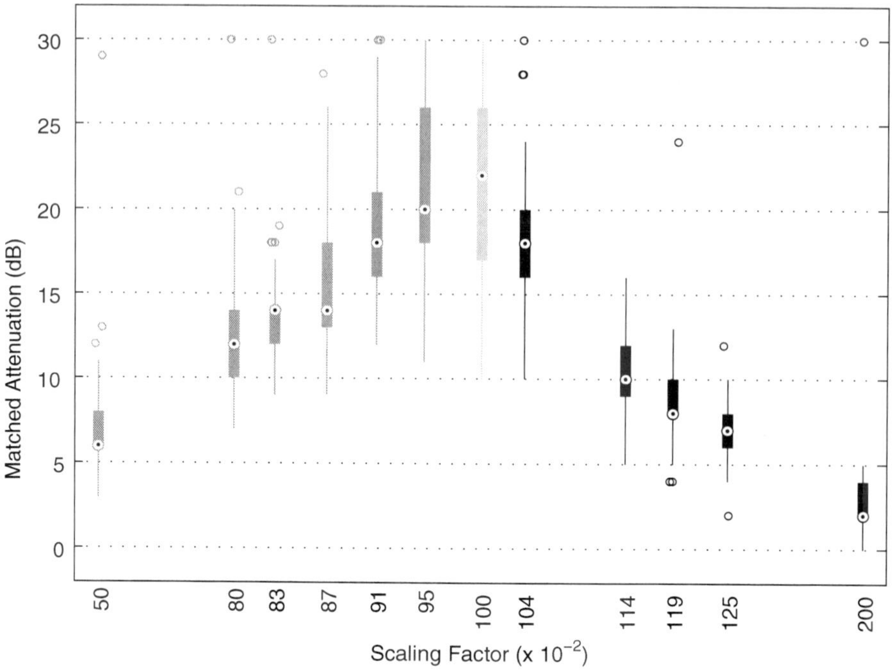

Fig. 17.2 Boxplots of the matched attenuation of the odd harmonics of the comparison stimulus as a function of the scaling factor

harmonics did not seem familiar as the daily musical experience of the listeners. In the next section, a simulation based on an auditory model is performed to provide a reasonable predictor for this pitch shift effect.

4 Auditory Simulation

The auditory simulation was accomplished based on the "auditory image model (AIM)" (Patterson et al. 1995). AIM can provide stabilized auditory images (SAIs) for acoustic input signals. SAIs are time interval histograms for multiple tonotopic channels. It can be assumed that the pitch will be determined by the dominant time interval that might correspond to the peak position in the summary SAI. There will generally be multiple local peaks in the summary SAI for harmonic complex tones, and two major peaks appear at 8 and 16 ms for the test and comparison stimuli in Experiment 2. The ratios between the heights of these two peaks were calculated for each stimulus, and boxplots were replotted by converting the scaling factor and the odd-harmonic attenuation to the corresponding peak ratio in Fig. 17.3. The broken line in Fig. 17.3 depicts the equilibrium between the peak ratios for the test stimuli

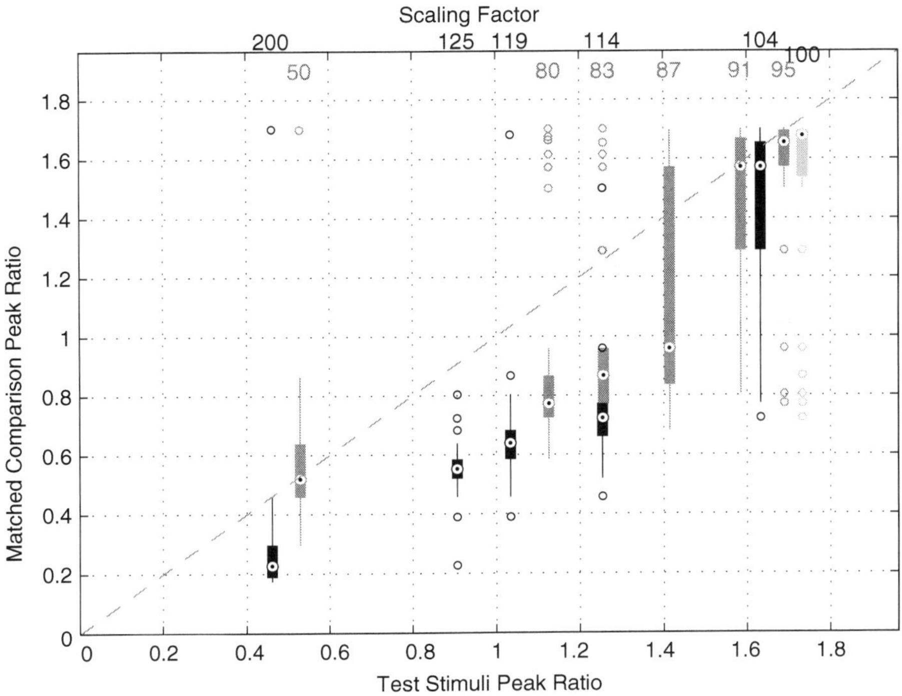

Fig. 17.3 Boxplots of the peak ratio in the summary SAIs of the pitch-matched comparison stimulus as a function of the peak ratio for the test stimulus

and those for the matched comparison stimuli. Although the proposed decision statistics do not seem to be misdirected, many points are below the line. This might indicate that some other factors must be considered.

5 Summary

The pitch of SAWSs shifted downwards by an octave when the scaling factor exceeded a certain amount. Although this octave shift appeared discontinuous along the pitch chroma continuum, it could be regarded as a continuous change in terms of the dominance of time intervals and could be approximately explained by the ratio between the first and second peaks in summary SAIs.

Acknowledgments This study was supported by JSPS KAKENHI (Grants-in-Aid for Scientific Research) B, No. 20346331, and JSPS KAKENHI (Grants-in-Aid for Young Scientists) B, No. 23730715. The authors are thankful to Dr. Roy D. Patterson, Cambridge University, for his useful advice. They also thank Ms. Sawa Hanada for her helps for running Experiment 2.

References

Kawahara H, Morise M, Takahashi T, Nishimura R, Irino T, Banno H (2008) Tandem-STRAIGHT: a temporally stable power spectral representation for periodic signals and applications to interference-free spectrum, F0, and aperiodicity estimation. In: ICASSP 2008: Acoust., Speech Signal Processing, Las Vegas, 2008, pp 3933–3936

Patterson RD, Milroy R, Allehand M (1993) What is the octave of a harmonically rich note? Contemp Music Rev 9:69–81

Patterson RD, Allerhand MH, Giguère C (1995) Time-domain modeling of peripheral auditory processing: a modular architecture and a software platform. J Acoust Soc Am 98:1890–1894

Titze IR (2008) Nonlinear source–filter coupling in phonation: theory. J Acoust Soc Am 125: 2733–2749

Tsuzaki M, Irino T, Takeshima C, Matsui T (2012) Effects of the correlation between the fundamental frequencies and resonance scales as a cue for the auditory stream segregation. ARO midwinter research meeting, San Diego, 2012

Chapter 18
Putting the Tritone Paradox into Context: Insights from Neural Population Decoding and Human Psychophysics

Bernhard Englitz, S. Akram, S.V. David, C. Chambers, Daniel Pressnitzer, D. Depireux, J.B. Fritz, and Shihab A. Shamma

Abstract The context in which a stimulus occurs can influence its perception. We study contextual effects in audition using the tritone paradox, where a pair of complex (Shepard) tones separated by half an octave can be perceived as ascending or descending. While ambiguous in isolation, they are heard with a clear upward or downward change in pitch, when preceded by spectrally matched biasing sequences. We presented these *biased Shepard pairs* to awake ferrets and obtained neuronal responses from primary auditory cortex. Using dimensionality reduction from the neural population response, we decode the perceived pitch for each tone. The bias sequence is found to reliably shift the perceived pitch of the tones away from its central frequency. Using human psychophysics, we provide evidence that this shift in pitch is present in active human perception as well. These results are incompatible with the standard absolute distance decoder for Shepard tones, which would have predicted the bias to attract the tones. We propose a relative decoder that takes the stimulus history into account and is consistent with the present and other data sets.

B. Englitz, PhD (✉) • S. Akram • D. Depireux • J.B. Fritz
Institute for Systems Research, University of Maryland,
A.V. Williams Bldg., College Park, MD 20742, USA
e-mail: benglitz@gmail.com

S.V. David
Institute for Systems Research, University of Maryland,
A.V. Williams Bldg., College Park, MD 20742, USA

Oregon Hearing Research Center,
Oregon Health & Science University, Portland, OR, USA

C. Chambers • D. Pressnitzer
Institute for Systems Research, Equipe Audition, Ecole Normale Superieure, Paris, France

S.A. Shamma
Department of Electrical and Computer Engineering,
Institute for Systems Research, University of Maryland, College Park, MD, USA

B.C.J. Moore et al. (eds.), *Basic Aspects of Hearing*,
Advances in Experimental Medicine and Biology,
DOI 10.1007/978-1-4614-1590-9_18,

1 Introduction

In the present study, we address the physiological basis of the auditory 'tritone paradox' (Deutsch et al. 1986), in which a sequence of two tonal stimuli can be perceived as ascending or descending in pitch. Various factors, including linguistic background (Deutsch 1991, 1994), amplitude envelope (Repp 1997), and stimulus context (Chambers and Pressnitzer 2011), have been shown to influence the perceived direction. The present focus is the influence of stimulus history on the directionality percept. We address this question by preceding the ambiguous pair with a sequence of tonal stimuli ('bias') in a limited spectral range, which has been demonstrated to reliably influence human perception (Chambers and Pressnitzer 2011). Identical stimuli are presented to ferrets in physiological experiments and humans in psychophysical experiments.

We find the bias to influence the perceived pitch of the tones in the ambiguous pair in a repulsive manner, i.e. the pitch of the tones shifts away from the bias. These results from neuronal population decoding are supported by human psychophysics.

These results allow us to evaluate different decoders for directionality in pitch. The most intuitive decoder, which relies on the minimal—ascending or descending—distance between the two tonal stimuli, is not consistent with the bias-induced increase in distance on the biased side of the pitch circle. We propose a more general decoder, which takes the stimulus history into account and allows for a transitive conjunction of two relative directionality judgements to provide an explanation for the central findings on the tritone paradox and its context dependence.

2 Methods

2.1 *Physiology and Psychophysics: Stimulation and Recording*

Both human and animal experiments were conducted according to the ethical guidelines of the University of Maryland. Stimuli were presented in sound-attenuated chambers over calibrated speakers. Extracellular neuronal recordings were collected from seven awake ferrets using chronically implanted multielectrode arrays. In human experiments, an optimized sampling technique was applied to rapidly estimate the psychophysical curve with high precision (Poppe et al. 2012) (Fig. 18.1).

2.2 *Acoustic Stimuli*

All stimuli were composed of sequences of Shepard tones (Shepard 1964). A Shepard tone is a complex tone built as the sum of octave-spaced pure tones (a flat spectral envelope was used here). A Shepard tone can be characterized by its position

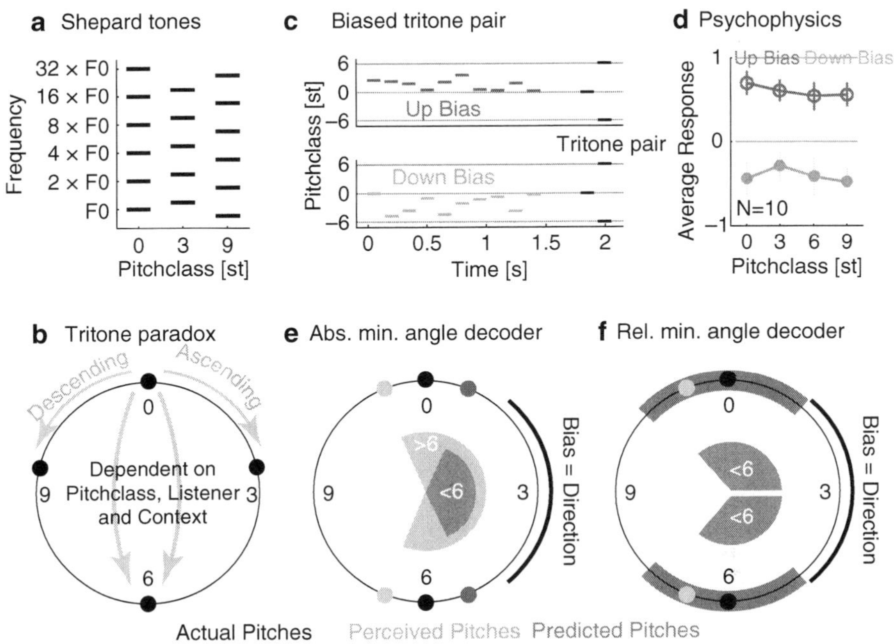

Fig. 18.1 The tritone paradox and contextual modulation of the perceived change in pitch. (**a**) The ambiguity in the tritone paradox stems from the Shepard tones' repetitive structure (see Sect. 2). (**b**) Steps of up to almost 6 st are perceived as up/down steps, whereas the half-octave (tritone) step is ambiguous and influenced by various factors (acoustic context, pitch class, etc.). (**c**) Preceding an ambiguous Shepard pair by a sequence of Shepard tones within the half octave above ('up bias', *dark grey*) or below ('down bias', *light grey*), the first tone (**d**) influences the percept of directionality for human listeners. (**e**) As indicated in the Sect. 1 and detailed in the Sect. 4," the standard decoder assumes the absolute minimal angle between the Shepard pair to determine the perceived direction of pitch change (perceived = *light grey*, prediction = *dark grey*). We show that the perceived pitches move away (i.e. >6 st) from the bias (actual = *black*), thus necessitating a different decoder. (**f**) A relative decoder takes the acoustic context into account and makes relative judgements from the acoustic history, consistent with the present data set. The *dark grey* areas indicate acceptable perceived pitches of the tones to still be heard according to the bias for the relative decoder

in an octave, termed pitch class (in units of semitones), w.r.t. a base tone. Across the entire set of experiments the duration of the Shepard tones was 0.1 s and the amplitude 70 dB SPL.

The *biased Shepard pairs* consisted of a bias sequence ('bias') followed by an ambiguous, i.e. 6 st separated, Shepard pair. The bias precedes the pair at various temporal separations ([0.05,0.2,0.5,1]s) and consists of a sequence of Shepard tones (lengths, 5 and 10 stimuli), which are within 5 semitones above or below the first Shepard tone in the pair. These biases are called 'up' and 'down' bias, respectively, as they bias the perception of the ambiguous pair to be 'ascending' or 'descending', respectively, in pitch (Chambers and Pressnitzer 2011). Altogether we presented 32 conditions (4 base pitch classes ([0,3,6,9]st), 2 randomization, 2 bias lengths ([5,10]

stimuli), 'up'/'down bias) and different bias sequences, which in total contained 240 distinct Shepard tones, finely covering one octave. In the present study, we use one of the simpler versions of this contextual influence—more detailed psychophysics will be described in a forthcoming study by CC, SAS and DP.

Further, we used pitch comparison sequences in the psychophysical studies. The pitch comparison sequences consisted of a bias, a reference Shepard tone, and a target Shepard tone. The bias that preceded the reference was followed by a target (drawn from the set [−3,−2.9,…,2.9,3]st) 3 s later. Subjects were asked to report whether the target's pitch was higher or lower than the reference's.

2.3 Population Decoding

The perceived stimuli in the ambiguous pair were estimated from the neural responses by training a decoder on the biasing sequences and then applying the decoder to the neural response of the pair. We first build a matrix of responses which had the (240) different Shepard tones occurring in the bias running along one dimension and the neurons along the other dimension. The PCA (Principle Component Analysis) decoder performed a linear dimensionality reduction, interpreting the stimuli as examples and the neurons as dimensions of the representation. The data was projected to the first three dimensions, which represented the pitch class as well as the position in the sequence of stimuli. To assign a pitch class to the decoded stimuli of the test pair, we projected them onto the 'pitch circle' formed by the decoded stimuli from the bias sequences. More precisely we estimated a smoothed trajectory through the set of bias tones which was assigned a pitch class at every point, by averaging the pitch classes of the closest 10 bias stimuli weighted by their distance to the point. The pitch class of the test tone was set to the pitch class of the closest point on the trajectory.

2.4 Statistical Analysis

Nonparametric tests were used throughout the study to avoid assumptions regarding distributional shape.

3 Results

We obtained single-unit recordings from 555 neurons in the primary auditory cortex of seven awake ferrets and conducted psychophysical experiments with ten subjects under various stimulus conditions.

3.1 *Neurons in Auditory Cortex Exhibit Tuning to Shepard Tones*

A considerable subset of neurons in auditory cortex responded to the presentation of Shepard tones with a significant change in response rate compared to spontaneous rate (55, 43 % increased, 12 % decreased; $p<0.05$), while 41 % of the neurons also had a significantly tuned response. A well-tuned unit is shown in Fig. 18.2a, where the firing rate varies as a function of the pitch class of the Shepard tone. Neurons typically exhibited a single peak of varying width, although multipeaked tuning curves existed as well (~30 % out of the tuned cells). Overall, the median tuning width was 2.06 [25 %, 0.82; 75 %, 6.44]st (2 SD of a Gaussian fit to the tuning). Neurons exhibited strongest tuning in onset, sustained, and offset responses in similar proportions (onset, 38 %; sustained, 33 %; offset, 29 %).

Many cells (39 %) showed significant changes in response strength over the sequence of presented stimuli (Friedman test, Fig. 18.2f), where 33 % showed a decrease (rel. to 1 st stimulus, median 31 %, range 14–45 %) over time and 6 % an increase (median 11 %, range 5–31 %). Suppression/facilitation was assessed by fitting a single time-constant exponential decay to the average PSTH of all biasing stimuli.

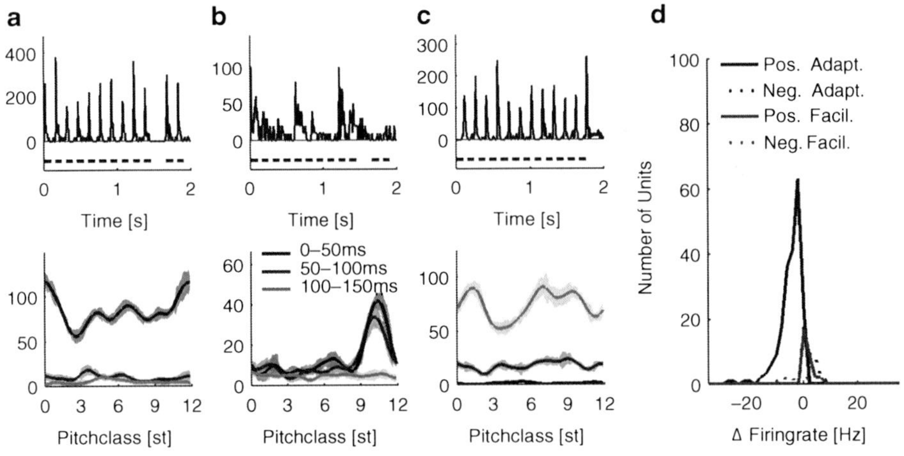

Fig. 18.2 Response of auditory cortex neurons to Shepard tones is typically tuned and adapting. Three example cells are shown in (**a–c**), an onset tuned cell (**a**), a sustained tuned cell (**b**) and an offset tuned cell (**c**). *Top panels* show an average over 10 trials to one bias sequence and the ambiguous pair after a pause of 0.2 s. Bottom panels show the responses as a function of all Shepard tones in the bias sequences (smoothed with a Gaussian fit, 0.5 st), with different grey values indicating the tuning at different times of the response. (**d**) The large majority of significantly responding cells adapted their response over the duration of the bias (*black*), while only a small minority facilitated to a smaller degree (*grey*). Positive and negative deviations from spontaneous rate are considered separately here (see text)

3.2 Bias Stimulus Repels the Tritone Pair in Pitch

The pitch class of a stimulus is encoded in the population response, via the various response properties of different neurons. We used a dimensionality reduction-based decoder to transition from the high-dimensional space of neural responses to only two dimensions. Stimuli of similar pitch class (Fig. 18.3a, different colours) occupy close regions in space. Globally, the stimuli form a slightly distorted torus, along which the pitch class changes in an orderly fashion. This circular, closed shape is expected, given that the Shepard tones form a circular stimulus space themselves.

The represented pitch class of the tones in the tritone pair generally falls into their respective regions on the pitch circle; however, they are shifted in different directions along the circle depending on which bias preceded them (Fig. 18.3b). If the bias is above the first tone in the pair, the first tone is shifted down, while the second tone shifted up, and conversely for a bias below the first tone. More simply put, the tones in the ambiguous pair shift away in pitch class from the bias sequence (127/128 comparisons between the positions of the tones across all conditions (see Sect. 2); Wilcoxon test, $p < 10$–22). The bias-induced shift was 0.62 st for the first and 0.49 st for the second tone and returned to baseline with an exponential time constant of 0.56 s (estimated based on the recovery of the spontaneous rate after the tritone pair).

The presence of the bias hence increases the distance between the tritone pair along the side of the pitch circle, which humans judge to be the direction of pitch change.

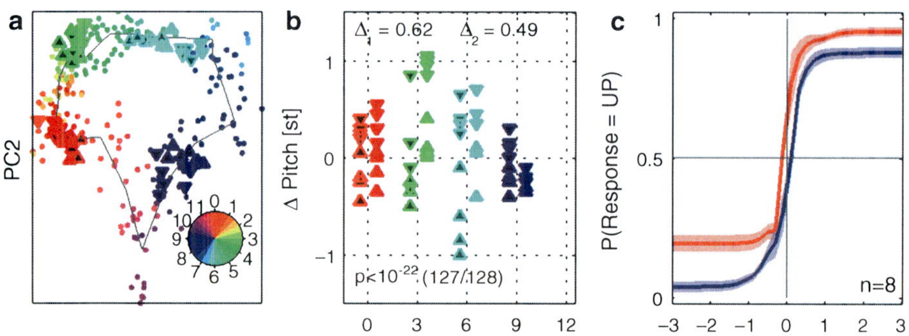

Fig. 18.3 Neural population decoding shows the bias to repel the tritone pair in pitch. (**a**) After dimensionality reduction to three dimensions, the neuronal responses to Shepard tones spanning the whole octave exhibit a circular progression of pitch classes (*colours*) in dimensions 1 and 2 (the *grey line* connects local averages every 0.5 st). (**b**) The Shepard tones in the tritone pairs (*triangles*) are mapped to their pitch classes; however, their precise position depends on the preceding bias, i.e. whether it is above (*triangle pointing up*) or below (*triangle pointing down*) the respective tone. The bias repels the tones in pitch, such that the same tone is perceptually separated by ~0.5 st between the two bias conditions (expected pitch class subtracted here to show differences). (**c**) Psychophysical results from humans exhibit a similar repulsion in perceived pitch in a pitch comparison paradigm with a 3 s delay separating reference and target (*red* = preceded by UP bias, *blue* = preceded by DOWN bias)

3.3 *Human Perception in Biased Pitch Discrimination Mirrors Physiological Pitch Shifts*

Since this result is unexpected, given the standard decoder, we tested whether it holds for human subjects. We modified the paradigm to test pitch perception more explicitly (rather than a local pitch change), by separating the tone pair temporally, in which case we hypothesize the bias to only influence the first—reference—tone. More precisely, subjects were asked to compare two Shepard tones which were separated by a long pause or noise (3 s) and preceded by the same bias sequences as above. The second Shepard tone was drawn from the range [−3…3]st relative to the first tone.

Pitch comparison judgements exhibited a behavior consistent with the neural decoding results. For Shepard tones preceded by *up* biases, the psychophysical curve shifted (by ~0.5 st on average) to lower pitch classes and the uncertainty for judgements on the low side increased (Fig. 18.3f, red curve, vice versa for *down* biases, blue curve). This is consistent with shifts in the same direction but variable size, which is predicted by the random nature of the bias sequence.

4 Discussion

How does an observer arrive at a judgement of whether the sequence of two Shepard tones is ascending or descending in pitch? Multiple hypotheses, i.e. decoders, are compatible with the previous psychophysical data. Using neural population analysis, we provide some physiological data to test whether the predictions of the classical decoder are compatible with human perception of the ambiguous Shepard pair in the context of a bias.

The classical decoder chooses the smaller angle between two Shepard tones on the pitch circle as the direction of pitch change, i.e. steps from 0 to up to 5.9 st are perceived as ascending, 0 to down to −5.9 st as descending, and the half-octave step remains ambiguous. For a biasing stimulus preceding a tritone pair, it predicts the corresponding angle to decrease, e.g. a 0–6 pair preceded by an 'up' bias could become a 1–5 pair and thus ascending (Fig. 18.1e).

A different hypothesis would be a relative decoder, which assesses a stimulus relative to the stimulus context (Fig. 18.1f), i.e. if the stimulus history is centred around 3 st, the ambiguous pair 0–6 is perceived as ascending, since 0 is relatively below and 6 relatively above the stimulus history, and by transitivity 6 is above 0. Hence, for a biasing stimulus preceding the tritone pair, it predicts an extended range of directional judgements in the direction of the bias. While it does not make a prediction about the absolute perceived location of the ambiguous pair, it is consistent with both decreases and increases in distance between them over a limited range. In the absence of a biasing stimulus, the first tone in the pair becomes the context for the second, and this decoder reduces to the classical one. A neural correlate of the relative decoder could be frequency-change selective cells, which could facilitate/adapt their response based on the stimulus history.

Our decoding results demonstrate that the angle between the two Shepard tones increases in the direction of the bias and are thus incompatible with the classical decoder, but consistent with the relative decoder. Further, as demonstrated by Chambers et al. (2009), the range of judgements following the bias exceeds 6 st, consistently with the relative decoder.

More generally, a repulsive effect of a spectrally limited stimulus history has been described in other contexts before, e.g. Holt (2005, *contrastive effect*). Our findings also relate to findings in the visual system (Seriès et al. 2009), where the overestimation of small differences between oriented gratings was linked to local adaptation and the 'homunculus' was also unaware of its own adaptation. Clearly, the 'Rosetta stone' to decode the 'tritone paradox' would be recordings from behaving animals, where perception and neural activity are comparable on a trial-by-trial basis.

Acknowledgment We thank Barak Shechter, Timm Lochmann, and Paul Watkins for interesting discussions.

References

Chambers C, Pressnitzer D (2011) The effect of context in the perception of an ambiguous pitch stimulus. ARO Abstract #1025

Chambers C, Park-Thompson V, Pressnitzer D (2009) Biasing perception of ambiguous pitch stimuli. BSA Short papers meeting, Liverpool, UK

Deutsch D (1991) The tritone paradox: an influence of language on music perception. Music Percept 8(4):335–347

Deutsch D (1994) The tritone paradox: some further geographical correlates. Music Percept 12(1):125–136

Deutsch D, Kuyper WL, Fisher Y (1986) The tritone paradox: its presence and form of distribution in a general population. A musical paradox, music perception, 3(3):275–280.

Holt LL (2005) Temporally nonadjacent nonlinguistic sounds affect speech categorization. Psychol Sci 16(4):305–312

Poppe S, Benner P, Elze T (2012) A predictive approach to nonparametric influence for adaptive sequential sampling of psychophysical experiments. J Math Psychol 56:179–195

Repp BH (1997) Spectral envelope and context effects in the tritone paradox. Perception 26(5):645–665

Seriès P, Stocker AA, Simoncelli EP (2009) Is the homunculus "aware" of sensory adaptation? Neural Comput 21(12):3271–3304

Shepard RN (1964) Circularity in judgments of relative pitch. J Acoust Soc Am 36(12):2346

Part III
Enhancement and Perceptual Compensation

Chapter 19
Spectral and Level Effects in Auditory Signal Enhancement

Neal F. Viemeister, Andrew J. Byrne, and Mark A. Stellmack

Abstract Auditory enhancement refers generally to the increased perceptual salience of a spectral region when that region is preceded by its spectral complement, e.g., reinserting a missing component in a harmonic complex makes that component "pop out." One manifestation of enhancement is the increased detectability of a signal in certain spectro-temporal configurations. In the present experiments, detection thresholds were measured for a 2-kHz signal that was masked by an inharmonic complex with a spectral notch centered at 2-kHz. When the masker was preceded by a precursor/adaptor with a spectral gap identical to that of the masker, detection thresholds were lowest when the gap width was 0.6 octave. The amount of signal enhancement, the difference in thresholds between the no-precursor and precursor conditions, decreased for smaller and larger gap widths. In addition, this general result was robust for precursors such as band-reject noise and harmonic complexes that were different in perceptual quality from the masker. This suggests that grouping/segregation processes do not mediate enhancement as assessed here. Similarly, significant enhancement was observed with precursor-masker level differences over a 40-dB range. Overall, these results further indicate that frequency resolution is a dynamic process that depends on spectro-temporal context. They also are consistent with a mechanism involving adaptation of inhibition that likely occurs at low levels in the auditory system.

1 Introduction

The term "enhancement" has been used in many different ways in audition, ranging from signal processing schemes to spectral context effects in speech perception (e.g., Holt et al. 2001). Whether and how these aspects are related are open questions.

N.F. Viemeister (✉) • A.J. Byrne • M.A. Stellmack
Department of Psychology, University of Minnesota,
75 E River Rd, Minneapolis, MN 55455, USA
e-mail: nfv@umn.edu

B.C.J. Moore et al. (eds.), *Basic Aspects of Hearing*,
Advances in Experimental Medicine and Biology,
DOI 10.1007/978-1-4614-1590-9_19,

Here, the focus is on a class of phenomena that appear to be related and that may reflect common, basic properties of auditory processing. Enhancement, as we refer to it, is essentially a perceptual effect in which a spectral region in a complex sound "pops out" when that region is preceded by its spectral complement. The well-known ASA demonstration "cancelled harmonics" is an example: reinserting a deleted component in an equal-amplitude harmonic complex makes that component perceptually salient (see Hartmann and Goupell 2006). The general question concerns the mechanisms/processes that underlie this rather striking phenomenon. In an effort to quantify the salience of this enhancement, Viemeister (1980) showed that the masked threshold for a target component in a harmonic complex was reduced by 10 dB when that complex was preceded by the same complex but with the target component deleted. It was also shown that the threshold reduction is robust and, specifically, does not require harmonic complexes, e.g., the detection threshold for a band of noise masked by a surrounding band-reject noise is comparably reduced when the band-reject noise is a precursor/adaptor/conditioner. Following Wright et al. (1993), we will refer generally to the reduction in detection threshold as "signal enhancement." Signal enhancement is the focus of the experiments in this chapter.

Although the research has been intermittent, there has been considerable effort towards further exploring enhancement and enhancement-related phenomena. A brief, selected review of the psychophysical literature follows. Viemeister and Bacon (1982) showed that an enhanced component in a harmonic complex produces more forward masking than when there is no precursor ("masker enhancement"). They proposed that the results indicate adaptation-of-suppression, i.e., the effective level of the enhanced forward masker is increased because suppression by neighboring components is reduced by adaptation of those components by the precursor. Summerfield et al. (1987) showed that vowel identification was possible for a spectrally flat harmonic complex when that complex was preceded by the spectral complement of the target vowel. This is similar to the auditory afterimage effect reported by Wilson (1970). Carlyon (1989) examined the effects of precursor level and showed that, over the range of levels they examined, there was little effect on the amount of signal enhancement. They proposed that signal enhancement reflects a grouping/segregation process rather than adaptation. Thibodeau (1991) showed that listeners with cochlear impairment showed no masker enhancement when the masker and probe were presented in frequency regions of hearing loss. Presumably these listeners also showed no psychophysical suppression in those regions. Similarly, Wright et al. (1993) and Wright (1996) thoroughly examined the possible role of suppression in normal-hearing listeners and showed that the amounts of suppression and signal enhancement were not significantly correlated. Byrne et al. (2011) examined enhancement using a binaural centering task and showed that lateralization of a target component in an inharmonic complex was shifted to the ear in which a precursor (without the target component) was presented. They also showed that the amount of enhancement was similar for signal enhancement, masker enhancement, and centering.

There have been several physiological studies that have pursued enhancement-like effects. Using recordings of single auditory nerve fibers in anesthetized guinea pigs, Palmer et al. (1995) showed that the response of single auditory nerve fibers does not show enhancement-like effects. Nelson and Young (2010), however, showed strong correlates of enhancement in single-unit recording in the inferior colliculus (IC) of awake marmoset monkeys. They proposed, similar to the adaptation-of-suppression hypothesis, that the increased response reflected adaptation of inhibition.

Overall, the published data strongly suggest that enhancement, as we use the term, is an auditory phenomenon that reflects a basic property of audition, namely, that functional frequency selectivity, and frequency resolution is a dynamic process, one that is determined by the spectro-temporal context of the input. The existing data also suggest that enhancement reflects "low-level" auditory processing, probably resulting from adaptation of inhibition/suppression.

Two experiments on signal enhancement are presented. The first examines the effects of varying the notch width of the precursor and masker. The second examines the effects of precursor-masker level differences and the type of precursor. The motivation for these experiments is to further describe the properties of enhancement as they relate to frequency selectivity, to relate them to the physiological data, and to address issues related to processes such as stream segregation.

2 General Methods and Procedure

In both experiments, detection thresholds for a 2-kHz signal were obtained using a 2IFC tracking procedure that estimated the signal level necessary for 70.7 % correct responses (see Byrne et al. 2011, for further details). The masker was an inharmonic complex that ranged from 0.5 to 8 kHz with a component spacing of 0.1 octave. The phases of the components were random over presentations. The masker and the signal were gated synchronously with 10-ms raised-cosine ramps.

The precursors, levels, and durations were different for the two experiments and will be described later. For both experiments, the precursor lasted 500 ms including 10-ms cosine ramps. The delay between the precursor and masker was 20 ms between the beginning of the off ramp for the precursor and the termination of the on ramp for the masker.

Four listeners with normal hearing participated in both experiments. Stimuli were generated with MATLAB on a PC using a 24-bit sound card and were presented monaurally in a sound attenuating chamber via Sony MDR-V6 headphones (levels are specified based on the output of the headphones at 1 kHz).

Correct-answer feedback was provided in all conditions. Listeners were trained to stable performance and were paid for their participation. Final thresholds are based on four runs for each condition. The order of the conditions was random over the multiple 1–2-h sessions. There was little across-listener variability in thresholds and so only averaged thresholds are presented.

3 Experiment 1: Effects of Notch Width

The primary motivation for this experiment was to compare, using similar stimuli, the human behavioral results with those obtained from the unit recordings in IC by Nelson and Young (2010). They varied the width of the spectral notch in an inharmonic complex with 0.1-octave spacing with the notch centered in octaves about the best frequency of the unit. The present experiment was similar except that the signal frequency was fixed and detection thresholds rather than response magnitude (firing rate) were measured.

As in Nelson and Young (2010), the notch width of the masker and the precursor were identical and varied from 0.2 to 1.8 octaves in 0.4-octave steps. The notch was symmetrical in octaves about 2 kHz. In a condition in which there was no notch, the signal was added in 90° phase to the masker component at 2 kHz. The levels of the precursor and masker were 70 dB SPL per component and the masker and probe durations were 250 ms.

Figure 19.1 shows thresholds as a function of masker/precursor notch width (left panel). As expected based on the many experiments using notched noise to estimate frequency selectivity, there is a monotonic decrease in threshold with increasing notch width for both the enhanced condition (open circles) and the reference (unenhanced condition, filled squares). Importantly, there is a significant effect of notch width on enhancement [$F(5,15)=10.72$, $p<0.001$]. This is clearly indicated in the right panel where the amount of enhancement, the difference in thresholds between the unenhanced and enhanced thresholds, shows a clear dependence on notch width:

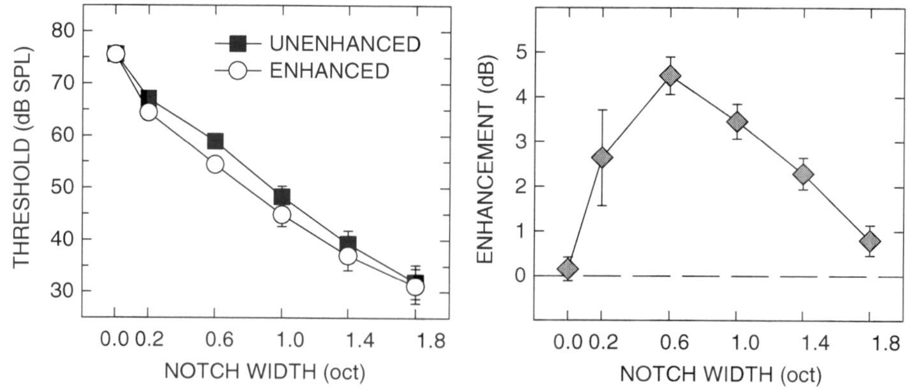

Fig. 19.1 *Left panel*: mean thresholds across four listeners as a function of precursor/masker notch width for the unenhanced (*filled squares*) and enhanced (*open circles*) conditions. *Right panel*: threshold differences (unenhanced-enhanced) vs. notch width. *Error bars* represent the standard errors

at 0.6 octave, the mean threshold in the enhanced condition is 4.5 dB lower than in the unenhanced condition.

These results are in good agreement, qualitatively, with those of Nelson and Young (2010), which showed maximum enhancement for a notch width of 0.5 octave. The agreement between the IC data and the present data further suggests that signal enhancement as observed in humans is a low-level process and as such may reflect a basic property of hearing.

The present data also indicate a strong increase in frequency selectivity under conditions of enhancement. Although the data do not permit a compelling derivation of "auditory filter shapes" and bandwidths, they indicate, according to the assumptions underlying the notched-noise technique (see Patterson and Moore 1986), a reduced effective bandwidth under conditions of enhancement: at 0.6 octave, there is 4.5-dB reduction in the effective excitation being passed by the filter, implying an increase in frequency selectivity. Carlyon (1989) using 5-ms, 1-kHz signals also showed significant enhancement effects with noise precursors of varying notch widths. His data are highly variable but are consistent in showing increased frequency selectivity under conditions of enhancement. In contrast, Moore et al. (1987) argued that the derived auditory filter shape and bandwidth did not change significantly with delay for a 20-ms, 1-kHz signal presented at the onset and in the temporal center of a notched-noise masker. Their conditions were similar to those of the present experiment, and indeed their data show approximately 5-dB-lower thresholds for signals presented in the temporal center of the masker (likely an enhancement condition) when the notch width was approximately 0.6 octave. Since the thresholds appear to be identical at the onset and center of the masker when there is no notch, their results are consistent with the present results and thus imply a reduction in filter bandwidth according to the power summation assumptions that underlie critical band/auditory filter estimates. Why their derived filter shapes and bandwidths are similar under conditions of enhancement is unclear. However, their data, the data of Carlyon (1989), and the present data are consistent with an explanation based on adaptation of suppression/inhibition.

4 Experiment 2: Effects of Precursor Level and Precursor Type

In addition to providing additional parametric data, this experiment addresses an issue that has been raised in several contexts, namely, the role of grouping and segregation in enhancement. The argument, essentially, is that the explanation for enhancement is that the precursor and masker form one stream and that when a new component is introduced it is segregated and thus is more salient, shows a lower threshold, etc. In our opinion, this is an inadequate explanation and merely describes the phenomenon. For example, it seems inconsistent with data indicating that a temporal gap between the precursor and masker that clearly segregates the two still

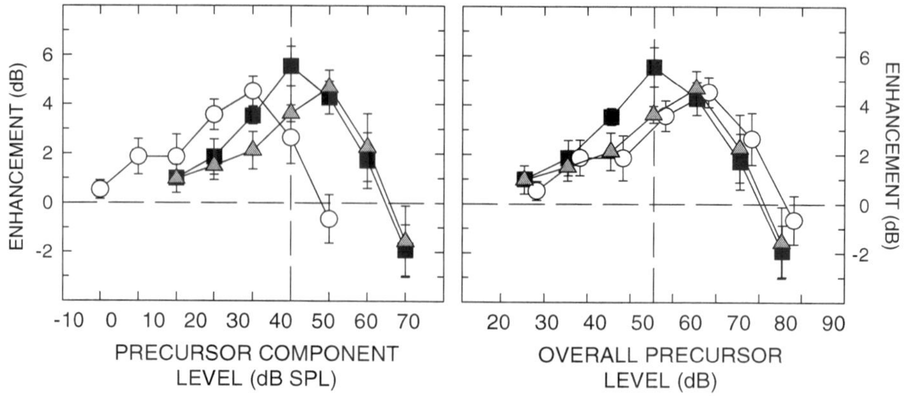

Fig. 19.2 *Left panel*: mean enhancement across five listeners as a function of precursor component level with inharmonic (*filled squares*), harmonic (*shaded triangles*), and notched-noise (*open circles*) precursors. *Error bars* represent the standard error of the mean, and the *vertical dashed line* indicates the level of the inharmonic masker. *Right panel*: the data from the left panel replotted in terms of the overall precursor level

produces substantial enhancement. In any case, in the present experiment, two of the precursors were perceptually different from the masker and certainly were not perceived as being grouped with the masker.

In this experiment, the precursors were either (1) an inharmonic complex spectrally identical to the masker; (2) a harmonic complex with a fundamental frequency of 200 Hz, spanning 0.6 to 8 kHz; or (3) a notched noise, spanning 0.5 to 8 kHz. The notch width for all precursors was 0.6 octave centered at 2 kHz. The levels of the precursors were varied over a 60-dB range but fixed within a block of trials. The masker level was fixed at 40 dB SPL per component and the masker and probe durations were 100 ms. An additional listener was employed for this experiment.

The left panel of Fig. 19.2 shows the amount of enhancement as a function of level. For the notched-noise precursor, the levels are the spectrum levels (1-kHz equivalents) in the passbands. The differences between the functions shown in the left panel are reduced by equating for overall level, as shown in the right panel of Fig. 19.2.

The general characteristics of the data are similar for all three precursors: there is a fairly broad maximum in enhancement at precursor levels close to that of the masker. At very low precursor levels, those approaching a no-precursor situation, there is little enhancement. At high precursor levels, the precursor forward masks the signal and eventually results in negative enhancement.

These results are roughly comparable to those of Carlyon (1989), who examined enhancement-like effects over a narrower range of precursor levels (30 dB). They concluded that because there were no strong level effects, an explanation based on adaptation was implausible and that an explanation based on grouping was more plausible. The logic of their argument remains elusive.

The other notable general characteristic of these data is that the amount of enhancement is similar for all three conditions. It could be argued that when the precursor and masker were spectrally identical, they formed a stream and thus enhanced the salience of the (new) signal/object. This seems unlikely considering that there was a brief gap between the precursor and masker that listeners reported made them temporally distinct. More directly, the harmonic and the band-reject noise precursors were perceptually very dissimilar from the masker and, consistent with listeners reports, almost certainly were not grouped together.

5 Discussion

These results suggest that signal enhancement results from processes that occur at early stages in auditory processing and manifest themselves as a dynamic increase in frequency selectivity. The effects of varying precursor/masker notch width (Experiment 1) agree qualitatively with unit recordings from the IC and are consistent with a process involving adaptation of inhibition/suppression. The increase in frequency selectivity under conditions of enhancement can be substantial: at a notch width of 0.6 octave, the effective level of the masker is reduced by 4.5 dB, implying a narrowing of the auditory filter. The results from Experiment 2 indicate the robustness of signal enhancement and suggest that signal enhancement does not result from processes such as grouping/segregation and attention. More likely, in our opinion, segregation can, as in the familiar "pop-out" demonstrations, result from low-level processes.

Although it appears that enhancement is a low-level, fundamental aspect of audition, this does not exclude the possibility of top-down influences. Indeed, the recent psychophysical research on overshoot (Strickland 2001; Jennings et al. 2009), a well-known phenomenon that is likely related to signal enhancement, suggests an involvement of cochlear efferents.

Acknowledgment This work was supported by Research Grant No. R01 DC 00683 from the National Institute on Deafness and Communication Disorders, National Institutes of Health.

References

Byrne AJ, Stellmack MA, Viemeister NF (2011) The enhancement effect: evidence for adaptation of inhibition using a binaural centering task. J Acoust Soc Am 129:2088–2094

Carlyon RP (1989) Changes in the masked thresholds of brief tones produced by prior bursts of noise. Hear Res 41:223–236

Hartmann WM, Goupell MJ (2006) Enhancing and unmasking the harmonics of a complex tone. J Acoust Soc Am 120:2142–2157

Holt LL, Lotto AJ, Kluender KR (2001) Influence of fundamental frequency on stop consonant voicing perception: a case of learned covariation or auditory enhancement? J Acoust Soc Am 109:764–774

Jennings SG, Strickland EA, Heinz MG (2009) Precursor effects on behavioral estimates of frequency selectivity and gain in forward masking. J Acoust Soc Am 125:2172–2181

Moore BCJ, Poon PWF, Bacon SP, Glasberg BR (1987) The temporal course of masking and the auditory filter shape. J Acoust Soc Am 81:1873–1880

Nelson PC, Young ED (2010) Neural correlates of context-dependent perceptual enhancement in the inferior colliculus. J Neurosci 30:6577–6587

Palmer AR, Summerfield Q, Fantini DA (1995) Responses of auditory-nerve fibers to stimuli producing psychophysical enhancement. J Acoust Soc Am 97:1786–1799

Patterson RD, Moore BCJ (1986) Auditory filters and excitation patterns as representations of frequency resolution. In: Moore BCJ (ed) Frequency selectivity in hearing. Academic, London

Strickland EA (2001) The relationship between frequency selectivity and overshoot. J Acoust Soc Am 109:2062–2073

Summerfield Q, Sidwell A, Nelson T (1987) Auditory enhancement of changes in spectral amplitude. J Acoust Soc Am 81:700–708

Thibodeau LM (1991) Performance of hearing-impaired persons on auditory enhancement tasks. J Acoust Soc Am 89:2843–2850

Viemeister NF (1980) Adaptation of masking. In: van den Brink G, Bilsen FA (eds) Psychophysical, physiological and behavioural studies in hearing. Delft University Press, Noordwijkerhout, pp 190–199

Viemeister NF, Bacon SP (1982) Forward masking by enhanced components in harmonic complexes. J Acoust Soc Am 71:1502–1507

Wilson JP (1970) An auditory after-image. In: Plomp R, Smoorenburg GF (eds) Frequency analysis and periodicity detection in hearing. Sijthoff, Leiden, pp 303–318

Wright BA (1996) Correlated individual differences in conditions used to measure psychophysical suppression and signal enhancement. J Acoust Soc Am 100:3295–3303

Wright BA, McFadden D, Champlin CA (1993) Adaptation of suppression as an explanation of enhancement effects. J Acoust Soc Am 94:72–82

Chapter 20
Enhancement of Increments in Spectral Amplitude: Further Evidence for a Mechanism Based on Central Adaptation

Samuele Carcagno, Catherine Semal, and Laurent Demany

Abstract The threshold for detecting a tone in a multitone masker is lower when the masker-plus-signal stimulus is preceded by a copy of the masker. One potential explanation of this "enhancement" phenomenon is that the precursor stimulus acts as a "template" of the subsequent masker, thus helping listeners to segregate the signal from the masker. To assess this idea, we measured enhancement for precursors that were perceptually similar to the masker and for precursors that were made dissimilar to the masker by gating their components asynchronously. We found that the two types of precursor produced similar amounts of enhancement. This was true not only when the precursor and the subsequent test stimulus were presented to the same ear but also when they were presented to opposite ears. In a second experiment, we checked that the precursors with asynchronously gated components were perceptually poor templates of the subsequent maskers. Listeners now had to discriminate between test stimuli containing the same components as the precursor and test stimuli containing all but one of the precursor components. We found that in this experimental situation, where enhancement could play no role, gating the precursor components asynchronously disrupted performance. Overall, our results are inconsistent with the hypothesis that precursors producing enhancement are beneficial because they are used as perceptual templates of the masker. Our results are instead consistent with an explanation of enhancement based on selective neural adaptation taking place at a central locus of the auditory system.

S. Carcagno, PhD (✉) • C. Semal • L. Demany, PhD
Institut de Neurosciences Cognitives et Intégratives d'Aquitaine (UMR CNRS 5287), Université de Bordeaux, 146 rue Léo Saignat,
Bordeaux F-33076, France
e-mail: samuele.carcagno@u-bordeaux2.fr

B.C.J. Moore et al. (eds.), *Basic Aspects of Hearing*,
Advances in Experimental Medicine and Biology,
DOI 10.1007/978-1-4614-1590-9_20,

1 Introduction

In audition, a contrast effect known as "enhancement" occurs when a test sound consisting of simultaneous pure tones is preceded by the same sound with one component deleted: the component that was deleted in the precursor sound perceptually "pops out" in the test sound (Viemeister and Bacon 1982), and its detection threshold within the test sound is lower than in the absence of the precursor (Viemeister 1980). Enhancement is observable even when the precursor and test sounds are separated by several seconds (Viemeister 1980) or are presented to opposite ears (Erviti et al. 2011).

The enhancement phenomenon may result from neural adaptation: following the precursor, adaptation is expected to be weaker for neurons that will respond to the target of enhancement than for neurons responding to the other components of the test sound; this should reduce the masking effect of the latter components. However, non-sensory phenomena may also be involved. In a signal-detection task, for example, the precursor may provide a reference "template" that could help listeners segregate the masker from the signal. If that is the case, then the advantage provided by the precursor should be eliminated or reduced when the precursor is made perceptually dissimilar to the following masker while remaining similar to it in terms of power spectrum. Contrary to this hypothesis, Summerfield et al. (1987) found that for a harmonic test stimulus, the enhancement produced by a harmonic precursor was equivalent to the enhancement produced by a noise precursor with the same spectral envelope. Also, Byrne et al. (2011) showed that enhancement is still present when the target tone in the test sound is physically more intense than the nontarget tones, so that it is already perceptually segregated from them. In the latter two studies, however, the test sound was presented *immediately* after the precursor. As a consequence, enhancement could have been due entirely to *peripheral* adaptation. Perceptual similarity between the precursor and test sounds may play a more important role when the interstimulus interval (ISI) is longer. In support of this hypothesis, Kidd et al. (2011), using a 250-ms ISI, found that enhancement was reduced when the precursor, rather than being an exact copy of the tonal masker, consisted of a noise with a spectral notch centred on the signal frequency. The enhancement produced by a precursor presented contralaterally to the test sound (i.e. to the opposite ear) cannot be explained by peripheral adaptation and may also depend crucially on the perceptual similarity between the two sounds. The aim of the study reported here was to assess the role played by the perceptual similarity between precursor and test in ipsilateral and contralateral enhancement. To this end, in the main experiment, we measured the enhancement produced by precursor bursts that were either perceptually similar to the test sound or were made perceptually dissimilar to the test sound by gating their components asynchronously.

2 Experiment 1

2.1 Method

Eight normally hearing listeners, including author SC, were tested. They ranged in age between 19 and 29 years (mean = 23). Enhancement was measured using a simultaneous masking paradigm. Listeners had to detect a 100-ms pure tone gated synchronously with a 100-ms masker in a two-interval, two-alternative forced-choice task (see Fig. 20.1). The two observation intervals were separated by a 500-ms ISI and were marked by lights on a computer screen. Visual feedback was provided at the end of each trial.

The masker consisted of a lower and an upper frequency band. These bands were composed of three pure tones each, and it was placed symmetrically around the signal frequency. The spacing between the three tones in each masker band was 100 cents (1 cent = 1/1,200 octave), while the distance between the signal and the masker components closest to it was 350 cents. The level of each masker component was 50 dB SPL. The signal was roved in frequency between 600 and 2,400 Hz. In order to investigate the effect of frequency region, we used three interleaved adaptive tracks estimating thresholds separately in a LOW (600–952 Hz), a MID (952–1,512 Hz) and a HIGH (1,512–2,400 Hz) frequency region. In each track, the level of the signal was initially set at 60 dB SPL and was then varied adaptively using a 2-down 1-up rule targeting the 70.7 % correct point on the psychometric function (Levitt 1971). Track selection was pseudorandom, with a maximum of

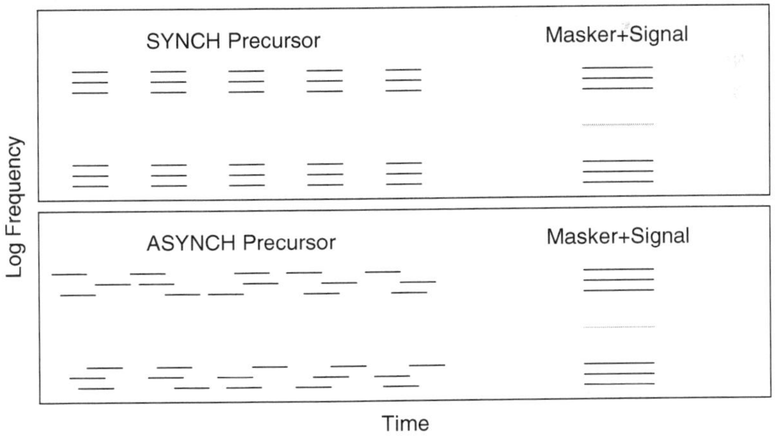

Fig. 20.1 Stimuli used in Experiment 1. Listeners had to detect a 100-ms signal (*grey line*) embedded in a 100-ms masker. The masker + signal stimulus could be preceded by a synchronous precursor (*top panel*), an asynchronous precursor (*bottom panel*) or silence (not shown)

three consecutive trials per track permitted. The step size was 4 dB for the first four turnpoints and 2 dB thereafter. A block of trials was terminated when at least 16 turnpoints per track had occurred.

Each trial began with the presentation of a copy of the signal, at 50 dB SPL, in order to eliminate uncertainty about the signal frequency. The first observation interval started 500 ms after the offset of this signal cue. There were three precursor types: SYNCH, ASYNCH and SILENT. In the SYNCH conditions, the precursor was an exact copy of the masker, except that its duration was 50 ms. It was presented five times before the test sound, with an ISI of 62.5 ms between precursor bursts. In the ASYNCH conditions, the components of each precursor burst, rather than being gated simultaneously, had a 12.5-ms onset asynchrony; there was no ISI between the precursor bursts. The order in which the six precursor components were successively gated in each ASYNCH burst was random. For both the SYNCH and the ASYNCH conditions, the time interval between the middle time point of the last precursor burst and the onset of the test sound was 256.25 ms. As a consequence, the ISI between the offset of the last precursor burst and the onset of the test sound was 231.25 ms in the SYNCH conditions and 200 ms in the ASYNCH conditions. In the SILENT conditions, the test sound was separated from the beginning of the observation interval by 793.75 ms of silence. All tones were gated with 10-ms raised-cosine ramps.

In the SYNCH and ASYNCH conditions, the precursor was presented either to the same ear as the test sound ("*Ipsi*") or to the opposite ear ("*Contra*"), while the initial signal cue was always presented to the same ear as the test sound. Listeners completed twelve sessions. During each session, they completed one block of trials for each precursor type (SILENT, *Ipsi* SYNCH, *Contra* SYNCH, *Ipsi* ASYNCH and *Contra* ASYNCH); these five blocks were randomly ordered. The first two sessions were considered as practice sessions, and the final thresholds were computed as the arithmetic average of the remaining ten threshold estimates, for each precursor type and frequency region. The stimuli were presented via TDH-39 headphones fitted with audiocups that ensured no interaural crosstalk at the presentation levels we used.

2.2 Results

Figure 20.2 displays enhancement magnitude, defined as the difference in threshold between a given condition with a nonsilent precursor and the corresponding SILENT condition. Overall, enhancement magnitude was about 4 dB in the *Ipsi* case and 2 dB in the *Contra* case. Averaged across frequency regions, enhancement was significantly greater than zero for each precursor type [*Ipsi* SYNCH $t(7)=8.66$, $p<0.001$; *Ipsi* ASYNCH $t(7)=9.08$, $p<0.001$; *Contra* SYNCH $t(7)=4.65$, $p=0.002$; *Contra* ASYNCH $t(7)=3.72$, $p=0.007$]. Thus, some enhancement was obtained even when the precursor and test sounds were perceptually dissimilar and/or presented to opposite ears.

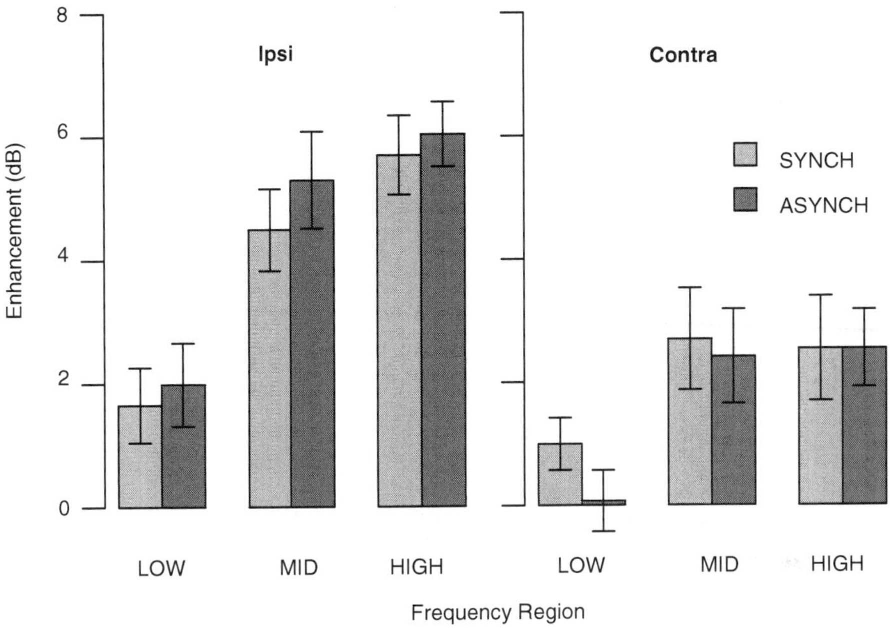

Fig. 20.2 Mean enhancement magnitude in Experiment 1 (± 1 s.e. of the mean)

The enhancement data were entered in a repeated-measures ANOVA with laterality (*Ipsi* vs. *Contra*), synchronicity (SYNCH vs. ASYNCH) and frequency region (LOW, MID or HIGH) as within-subject factors. The ANOVA revealed a significant main effect of laterality [$F(1, 7)=52.5$, $p<0.001$] and frequency region [$F(2, 7)=10.24$, $p=0.002$], but no main effect of synchronicity [$F(1, 7)=0.01$, $p=0.92$]. The interaction between laterality and synchronicity was significant [$F(1, 7)=9.64$, $p=0.02$], as well as the interaction between laterality and frequency region [$F(2, 14)=14.56$, $p<0.001$], while the other interactions were not. These results indicate that enhancement was overall stronger in the *Ipsi* than in the *Contra* conditions and that the effects of synchronicity and frequency region were dependent on the laterality factor.

In order to investigate these interactions, we performed separate ANOVAs for the *Ipsi* and *Contra* conditions. For the *Ipsi* conditions, there was no significant effect of synchronicity [$p=0.38$], while the effect of frequency region was highly significant [$F(2, 7)=18.45$, $p<0.001$]. The interaction between synchronicity and frequency region was not significant [$p=0.75$]. Follow-up *t*-tests (corrected with the Holm procedure) indicate that enhancement was significantly less strong in the LOW frequency region than in both the MID [$t(7)=-4.99$, $p=0.003$] and the HIGH [$t(7)=-5.35$, $p=0.003$] frequency regions, which did not significantly differ from each other [$p=0.2$]. In the *Contra* conditions, as in the *Ipsi* conditions, there was no main effect of synchronicity [$p=0.27$] and no significant interaction of synchronicity and frequency region [$p=0.44$], but a significant main effect of frequency region

[$F(2,14)=4.34$, $p=0.03$]. Follow-up t-tests indicate that enhancement was significantly less strong in the LOW region than in the MID region [$t(7)=-3.22$, $p=0.042$]; the other contrasts were not significant.

Overall, therefore, similar patterns of results were obtained in the *Ipsi* and *Contra* conditions, despite the significant interactions found in the main ANOVA. The significant interaction between laterality and synchronicity reflects the fact that whereas in the *Ipsi* conditions there was a trend for greater enhancement with asynchronous than with synchronous precursors, the opposite was true in the *Contra* conditions.

3 Experiment 2

The rationale of Experiment 1 was based on the assumption that while a synchronous precursor could be easily used as a template of the following masker, this was not the case for an asynchronous precursor. The validity of this assumption was tested in Experiment 2. Whereas, in Experiment 1, the listeners' task was to detect the *addition* of a tone to a copy of the precursor, the task in Experiment 2 was to detect the *subtraction* of a tone from a copy of the precursor. In the latter task, enhancement could play no role (Summerfield et al. 1984). We reasoned that if, contrary to our assumption, synchronous and asynchronous precursors could be used equally well as templates of the following masker in Experiment 1, then they should also be equivalent in Experiment 2.

3.1 Method

Seven of the listeners who had taken part in Experiment 1 were tested. As in Experiment 1, listeners were presented with five bursts of a synchronous or an asynchronous precursor, followed by a synchronous test sound. However, each precursor burst now had the same frequency components as a *test* sound of Experiment 1; therefore each precursor burst now had seven frequency components. The test sound either had the same seven components or did not contain the central one. On each trial, a single precursor-test sound sequence was presented, and listeners had to judge whether the precursor and test contained the same frequency components or not. Feedback was provided on each trial.

Again, synchronous and asynchronous precursors were used in different blocks of trials. However, all stimuli were now presented diotically. The duration of each precursor and test sound component was the same as in Experiment 1. As each precursor burst now had seven components, its duration in the ASYNCH condition was 12.5 ms longer than in Experiment 1. Therefore, the inter-burst ISI in the SYNCH condition was also increased by 12.5 ms. In the ASYNCH condition, the ISI between the offset of the last precursor component and the onset of the test sound was 200 ms.

This ISI was set to 237.5 ms in the SYNCH condition, so as to equalize, for the two conditions, the time interval between the middle point of the last precursor burst and the onset of the test sound. The central component of the precursor and of the test sound (when the test sound had seven components) had the same intensity level; this level was adjusted for each listener in a preliminary phase of the experiment so as to avoid floor or ceiling effects. The level of all the other precursor and test sound components was set to 50 dB SPL. Listeners completed eight blocks of 50 trials for each condition, in random order.

3.2 Results

The average d' in the SYNCH condition was 2.1, while in the ASYNCH condition it was 1.5. This difference was statistically significant [$t(6)=3.01$, $p=0.02$]. Six out of the seven listeners performed better in the SYNCH condition than in the ASYNCH condition. The remaining listener showed a weak trend in the opposite direction.

4 Discussion

In Experiment 1, the enhancement produced by the asynchronous precursors was very similar in magnitude to that produced by the synchronous precursors. Yet, as shown by Experiment 2, it was more difficult to compare the frequency contents of the precursor and test sounds when the precursors were asynchronous; perceptually, the asynchronous precursors were thus poorer templates of the following maskers. Our study therefore indicates that enhancement is not critically dependent on the perceptual similarity between the precursor and test sounds, even when these two sounds are presented to opposite ears. In other words, our data are at odds with the idea that enhancement can be understood as the consequence of a comparison between sounds (the precursor and the test) which are processed as integrated "auditory objects".

What is, then, the origin of enhancement? One possibility is that enhancement is due to the activation of the medial olivocochlear efferent reflex (MOCR) by the precursor sound (Strickland 2004). The MOCR would cause a frequency-specific reduction in the gain of the cochlear amplifier, thus decreasing the ability of the nontarget components of the test sound to mask the target component. If, as some evidence suggests (Hicks and Bacon 1999), the cochlear amplifier has a weaker action at low frequencies than at high frequencies, the MOCR hypothesis could explain the increase in enhancement that we observed in the higher-frequency regions. However, recent studies that have investigated the frequency tuning of the MOCR in humans using otoacoustic emissions do not provide much evidence that the MOCR is frequency specific (Guinan 2011). Therefore, the MOCR is unlikely to explain enhancement.

Another possibility is that enhancement results from some form of adaptation (which could be "adaptation of inhibition"; Byrne et al. 2011). Importantly, as we measured enhancement at relatively long ISIs (≥200 ms) and across the ears, the adaptation in question should take place centrally. A recent neurophysiological study (Nelson and Young 2010) indicates that enhanced neural responses can be observed at the level of the inferior colliculus. These responses were measured only at a 0-ms ISI between the precursor and test sounds, and it is not obvious that they could also be obtained at long ISIs (in the order of seconds), for which psychophysical enhancement is still present (Viemeister 1980). However, stimulus-specific adaptation with long time constants has been recently observed at several levels of the auditory system, from the inferior colliculus (Malmierca et al. 2009) to the auditory cortex (Ulanovsky et al. 2004). These adaptation effects could potentially account for the enhancement observed psychophysically at long ISIs.

Acknowledgements This work was supported by a grant from the Agence Nationale de la Recherche (LEAP, Programme Blanc 2010).

References

Byrne AJ, Stellmack MA, Viemeister NF (2011) The enhancement effect: evidence for adaptation of inhibition using a binaural centering task. J Acoust Soc Am 129:2088–2094

Erviti M, Semal C, Demany L (2011) Enhancing a tone by shifting its frequency or intensity. J Acoust Soc Am 129:3837–3845

Guinan JJ (2011) Physiology of the medial and lateral olivocochlear systems. In: Ryugo DK, Fay RR (eds) Auditory and vestibular efferents. Springer, New York, pp 39–81

Hicks ML, Bacon SP (1999) Psychophysical measures of auditory nonlinearities as a function of frequency in individuals with normal hearing. J Acoust Soc Am 105:326–338

Kidd G, Richards VM, Streeter T, Mason CR, Huang R (2011) Contextual effects in the identification of nonspeech auditory patterns. J Acoust Soc Am 130:3926–3938

Levitt H (1971) Transformed up-down methods in psychoacoustics. J Acoust Soc Am 49:467–477

Malmierca MS, Cristaudo S, Pérez-González D, Covey E (2009) Stimulus-specific adaptation in the inferior colliculus of the anesthetized rat. J Neurosci 29:5483–5493

Nelson PC, Young ED (2010) Neural correlates of context-dependent perceptual enhancement in the inferior colliculus. J Neurosci 30:6577–6587

Strickland EA (2004) The temporal effect with notched-noise maskers: analysis in terms of input–output functions. J Acoust Soc Am 115:2234–2245

Summerfield Q, Haggard M, Foster J, Gray S (1984) Perceiving vowels from uniform spectra: phonetic exploration of an auditory aftereffect. Percept Psychophys 35:203–213

Summerfield Q, Sidwell A, Nelson T (1987) Auditory enhancement of changes in spectral amplitude. J Acoust Soc Am 81:700–708

Ulanovsky N, Las L, Farkas D, Nelken I (2004) Multiple time scales of adaptation in auditory cortex neurons. J Neurosci 24:10440–10453

Viemeister NF (1980) Adaptation of masking. In: van den Brink G, Bilsen FA (eds), Psychophysical, physiological and behavioural studies in hearing. Delft University Press, Delft, pp 190–199

Viemeister NF, Bacon SP (1982) Forward masking by enhanced components in harmonic complexes. J Acoust Soc Am 71:1502–1507

Chapter 21
The Role of Sensitivity to Transients in the Detection of Appearing and Disappearing Objects in Complex Acoustic Scenes

Francisco Cervantes Constantino, Leyla Pinggera, and Maria Chait

Abstract We report a series of psychophysics experiments that investigated listeners' sensitivity to changes in complex acoustic scenes. Specifically, we sought to test the hypothesis that change detection is supported by sensitivity to change-related transients (an abrupt change in stimulus power within a certain frequency band, associated with the appearance or disappearance of a scene element). This hypothesis, in the context of natural scenes, is commonly dismissed on account that the elements of the scene may themselves be characterized by on-going energy fluctuations that would mask any genuine change-related transients. We created artificial 'scenes' populated by multiple pure-tone components. Tones were modulated (by a square wave at a distinct rate) so as to mimic the fluctuation properties of complex sounds. "Change" was defined as the appearance or disappearance of one such element. Importantly, such scenes lack semantic attributes, which may have been a limiting factor in interpreting previous auditory change-detection studies, thus allowing us to probe the low-level, pre-semantic, processes involved in auditory change perception. In Experiment 1 we measured listeners' ability to detect item appearance and disappearance in conditions where change-related transients are masked by a silent gap. In Experiment 2, we investigated the effect of an acoustic distractor – a brief signal that occurs at the time of change, but does not mask any scene components. The data show that gaps adversely affected the processing of item appearance but not disappearance. However, distractors reduced both

F.C. Constantino
Department of ECE, University of Maryland,
College Park, MD 20742, USA

L. Pinggera
University Clinic for Ear, Nose and Throat Medicine,
Anichstrasse 35, Innsbruck 6020, Austria

M. Chait (✉)
UCL Ear Institute, 332 Gray's Inn Rd,
London WC1X 8EE, UK
e-mail: m.chait@ucl.ac.uk

B.C.J. Moore et al. (eds.), *Basic Aspects of Hearing*,
Advances in Experimental Medicine and Biology,
DOI 10.1007/978-1-4614-1590-9_21, © Springer Science+Business Media New York 2013

appearance and disappearance detection. Together our results suggest a role for sensitivity to transients in the process of auditory change detection, similar to what has been demonstrated for visual change detection.

1 Introduction

The ability to detect abrupt changes in our surroundings has serious implications for survival. These processes have received considerable attention in vision (Rensink 2000; Simons and Rensink 2005) but remain poorly understood in the context of acoustic scenes, despite the fact that the auditory system is commonly assumed to be the brain's early warning device, rapidly directing attention to new events in the scene (Spence and Driver 1997). We are aware of only a handful of studies that have directly examined listeners' sensitivity to changes in scene contents (Eramudugolla et al. 2005; Gregg and Samuel 2008; Pavani and Turatto 2008). These investigations used 'acoustic scenes' comprised of concurrently presented naturalistic sounds (animal or human vocalisations, environmental sounds, etc.). Participants were instructed to detect changes, manifested as the appearance, disappearance or switch in the location of an object within the scene. Curiously, performance was poor, even for very small scenes (4 objects), unless the subject knew the identity of the change in advance, suggesting, rather surprisingly, that the auditory system is not as sensitive to change in the structure of scenes as often assumed.

However, these studies are limited by the use of easily identifiable natural sounds. The use of natural sounds in the laboratory poses several problems. First, it is hard to control their physical parameters (indeed they were not controlled in previous work) and therefore impossible to relate specific effects to underlying stimulus properties. Second, the sounds were familiar and associated with semantic labels (in fact in most previous studies, subjects were explicitly encouraged and trained to 'label' the sounds, e.g. 'dog', 'train'). It is thus difficult to exclude the possibility that the observed performance limits may be reflecting limits of general working memory rather than a specific auditory change-detection system. To address these issues, we have been using a new paradigm (Cervantes Constantino et al. 2012), based on artificial acoustic scenes, designed to model the dynamics of natural 'soundscapes' but devoid of semantic attributes (Fig. 21.1). This paradigm taps predominantly low-level (pre-semantic) processes involved in detecting a change in scene contents.

This chapter is focused on examining sensitivity to change-related transients. While such sensitivity has been shown to underlie change detection in vision (Yantis and Jonides 1984; O'Regan et al. 1999), in audition, the role of sensitivity to change-evoked transients is generally dismissed on account that natural sounds are inherently characterised by energy fluctuations that would mask any genuine change-related transients. To address this issue, we designed our stimuli such that scene elements contain numerous onsets and offsets, mimicking the fluctuation properties of complex sounds. Physically, appearance and disappearance of a component are associated with

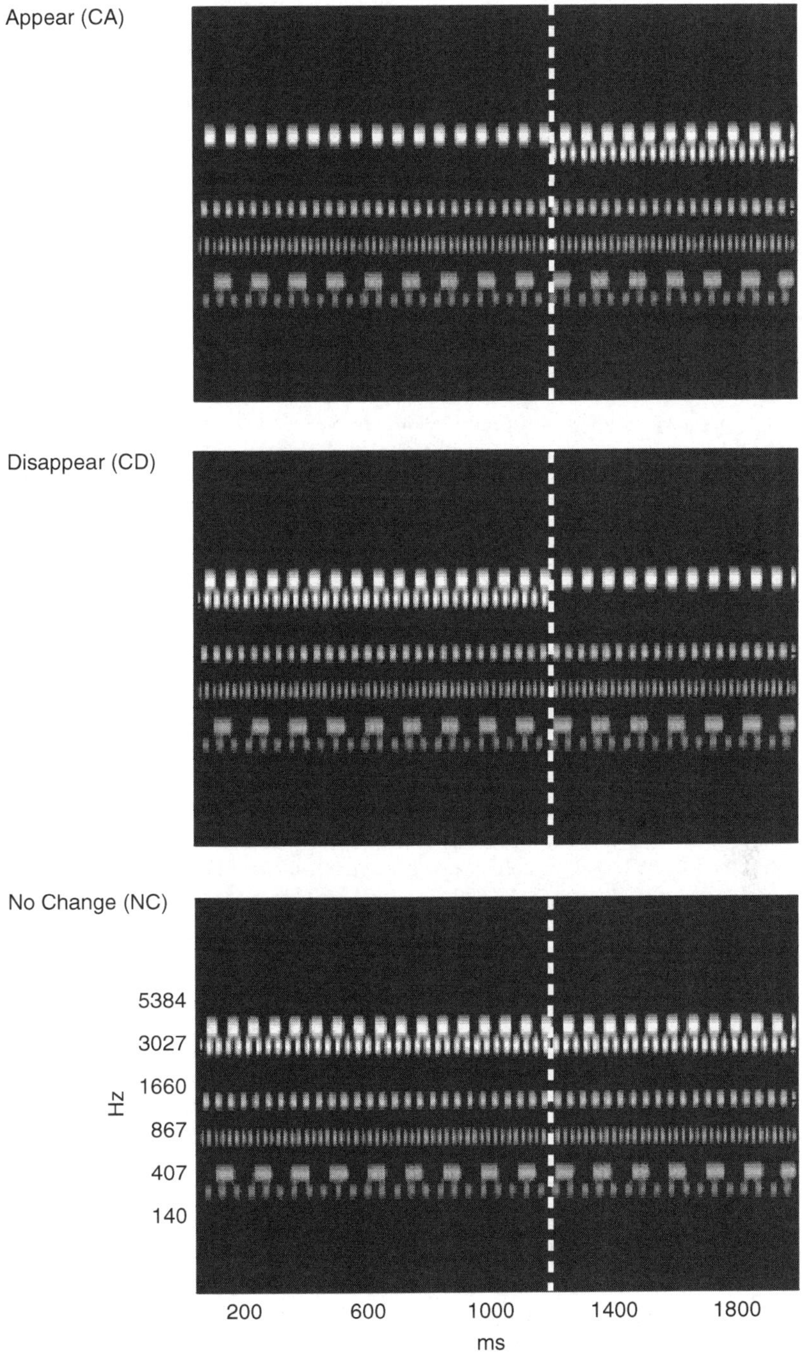

Fig. 21.1 Example of the 'artificial scene' stimuli used. See text for details

a 'local transient' – an abrupt change in stimulus power within a certain frequency band, resulting in a small number of neighbouring cells sharply changing their firing pattern, while the statistics of the activity in the rest of the array are unaltered. A mechanism able to detect such local transients from within the numerous noninformative transients characterising scene elements might support change detection since it is able to indicate the time, the frequency region and the nature of the change (appearance vs. disappearance).

Computationally, detection of item appearance should be comparatively easy, as it is associated with appearance of energy within a frequency band that was previously silent. Disappearance, on the other hand, is not easily distinguished from the many offsets that occur due to ongoing modulation. Disappearance detection requires a smarter, 'second-order transient' detection mechanism, capable of acquiring the statistical rules of the ongoing sound and signalling when those rules are violated, e.g. when an expected tone pip fails to arrive. The existence of such 'smart' offset detection mechanisms (albeit in the context of a single sequence rather than several concurrent sources), operating automatically irrespective of listeners' attentional focus, has been demonstrated in several recent human brain imaging studies (Chait et al. 2008; Yamashiro et al. 2009; Fujioka et al. 2009), and it has been hypothesised that they might play a role in scene change detection.

2 Stimuli

We use artificial 'scenes' (Fig. 21.1) populated by multiple streams of pure tones that are designed to model acoustic sources (scene size of 4, 8 and 14 sources). Each source is characterised by a different carrier frequency (drawn from a pool of fixed values spaced at 2*ERB) and is furthermore amplitude modulated (AM) by a square wave (the source can be seen as a stream of tone pips) at a distinct rate (between 3 and 35 Hz). The AM mimics temporal properties found across many natural sounds and ensures that, similarly to natural scenes, the stimuli are perceived as a composite 'soundscape' that is perceptually separable, so that each stream can be attended to individually (as verified in a control experiment).

We refer to scenes in which each source is active throughout the stimulus, as 'no change' stimuli (NC). Additionally, versions in which a single component is removed partway through the scene ('change-disappear', CD, stimuli) and versions in which the same single component is added to the scene ('change-appear', CA, stimuli) are created. The timing of change varies randomly. For appearing components, the nominal time of change is set at the introduction of the first non-zero sound sample to the scene; for disappearing components, the time of change is the time at which the next tone burst is expected to appear (dashed line in Fig. 21.1). The choice of frequencies and AM rates is random for each scene, but to enable a controlled comparison between CA and CD, the stimuli are generated as NC/CD/CA triplets (as in Fig. 21.1). These are presented in random order during the experiment (blocked by change type, NC/CA and NC/CD).

3 Experiment 1

In Experiment 1 (part of the data reported in Cervantes Constantino et al. 2012), we tested listeners' ability to detect sudden scene changes, manifested as the appearance or disappearance of an element. We also assessed the extent to which performance is affected by interrupting the scenes (at the time of change) with a silent gap.

3.1 *Materials and Methods*

Ten subjects participated in the experiment (5 female; mean age = 23.8 years).

The stimulus set included two conditions: (a) 'continuous' stimuli constructed as described in Sect. 2, above, and (b) 'silent-gap' stimuli with a 200 ms silent gap inserted at the time of change; gap duration (200 ms) was chosen to be as short as possible so as to minimise reliance on memory capacities but longer than the longest inter-pulse interval (corresponding to the slowest AM rate used) in order to introduce a detectable gap for all scene components. The signals before and after the gap were ramped with a 10 ms cosine-squared ramp. For each gap condition, signals were generated as NC/CD/CA triplets such that NC signals also contained a gap at the same time as their matching CA and CD scenes. Stimulus presentation was blocked by change type and gap type (no gap/silence). The proportion of change events was 50 % in each block. Block order was randomised across listeners. Experimental sessions lasted about 2 h and consisted of a short practice session with feedback, followed by the main experiment with no feedback, divided into runs of about 10 min. Subjects were instructed to fixate at a cross presented on the computer screen and press a keyboard button as soon as possible when they detected a change in the ongoing scene stimulus.

3.2 *Results and Discussion*

The results are presented in Fig. 21.2. Panel a shows data for continuous scenes – listeners are very sensitive to source appearance but have difficulty detecting disappearing sources (the performance advantage is also reflected in response times; not shown). This finding is likely related to previously reported 'enhancement' effects (e.g. Erviti et al. 2011, for review; see also Chap. 19) which refer to the finding that local power change, e.g. an increase in the power of one component within a pure-tone chord, results in perceptual pop-out of the associated component away from the rest of the mixture (see also Bregman et al. 1994). Investigations of enhancement phenomena have usually used much simpler stimuli (pure-tone chords) than those used here, and the present demonstration thus suggests that appearance pop-out is a wide-ranging phenomenon, unaffected by the multitudes of onsets and offsets in our scene components.

The CA advantage likely receives contribution from at least two different low-level neural mechanisms: (a) adaptation effects – changes to the sustained neuronal firing rate, which follows the signal while it is present in the scene. Adaptation

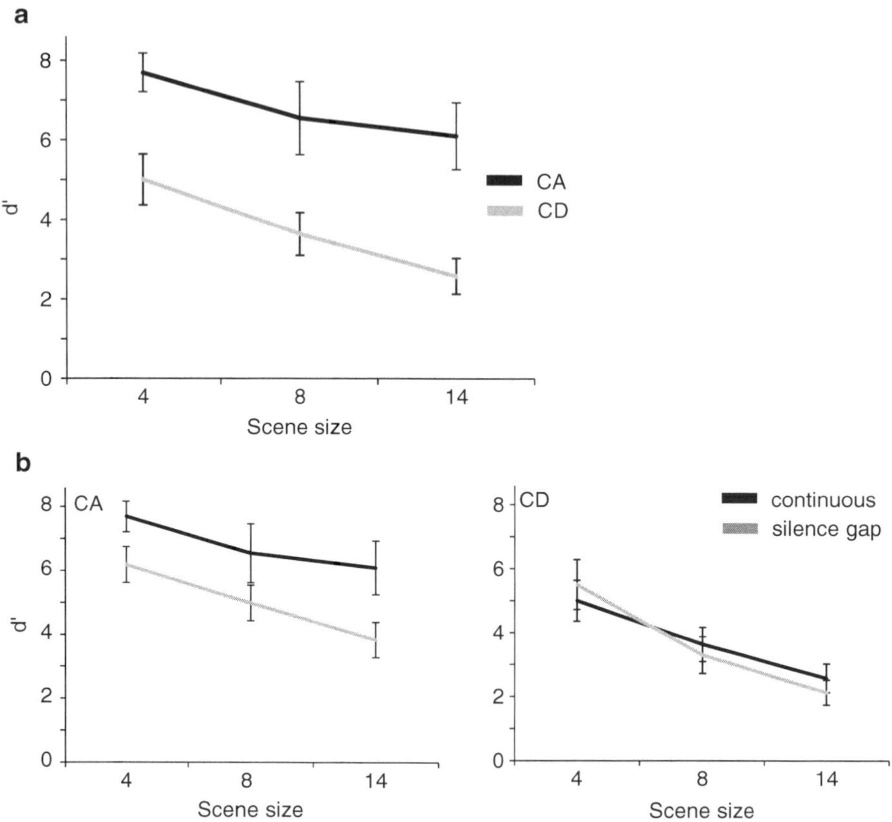

Fig. 21.2 Results of Experiment 1 (**a**) Continous scenes (**b**) comparison between performance oncontinous scenes and silent-gap interrupted scenes

could thus contribute to onset detection by reducing responses to the ongoing (non-changing) scene components. (b) Local transients generated at signal onset – auditory onset-/offset-tuned cell populations are characterised by markedly different properties: offset-tuned cells are fewer in number, and their responses tend to be of longer latency and smaller amplitude (Scholl et al. 2010; Phillips et al. 2002) thus leading to larger, and earlier, on- than off-transient responses. These differences are consistent with our finding that appearance events are overall more detectable.

To assess the degree to which change detection is supported by sensitivity to local transients, we measured change-detection performance in conditions where the scenes are interrupted by a silent gap at the time of change. The silent gap serves as a 'global transient' – disturbing activity in all channels simultaneously and thus masking any change-related local transients. The data from this condition are compared to the continuous condition in Fig. 21.2b. A repeated measures ANOVA showed a main effect of gap ($F(1,9)=16.9$, $p=0.003$), change type ($F(1,9)=147.62$, $p<0.0001$), scene size ($F(2,18)=57.311$, $p<0.001$) as well as an interaction between gap and change type ($F(1,9) = 25.96, p = 0.001$). As is clear from the figure, the silent gap significantly reduced CA detection while performance on CD remained intact (and above floor) in all tested scene sizes. It is also noteworthy that even after introducing the gap, detection of CA remained superior (a difference of $d' = 1.5$) to detection of CD, possibly implying additional effects of ongoing adaptation.

4 Experiment 2

It is possible that the results of Experiment 1 reflect the effects of adaptation rather than sensitivity to transients. The gaps were meant to serve as maskers for the change-related local transient but they might also have caused a resetting of adaptation across the channels thereby reducing CA pop-out. In Experiment 2, we investigate the possible role of sensitivity to change-related transients by using another method for masking the onset/offset transients associated with item appearance or disappearance (inspired by the 'mud-splash' experiments of O'Regan et al. 1999). Instead of a global transient, the scene stimuli in Experiment 2 include a brief interruption that occurred at the time of change but did not mask any scene components. Neural adaptation to the ongoing scene elements is therefore not disturbed; rather, the interruption 'perceptually' masked the change-related transient by temporarily attracting attention.

4.1 Materials and Methods

Ten subjects participated in the experiment (5 female; mean age = 27.4 years). The stimulus set included three conditions: (a) 'continuous stimuli (as in Sect. 3, above), (b) 'silent-gap' stimuli (as in Sect. 3, above) and (c) 'splash' scenes (nomenclature inspired by O'Regan et al. 1999) in which the scene was interrupted by an acoustic distractor ('splash'). The 'splashes' are chords of four, concurrently presented, 200 ms pure tones. Tone frequencies were selected out of the remaining values from the frequency bank used for the scene components (this strategy ensures that 'splash' components did not mask any scene elements) and thus varied from trial to trial. To make the 'splash' components 'stand out' as different from the scene elements, they were additionally amplitude modulated at 100 Hz (depth of 0.5).

Scenes in this experiment were comprised of 4, 6 or 10 components. Procedure and other parameters were as in Experiment 1.

4.2 Results and Discussion

The results of Experiment 2 are presented in Fig. 21.3 for CA and CD changes separately.

Fig. 21.3 Results of Experiment 2

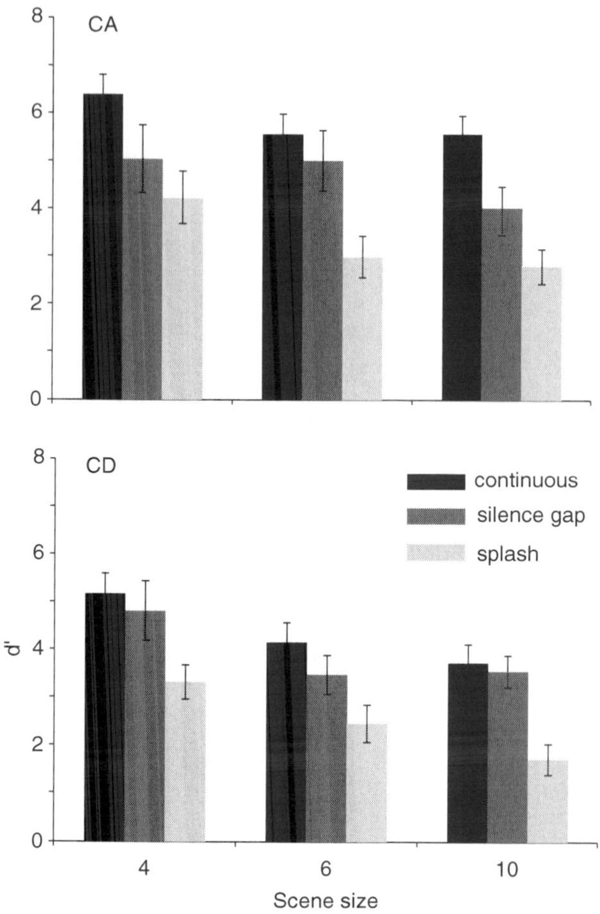

For CA, a repeated measures ANOVA with interruption type and scene size as factors revealed a main effect of interruption ($F(2,18)=20.93$, $p<0.0001$) and a marginally significant main effect of scene size ($F(2,18)=4.09$, $p=0.048$). A post hoc (Bonferroni corrected) pairwise comparison test indicated a significant difference between the 'continuous' and 'splash' conditions ($p<0.0001$) and 'continuous' and 'gap' conditions ($p=0.025$), but not between then 'gap' and 'splash' conditions ($p=0.055$). The data suggest, therefore, that performance was equally affected by 'splash' and 'gap' stimuli.

In the case of CD, a repeated measures ANOVA revealed a main effect of interruption ($F(2,18)=23.69$, $p<0.0001$) and an interaction between interruption and scene size ($F(4,36)=0.33$, $p=0.036$). A post hoc (Bonferroni corrected) pairwise comparison test indicated a significant difference between the 'continuous' and 'splash' conditions ($p<0.0001$) and the 'gap' and 'splash' conditions ($p<0.0001$) but no difference between 'continuous' and 'gap' ($p=0.75$).

The performance on the 'continuous' and 'silent-gap' conditions replicates the results of Experiment 1: gaps produced a decline in CA performance but did not affect CD performance. In contrast, 'splashes' significantly reduced performance on both CA and CD. This pattern of results suggests therefore that gaps and 'splashes' (of the same duration) have differing effects on performance and that 'splashes' are overall more detrimental, potentially because they constitute a new event that might be more effective at capturing attention away from the change events.

Because the 'splashes' were brief and did not physically mask any of the scene components, it is unlikely that this drop in performance is caused by resetting of some form of memory representation of the ongoing scene. Instead, our results provide evidence that despite the multitudes of onset and offset transients characterising the ongoing components, the auditory system is able to identify specific onset/offset events associated with the addition or deletion of scene elements. In the case of appearance detection, this is rather simple in that the system only needs to be able to identify a new event in a channel that was previously inactive. In contrast, in the case of CD, what is required is a mechanism that is able to learn the activation pattern within each channel (duration of tone pips and silence intervals) and respond when an expected tone fails to arrive. While performance is overall rather impaired on this condition (in the larger scene sizes, listeners tend to miss about 50 % of the CD changes; relatively high d' values are produced because they also make few false positives), our evidence strongly indicates that the auditory system, to some extent, is able to make use of such 'smart' disappearance detection mechanisms.

Acknowledgements This work was supported by a Deafness Research UK Fellowship and a Wellcome Trust project grant to MC.

References

Bregman AS, Ahad PA, Kim J (1994) Resetting the pitch-analysis system. 2. Role of sudden onsets and offsets in the perception of individual components in a cluster of overlapping tones. J Acoust Soc Am 96:2694–2703

Constantino CF, Pinggera L, Paranamana S, Kashino M, Chait M (2012) Detection of appearing and disappearing objects in complex acoustic scenes. PLoS ONE 7(9):e46167

Chait M, Poeppel D, Simon JZ (2008) Auditory temporal edge detection in human auditory cortex. Brain Res 1213:78–90

Eramudugolla R, Irvine DR, McAnally KI, Martin RL, Mattingley JB (2005) Directed attention eliminates 'change deafness' in complex auditory scenes. Curr Biol 15:1108–1113

Erviti M, Semal C, Demany L (2011) Enhancing a tone by shifting its frequency or intensity. J Acoust Soc Am 129:3837–3845

Fujioka T, Trainor LJ, Large EW, Ross B (2009) Beta and gamma rhythms in human auditory cortex during musical beat processing. Ann N Y Acad Sci 1169:89–92

Gregg MK, Samuel AG (2008) Change deafness and the organizational properties of sounds. J Exp Psychol Hum Percept Perform 34:974–991

O'Regan JK, Rensink RA, Clark JJ (1999) Change-blindness as a result of 'mudsplashes'. Nature 398:34

Pavani F, Turatto M (2008) Change perception in complex auditory scenes. Percept Psychophys 70:619–629

Phillips DP, Hall SE, Boehnke SE (2002) Central auditory onset responses, and temporal asymmetries in auditory perception. Hear Res 167:192–205

Rensink RA (2000) Seeing, sensing, and scrutinizing. Vision Res 40:1469–1487

Scholl B, Gao X, Wehr M (2010) Nonoverlapping sets of synapses drive on responses and off responses in auditory cortex. Neuron 65:412–421

Simons DJ, Rensink RA (2005) Change blindness: past, present, and future. Trends Cogn Sci 9: 16–20

Spence C, Driver J (1997) Audiovisual links in exogenous covert spatial orienting. Percept Psychophys 59:1–22

Yamashiro K, Inui K, Otsuru N, Kida T, Kakigi R (2009) Automatic auditory off- response in humans: an MEG study. Eur J Neurosci 30:125–131

Yantis S, Jonides J (1984) Abrupt visual onsets and selective attention: evidence from visual search. J Exp Psychol Hum Percept Perform 10:601–621

Chapter 22
Perceptual Compensation When Isolated Test Words Are Heard in Room Reverberation

Anthony J. Watkins and Andrew P. Raimond

Abstract Room reverberation usually degrades speech reception, such as when listeners identify test words from a 'sir'-to-'stir' continuum. Here, substantial reverberation introduces a 'tail' from the [s], which tends to fill the gap that cues the [t], and a degradation effect arises as listeners report correspondingly fewer 'stir' sounds. This effect is particularly clear when test words are preceded by a precursor phrase (e.g. 'next you'll get…') that contains much less reverberation than the test word. When the precursor's reverberation is increased to be the same as in the test word, the degradation diminishes as more 'stir' sounds are heard once again. This last effect has been attributed to a perceptual compensation mechanism that is informed by the precursor's reverberation level. However, a recent claim is that the degradation is caused by 'modulation masking' from precursors with a low level of reverberation. Such masking is likely to diminish when the precursor's reverberation level is raised, because reverberation acts as a low-pass modulation filter. Support for this hypothesis comes from results in conditions where degradation effects seem to be entirely absent, despite substantial reverberation. In these conditions, test words were played in isolation, with no precursor, and reverberation was kept at the same level in the test words of every trial. The experiments reported here have conditions that are similar, except that reverberation in test words is varied unpredictably from trial to trial, so that substantial-level trials are interspersed with trials that have a much lower level of reverberation. The result is that under these conditions, the degradation effect is entirely restored, allowing rejection of the modulation-masking hypothesis. An alternative is that some perceptual compensation comes from reverberation information within test words, and its effects accumulate over sequences of trials as long as the test word's reverberation level stays the same from trial to trial.

A.J. Watkins (✉) • A.P. Raimond
Department of Psychology, University of Reading, Reading RG6 6AL, UK
e-mail: syswatkn@gmail.com

B.C.J. Moore et al. (eds.), *Basic Aspects of Hearing*,
Advances in Experimental Medicine and Biology,
DOI 10.1007/978-1-4614-1590-9_22,

1 Introduction

When a speech message is played at different distances from a listener in a room, the different amounts of reflected sound from the room's surfaces make the temporal envelopes of the signals very different. Nevertheless, with moderate levels of reverberation, these sounds are generally heard to have very similar phonetic content at diverse distances, suggesting that there is a 'perceptual constancy' operation in hearing (Watkins and Makin 2007). This constancy would appear to arise from a mechanism that takes account of the amount of reflected sound in the surrounding context, and it effects some perceptual compensation for the degradation that room reverberation might otherwise have on speech reception. This degradation includes the self- and overlap-masking effects apparent in higher levels of reverberation when listeners identify the consonant at the beginning or the end of a syllable (Nábělek et al. 1989).

One degradation effect from moderate levels of reverberation occurs when listeners identify test words from a 'sir'-to-'stir' continuum (Watkins 2005). Here, reverberation introduces a 'tail' from the [s], which will tend to fill the gap that cues the [t], and a degradation effect arises when the level of reverberation is increased. As this degradation becomes more substantial, listeners report correspondingly more 'sir' sounds, as if they no longer hear the [t] in sounds that were previously heard as 'stir'. This effect is particularly clear when test words are preceded by a precursor phrase (e.g. 'next you'll get…') that contains much less reverberation than the test word. When the precursor's reverberation is increased to be the same as in the test word, the degradation diminishes as more 'stir' sounds are heard once again. This last effect was attributed to a perceptual compensation mechanism that is informed by the precursor's reverberation level.

In a recent study (Nielson and Dau 2010), degrading effects of reverberation on speech reception were not apparent, even though test words from a 'sir'-to-'stir' continuum were used and the level of reverberation was reasonably substantial. However, other aspects of this study were rather more unusual; for example, test words were heard in isolation, with no precursor, and reverberation was kept at the same level in the test words of every trial throughout the experiment. The present experiment tests three hypotheses about why degradation might not be apparent in such conditions:

1. Nielson and Dau's hypothesis was that there are no degrading effects of reverberation for these test words. Rather, effects of different precursors are brought about by a modulation-masking effect (Wojtczak and Viemeister 2005). Such an effect was thought to be more substantial when there is *less* reverberation in the precursor, as reverberation reduces the signal's modulation (Houtgast and Steeneken 1973). These are forward-masking effects, so they should be less substantial when test words are played in isolation than when test words have a precursor.
2. An alternative hypothesis concerns the special nature of conditions in which the test word's reverberation does not vary. This may somehow have led to an amelioration of the effects of reverberation, so it might be possible to restore the

degradation by varying the level of the test word's reverberation in an unpredictable way from trial to trial.

3. A further possibility concerns the special nature of isolated test words. It may be that there is a source of information about reverberation in such sounds, with effects that only become apparent when there is no overriding information from a precursor. Such a source could be the 'tails' that reverberation introduces after the offset of the test word's vowel. These tails can be removed by a simple 'gating' operation, and so it might be possible to restore the degradation in such a way.

2 Method

2.1 *Speech Precursors and the Test-Word Continuum*

The methods described by Watkins (2005) were used to obtain phrases with a test word that was drawn from a continuum between 'sir' and 'stir', each with a preceding 'precursor' phrase, 'next you'll get _'. This method used the speech of one of the authors (AW) recorded with 16-bit resolution at a 48-kHz sampling rate using a Sennheiser MKH 40 P48 cardioid microphone in an IAC 1201 double-walled booth, giving 'dry' speech. The precursor was originally such a recording of the phrase 'next you'll get sir to click on', with the 'sir' test word and 'to click on' parts excised using a waveform editor. A recording of a 'stir' test word was also obtained in this context. The duration of the precursor was 685 ms, and the original recordings of the test words were both 577 ms long.

To form a test-word continuum, the wideband temporal envelopes of 'sir' and 'stir' were obtained by full-wave rectification followed by a low-pass filter with a 50-Hz corner frequency. The envelope of 'stir' was then divided (point-wise) by the envelope of 'sir' to give a modulation function, and clear 'stir' sounds were obtained by amplitude modulating the waveform of 'sir' with this function. The original 'sir' along with the 'stir' produced by the modulation were the 11-step continuum's end points, nominally steps 0 and 10, respectively. The intermediate steps were produced from the recording of 'sir' using appropriately attenuated versions of the modulation function.

In conditions with a precursor, test words were re-embedded into the context parts of the original utterance. This re-embedding was performed by adding the precursor's waveform to the test word's waveform. Before the addition, silent sections were added to preserve temporal alignment and to allow different reflection patterns to be separately introduced into the test word and the precursor.

2.2 *Category Boundaries*

When room reflections oppose the amplitude modulation that forms the continuum, they obscure cues to the presence of a [t] in test words, so more of the continuum's steps are identified as 'sir'. To indicate differences between conditions in the number of steps that are identified as 'sir', listeners' category boundaries were compared. The boundary is the step, or point between steps, where listeners switch from predominantly 'sir' to predominantly 'stir' responses.

For each of the continua used in the experiment, listeners were asked to identify four presentations of each step, and category boundaries were found from the total number of 'sir' responses across all 11 steps. This total was divided by 4 before subtracting 0.5, to give a boundary step number between –0.5 and 10.5.

2.3 Room Reflections

The methods described by Watkins (2005) were also used to introduce room reflections into the dry precursors and test words by convolution with room impulse responses. This gives the effect of monaural real-room listening over headphones. The monaural impulse responses were obtained in rooms using dummy-head transducers (a loudspeaker in a Bruel and Kjaer 4128 head and torso simulator and a Bruel and Kjaer 4134 microphone in the ear of a KEMAR mannequin), so that they incorporate the directional characteristics of a human talker and a human listener. To obtain signals at the listener's eardrum that match the signal at KEMAR's ear, the frequency-response characteristics of the dummy-head talker and of the listener's headphones were removed using appropriate inverse filters.

The room impulse responses were obtained in a disused office that was L-shaped with a volume of 183.6 m^3. The transducers faced each other, while the talker's position was varied to give distances from the listener of 0.32 or 10 m. This gave different levels of reflected sound, as indicated by the ratio of early (first 50 ms) to late energy in the impulse response (C_{50}, ISO 3382 1997), which was 8 dB at 0.32 m, descending to 2 dB at 10 m.

The processing for 'gated' conditions deleted the 'tails' that arise as room reflections extend test words beyond the dry vowel's offset. From this offset point, a short segue to the subsequent deletion was effected with a 1-ms half-Hann offset ramp, which obviated any clicks at the ends of gated sounds. The resulting effects on test words are shown in the spectrograms of Fig. 22.1.

2.4 Design

The distance of the test word's reflection pattern was 0.32 or 10 m for the 'near' and 'far' continua that were both presented in each of the four conditions. In one of the two conditions without a precursor, test words were gated to remove reverberant tails. The distance of the precursor's reflection pattern was 0.32 or 10 m for near- and far-precursor conditions, where test words were not gated. Series of trials that each had one of the test words were presented to the six listeners in individual

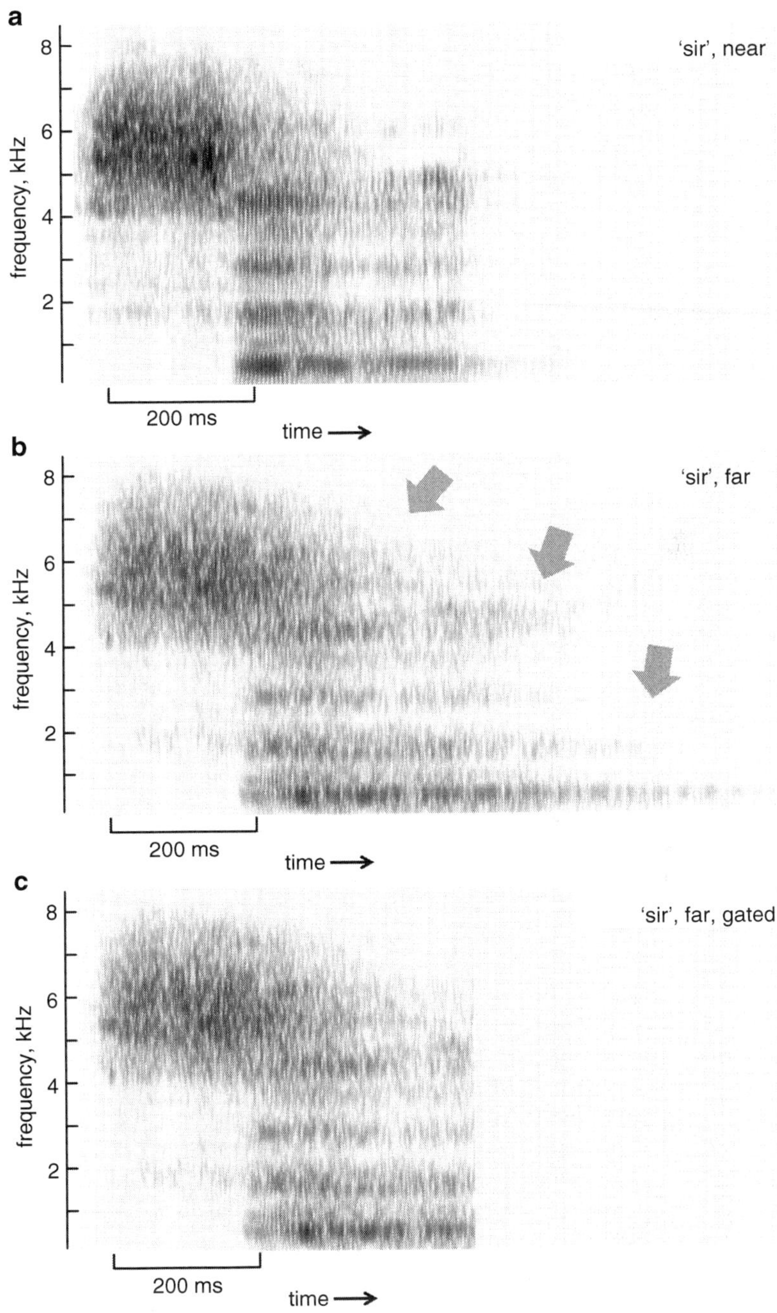

Fig. 22.1 Spectrograms of the 'sir' end point of the test-word continuum showing the near sound (**a**) and the effects of increasing its reverberation with the 10-m room impulse response (**b**) which adds tails (*arrowed*) at the end of the vowel and at the transition from the [s] to the vowel. The effect of gating (**c**) is to delete the parts of these tails that extend beyond the end of the vowel

sessions, with 11 steps × 4 repeats × 2 test-word distances = 88 trials for each of the four conditions, all in a different randomized ordering for each listener.

The dependent variable is the difference between category boundaries for the near and far continua of a condition. This 'near-far difference' is reduced when a condition gives more perceptual constancy. This quantity is compared pairwise between conditions for the hypothesis tests.

2.5 Procedure

Sounds were presented to listeners at a peak level of 48 dB SPL through the left earpiece of a Sennheiser HD480 headset in the otherwise quiet conditions of the IAC booth. Before the experimental trials, listeners were informally given a few randomly selected practice trials to familiarize them with the sounds and the setup. Trials were administered to listeners in individual sessions by an Athlon 3500 PC computer with Matlab 7.1 software and with an M-Audio FireWire 410 sound card. On each of these trials, a context with an embedded test word was presented. Listeners then identified the test word with a click of the computer's mouse, which they positioned while looking through the booth's window at the 'sir' and 'stir' alternatives displayed on the computer's screen. This click also initiated the following trial.

3 Results

For each condition, category boundaries for the 'near' and the 'far' continua were pooled along with the near-far difference across the 6 listeners, and the resulting means are shown with their standard errors in Fig. 22.2.

Results with near and far precursors replicate the constancy effects reported in earlier work (e.g. Watkins 2005; Nielson and Dau 2010; Watkins et al. 2011). When the precursor is nearby, increasing the test word's distance to the far value of 10 m causes more of the continuum's members to be heard as 'sir', and there is a corresponding difference between the category boundaries for near and far test words. However, when the context's distance is also increased to 10 m, there is more constancy, as the difference between category boundaries for near and far test words is markedly reduced. The mean of the near-far differences in far-precursor conditions is significantly less than it is in near-precursor conditions, where $t(5) = 14.8$ and $p < 0.00003$. This and subsequent t-tests are two tailed with Bonferroni corrections for three comparisons.

Results in far- and no-precursor conditions allow rejection of the modulation-masking hypothesis proposed by Nielson and Dau (2010). According to this idea, any effect of removing the precursor should be a decrease in the masking of subsequent sounds, or perhaps it might have no effect at all if masking is insubstantial.

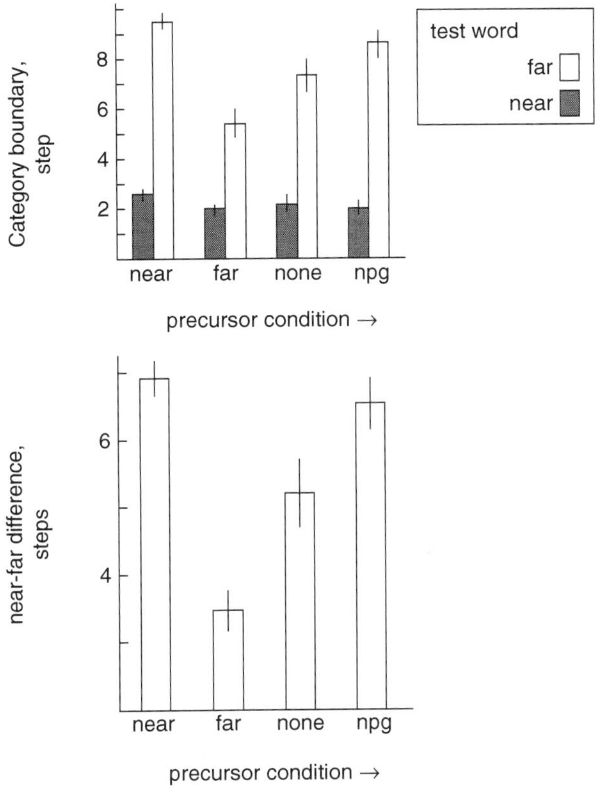

Fig. 22.2 The *upper* histogram shows means and standard errors of category boundaries for the near and far test words in each of the experiment's precursor conditions. The *lower* histogram plots the difference between the near and far category boundaries in these precursor conditions. When this near-far difference is small, constancy is greater, so, moving from left to right, the data show constancy increasing for 'far' precursors, then decreasing when the precursor is removed, and then decreasing further when the test word's tails are removed by gating in the 'no-precursor gated' (*npg*) condition

The data here contradict both versions of the masking idea as the near-far difference actually increases when the far precursor is removed. This increase is considerable; the mean of the near-far differences in the no-precursor condition is significantly higher than it is in far-precursor conditions, where $t(5)=4.84$ and $p<0.005$. It would appear that degradation from reverberation has been restored in these no-precursor conditions, as the perceptual compensation from the precursor is removed. So the absence of degradation in Nielson and Dau's experiment is probably due to their keeping the test word's reverberation at the same constant level in every trial of their experiment.

Effects of gating the test word, in the no-precursor gated ('npg') condition, indicate that the tails that reverberation adds to test words seem also to have a compensatory influence. When these tails are removed, by the gating operation, there is reduced constancy as the mean of the near-far differences in the gated condition is significantly greater than it is in the condition where these isolated test words are not gated. For this comparison, $t(5)=5.19$ and $p<0.004$.

In summary, degrading effects of adding reverberation to isolated test words are apparent in the present experiment, where there is unpredictable variation in the level of reverberation from trial to trial. 'Far' precursors appear to have a compensatory influence, but when this influence is removed by isolating test words, the degrading effect of reverberation emerges. Some compensation is also apparent for isolated test words, where it seems to be effected by the reverberant tails at the end of the sound.

4 Discussion

In the present experiment, the degradation from reverberation is substantial for isolated test words. This degradation becomes much less apparent when test words are preceded by a precursor phrase but only when it has the same level of reverberation as the test word. This supports the idea that there is a perceptual compensation for the degrading effects of reverberation for the 'far' test words and that the mechanism involved is informed by the precursor's reverberation level (Watkins 2005). At the same time, the present results allow rejection of Nielson and Dau's (2010) alternative hypothesis, concerning the putative effects of modulation masking in these sounds.

The constancy mechanism seems to be independent of the sounds' modulation characteristics in other important respects. For example, Watkins (2005) used test words in speech contexts that had reversed reverberation and found that compensation was only obtained with reverberation in its characteristically forward time direction. Compensation was not obtained with speech contexts containing reversed reverberation, even though the modulation attenuation in such sounds is much the same as it is with forward reverberation (Longworth-Reed et al. 2009).

Modulation masking does not appear to be an important consideration in the present type of experiment, which is probably because the salient modulation rates for speech distinctions are relatively low, as in the difference between temporal envelopes of the 'sir' and 'stir' test words used here. Whereas, a typical conclusion from studies of modulation masking is that for '… modulation rates that are dominant in speech (below 16 Hz), one cycle of modulation approaches the time interval over which the auditory system recovers from forward masking, and thus the AM forward masking may not appreciably affect their perception' (Wojtczak and Viemeister 2005, p 3206).

Effects from precursors are 'extrinsic' in that they involve information from beyond the test word's syllable (Watkins and Makin 1994), but other aspects of the present results suggest compensation effects that have 'intrinsic', within-syllable origins. This can be seen in conditions where test words are gated to remove the tails from reverberation that extend beyond the end of the sound's vowel, with the effect that the degradation from reverberation in isolated test words becomes more apparent. It would seem from this that these test-word tails can also bring about a compensation effect, even though they arise some time after the crucial fricative part that distinguishes the test words.

Although the intrinsic effects seen here are small by comparison with effects from precursors, effects of the test word's tails may have been more substantial in Nielson and Dau's (2010) experiment. In those conditions, the lack of variation in the test word's reverberation level from trial to trial would have consistently made the test word on a preceding trial a 'far precursor' for the subsequent trial's test word. This means that in these authors' experiment, there were such 'far precursors' for every trial, except only for the very first one. Any consequent compensation effects might thus have been mistaken for a lack of degradation from reverberation in test words.

Acknowledgements This work was supported by a grant to the first author from EPSRC. We are grateful to Amy Beeston, Guy Brown, Peter Derleth, Kalle Palomäki and Hynek Hermansky for discussions.

References

Houtgast T, Steeneken HJM (1973) The modulation transfer function in acoustics as a predictor of speech intelligibility. Acustica 28:66–73

ISO 3382 (1997) Acoustics - measurement of the reverberation time of rooms with reference to other acoustical parameters. International Organization for Standardization, Geneva

Longworth-Reed L, Brandewie E, Zahorik P (2009) Time-forward speech intelligibility in time-reversed rooms. J Acoust Soc Am 125:EL13–EL19

Nábělek AK, Letowski TR, Tucker FM (1989) Reverberant overlap- and self-masking in consonant identification. J Acoust Soc Am 86:1259–1265

Nielson JB, Dau T (2010) Revisiting perceptual compensation for effects of reverberation in speech identification. J Acoust Soc Am 128:3088–3094

Watkins AJ (2005) Perceptual compensation for effects of reverberation in speech identification. J Acoust Soc Am 118:249–262

Watkins AJ, Makin SJ (1994) Perceptual compensation for speaker differences and for spectral-envelope distortion. J Acoust Soc Am 96:1263–1282

Watkins AJ, Makin SJ (2007) Perceptual compensation for reverberation in speech identification: effects of single-band, multiple-band and wideband contexts. Acta Acust United Acust 93:403–410

Watkins AJ, Raimond AP, Makin SJ (2011) Temporal-envelope constancy of speech in rooms and the perceptual weighting of frequency bands. J Acoust Soc Am 130:2777–2788

Wojtczak M, Viemeister NF (2005) Forward masking of amplitude modulation: basic characteristics. J Acoust Soc Am 118:3198–3210

Chapter 23
A New Approach to Sound Source Segregation

Robert A. Lutfi, Ching-Ju Liu, and Christophe N.J. Stoelinga

Abstract We rely critically on our ability to 'hear out' (*segregate*) individual sound sources in a mixture. Yet, despite its importance, little is known regarding this ability. Perturbation analysis is a psychophysical method that has been successfully applied to related problems in vision [Murray, R.F. 2011. J. of Vision 11, 1–25]. Here the approach is adapted to audition. The application proceeds in three stages: First, simple speech and environmental sounds are synthesized according to a generative model of the sound-producing source. Second, listener decision strategy in segregating target from non-target (noise) sources is determined from decision weights (regression coefficients) relating listener judgments regarding the target to lawful perturbations in acoustic parameters, as dictated by the generative model. Third, factors limiting segregation are identified by comparing the obtained weights and residuals to those of a maximum-likelihood (ML) observer that optimizes segregation based on the equations of motion of the generating source. Here, the approach is applied to test between the two major models of sound source segregation; target enhancement versus noise cancellation. The results indicate a tendency of noise segregation to preempt target enhancement when the noise source is unchanging. However, the results also show individual differences in segregation strategy that are not evident in the measures of performance accuracy alone.

1 Introduction

We possess a remarkable ability to 'hear out' and identify individual sound sources in everyday, multisource acoustic environments. Understanding this ability has been a preoccupation of psychoacoustic research, but it is a tremendous challenge

R.A. Lutfi (✉) • C.J. Liu • C.N.J. Stoelinga
Auditory Behavioral Research Laboratory,
Department of Communicative Disorders, University of Wisconsin,
Madison, WI 53706, USA
e-mail: ralutfi@wisc.edu

B.C.J. Moore et al. (eds.), *Basic Aspects of Hearing*,
Advances in Experimental Medicine and Biology,
DOI 10.1007/978-1-4614-1590-9_23, © Springer Science+Business Media New York 2013

(cf. Bregman 1990). The pressure waveform arriving at the ear is a confluence of information from different sound sources, so determining precisely how a listener is able to identify each source in the mixture requires sophisticated methods. The approach to date has largely been based on a traditional *performance model* – this is the view that changes in performance accuracy or threshold in different conditions can be used to infer the listener's decision processes in those conditions. The approach is widely practiced, but it can easily lead to false conclusions regarding the data. Recent studies involving the identification of simple sound sources in quiet have shown that listeners can and regularly do adopt different listening strategies that yield the same performance accuracy and, conversely, that the same listening strategy often produces very different levels of performance (Lutfi and Liu 2007, 2011a, b). These studies call into question the conclusions that can be drawn regarding listening strategy from performance measures alone, and they underscore the challenge of understanding listening in multisource acoustic environments. Here we undertake an alternative approach that overcomes some of the limitations of the performance model.

2 Perturbation Analysis

The approach amounts to a variation in methods that have been used to determine listener decision strategy in multitone pattern discrimination studies (Berg 1989). The procedure, referred to as *perturbation analysis*, involves introducing small random perturbations in the tones from trial to trial along the dimension to be discriminated. Listener decision strategy is then assessed from estimates of decision weights on the tones given by regression coefficients in a general linear model for which the perturbations are predictor variables for the listener's trial-by-trial response. The present application proceeds in much the same way, but with two important differences. First, the stimulus is taken to be comprised of only two elements, target and noise. The listener's decision strategy for segregating target from noise is determined from the relative decision weights given to the different material, geometric and/or driven properties that serve to distinguish target from noise. Second, the target and noise are either speech sounds or the impact sounds of simple resonant sources. The perturbations in acoustic parameters are applied indirectly by perturbing the physical parameters of a mechanical model of these sources used to synthesize the sounds (Klatt 1980; Morse and Ingard 1968; pp. 175–221). This modification provides the variation in acoustic parameters necessary to derive listener decision weights without violating the lawful relations that exist among these parameters. It also prevents the listener from performing better than chance by simply discriminating a change in a fixed acoustic waveform, what would otherwise be described as a simple discrimination or detection task.

The goal of the approach is to determine precisely *how* the listener makes use of the information in sound to segregate target from noise. This is achieved by formulating a general decision model of listener's trial-by-trial judgments regarding some physical property (or properties) of the target and then estimating the parameters of this model by regression. Here we demonstrate using a simple example from a previously published study (Lutfi and Liu 2011b). The listener's task was to identify as target the impact sound produced by the larger of two circular stretched membranes (the mechano-acoustic analogy to speech would be judging the gender of a speaker by vocal tract length; cf. Patterson et al. 2008). The noise was the impact sound of a variable-size, circular plate presented simultaneously with the target. Let p_T and p_N denote the perturbation in size on each trial, respectively, for target and noise (normally distributed with zero mean). A general decision rule for this task is then as follows: Respond 'larger target' if and only if

$$w_T(\mu_T + p_T + e_T) + w_N(\mu_N + p_N + e_N) > \beta + e_\beta \tag{23.1}$$

where w_T and w_N are, respectively, the decision weights on the target and the noise; μ_T and μ_N denote the nominal sizes of target and noise (before perturbation); β is a response criterion; and e_T, e_N, and e_β are zero-mean normal deviates representing various imperfections in the decision process (internal noise). The internal noise terms are taken to be independent of p_T and p_N and are so pooled to yield a single error term, e (a step justified by the fit of the model). The parameters of the decision rule are then estimated by regressing the listener's trial-by-trial response on the values of p_T and p_N, where the influence of internal noise is estimated from the regression error, σ_e. In practice, the regression is performed using the glmfit routine of MATLAB v.7.0.1 with logistic link function,

$$\text{logit}[P(\text{Larger})] = c_0 + c_1 p_T + c_2 p_N + e, \tag{23.2}$$

where P(Larger) is the probability that the target presented is identified as the larger of the two and c_0 is an estimate of $-\beta$. (Note that the regression is performed on the type of response, not on whether the response is correct or incorrect.) The regression coefficients c_1 and c_2 are finally used to estimate the decision weight on the noise *relative* to that of the target,

$$\hat{w}_N = \frac{c_2}{c_1 + |c_2|} \approx \frac{w_N}{w_T + |w_N|}, \tag{23.3}$$

where $c_1 > 0$. The approach allows for separate evaluation of the effects of the decision weights and internal noise on performance and so provides a stronger test of listener decision strategy than an overall performance metric alone can provide. It also allows the decision weights to be evaluated in different conditions relative to that of a maximum-likelihood (ML) observer that maximizes performance accuracy. For the example described here, the decision weight on the noise for the ML observer is $w_N = 0$.

3 Application: Target Enhancement Versus Noise Cancellation

Here we consider a specific application of the approach designed to test different models of sound source segregation in noise, noise in this case referring to any sound source other than the target. These models generally fall into two classes: those based on enhancement of the target and those based on cancellation of the noise (Bregman 1990; de Cheveigné et al. 1995; Durlach 1963; Meddis and Hewitt 1992; Piechowiak et al. 2007). Much of the work aimed at testing these models has been done on the identification of concurrent pairs of harmonic and inharmonic vowels (see de Cheveigné et al. (1995) for a review). This work would seem to support a predominant role of noise cancellation; however, the tests are based on measures of performance accuracy which, as noted, may not best reflect the processes underlying segregation. In the present study, the relative contribution of target enhancement and noise cancellation is determined by the sign and magnitude of listener weights on the noise.

The listener's task was size discrimination as in the example described above. A two-interval forced-choice procedure with feedback was used. The impact sounds of the membrane (target) and plate (noise) were synthesized using first-order analytic equations for the motion of these sources (Morse and Ingard 1968). The result in each case was a sum of exponentially damped sinusoids. For the membrane the frequencies were 500, 797, 1,068, 1,148, 1,327, and 1,459 Hz; for the plate they were 250, 700, 1,287, 1,495, 2,437, and 3,522 Hz. The amplitudes and decay moduli decreased proportionally with frequency beginning with a decay modulus of 0.2 s for the membrane and 0.4 s for the plate. The frequencies were perturbed from one presentation to the next as would correspond to changes in size. The perturbations were normally distributed in just-noticeable (jnd) units, with one jnd being equal to log(1.002) (Wier et al. 1977). The standard deviation of perturbations for the plate was fixed at 10 jnds; for the membrane it took on values $\sigma_T = 10$–80 jnds, somewhat different for each listener. Sounds were played at a 44,100-Hz sampling rate with 16-bit resolution and were calibrated to be approximately 70 dB SPL at the eardrum (see Lutfi et al. 2008). They were delivered diotically over Beyerdynamic DT 990 headphones to listeners seated individually in a double-walled, IAC sound-attenuated chamber. Listeners were five students of the University of Wisconsin-Madison, aged 24–36 years. All had normal hearing sensitivity (ANSI S3.6-2004) and extensive previous experience with the task.

Now consider the predictions of the two classes of models for the noise weights in Eq. (23.1). Fundamentally, all noise cancellation models detect or recognize the target as a change from a baseline established by the noise. Probably, the simplest example is the old-versus-new (figure-ground) heuristic of Bregman (1990). Models that implement this process computationally involve some form of noise equalization followed by subtraction (e.g., Durlach 1963). For the present application, this can be shown algebraically to be equivalent to giving a negative weight to the noise in Eq. 23.1. We expect then that $-1 < \hat{w}_N < 0$. Target enhancement models, by com-

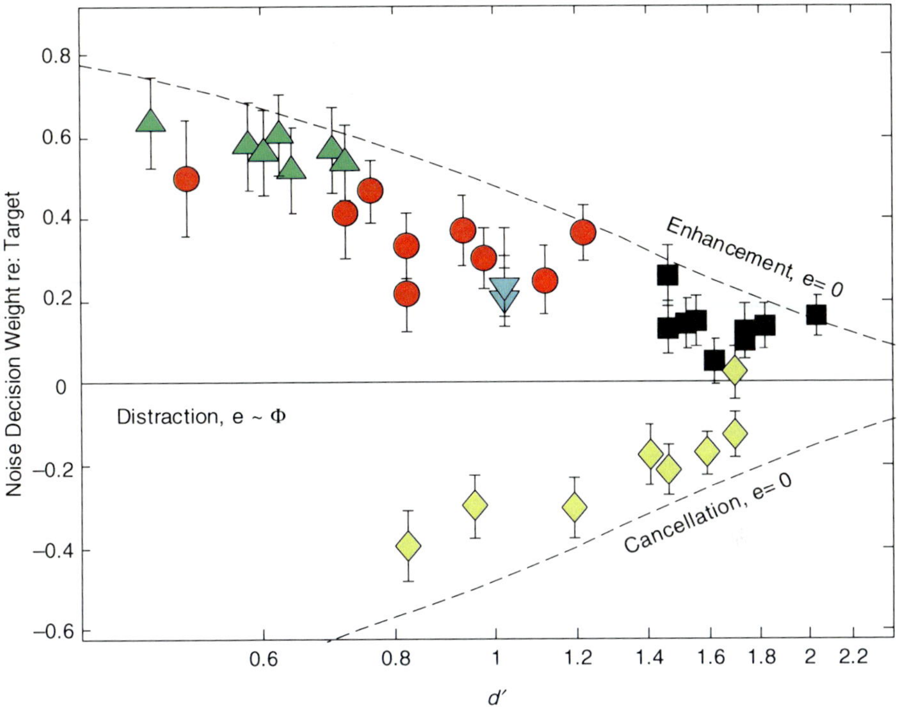

Fig. 23.1 Noise decision weights $\hat{w}_N$ versus d' for five listeners (symbols with 95 % confidence intervals) judging membrane size in the presence of a variable-size plate. Curves give predictions for three models of interference. Repeated symbols represent different values of σ_T

parison, attribute noise interference to imperfections in the enhancement process that result in partial enhancement of the noise. A common example is the auditory filter model for tone-in-noise detection (Patterson 1976; Patterson and Nimmo-Smith 1980). These models predict the weight on the noise to be positive, so that we expect $0< \hat{w}_N <1$. (The analytic proof for both predictions is given fully in the recent publication of Lutfi and Liu (2011b).) Note that the decision model given by Eq. (23.1) also allows for the evaluation of a third alternative in which the noise simply serves to distract attention away from the target, without itself being given any weight (cf. Lutfi RA and Wightman FL 1996; Carlyon and Moore 1986; Werner and Bargones 1991). In this case $\hat{w}_N$ is expected to be near the optimal value of zero with less than optimal performance dictated by an increase in internal noise e resulting from the distraction.

Figure 23.1 gives the estimated noise decision weight, $\hat{w}_N$, for each listener (symbol type) as a function of performance level in d', each point representing 400 trials. Repeated symbols are for different values of σ_T. The horizontal and dashed lines are the predictions of models as labeled. For four of five listeners, the weights are positive, indicating target enhancement; however, for the remaining listener,

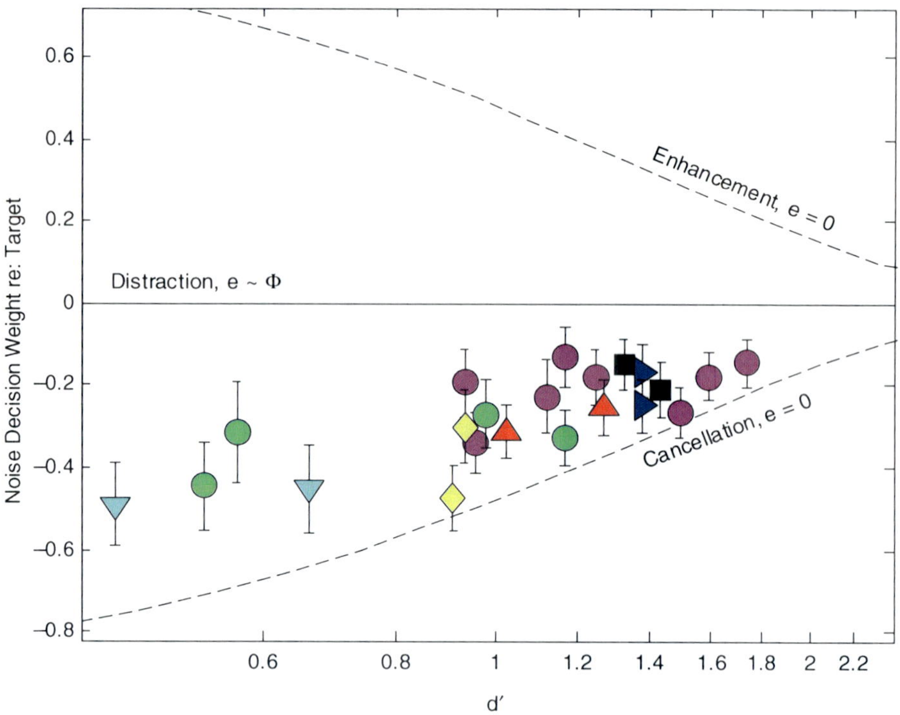

Fig. 23.2 Same as Fig. 23.1 except task is to judge force of membrane impact in the presence of a plate hit with variable force

they are negative, indicating noise cancellation. In a second condition, involving a different group of listeners, the task was to choose the membrane hit with harder force; the noise was a plate hit with variable force. The noise weights for all listeners in this condition indicated noise cancellation (Fig. 23.2). With the exception of the one listener in Fig. 23.1, the results from Figs. 23.1 and 23.2 suggest that noise cancellation occurs whenever the noise source is unchanging (force of impact condition). This would make sense since for a constant noise source, only the amplitudes of modes change; the frequencies of the modes are fixed. Such conditions would be expected to allow for the most effective noise equalization and cancellation, perhaps through adaptation or inattention. We will refer to this as the *fixed-noise hypothesis*; it is fundamentally a version of the old-versus-new heuristic described by Bregman (1990).

A subsequent study was undertaken to further test this hypothesis. The impact sounds were replaced with synthesized vowel sounds (Klatt 1980). A new group of listeners was asked to judge the size of a speaker (length of the vocal tract) uttering the vowel /ae/ in the presence of a variable-size speaker uttering the vowel /a/. The frequencies of the spectral prominences of the vowels in this case varied in much the same way as they did for the variable-size membrane and plate of the first study (cf. Patterson et al. 2008). The results shown in Fig. 23.3 supported target enhancement, as expected. In a second task listeners were asked to judge the force on the vocal folds

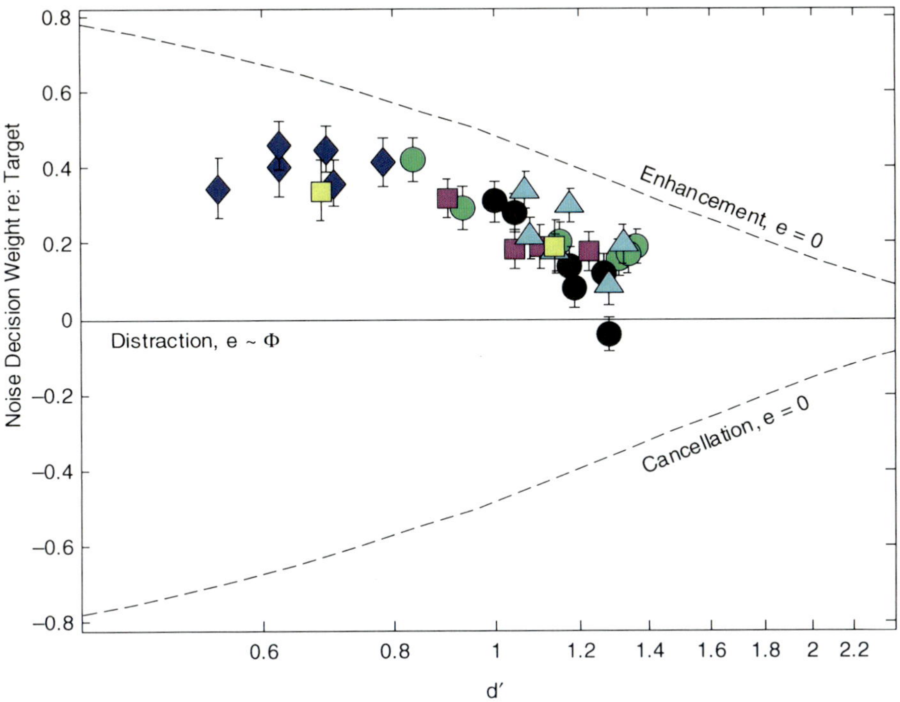

Fig. 23.3 Same as Fig. 23.1 except task is to judge size of speaker uttering the vowel /ae/ in the presence of a variable-size speaker uttering the vowel /a/

for /ae/ ('loudness') in the presence of /a/ spoken with variable force. The levels of the spectral prominences in this case varied in much the same way as they did for the membrane and plate struck with variable force. The results, given in Fig. 23.4, show listeners in this case to be split between target enhancement and noise cancellation.

The results fail to provide consistent support for the fixed-noise hypothesis, but in so doing, they underscore the difficulty in inferring listener decision strategy from performance accuracy alone. Note, in particular, that in Fig. 23.4, where the performance of two listeners is nearly identical (cyan and red squares), the decision weights reflect clearly different listening strategies ($\hat{w}_N$ for the two listeners being different in sign). Conversely, where the performance of two other listeners is quite different (yellow and blue diamonds), the decision weights indicate similar decision strategies ($\hat{w}_N$ for the two listeners being both positive in sign).

4 Summary Comments

We conclude by mentioning two other unexpected outcomes in application of this approach involving the identification of single sound sources in isolation. The first is *level dominance*. This is the finding that the level of a partial, even in the absence

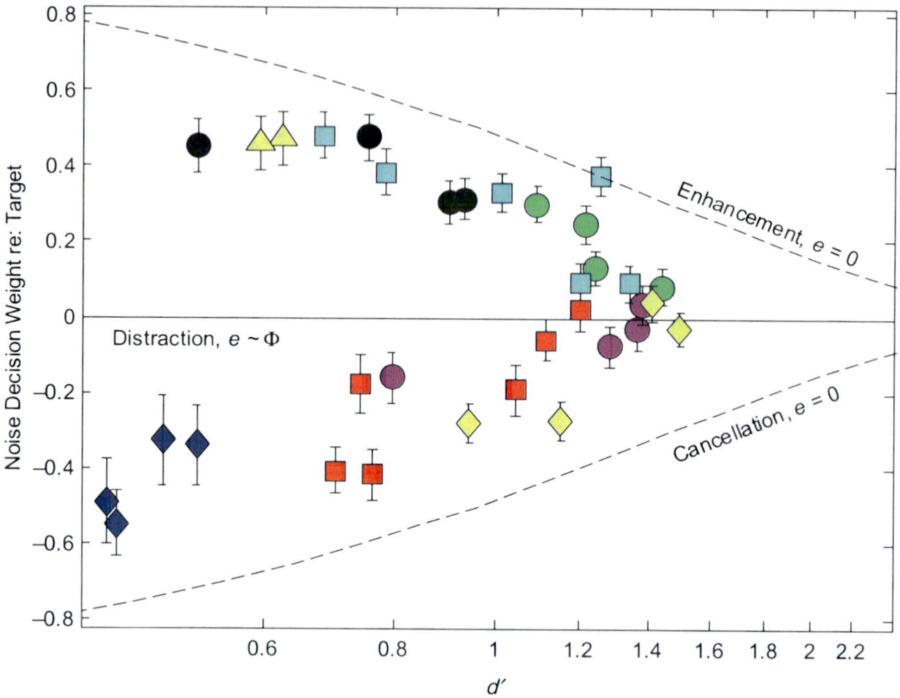

Fig. 23.4 Same as Fig. 23.3 except task is to judge spoken force ('loudness') of vowel /ae/ in the presence of a vowel /a/ spoken with variable force

of any significant masking, is often more predictive of the listener's decision weight on that partial than the amount of information it conveys for identification (Lutfi et al. 2008). The result is surprising because it is exactly opposite to that expected from a maximum-likelihood observer. Level dominance has not, to the authors' knowledge, been studied in the hard of hearing, but it might be expected to have an influence on the benefit of amplification provided by hearing aids. The second unexpected finding is that the acoustic information for basic source attributes, such as material and size, is highly vulnerable to information loss at the cochlea (Lutfi 2001; Lutfi and Stoelinga 2010). This too would seem to have implications for understanding the impact of hearing loss on everyday listening when hearing sensitivity is reduced. Taken together, it is hoped that the results of these and future studies applying this approach can be used to advance development of a computational model of normal listening in multisource acoustic environments, one that can ultimately be used to generate new and testable hypotheses regarding the problem of listening in such environments for the hard of hearing.

Acknowledgments Figure 23.1 was reproduced from Lutfi and Liu (2011b) with permission from the Journal of the Acoustical Society of America. This research was supported by NIDCD grants R01DC006875 and R01DC01262.

References

ANSI (2004) ANSI S3.6-2004. Specification for audiometers. American National Standards Institute, New York

Berg BG (1989) Analysis of weights in multiple observation tasks. J Acoust Soc Am 86:1743–1746

Bregman AS (1990) Auditory scene analysis: the perceptual organization of sound. Bradford Books/MIT Press, Cambridge, MA

Carlyon RP, Moore BCJ (1986) Continuous versus gated pedestals and the severe departure from Weber's law. J Acoust Soc Am 79:453–460

de Cheveigné A, McAdams S, Laroche J, Rosenberg M (1995) Identification of concurrent harmonic and inharmonic vowels: a test of the theory of harmonic cancellation and enhancement. J Acoust Soc Am 97:3736–3748

Durlach N (1963) Equalization and cancellation theory of binaural masking-level differences. J Acoust Soc Am 35:1206–1218

Klatt DH (1980) Software for a cascade/parallel formant synthesizer. J Acoust Soc Am 67:971–995

Lutfi RA (2001) Auditory detection of hollowness. J Acoust Soc Am 110:1010–1019

Lutfi RA, Liu CJ (2007) Individual differences in source identification from synthesized impact sounds. J Acoust Soc Am 122:1017–1028

Lutfi RA, Liu CJ (2011a) A method for evaluating the relation between sound source segregation and masking. J Acoust Soc Am 129:EL34–38

Lutfi RA, Liu CJ (2011b) Target enhancement and noise cancellation in the identification of a rudimentary sound source in noise. J Acoust Soc Am 129:EL52–56

Lutfi RA, Stoelinga CNJ (2010) Sensory constraints on the identification of the geometric and material properties of struck bars. J Acoust Soc Am 127:350–360

Lutfi RA, Liu CJ, Stoelinga CNJ (2008) Level dominance in sound source identification. J Acoust Soc Am 124:3784–3792

Lutfi RA, Wightman FL (1996) Guessing or confusion?: Analytic predictions for two models of target-distracter interference in children. Abstracts of the Assoc. for Research in Otolaryngology 19:142

Meddis R, Hewitt MJ (1992) Modeling the identification of concurrent vowels with different fundamental frequencies. J Acoust Soc Am 91:233–245

Morse PM, Ingard KU (1968) Theoretical acoustics. Princeton University Press, Princeton, pp 175–121

Patterson RD (1976) Auditory filter shapes derived with noise stimuli. J Acoust Soc Am 59:640–654

Patterson RD, Nimmo-Smith IN (1980) Off-frequency listening and auditory-filter asymmetry. J Acoust Soc Am 67:229–245

Patterson RD, Smith DRR, van Dinter R, Walters TC (2008) Size information in the production and perception of communication sounds. In: Yost WA, Popper AN (eds) Springer handbook of auditory research: auditory perception of sound sources. Springer, New York, pp 43–76

Piechowiak T, Ewert SD, Dau T (2007) Modeling comodulation masking release using an equalization-cancellation mechanism. J Acoust Soc Am 121:2111–2126

Werner LA, Bargones JY (1991) Sources of auditory masking in infants: distraction effects. Percept Psychophys 50:405–412

Wier CC, Jesteadt W, Green DM (1977) Frequency discrimination as a function of frequency and sensation level. J Acoust Soc Am 61:178–184

Part IV
Binaural Processing

Chapter 24
Maps of ITD in the Nucleus Laminaris of the Barn Owl

Catherine Carr, Sahil Shah, Go Ashida, Thomas McColgan, Hermann Wagner, Paula T. Kuokkanen, Richard Kempter, and Christine Köppl

Abstract Axons from the nucleus magnocellularis (NM) and their targets in nucleus laminaris (NL) form the circuit responsible for encoding interaural time differences (ITDs). In barn owls, NL receives bilateral inputs from NM such that axons from the ipsilateral NM enter NL dorsally, while contralateral axons enter from the ventral side. These afferents and their synapses on NL neurons generate a tone-induced local field potential, or neurophonic, that varies systematically with position in NL. From dorsal to ventral within the nucleus, the best interaural time difference (ITD) of the neurophonic shifts from contralateral space to best ITDs around 0 μs. Earlier recordings suggested that in NL, iso-delay contours ran parallel to the dorsal and ventral borders of NL (Sullivan WE, Konishi M. Proc Natl Acad Sci U S A 83:8400–8404, 1986). This axis is orthogonal to that seen in chicken NL, where a single map of ITD runs from around 0 μs ITD medially to contralateral space laterally (Köppl C, Carr CE. Biol Cyber 98:541–559, 2008). Yet the trajectories of the NM axons are similar in owl and chicken (Seidl AH, Rubel EW, Harris DM, J Neurosci 30:70–80, 2010). We therefore used clicks to measure conduction time in NL and made lesions to mark the 0 μs iso-delay contour in multiple penetrations along an isofrequency slab. Iso-delay contours were not parallel to the dorsal

C. Carr, PhD (✉) • S. Shah • G. Ashida, PhD
Department of Biology, University of Maryland,
Stadium Drive, College Park, MD 20742, USA
e-mail: cecarr@umd.edu

T. McColgan • H. Wagner
Institute for Biology II, RWTH Aachen, Mies-van-der-Rohe-Str. 15, Aachen, Germany 52056

P.T. Kuokkanen • R. Kempter
Department of Biology, Institute for Theoretical Biology,
Humboldt-Universität zu Berlin, Invalidenstr. 43, Berlin, Germany D-10115

C. Köppl
Institute for Biology and Environmental Sciences, and Research
Center Neurosensory Science, Carl von Ossietzky University,
Carl von Ossietzky Str. 9-11, Oldenburg, Germany 26129

B.C.J. Moore et al. (eds.), *Basic Aspects of Hearing*,
Advances in Experimental Medicine and Biology,
DOI 10.1007/978-1-4614-1590-9_24,

and ventral borders of NL; instead the 0 μs iso-delay contour shifted systematically from a dorsal position in medial NL to a ventral position in lateral NL. Could different conduction delays account for the mediolateral shift in the representation of 0 μs ITD? We measured conduction delays using the neurophonic potential and developed a simple linear model of the delay-line conduction velocity. We then raised young owls with time-delaying earplugs in one ear (Gold JI, Knudsen EI, J Neurophysiol 82:2197–2209, 1999) to examine map plasticity.

1 Introduction

The barn owl's nucleus laminaris remains an outstanding preparation for analysing ITD coding. A major advantage is its large size and accessibility. The nuclei of the auditory brainstem occupy most of the floor of the fourth ventricle, allowing for precise mapping of ITDs in vivo. NL also is a useful preparation for the experimental manipulation of developmental events, because the barn owl's head size more than doubles in size during the first month posthatch (Köppl et al. 2005). We have therefore used the neurophonic potential to measure the conduction times needed to create maps of ITD and developed a simple linear model of the delay-line conduction velocity.

Neurophonics are extracellular potentials well correlated with auditory stimuli and are found in mammals and birds (Köppl and Carr 2008; Sullivan and Konishi 1986; McLaughlin et al. 2010; Schwarz 1992). Although its precise source(s) is still uncertain and may differ between species (Köppl and Carr 2008; Mc Laughlin et al. 2010; Schwarz 1992), the neurophonic is generally accepted as a valid reflection of local activity within the NL.

2 Materials and Methods

The experiments were conducted at the Departments of Biology of the University Maryland and University of Oldenburg. Eighteen barn owls (*Tyto alba*) were used to collect the data presented in this and other studies (Wagner et al. 2005, 2009; Kuokkanen et al. 2010). Procedures conformed to NIH Guidelines for Animal Research and were approved by the Animal Care and Use Committee of the Universities of Maryland and Oldenburg. Anaesthesia was induced by intramuscular injections of 10–20 mg/kg ketamine hydrochloride and 3–4 mg/kg xylazine. Supplementary doses were administered to maintain a suitable plane of anaesthesia. Body temperature was maintained at 39 °C by a feedback-controlled heating blanket. For mapping experiments, electrodes were moved in defined amounts in the rostrocaudal and mediolateral axes. More details may be found in Wagner et al. (2009).

Recordings were made in a sound-attenuating chamber (IAC, New York). Tungsten electrodes were used, with impedances between 5 and 20 MΩ. A grounded silver chloride pellet, placed under the animal's skin around the incision, served as the reference electrode. Amplified electrode signals were passed to a threshold discriminator (SD1 (Tucker-Davis Technologies (TDT) Gainesville, FL)) and an analogue-to-digital converter (DD1 (TDT)) connected to a personal computer via an optical interface (OI (TDT)). Acoustic stimuli were digitally generated by custom-made software ("Xdphys" written in Dr. M. Konishi's lab at Caltech) driving a signal-processing board (DSP2 (TDT)). Acoustic signals were fed to miniature earphones and inserted into the owl's left and right ear canals, respectively. At a given recording site, we measured frequency tuning, then tuning to ITD, and responses to monaural clicks. For most penetrations, an electrolytic lesion (1–10 μA, 3–10 s) was made at or near the response to a best ITD of 0 μs. After a survival time of 5–14 days, owls were perfused transcardially with saline, followed by 4 % paraformaldehyde in phosphate buffer. Brains were blocked in the same stereotaxic apparatus as for in vivo recordings, and sections through NL were reconstructed with the aid of a Neurolucida system.

Young owls were raised from about posthatching day 20 with earplugs designed by Gold and Knudsen (1999) to introduce a time delay in the 3–5 kHz inputs from one ear. Maps of ITD were investigated as above in owls raised with earplugs and lesions made as above.

3 Results

NL receives bilateral inputs from the nucleus magnocellularis (NM) such that axons from the ipsilateral NM enter NL dorsally, while contralateral axons enter from the ventral side. These afferents, and their synapses on NL neurons, generate the neurophonic, which varies systematically with position in NL (Fig. 24.1; Kuokkanen et al. 2010; Sullivan and Konishi 1986). From dorsal to ventral within NL, the best ITD of the neurophonic shifts from contralateral space to best ITDs around 0 μs and

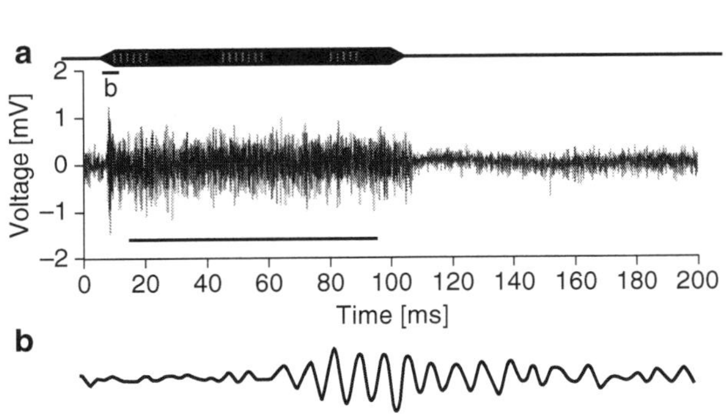

Fig. 24.1 (**a**) 5 kHz tone stimulus (*top*) and neurophonic response (*bottom*). (**b**) 5 ms interval of the neurophonic at tone onset (Modified from Kuokkanen et al. 2010)

on to ipsilateral ITDs. Neurophonics were used to map best ITD, accompanied by small electrolytic lesions, generally at the location of the response to 0 μs ITD. Based on our 3D reconstructions of NL architecture and our measures of delay, we generated an explanation of how conduction velocity of the incoming delay-line axons from NM creates these maps and used modeling to constrain measurements of conduction velocity.

3.1 Dorsoventral Maps of ITD in Owls Derived from Tone-Induced Responses

Within NL, binaurally evoked neurophonics vary with interaural phase difference (IPD), as do ITD-sensitive responses recorded from NL neurons (Carr and Konishi 1990; Peña et al. 2001). The ITD that evokes the maximum neurophonic response changes systematically, always shifting towards ipsilateral space with increasing depth (Sullivan and Konishi 1986); Fig 24.2a). These location-dependent shifts in best ITD occur only within NL; lesions marking the onset of the systematic shift are found on the dorsal edge of the nucleus, and those marking the end of the phase shift are on the ventral edge (Sullivan and Konishi 1986).

We measured the changes in the neurophonic obtained when traversing the short, dorsoventral axis of NL. From dorsal to ventral within NL, the phase delay of a contralaterally elicited potential decreased and that of its ipsilateral counterpart increased (Sullivan and Konishi 1986). This neurophonic delay reflects the delay of phase-locked spikes originating from interdigitating ipsi- and contralateral axons from NM and generates an orderly representation of delay disparities.

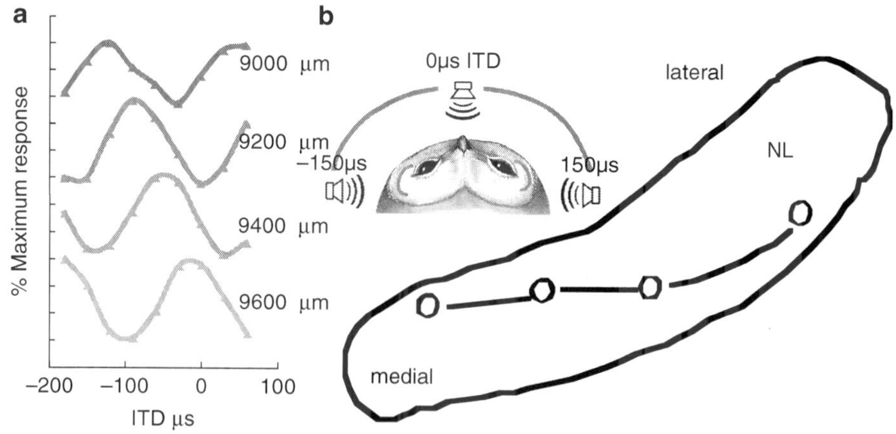

Fig. 24.2 (**a**) Neurophonic responses recorded at 200 μm intervals in a single penetration through NL. (**b**) Reconstructed section through NL, showing location of lesions placed at 0 μs ITDs in multiple penetrations and revealing that iso-ITD lines were not parallel to the borders of NL. *Inset*: schematic owl head showing range of ITDs (von Campenhausen and Wagner 2006)

NL is tonotopically organized, and its isofrequency laminae are oriented along a plane that is inclined about 45° to the midline. We quantified the distribution of ITDs within a single isofrequency lamina by stereotactically sampling ITD at multiple locations within this plane. Each isofrequency lamina contained a map of optimal binaural delays. The map is best illustrated by iso-delay contours (Fig. 24.2b).

Medial portions of each tonotopic lamina mapped best IPDs and ITDs near 0 μs dorsally in NL, with a representation of ipsilateral space below (Fig. 24.2b). The central portions of the tonotopic lamina contained maps of best ITD that were centered on frontal space. The most lateral lesions were characterized by representations of contralateral space, and the most extreme lateral positions did not include a representation of 0 μs. Thus, there was a steady shift in the mapping of 0 μs from dorsal in medial NL to ventral in lateral NL.

3.2 ITD at 0 μs Can Be Derived from Latencies of Click-Induced Responses

Since the representation of frontal space shifted systematically to more ventral locations with mediolateral progression within each tonotopic band, we measured delay latency at different mediolateral positions to determine the basis of the shift and to show how regulation of conduction velocity could underlie map formation. Click stimuli provide a useful measure of conduction delay, since click phase is precise (Wagner et al. 2005). We compared the neurophonic responses to ipsi- and contralateral clicks at best ITDs at or near 0 μs and observed overlaid, phase-locked but not identical ipsi- and contralateral click responses (Fig. 24.3a).

3.3 Models of Conduction Velocity and Maps of ITD

We developed a simple linear model of NL that incorporated click delays (Fig. 24.3b). Conduction velocities along the delay lines inside (d_1–d_4) and outside NL (L_1–L_4) were assumed to be different but constant, consistent with our measures of axon diameter and internodal distance (Carr and Konishi 1990; Carr and Boudreau 1993). Also, we assumed a common path from NM to the midline, followed by a measured path from the midline to the edges of NL. Axon lengths were estimated using a predefined straight-line geometry, derived from reconstructions of axon paths. The parameters of this geometry were fit to three-dimensional reconstructions of the NL borders and the locations of lesions reconstructed from counterstained sections.

The model was used to evaluate possible conduction velocities along the delay lines in NL. The model yielded a range of possible conduction velocities lying between 1 and 10 m/s with velocities within the nucleus being smaller than those outside NL. The model also showed that iso-ITD lines cannot run parallel to the dorsal and ventral borders of NL in the mediolateral direction (Sullivan and Konishi

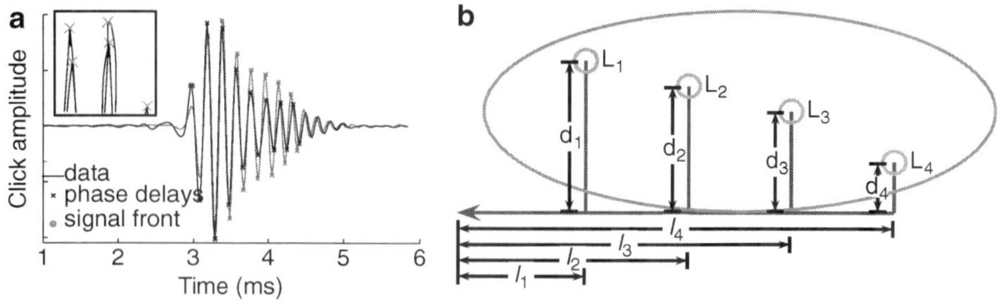

Fig. 24.3 (**a**) Left and right neurophonic click responses recorded at 0 μs best ITD. (**b**) Model of neurophonic conduction velocity accounts for observed shifts in best ITD. Mediolateral lengths are marked l1-4, while dorsoventral lengths are marked d1-4

1986), because parallel iso-ITD lines yielded biologically unrealistic conduction velocities, which is consistent with our experimentally observed iso-ITD contours.

3.4 *Plasticity in Maps of ITD*

Young owls (from P21) were unilaterally fitted with an ear-canal insert designed to introduce a time delay to inputs from that ear (Gold and Knudsen 1999). After growing up with this altered auditory experience, the representation of ITD in NL was mapped using neurophonic responses and tested for any differences to normal. Our hypothesis was that if NL adjusts for the temporally altered input, the maps of ITD would be shifted compared to normal. In addition, complementary shifts are expected in the NL of both sides. Specifically, on the side ipsilateral to the ear insert, the point of zero ITD should shift ventrally and dorsally on the contralateral side. Preliminary analysis did not reveal any shifts in the predicted, adaptive direction. This suggests that nucleus laminaris does not show experience-dependent plasticity, at least not after P20–30. The data are consistent with the observation that myelination of the delay lines peaks at the third week posthatch (Cheng and Carr 2007).

4 Discussion

Based on NL architecture and our results, we have generated an explanation of how conduction velocity of the incoming delay-line axons from NM affects maps of ITD in NL, and we have used a computational model to constrain estimates of NM axonal conduction velocity.

In the barn owl and chicken, the bilateral projection from NM to NL conforms to the requirements of the Jeffress model. In the barn owl, NM axons innervate dorsoventral arrays of neurons in NL in a sequential fashion. For each frequency band, recordings from these interdigitating ipsilateral and contralateral axons show regular changes in delay with depth in NL (Carr and Konishi 1990). These conduction delays are similar to the ±250 μs range of ITDs available to the barn owl (von Campenhausen and Wagner 2006).

The delay-line circuit in the chicken is organized on the same plan as the circuit described for the barn owl, except that the chicken NL is not a large nucleus, but a monolayer of bipolar cells oriented in the mediolateral dimension of the brainstem (Parks and Rubel 1975; Young and Rubel 1983; Seidl et al. 2010). In the barn owl and chicken, NL cells receive input from the ipsilateral NM axons, which splay out to innervate NL neurons with approximately equal lengths to each cell, so that ipsilateral inputs arrive fairly simultaneously along the mediolateral extent of NL. The contralateral axons act as delay lines; each axon runs along the ventral surface of NL, giving off collateral branches in the nucleus. Patterns of axons in the cat medial superior olive are very similar to those of the chicken, in that axons from the contralateral ventral cochlear nucleus project across the rostrocaudal axis of the nucleus, while the ipsilateral axons form a less organized projection (Smith et al. 1993; Beckius et al. 1999). The advantage of electrophysiology in birds is that NL is fairly straight and delays can therefore be mapped within a single tonotopic band. We therefore were able to measure delays with a high enough precision to determine how continuous maps of IPD are formed. We found that the three-dimensional network of delays representing response latencies is well suited for the derivation of IPDs by coincidence detection, with long delays in the dorsoventral dimension and short delays in the mediolateral dimension.

Acknowledgments This research was sponsored by NIH DC00436 to CEC, by NIH P30 DC04664 to the University of Maryland Center for the Comparative and Evolutionary Biology of Hearing, by the German Research Foundation (DFG, Wa-606/12, Ke-788/1-3, 4), by the Bundesministerium für Bildung und Forschung (BMBF, Bernstein Collaboration Temporal Precision, 01GQ07101 to HW and 01GQ07102, 01GQ1001A and 01GQ0972 to RK) and by fellowships from the Humboldt Foundation and the Hanse-Wissenschaftskolleg to CEC and GA.

References

Beckius GE, Batra R, Oliver DL (1999) Axons from anteroventral cochlear nucleus that terminate in medial superior olive of cat: observations related to delay lines. J Neurosci 19:3146–3161

Carr CE, Boudreau RE (1993) Organization of the nucleus magnocellularis and the nucleus laminaris in the barn owl: encoding and measuring interaural time differences. J Comp Neurol 334:337–355

Carr CE, Konishi M (1990) A circuit for detection of interaural time differences in the brain stem of the barn owl. J Neurosci 10:3227–3246

Cheng S-M, Carr CE (2007) Functional delay of myelination of auditory delay lines in the nucleus laminaris of the barn owl. Dev Neurobiol 67:1957–1974

Gold JI, Knudsen EI (1999) Hearing impairment induces frequency-specific adjustments in auditory spatial tuning in the optic tectum of young owls. J Neurophysiol 82:2197–2209

Köppl C, Carr CE (2008) Maps of interaural time difference in the chicken's brainstem nucleus laminaris. Biol Cyber 98:541–559

Köppl C, Futterer E, Nieder B, Sistermann R, Wagner H (2005) Embryonic and posthatching development of the barn owl (Tyto alba): reference data for age determination. Dev Dyn 233:1248–1260

Kuokkanen PT, Wagner H, Ashida G, Carr CE, Kempter R (2010) On the origin of the extracellular field potential in the nucleus laminaris of the barn owl (Tyto alba). J Neurophysiol 104:2274–2290

Mc Laughlin M, Verschooten E, Joris PX (2010) Oscillatory dipoles as a source of phase shifts in field potentials in the mammalian auditory brainstem. J Neurosci 30:13472–13487

Parks TN, Rubel EW (1975) Organization and development of brain stem auditory nuclei of the chicken: organization of projections from n. magnocellularis to n. laminaris. J Comp Neurol 164:435–448

Peña JL, Viete S, Funabiki K, Saberi K, Konishi M (2001) Cochlear and neural delays for coincidence detection in owls. J Neurosci 21:9455–9459

Schwarz DW (1992) Can central neurons reproduce sound waveforms? An analysis of the neurophonic potential in the laminar nucleus of the chicken. J Otolaryngol 21:30–38

Seidl AH, Rubel EW, Harris DM (2010) Mechanisms for adjusting interaural time differences to achieve binaural coincidence detection. J Neurosci 30:70–80

Smith PH, Joris PX, Yin TCT (1993) Projections of physiologically characterized spherical bushy cell axons from the cochlear nucleus of the cat: evidence for delay lines to the medial superior olive. J Comp Neurol 331:245–260

Sullivan WE, Konishi M (1986) Neural map of interaural phase difference in the owl's brainstem. Proc Natl Acad Sci U S A 83:8400–8404

von Campenhausen M, Wagner H (2006) Influence of the facial ruff on the sound-receiving characteristics of the barn owl's ears. J Comp Psychol 192:1073–1082

Wagner H, Brill S, Kempter R, Carr CE (2005) Microsecond precision of phase delay in the auditory system of the barn owl. J Neurophysiol 94:1655–1658

Wagner H, Brill S, Kempter R, Carr CE (2009) Auditory responses in the barn owl's nucleus laminaris to clicks: impulse response and signal analysis of neurophonic potential. J Neurophysiol 102:1227–1240

Young S, Rubel EW (1983) Frequency-specific projections of individual neurons in chick brainstem auditory nuclei. J Neurosci 3:1373–1378

Chapter 25
The Influence of the Envelope Waveform on Binaural Tuning of Neurons in the Inferior Colliculus and Its Relation to Binaural Perception

Mathias Dietz, Torsten Marquardt, David Greenberg, and David McAlpine

Abstract Recently, Klein-Hennig et al. (J Acoust Soc Am 129:3856–3872, 2011) suggested a design for envelope waveforms that allows for independent setting of the duration of the four segments of an envelope cycle – pause, attack, sustain, and decay. These authors conducted psychoacoustic experiments to determine the threshold interaural time differences (ITDs) for different waveforms and revealed that a steep attack flank and at least 4 ms of pause duration prior to the attack are optimal for discrimination performance, whilst sustained and decay durations were of only minor influence. The current study tests the sharpness of rate-ITD-functions recorded in the inferior colliculus of guinea pigs in response to a similar set of waveforms, examining their relationship to the psychoacoustic data. Particular focus is applied to temporally asymmetric envelope waveforms: a long 15-ms attack and a short 1.5-ms decay envelope and the temporally inverted envelope with a short 1.5-ms attack and a long 15-ms decay.

1 Introduction

Perrott and Baars (1974) reported that threshold interaural time differences (ITDs) are much smaller for the stimulus onset than for the stimulus offset. However, until recently it was not clear if the same holds for the attack and decay of an ongoing modulation period. Klein-Henning et al. (2011) employed slightly different envelopes in the two ears to demonstrate that even in the ongoing envelope, sensitivity to attack ITD is much higher than sensitivity to decay ITDs. Consequently, stimuli with short, steep attacks and long, shallow decay flanks (short/long) elicit much lower threshold ITDs than do the reverse (i.e. long/short)

M. Dietz (✉) • T. Marquardt • D. Greenberg • D. McAlpine
Ear Institute, University College London,
London, UK
e-mail: mathias.dietz@uni-oldenburg.de

B.C.J. Moore et al. (eds.), *Basic Aspects of Hearing*,
Advances in Experimental Medicine and Biology,
DOI 10.1007/978-1-4614-1590-9_25,

stimuli. This is difficult to explain with existing models of binaural interaction, as their performance often depends on the spectrum of the preprocessed signal (e.g. Bernstein and Trahiotis 1996; Dietz et al. 2012), and at least in the absence of any peripheral adaptation, the spectrum is identical for the two stimuli. Even when incorporating adaptation in the preprocessing stage, Klein-Hennig et al. were unable to account for the large observed difference in ITD sensitivity when employing a binaural model based on the interaural cross-correlation coefficient (Bernstein and Trahiotis 1996).

The goal of the current study is to measure the ITD tuning of binaurally sensitive neurons in the inferior colliculus (IC) of guinea pigs to an extensive set of envelope waveforms and to analyse whether these physiological data can account for the behavioural ITD discrimination thresholds obtained by Klein-Henning et al. (2011).

2 Methods

Recording equipment and methods were similar to those used by Griffin et al. (2005). In brief, all experiments were carried out in accordance with the Animal (Scientific Procedures) Act of 1986 of Great Britain and Northern Ireland. Single-neuron recordings were made from 28 neurons in the right IC of 8 adult tri-coloured guinea pigs anaesthetised with urethane (20 % solution, 1.5 g/kg dosage, IP). Administered subcutaneously at the beginning of each experiment were Rimadyl (50 mg/ml solution, 2 mg constant dosage), a nonsteroidal anti-inflammatory drug that also acts as an analgesic; Buprenorphine (0.3 mg/ml solution, 0.05 mg/kg dosage) an opioid with which supplementary doses were administered when an experiment continued beyond 8 h; and Colvasone (2 mg/ml solution, 2 mg/kg dosage), a corticosteroid with an anti-inflammatory action.

Sounds were produced using Tucker-Davis Technologies (TDT, Alachua, FL) digital signal processing hardware. TDT Brainware, Real-Time Processor Visual Design Studio (RPvdsEx), and system III hardware were used to generate the stimuli (48,828-kHz sampling rate). Stimuli were generated and scaled such that their peak voltages were at 5 V of the D/A converters (DACs). The outputs were attenuated to achieve the desired level for the experiments using PA5 (system III) modules (TDT). Sounds were delivered by Etymotic ER-4S headphones with the common reference wire separated in order to abolish crosstalk.

Each cycle of the stimulus envelope was constructed from four segments, identical to those used by Klein-Hennig et al. (2011): (1) a pause segment with duration T_p and zero amplitude, (2) an attack segment identical to the rising portion of a squared sinusoid and duration T_a, (3) a sustain segment with duration T_s and maximum amplitude, and (4) a decay segment identical to the falling portion of a squared sinusoid and duration T_d. All stimuli were fully modulated. The stimulus duration was always an integer multiple of the cycle duration (starting and stopping in the modulation trough) and as close to 1 s as possible (990–1,010 ms). The interstimulus interval was at least 300 ms. Responses to 17 different envelope shapes with an average of 38 different ITDs were recorded at least 6 times (mean 8.2), resulting in a minimum recording duration of 84 min. The stimulus presentation order was chosen so as to record conditions with identical envelope shapes en bloc. Within each block, the ITD was successively increased. Thereafter, for the next run, the presentation order was reversed. Stimuli were presented with a constant maximum amplitude approximately 20 dB above the respective pure-tone threshold for the neuron. The ITD was applied after multiplying the carrier with the envelope resulting in a full waveform shift. Characteristic frequencies (CFs) of all recorded neurons were >2.5 kHz (i.e. beyond the presumed upper limit for sensitivity to carrier ITD). Nevertheless, half of the runs were recorded by inverting the carrier in one channel in order to cancel out potential residual tuning to the carrier IPD (SUMCOR, Joris 2003).

Electrical signals from the electrode were current amplified by a headstage (RA16AC, TDT) and further amplified and digitised by a preamplifier (TDT Medusa RA16PA) at a 25-kHz sampling rate. The signal was conducted by a fibre-optic cable to the RX5 base station for filtering (300–3,500 Hz). Spike data were passed from the RA16 base station to TDT Brainware, and spikes were selected, according to manually adjusted spike characteristics, to ensure data were analysed from a single neuron.

3 Results

Rate-ITD-functions for 17 different envelope shapes of one example neuron are shown in Fig. 25.1. Envelope shapes are displayed in six different groups.

For the sinusoidally amplitude modulated (SAM) tones (group 1), this neuron clearly shows ITD tuning for modulation frequencies between 40 and 200 Hz. For pseudo-square-wave (PSW) envelopes (group 2), the rate-ITD-functions are generally steeper than for SAM. For a modulation frequency of 333 Hz, the PSW stimulus is identical to the SAM. If the duty cycle is varied at a constant 40-Hz modulation frequency (group 3), the response R of the neuron increases with a decrease in duty cycle, that is, with an increase in pause duration. The steepness of the rate-ITD-function remains constant in terms of the R^2-linearized ordinate for all but the $T_p = 0$ condition, where almost no tuning is observed. A similar pattern is observed for group 4, where the pause duration was varied with a constant sustain duration (and thus co-varying f_m). Here, however, the steepness of the ITD tuning starts to decrease again for long pause durations, once more demonstrating this neuron's preference for $f_m > 40$ Hz. For group 5, the sustain duration T_s was varied with a constant $T_p = 8$ ms, revealing a reduction in tuning steepness when the sustain duration is increased from 0.5 to 4 ms. If the sustain duration is further increased to 14 ms, the steepness stays constant but the overall rate decreases. In the final, but perhaps most critical, group of conditions, the isolated influence of the duration of the attack and the decay flank was investigated. This example neuron shows the steepest tuning if both attack and decay flank are short (short/

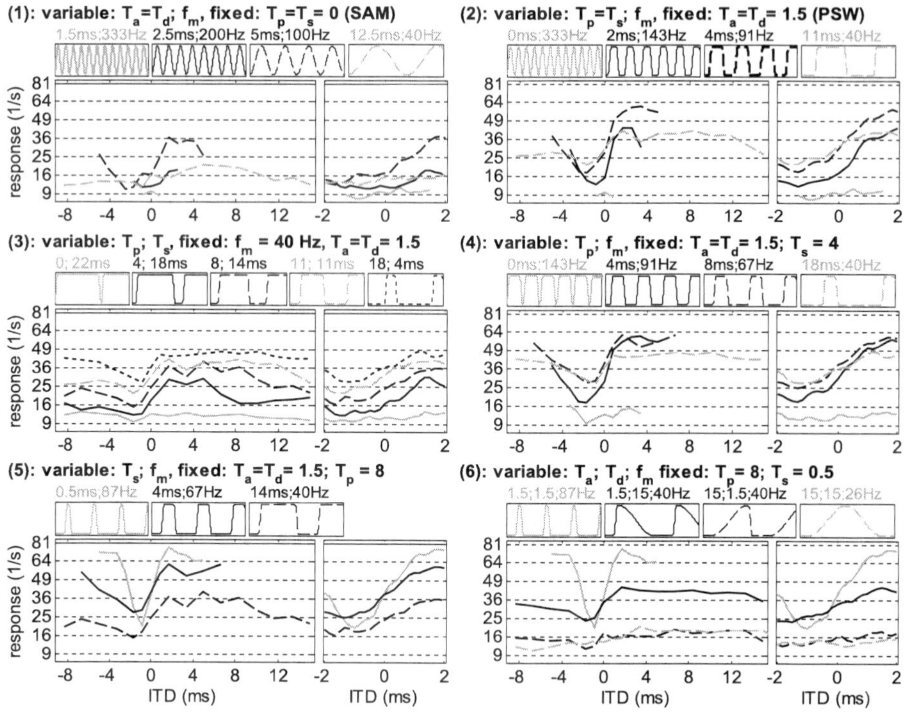

Fig. 25.1 Rate-ITD-functions of 17 different envelope shapes for an example neuron with $CF = 10{,}600$ Hz and a spontaneous rate of 15 s^{-1}. Conditions are displayed in six groups. Within one group, one segment duration (either T_p, T_a, T_s, or T_d) is varied, the other three segment durations and the modulation frequency f_m are either fixed or co-dependent. The only exception is group 6 where both T_a and T_d are varied independently. There is always at least one co-dependent variable because of the relation $T_p + T_a + T_s + T_d = 1/f_m$. Some envelope shapes are used in several groups. Upper subpanels of each group are the legend illustrating the envelope of the conditions in the line style of the corresponding rate-ITD-function. Numbers over each envelope are the variable parameters given in the same order (segment duration(s) in ms, modulation frequency in Hz). The fixed parameters are included in the title of each group. The main panel shows the rate-ITD-functions over the duration of one modulation period, recorded with an ITD spacing of 1.67 ms or less. The smaller panel on the right shows the central ITD range [–2 ms; +2 ms] recorded with a narrower spacing of 167 μs and plotted after weighted-3-point-averaging. Note that the ordinate is linear for the squared response rate. Explanation for this choice of ordinate is provided in Sect. 3

short condition). Relatively sharp ITD tuning is also observed for the short/long condition, but almost no tuning for the long/short and long/long conditions.

Rate-ITD-functions of a second example neuron are shown in Fig. 25.2. This neuron does not exhibit any ITD tuning for the SAM conditions. For the PSW, however, the rate-ITD-function is very steep around ITD = 0 for conditions with $f_m < 100$ Hz. The same holds for all conditions in groups 3, 4, and 5 as long as the pause duration is at least 4 ms. For group 6, the neuron shows a steep tuning function if the attack duration is short (1.5 ms). For the two conditions with long, shallow attack durations, the neural response is independent of the ITD.

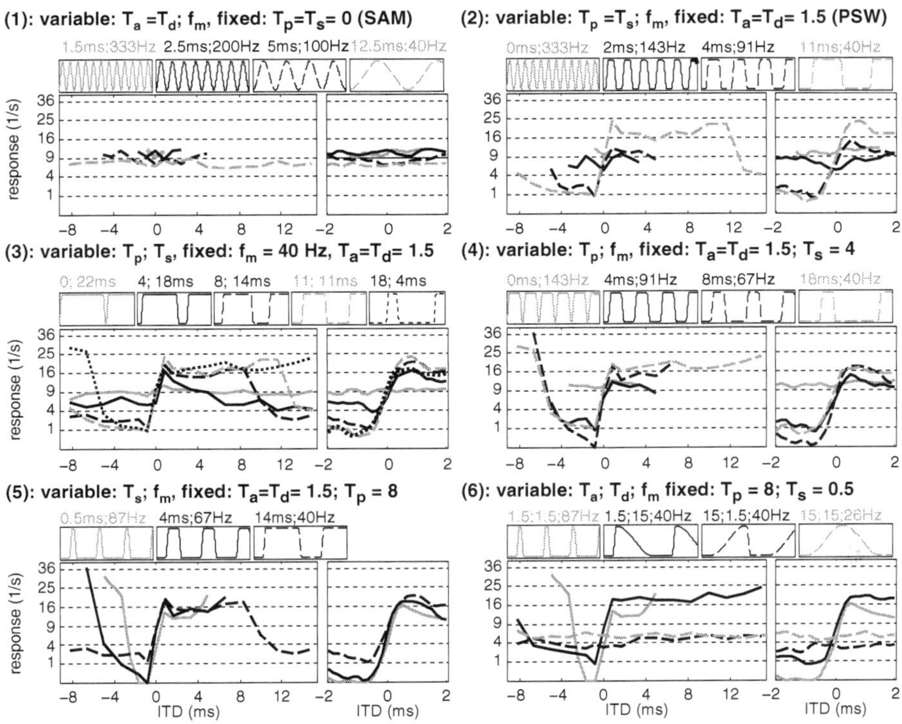

Fig. 25.2 Same format as Fig. 25.1 but for a second example neuron with $CF = 6{,}000$ Hz and a spontaneous rate of 0.1 s^{-1}

In Fig. 25.3, the conditions of group 6 are shown for an additional 12 neurons. The neurons were not selected as in Figs. 25.1 and 25.2 but were simply the first 12 neurons recorded that were sensitive to ITD in any of the four conditions. Each neuron is unique in the shape of its rate-ITD-functions even for this reduced number of only four conditions. The neural ITD discrimination thresholds were estimated only in the central ITD range [−2 ms; +2 ms]. It was defined as the smallest significant difference ($d' \geq 2$) in response rate R to any reference ITD:

$$d' = \frac{\Delta R}{\sigma} \cong \frac{\left|R(\mathrm{ITD}) - R(\mathrm{ITD}_{\mathrm{ref}})\right|}{\sqrt{R}} \cong \frac{\left|R(\mathrm{ITD}) - R(\mathrm{ITD}_{\mathrm{ref}})\right|}{\left(\sqrt{R(\mathrm{ITD})} + \sqrt{R(\mathrm{ITD}_{\mathrm{ref}})}\right)/2}.$$

The simplified assumption is that spike variance is proportional to mean spike rate (Poisson process). A mathematical approximation is to average the variance at $\mathrm{ITD}_{\mathrm{ref}}$ and at the test-ITD. The advantage of assuming a Poisson process is that on the R^2-axis employed in the figures, the steepness correlates directly with the discriminability, due to the variance normalisation. The distance between two neighbouring dashed lines corresponds to $d' = 2$ (e.g. $|25-16|/4.5 = 2$).

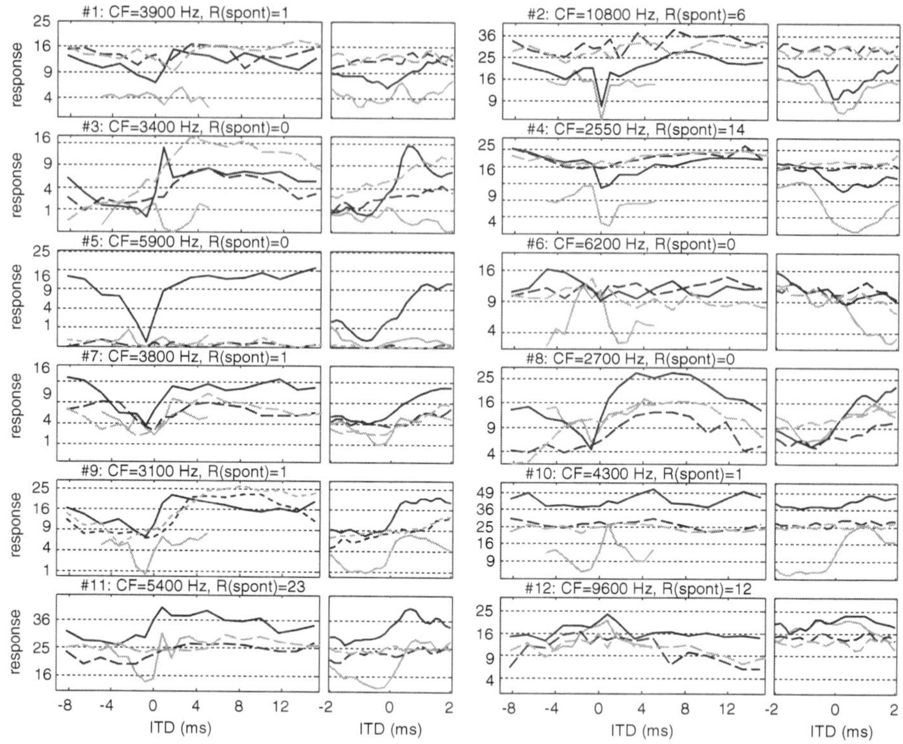

Fig. 25.3 Rate-ITD-functions of the conditions in group 6 for 12 additional neurons. Fixed parameters were $T_p = 8$ ms and $T_s = 0.5$ ms. Pairs of attack/decay duration (in ms) are 1.5/1.5 (*grey solid line*), 1.5/15 (*black solid line*), 15/1.5 (*black dashed line*), and 15/15 (*grey dashed line*). Focusing on the 1.5/15 and 15/1.5 conditions where the envelopes have the same modulation frequency of 40 Hz and are mutual temporal mirror images, it is notable that the condition with a steep attack (*black solid*) almost always shows a deeper and steeper modulated rate-ITD-function than the condition with the steep decay (*black dashed*)

For the short/long stimulus, an ITD discrimination threshold of $d' \geq 2$ was obtained for 21/28 neurons, but only for 8/28 in the long/short stimulus. Average thresholds for the best quartile (best 7 neurons) were 167 μs for the short/long and 1,000 μs for the long/short stimulus. For the short/short, 24/28 neurons had a sufficient response difference to establish a threshold (mean of best 7: 167 μs). For the long/long stimulus, just 6 neurons allowed for a threshold determination (mean 1,100 μs). Of all conditions and neurons that showed an optimal reference ITD in the interval (−2 ms; +2 ms), the average optimal reference ITD was +100 μs (contralateral leading).

4 Discussion

In line with the differences in psychoacoustic threshold ITDs (Klein-Hennig et al. 2011), rate-ITD-functions to the short/long stimuli are better modulated than those of the long/short stimuli. The estimated difference factor 6 of ITD discriminability is even larger than the factor 3.3 obtained by Klein-Hennig and colleagues, although with slightly different stimulus parameters. Irrespective of their binaural tuning, 21/28 neurons have a higher response to the short/long than to the long/short. This is also in agreement with a study of monaural IC responses to temporally asymmetric envelopes (Neuert et al. 2001). However, seven neurons have a higher response for the long/short stimuli (although this higher response does not result in a better binaural tuning).

On average, ITD tuning of the short/short was even a little better compared to the short/long. This does not necessarily mean that the short decay ramp as such improves ITD tuning, but only that spikes generated during the rising flank are temporally more precise. Timing of spikes generated during the long decay period is almost random and will therefore weaken ITD tuning. The observation that at least the trough of the short/long tuning curve has generally a higher spike rate than the short/short tuning curve supports this interpretation.

None of the cross-correlation function, the binaural envelope modulation filters (Dietz et al. 2012), or a squared difference detector (Breebaart et al. 2001) have the ability to account for the observed differences in ITD tuning and the consequent different psychoacoustic thresholds for the short/long, compared to the long/short, stimuli. Peripheral adaptation, as introduced in binaural models by Breebaart and colleagues, suggests a plausible mechanism by which these differences might arise, but the difference after a subsequent cross-correlation is too small to explain either the neural or the psychoacoustic data. More sophisticated binaural processors including timed inhibition, or low-threshold, voltage-gated potassium channels (e.g. Gai et al. 2009) would appear necessary to model neural ITD tuning and binaural perception with temporally asymmetric envelopes.

Rate-ITD-functions for the other conditions are shown here for only two example neurons (Figs. 25.1 and 25.2), and the data are generally in line with the psychoacoustic data of Klein-Hennig et al. (2011). Analyses of the whole data set, however, which could not be shown here in detail, reveal a wide variety of cell response patterns across conditions that are often, but not always, in line with the psychoacoustic data. One difference, for example, lies in the different ITD tuning for different sustain durations (group 5) shown by a subset of neurons, including that in Fig. 25.1. Despite this exception, however, the rate-ITD-functions of the neuron from Fig. 25.1 are almost perfectly in line with the trends in the psychoacoustic data: a maximum sensitivity for the SAM with $f_m = 100$ Hz (group 1), a relatively constant sensitivity for the PSW in the range 40–143 Hz (group 2), the strongly increasing sensitivity with longer T_p (from 0 to 4 ms) at constant f_m (group 3), and no further increase for even longer pause durations. Only in this last relationship did a minor difference exist, namely, that the flattening of the sensitivity was evident by $T_p = 4$ ms rather than 8.8 ms. The neuron in Fig. 25.2 does not show binaural tuning to any SAM tone (group 1) and to the long/short stimulus (group 6). All other conditions are consistent with the trends observed in the psychoacoustic data.

Neurons exist in which ITDs corresponding to the minimal (and maximal) response (worst ITD and best ITD, respectively) differ strongly between the short/long and the long/short stimuli (e.g. Fig. 25.3). For this example neuron, worst ITD is at −0.7 ms, best ITD at +0.7 ms for the short/long, but at −7 ms and +7 ms for the long/short. However, in order to characterise binaural properties of a neuron in terms of characteristic phase (CP) and characteristic delay (CD), it is necessary that for a given cycle duration (i.e. modulation frequency), the best ITD (or worst ITD) is constant and does not depend on a particular envelope shape. Short/long and long/short have the same cycle duration so the systematically larger best ITDs with longer attack duration suggests the possibility that attack duration is more important than cycle duration for the binaural characterisation of some neurons.

In summary, the attack segment of each modulation cycle is mainly influencing the binaural characteristics of the investigated IC neurons. A pause duration of 4–8 ms between AM cycles is usually necessary to elicit binaural sensitivity. Interpretation of the data as a rate code results in neural threshold ITDs similar to human psychoacoustic thresholds and with an optimal reference ITD close to 0 ms.

Acknowledgments This work was supported by the MRC. Mathias Dietz was supported by the Alexander von Humboldt-Foundation with a Feodor Lynen Fellowship.

References

Bernstein LR, Trahiotis C (1996) On the use of the normalized correlation as an index of interaural envelope correlation. J Acoust Soc Am 100:1754–1763

Breebaart J, van de Par S, Kohlrausch A (2001) Binaural processing model based on contralateral inhibition. I. Model structure. J Acoust Soc Am 110:1074–1088

Dietz M, Ewert SD, Hohmann V (2012) Lateralization based on interaural differences in the second-order amplitude modulator. J Acoust Soc Am 131:398–408

Gai Y, Doiron B, Kotak V, Rinzel J (2009) Noise-gated encoding of slow inputs by auditory brain stem neurons with a low-threshold K+ current. J Neurophysiol 102:3447–3460

Griffin S, Bernstein LR, Ingham N, McAlpine D (2005) Neural sensitivity to interaural envelope delays in the inferior colliculus of the guinea pig. J Neurophysiol 93:3463–3478

Joris PX (2003) Interaural time sensitivity dominated by cochlea-induced envelope patterns. J Neurosci 23:6345–6350

Klein-Hennig M, Dietz M, Hohmann V, Ewert SD (2011) The influence of different segments of the ongoing envelope on sensitivity to interaural time delays. J Acoust Soc Am 129: 3856–3872

Neuert V, Pressnitzer D, Patterson RD, Winter IM (2001) The responses of single units in the inferior colliculus of the guinea pig to damped and ramped sinusoids. Hear Res 159:36–52

Perrot DR, Baars BJ (1974) Detection of interaural onset and offset disparities. J Acoust Soc Am 55:1290–1292

Chapter 26
No Evidence for ITD-Specific Adaptation in the Frequency Following Response

Hedwig E. Gockel, Louwai Muhammed, Redwan Farooq, Christopher J. Plack, and Robert P. Carlyon

Abstract Neurons sensitive to interaural time differences (ITDs) in the fine structure of low-frequency signals have been found in binaurally responsive auditory nuclei in a wide range of species. The present study investigated whether the frequency following response (FFR) would show evidence for neurons "tuned" to ITD in humans. The FFR is a scalp-recorded measure of sustained phase-locked brainstem activity that has been shown to follow the frequency of low-frequency tones. The magnitude of the FFR often decreases over time for tones of long duration. The present study investigated whether this adaptation effect is ITD specific.

The FFR to a 100-ms, 80-dB SPL, 504-Hz target tone was measured for ten subjects. The target was preceded by a 200-ms, 80-dB SPL, 504-Hz adaptor. The target always led by 0.5 ms in the left ear. The adaptor led either in the left ear or in the right ear by 0.5 ms. Stimuli (adaptor + target = pair) were presented in alternating polarity at a rate of 1.81 Hz. We used a "vertical" montage (+Fz, – C7, ground = Fpz) for which the FFR is assumed to reflect phase-locked neural activity from rostral generators in the brainstem. The averaged FFR waveforms for each polarity were subtracted, to enhance temporal fine structure responses. The results showed significant adaptation effects in the spectral magnitude of the FFR. However, adaptation was not larger when the adaptor had the same ITD as the target than when the ITD of the adaptor differed from that of the target. Thus, the current data provide no evidence that the spectral magnitude of the scalp-recorded FFR provides a non-invasive indicator of ITD-specific neural activation.

H.E. Gockel (✉) • L. Muhammed • R. Farooq • R.P. Carlyon
MRC Cognition and Brain Sciences Unit,
15 Chaucer Rd, Cambridge CB2 7EF, UK
e-mail: hedwig.gockel@mrc-cbu.cam.ac.uk

C.J. Plack
School of Psychological Sciences,
The University of Manchester,
Manchester M13 9PL, UK

B.C.J. Moore et al. (eds.), *Basic Aspects of Hearing*,
Advances in Experimental Medicine and Biology,
DOI 10.1007/978-1-4614-1590-9_26,

1 Introduction

Interaural time differences (ITDs) provide a major cue for the localisation of low-frequency sounds in azimuth. Neurons sensitive to ITD in the fine structure of low-frequency signals have been found in binaurally responsive auditory nuclei in a wide range of species. However, the exact way in which ITD-sensitive neurons contribute to sound localisation is still under debate. The majority of physiological knowledge on ITD-sensitive neurons is based on animal studies. In these, electrophysiological recordings from one side of the brainstem or midbrain (mostly from individual neurons in the inferior colliculus, IC) indicate that the peaks of the functions relating neuronal discharge rate to ITD are distributed around a mean of 200–300 μs; for most neurons, the maximal discharge rate is found for a tone leading to the side contralateral to the recording side. There is evidence for adaptation in the IC specific to ITD that can be distinguished from adaptation in earlier monaural pathways (McAlpine et al. 2000; Ingham and McAlpine 2004). Ingham and McAlpine (2004) recorded from single IC neurons of the guinea pig. A given ITD-sensitive neuron was first adapted by a 1-s stimulus with its worst ITD, i.e. the ITD that produced the least firing. This was done so that all monaural components in the auditory pathway were adapted. When the ITD was suddenly changed to the neuron's best ITD (the ITD that produced the highest firing rate), the firing rate increased and then adapted again to a rate that was higher than the previous adapted rate. These results indicate that a change in ITD following an adaptation period can lead to an enhanced neural response in some neurons.

In humans, recording from individual neurons in the IC is rarely possible. The present study investigated whether ITD-specific adaptation would be evident in the scalp-recorded frequency following response (FFR) measured in humans. The FFR reflects sustained activity in a population of neurons that phase lock to stimulus-related periodicities (Marsh et al. 1975; Smith et al. 1975; Glaser et al. 1976). The contribution from various anatomical sources to the FFR depends on electrode positions. Here, a "vertical" electrode montage (see Sect. 2) was used. For this, the FFR is generally assumed to reflect sustained phase-locked neural activity from rostral generators in both hemispheres of the brainstem (IC and lateral lemniscus, LL). Hence, the FFR to our test sounds will have been driven by neurons with a wide range of locations throughout both ICs. Our experimental question was whether the FFR to a target sound with a given ITD would be reduced more by an adaptor that had the same, rather than a different, ITD. The rationale is based on the assumption that, as in other EEG adaptation studies (Näätänen et al. 1988), an adaptor will reduce the measured response more when it adapts those neurons responding most strongly to the target than when it adapts neurons that respond less strongly to the target.

2 Methods

2.1 Stimuli

The FFR to a 100-ms, 80-dB SPL, 504-Hz target tone was measured. The target was preceded by a 200-ms, 80-dB SPL, 504-Hz adaptor. The target always led by 0.5 ms in the left ear. The adaptor either led in the left ear (condition "LL") or the right ear (condition "RL") by 0.5 ms. The target followed the adaptor either immediately ("no-gap" condition) or after a silent gap corresponding to 10 cycles of the 504-Hz tone (19.841 ms; "with-gap" condition). All tone durations included 5-ms raised-cosine rise/fall times.

The target always had a starting phase of -0.4π (this is the phase that occurs 200 ms after the start of a 504-Hz tone starting with zero phase). A change of ITD between adaptor and target creates *within*-ear differences between the phase of the ringing on the basilar membrane (BM) after the end of the adaptor and the onset phase of the target, potentially leading to a "dip" in the response on the BM (Shailer and Moore 1987). Such a dip might affect the FFR in response to the target. To equate the dips across conditions, the starting phase of the adaptor was adjusted as follows. In condition RL, the adaptor started at zero phase in both ears (but was delayed by 0.5 ms in the left ear). In this condition, the change in ITD between adaptor and target resulted in unavoidable within-ear phase shifts at the transition between adaptor and target of $+\theta$ and $-\theta$ in the left and right ears, respectively; θ was equal to 0.504π (i.e. 2π ITD/period of the tone). In condition LL, the starting phase of the adaptor was $+\theta$ in both ears (but its onset was delayed by 0.5 ms in the right ear). The result was a phase shift at the transition between adaptor and target of $-\theta$ in both ears. Thus, assuming that phase shifts of $+\theta$ and $-\theta$ result in similar dips in the BM response (Shailer and Moore 1987), dips were equal across conditions.

Stimuli were generated with 16-bit resolution and a sampling rate of 40 kHz. They were played out through the digital-to-analogue converter included in the evoked potentials acquisition system (Intelligent Hearing Systems-Smart-EP, IHS) and presented binaurally through mu-metal shielded Etymotic Research ER2 insert earphones.

2.2 Recording

The FFR was recorded differentially between gold-plated scalp electrodes positioned at the midline of the forehead at the hairline (+, Fz) and at the seventh cervical vertebra (–, C7). A third electrode placed on the mid-forehead (Fpz) served as the common ground. Electrode impedances were less than 1 kΩ for all recordings. The FFR signal was recorded with a sampling rate of 8 kHz, amplified by a factor of 100,000, and band-pass filtered from 50 to 3,000 Hz (6 dB/octave roll-off, resistor-capacitor filter). Epochs with voltage changes exceeding 31 μV were automatically discarded and the trial repeated. The starting polarity of the stimuli (adaptor + target) was alternated for each presentation, and alternate-polarity sweeps were recorded and averaged in separate data buffers by the SmartEP system. Stimuli were played with a repetition rate of 1.81/s. The same stimulus condition was played in blocks of 1,500 (valid) trials; two blocks were run for each condition in randomized order across subjects.

2.3 Analysis

Offline processing was done using MATLAB (MathWorks, Natick, MA). First, the averaged FFR responses for original-polarity and for inverted-polarity stimuli were subtracted and the result divided by two, for each subject and condition. Subtraction of responses to alternating polarity stimuli enhances the representation of phase locking in response to temporal fine structure information and minimizes the representation of phase-locked activity to the envelope of the stimulus. The resulting waveform was high-pass and low-pass filtered at 150 and 2,000 Hz (eighth-order digital Butterworth filter; 3-dB down cutoff frequencies), respectively. The FFR was analysed and compared across five 50-ms time ranges: (1) from 12 to 62 ms after adaptor onset (A-start), (2) from 62 to 112 ms after adaptor onset (A-mid), (3) from 150 to 200 ms after adaptor onset (A-end), (4) from 12 to 62 ms after target onset (T-start), and (5) from 50 to 100 ms after target onset (T-end). For spectral analysis, for each subject, the 50-ms waveform was zero-padded symmetrically to make up a 1-s signal, and the magnitude spectrum was calculated via a Discrete Fourier Transform. The magnitude spectrum is specified in decibels re 0.01 μV. The dependent measure for the amount of phase-locked neural activity (the FFR strength) was defined as the highest magnitude present in the spectrum within a 12-Hz range centred at 504 Hz.

2.4 Subjects

Ten subjects (four male) participated in both experiments. They ranged in age from 20 to 24 years and had self-reported normal hearing. The 10 were selected from a pool of 13 subjects, because they showed clear FFR signals. The experiment was conducted in one session, lasting for about 3 h, including breaks.

3 Results

The latency of the unprocessed FFRs was between 7.5 and 10 ms, estimated visually as the time point relative to stimulus onset of the first occurrence of a major amplitude excursion followed by a regular pattern in the FFR traces. This is in good agreement with the range of latencies reported in the literature for FFRs (Glaser et al. 1976; Skoe and Kraus 2010) and is consistent with a generation site at the level of the IC or LL.

Figure 26.1 shows the FFR strength (the magnitude of the peak at 504 ± 6 Hz in the spectrum of the FFR) averaged across subjects and the corresponding standard error across subjects. First consider the no-gap conditions (panel (a), top). Importantly, adaptation over time is clearly visible in the FFR. To assess the significance of the reduction in FFR strength across time periods, four separate repeated measure two-way analyses of variances, ANOVAs (with factors time and adaptor ITD), were calculated. The first showed that the FFR strength was significantly larger during period A-start than during period A-mid [$F(1,9)=42.4$, $p<0.001$]. The second showed that the FFR strength was significantly larger during period A-mid than during period A-end [$F(1,9)=14.93$, $p<0.01$]. The third showed that during period T-start, the FFR strength was not significantly larger than during period A-end; the ANOVA showed no significant main effect or interaction. The fourth showed that the FFR strength was significantly larger during period A-start than during T-start [$F(1,9)=18.80$, $p<0.01$]. In all of these ANOVAs, the effect of adaptor ITD and its interaction with time period were both nonsignificant. The main effects of time period show that there was significant adaptation over time, which was not restricted to a short period after the onset of the stimulus and which was not abolished by the offset and onset ramps at the transition from adaptor to target.

To assess whether there was ITD-specific adaptation, the FFR strength during period T-start was compared across the conditions where the adaptor had the same ITD as the target (LL) and where it did not (RL). A paired-sample t-test showed no significant difference between the FFR strength for these two conditions [second group of two bars from the right; $t(9)=1.11$, $p=0.30$]. Thus, although the FFR response to the target was reduced due to the preceding adaptor, there was no evidence that this adaptation depended on whether the ITD of the adaptor matched that of the target.

Consider next the with-gap conditions (panel (b), bottom). Again, adaptation over time was visible in the FFR. However, the short silent gap between the adaptor and the target largely abolished the reduction in FFR strength during period T-start. The results of the ANOVAs were as follows. First, the FFR strength was significantly smaller during period A-mid than during period A-start [$F(1,9)=28.67$, $p<0.001$]. Second, the FFR strength was significantly smaller during period A-end than during period A-mid [$F(1,9)=5.57$, $p<0.05$]. Third, the FFR strength was significantly larger during period T-start than during period A-end [$F(1,9)=7.47$, $p<0.05$].

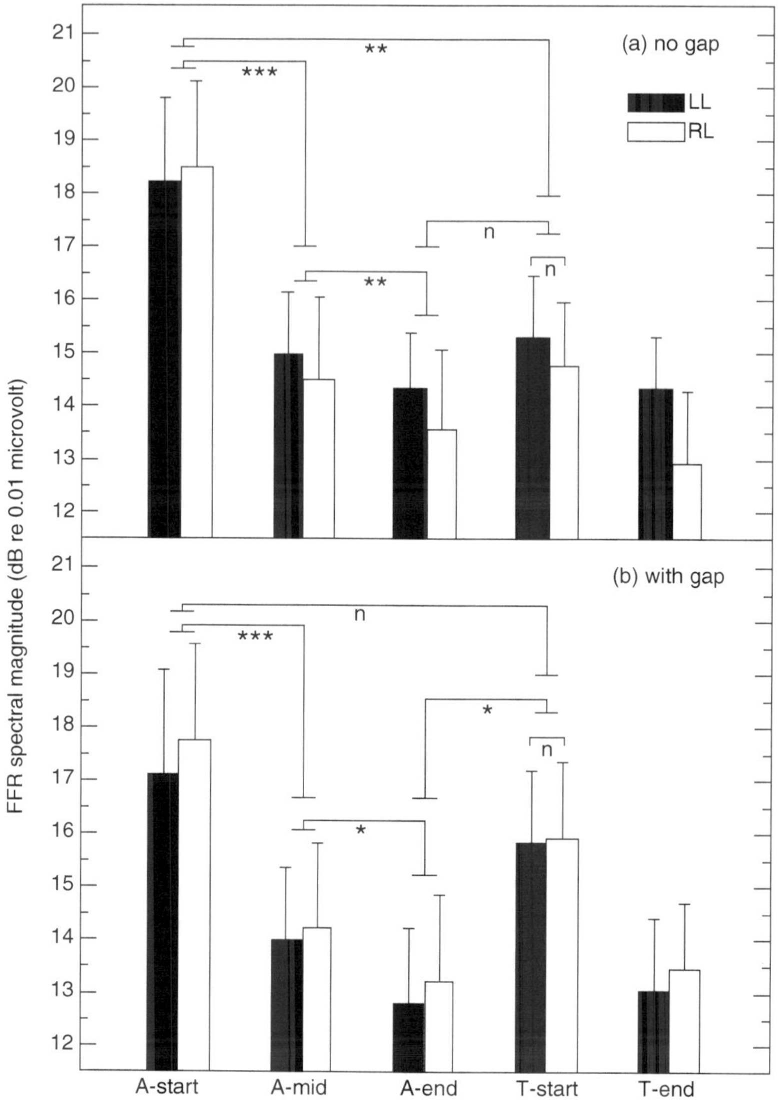

Fig. 26.1 Mean FFR spectral magnitude at 504 Hz and the corresponding standard errors across ten subjects. Panel **a** shows the data for the no-gap condition. The three left-hand groups of two bars show the FFR strength during three time ranges of the stimulation with the adaptor: 12–62 ms after onset (*A-start*), 62–112 ms after onset (*A-mid*), and 150–200 ms after onset (*A-end*). The two right-hand groups of two bars show the FFR strength during two time ranges of the stimulation with the target: 12–62 ms after onset (*T-start*) and 50–100 ms after onset (*T-end*). The *black* and the *white* bars show results when the adaptor led in the left and right ears, respectively. Panel **b** is as (a), but for the with-gap condition. $n = p > 0.05$; $* = p < 0.05$; $** = p < 0.01$; $*** = p < 0.001$

The effect of time period indicates significant recovery from adaptation. Fourth, the FFR strength was not significantly larger during A-start than during T-start [$F(1,9) = 2.13$, $p > 0.18$]. In all of these ANOVAs, the effect of adaptor ITD and its interaction with time period were both nonsignificant. Thus, while there was significant adaptation over the course of the adaptor, the larger FFR at T-start than at A-end meant that a silent gap of only ten cycles duration was sufficient to allow significant recovery from adaptation. In addition, a paired-sample t-test comparing FFR strength during T-start across conditions LL and LR showed no significant difference [$t(9)=-0.26$, $p=0.80$]. This was not entirely surprising, as the short silent gap before T-start allowed significant recovery from adaption.

4 Discussion and Conclusion

The FFR strength in response to the left-leading target was not smaller when it was preceded by the left-leading adaptor than when preceded by the right-leading adaptor. Thus, the observed adaptation effect of the adaptor on the response to the target could have been dominated by the adaptation of monaural components in the auditory pathway. Several (not mutually exclusive) factors may contribute to this finding.

1. Not all neurons in the IC, and not all neurons that contribute to the FFR, are ITD sensitive. The responses of such neurons would dilute a differential adaptation effect of the two adaptors.
2. One assumption underlying our rationale was that neurons responding to the left-leading target would have responded more strongly to the adaptor in condition LL than in condition RL. In animal studies, the majority of ITD-sensitive neurons were found to have peaks in the firing rate vs. ITD function between 100 and 300 μs; the best ITD varies across neurons. Given that we used ITDs of 500 μs, it is possible that some ITD-sensitive neurons would not have shown a large difference in response to the adaptors in condition LL vs. RL.

   A second assumption was that the FFR to the target would be reduced more when the adaptor preferentially excited neurons that were more sensitive to the target's ITD than when the adaptor preferentially excited neurons sensitive to the opposite ITD. Consider a simple model in which the FFR corresponds to the linear sum of activity from neurons tuned to left-leading and from neurons tuned to right-leading ITDs. The firing rate of neurons tuned to left-leading ITDs and responding to the left-leading target would be reduced more in condition LL than in condition RL. However, neurons tuned to right-leading ITDs may still have responded to the left-leading target, and for these neurons, the reduction in firing rate caused by the adaptor would have been greater in condition RL than in condition LL. If adaptation reduces the firing rate to the target by a certain amount (only depending on the response to the adaptor), then the two opposing effects could theoretically completely cancel each other. In practice, the fixed reduction in firing rate is limited by the fact that firing rate cannot drop below zero, and thus, complete cancellation would only occur if no or few neurons were subject to this "floor effect". Alternatively, if one assumes reduction by a certain factor (a multiplicative adaptation process), then the effects would not completely cancel. This follows because neurons tuned to the left-leading target have (by definition) a larger unadapted response to the target than neurons tuned to the opposite ITD, and so their firing rates would be reduced by a greater amount.

3. The FFR requires averaging over several thousand trials. Therefore, stimuli were played with a repetition rate of 1.81/s, giving an inter-target interval (end of target to start of next target) of 452.5 ms. It is possible that the interval between successive targets was not sufficiently long to allow ITD-sensitive neurons firing in response to the target to fully recover from adaptation due to the previous target (Ingham and McAlpine 2004). Hence, even in condition RL, there may have been some ITD-specific adaptation from the previous targets. However, this would presumably have been substantially smaller than the ITD-specific adaptation occurring in condition LL.

In conclusion, although we observed substantial and significant adaptation of the scalp-recorded FFR in humans, we found no evidence that this adaptation was ITD specific.

Acknowledgements This work was supported by Wellcome Trust Grant 088263. Thanks to Brian Moore for helpful comments.

References

Glaser EM, Suter CM, Dasheiff R, Goldberg A (1976) The human frequency-following response: its behavior during continuous tone and tone burst stimulation. Electroencephalogr Clin Neurophysiol 40:25–32

Ingham NJ, McAlpine D (2004) Spike-frequency adaptation in the inferior colliculus. J Neurophysiol 91:632–645

Marsh JT, Brown WS, Smith JC (1975) Far-field recorded frequency-following responses: correlates of low pitch auditory perception in humans. Electroencephalogr Clin Neurophysiol 38:113–119

McAlpine D, Jiang D, Shackleton TM, Palmer AR (2000) Responses of neurons in the inferior colliculus to dynamic interaural phase cues: evidence for a mechanism of binaural adaptation. J Neurophysiol 83:1356–1365

Näätänen R, Sams M, Alho K, Paavilainen P, Reinikainen K, Sokolov EN (1988) Frequency and location specificity of the human vertex N1 wave. Electroencephalogr Clin Neurophysiol 69: 523–531

Shailer MJ, Moore BCJ (1987) Gap detection and the auditory filter: phase effects using sinusoidal stimuli. J Acoust Soc Am 81:1110–1117

Skoe E, Kraus N (2010) Auditory brain stem response to complex sounds: a tutorial. Ear Hear 31:302–324

Smith JC, Marsh JT, Brown WS (1975) Far-field recorded frequency-following responses: evidence for the locus of brainstem sources. Electroencephalogr Clin Neurophysiol 39:465–472

Chapter 27
Interaural Time Difference Thresholds as a Function of Frequency

William M. Hartmann, Larisa Dunai, and Tianshu Qu

Abstract Different models of the binaural system make different predictions for the just-detectable interaural time difference (ITD) for sine tones. To test these models, ITD thresholds were measured for human listeners focusing on high- and low-frequency regions. The measured thresholds exhibited a minimum between 700 and 1,000 Hz. As the frequency increased above 1,000 Hz, thresholds rose faster than exponentially. Although finite thresholds could be measured at 1,400 Hz, experiments did not converge at 1,450 Hz and higher. A centroid computation along the interaural delay axis, within the context of the Jeffress model, can successfully simulate the minimum and the high-frequency dependence. In the limit of medium-low frequencies (f), where f . ITD $<< 1$, mathematical approximations predict low-frequency slopes for the centroid model and for a rate-code model. It was found that measured thresholds were approximately inversely proportional to the frequency (slope = –1) in agreement with a rate-code model. However, the centroid model is capable of a wide range of predictions (slopes from 0 to –2).

W.M. Hartmann (✉)
Department of Physics and Astronomy,
Michigan State University,
East Lansing, MI 48824, USA
e-mail: hartmann@pa.msu.edu

L. Dunai
Departamento de Ingeniería Gráfica,
Universitat Politècnica de València,
Camino de Vera, 46022 València, Spain

T. Qu
Key Laboratory on Machine Perception-Ministry of Education,
Peking University, Beijing 100871, China

B.C.J. Moore et al. (eds.), *Basic Aspects of Hearing*,
Advances in Experimental Medicine and Biology,
DOI 10.1007/978-1-4614-1590-9_27,

1 Introduction

The interaural time difference (ITD) is an important acoustical property that is used by humans and other animals to localize the sources of sound. This chapter studies the ability to discriminate between positive and negative ITDs for sine tones as a function of tone frequency. The chapter focuses on the functional dependence in high- and low-frequency limits with the goal of testing different models of binaural hearing.

2 High Frequencies

It is well known that human listeners are not able to detect interaural time differences (ITDs) in sine tones with frequencies greater than about 1,500 Hz (Zwislocki and Feldman 1956; Klumpp and Eady 1956). Those articles showed that although the smallest ITD thresholds occurred near 1,000 Hz, thresholds became unmeasurably high when the frequency increased above 1,300 Hz. The results point to a dramatic failure in ITD processing over a very small frequency range. Our high-frequency experiments explored this dramatic dependence in detail and determined the functional dependence of the high-frequency failure.

2.1 Methods

The listener heard two tones and was required to say whether the second tone appeared to be to the left or the right of the first. The tones were 460 ms in overall duration including a 140-ms rise duration and a 140-ms fall. The long rise time was intended to prevent the onset (identical for the two ears) from affecting ITD judgements. Tones were presented to the listener by headphones at a level of 60 dB SPL – the same in both ears. The level of the electrical signals was increased at lower frequencies to compensate for the headphone frequency response.

The application of ITDs was symmetrical about zero. For example, in a *left-right* trial with a nominal ΔITD of 20 μs, the tone led in the left ear by 10 μs during the first interval, and it led in the right ear by 10 μs during the second interval.

The experiment used a three-down, one-up adaptive staircase procedure with variable increments. The trials of an experimental run continued until the staircase had made 14 turnarounds, and the final 10 were averaged to obtain a ΔITD threshold for the run. Six runs were averaged to find the final threshold at any given frequency.

2.2 Results

There were four listeners in the experiment. Their thresholds appear in Fig. 27.1. The new information in Fig. 27.1 is that in the high-frequency limit, the threshold ΔITD grows faster than exponentially with increasing frequency.

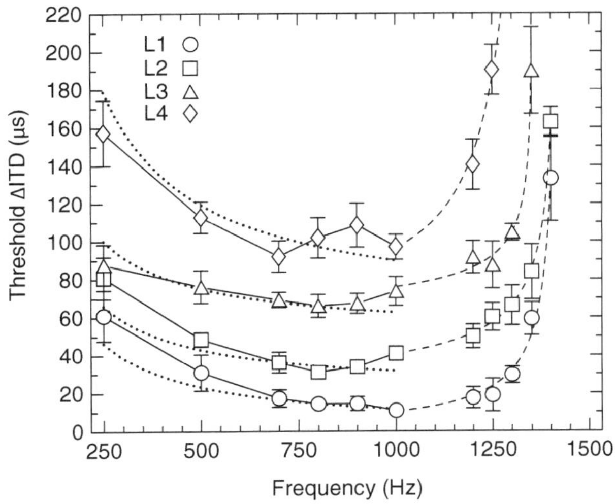

Fig. 27.1 ΔITD thresholds for four listeners are shown by symbols. Data for L2, L3, and L4 are offset vertically by 20, 50, and 60 μs, respectively. Error bars are two standard deviations in overall length. The *dotted line* is the maximum-likelihood fit to a 1/*f* law. The *dashed line* is the maximum-likelihood fit to the form $1/(f_c - f)^{\eta}$

It is possible to fit the dramatic lateralization failure at 1,450 Hz with a signal processing theory that extends the Jeffress (1948) model of the binaural system. The theory has two main parts. One part is an array of coincidence cells in the midbrain operating as cross-correlators, as observed physiologically in the medial superior olive (MSO) (e.g., Goldberg and Brown 1969; Yin and Chan 1990; Coffee et al. 2006). The second part is a hypothetical binaural display that is a nexus between the coincidence cells and a spatial representation that is adequate to determine laterality for a listener. The display is imagined to have a wide distribution of best delays with only a weak frequency dependence.

2.3 Centroid Theory

The centroid lateralization display was introduced by Stern and Colburn (1978) and applied to the lateralization of 500-Hz tones with interaural time and level differences. It was modified and extended to other frequencies by Stern and Shear (1996) to fit the lateralization data of Schiano et al. (1986). In this model display, a sine tone with angular frequency ω and an ITD of Δt excites midbrain cross-correlators represented by a cross-correlation function $c(\omega\tau)$, where τ is the lag and values of τ have a density distribution $p(\tau|\omega)$, centered on $\tau=0$.

The operative measure of laterality is the centroid of the density-weighted cross-correlation,

$$\bar{\tau} = \frac{\int d\tau\, p(\tau|\omega)\, \tau\, c(\omega\tau - \omega\Delta t)}{\int d\tau\, p(\tau|\omega)\, c(\omega\tau - \omega\Delta t)} \tag{27.1}$$

and the integrals are over the range of minus to plus infinity.

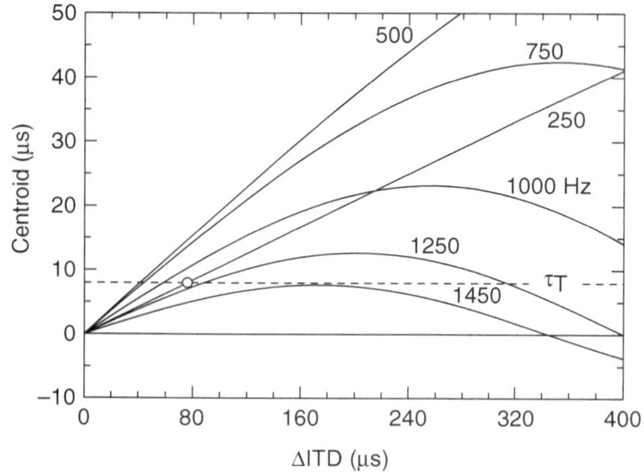

Fig. 27.2 Interaural delay centroid as computed in the centroid display model for six tone frequencies. An illustrative value of centroid threshold τ_T is shown at 8 ms. It predicts, for example, that for a 250-Hz tone, the ΔITD threshold is 76 μs, and that for 1,450 Hz, there is no threshold at all

Values of $\bar{\tau}$ can be computed given reasonable choices for $c(\omega\tau)$ and $p(\tau|\omega)$. A reasonable choice for $c(\omega\tau)$ is a cosine function, at least for frequencies of 750 Hz and higher,

$$c(\omega\tau) = 1 + m\cos(\omega\tau). \tag{27.2}$$

Parameter m ($m \le 1$) is the "rate-ITD modulation." The density of lag values can be modelled as a constant for very small τ followed by an exponentially decaying function of τ, independent of frequency as per Colburn (1977).

As pointed out by Stern and Shear, if the width of $p(\tau)$ is chosen correctly, then as the tone frequency increases, more and more cycles of $c(\omega\tau - \omega\Delta t)$ fit within the range of lags given by $p(\tau)$. This has the effect of preventing the centroid from increasing much as Δt increases because of partial cancellation of the positive lobes of $c(\omega\tau - \omega\Delta t)$ by the negative lobes. Because the centroid is the cue to laterality available to the listener, limiting the centroid in this way limits the perceived laterality. That limit could be a key to the failure to lateralize at 1,450 Hz and above.

Values of centroid $\bar{\tau}$ computed from the model are shown in Fig. 27.2 for $m = 0.4$ and for six different tone frequencies. Predictions for a threshold ΔITD can be made if it is assumed that there is a threshold value of centroid τ_T. For example, if it is assumed that the centroid threshold is $\tau_T = 8\ \mu s$, as shown by the dashed horizontal line in Fig. 27.2, then the model predicts that the threshold disappears altogether as the frequency approaches 1,450 Hz. However, in order to model the faster than exponential increase in threshold, the rate modulation m must decrease rapidly with increasing frequency.

3 Low Frequencies

All models for the neurophysiological processing of ITD depend on the cross-correlation between the neural inputs from the left and right ears. The cross-correlation function provides a measure of the difference the phase or timing of signals arriving at the two ears and thereby encodes the azimuth of a source. But although there is a general agreement about the importance of cross-correlation, there are differences of opinion about how it is applied functionally. The 1948 Jeffress model imagines a doubly tuned array of cross-correlators, tuned in best interaural delay and tuned in frequency. The tuning in best delay is normally thought to be influenced by the largest possible delay in free field given the head size, but is otherwise rather broad, enabling a *place* model for localization. Tones with different ITD cause different neurons in the central auditory system to light up.

An alternative model abandons the concept of place process. Physiological studies of single units in the inferior colliculus of guinea pigs show a strong correlation between the best interaural delay Δt and the best frequency f. The relationship is such that the phase angle $f\Delta t$ is in the neighborhood of an interaural phase of 45° (McAlpine et al. 2001). The two different models correspond to different mathematical forms for the density $p(\tau)$.

The goal of the low-frequency experiments reported here was to test models of ITD encoding against experimental measurements of just-detectable ITDs. The advantage of low-frequency experiments is that the synchrony of inputs to the cross-correlators becomes very high and stable (Joris et al. 1994) so that the remaining frequency dependence can be attributed to $p(\tau|\omega)$ and to differences between binaural display models.

3.1 Low-Frequency Theory

Binaural theories require three elements: a model cross-correlator c, a distribution function for cross-correlation units $p(\tau|\omega)$, and a model display. Function c is always a function of phase, which means that temporal parameters such as the ITD, Δt, always enter in the form $\omega\Delta t$.

In the low-frequency limit, the functional dependence on frequency can be extracted from c by expanding it in a Taylor series about $\omega\tau$. Then details of this cross-correlation function become unimportant, and attention can be focused on the distribution p and the display model. For our thresholds, the expansion in IPD, $\omega\Delta t$, is valid because the largest value obtained was always less than 0.07 cycles.

3.2 Centroid Model

The centroid model of Stern and Colburn (1978) and Stern and Shear (1996) is a model of the Jeffress type. A priori, it incorporates cells with a wide range of best delay at any frequency, though the effective extent of the array is limited by $p(\tau|\omega)$.

If the best delays, τ, scale with frequency such that $\omega\tau$ is constant, then density p can be written in terms of phase lag only, that is,

$$p(\tau \mid \omega)d\tau = p(\omega\tau)\mathrm{d}(\omega\tau) \tag{27.3}$$

Then because c and its derivative c' are also functions of phase, the low-frequency limit of ΔITD threshold is independent of the tone frequency.

A second simple case occurs when the distribution of best delays does not depend at all on frequency, as for the high-frequency calculations in the previous section. Then the predicted threshold becomes

$$\Delta t = \Delta \overline{\tau}_T / [2m < \tau^2 > \omega^2], \tag{27.4}$$

where $<\tau^2>$ is the second moment of $p(\tau|\omega)$. Therefore, the low-frequency limit of the ΔITD threshold varies inversely as the square of the tone frequency. When $p(\tau|f)$ includes parts that scale inversely with ω and parts that are independent of ω, as in the form investigated by Stern and Shear (1996), low-frequency dependences that are intermediate between flat and $1/\omega^2$ are possible.

3.3 Rate-Code Models

If function $p(\tau|\omega)$ is narrowly distributed on one side, [e.g., $p(\tau|\omega) \gg p(-\tau|\omega)$], the only available encoding for ITD is the difference in firing rates from left and right midbrain centers, $E_R - E_L$, presumably computed at a higher center.

A development parallel to the centroid model above, applied to the same symmetrical discrimination experiment, uses the change in $E_R - E_L$ between the two intervals, here defined as $\overline{\Delta}$, as the discrimination statistic. In the low-frequency limit,

$$\overline{\Delta} = 4\omega \Delta t \int_0^\infty d\tau \left[p(\tau|\omega) - p(-\tau|\omega)\right] |c'(\omega\tau)| \tag{27.5}$$

If $p(\tau|\omega)$ is a function of the product $\omega\tau$, as it appears to be in the guinea pig, the integral is independent of ω, and the frequency dependence of the ITD threshold is given by

$$\Delta t \propto \frac{\overline{\Delta}_T}{[4m\omega]}, \tag{27.6}$$

where $\overline{\Delta}_T$ is a constant, the threshold value of the discriminator. Therefore, the ΔITD is inversely proportional to the first power of the frequency.

If $p(\tau|\omega)$ is independent of ω, then

$$\Delta t = \frac{\overline{\Delta}_T}{[4m < \tau > \omega^2]}, \tag{27.7}$$

where $<\tau>$ is the first moment of $p(\tau|\omega) - p(-\tau|\omega)$. The conclusions of the calculations in the low-frequency limit are given in Table 27.1.

Table 27.1 Predicted low-frequency slopes of ΔITD as a function of frequency

Model	$p(\tau\|\omega)=p(\tau)$	$p(\tau\|\omega)=p(\omega\tau)$
Centroid	−2	0
Rate	−2	−1

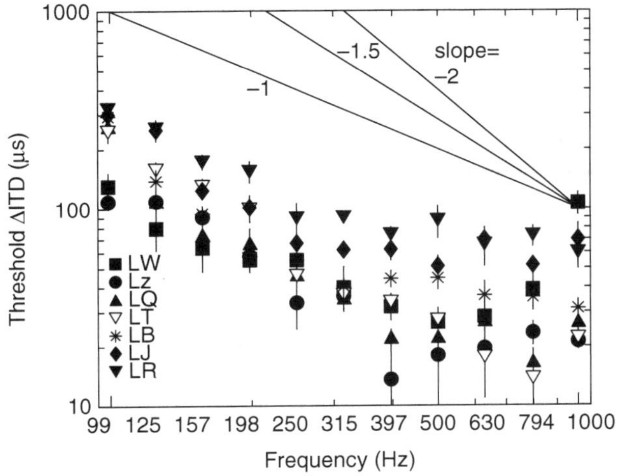

Fig. 27.3 Low-frequency thresholds for seven listeners. Three slopes are shown

3.4 Low-Frequency Experiment

Experimental stimuli were sine tones with 11 different frequencies spaced by one-third octave from 1,000 Hz down to 99 Hz. Methods were the same as for the high-frequency experiment. There were seven listeners in the experiment, with thresholds shown in Fig. 27.3. Maximum-likelihood slopes for the seven listeners were as follows: −0.97, −1.15, −1.43, −1.43, −1.08, −1.25, and −0.83 µs. The average is −1.16 and the standard deviation is 0.23. This result is most evidently consistent with the rate-code prediction with a distribution of the form $p(\tau|\omega)=p(\omega\tau)$, namely, a slope of −1. However, a final conclusion awaits the results of further experiments at still lower frequencies where the slopes appear to become steeper.

4 Conclusion

Mathematical models based on cross-correlation can successfully fit features of the mid- and high-frequency dependence of ITD thresholds, including the divergence at 1,450 Hz. They also predict the low-frequency limit of thresholds. The models gain simplicity by postulating thresholds internal to the binaural system. Incorporating variance, as in signal detection theory, can be expected to make the predictions fuzzier and more complicated.

Acknowledgments We are grateful to Les Bernstein, Constantine Trahiotis, and Richard Stern for the helpful conversations about the centroid model. LD was supported by The Vicerectorado de Profesorado y Ordenación Académica of the Universitat Politècnica de

Valeència (Spain). TQ was supported by grant 61175043 from the National Natural Science Foundation of China. The work was supported by the NIDCD grant DC-00181 and the AFOSR grant 11NL002.

References

Coffee CS, Ebert CS, Marshall AF, Skaggs JD, Falk SE, Crocker WD, Pearson JM, Fitzpatrick DC (2006) Detection of interaural correlation by neurons in the superior olivary complex, inferior colliculus, and auditory cortex of the unanesthetized rabbit. Hear Res 221:1–16

Colburn HS (1977) Theory of binaural interaction based on auditory-nerve data II. Detection of tones in noise. J Acoust Soc Am 61:525–533

Goldberg JM, Brown PB (1969) Response of binaural neurons of dog superior olivary complex to dichotic tonal stimuli: some physiological mechanisms of sound localization. J Neurophysiol 32:613–636

Jeffress LA (1948) A place theory of sound localization. J Comp Physiol Psychol 41:35–39

Joris PX, Carney LH, Smith PH, Yin TCT (1994) Enhancement of neural synchronization in the anteroventral cochlear nucleus. I. Responses to tones at the characteristic frequency. J Neurophysiol 71:1022–1036

Klumpp RB, Eady HR (1956) Some measurements of interaural time difference thresholds. J Acoust Soc Am 28:859–860

McAlpine D, Jiang D, Palmer AR (2001) A neural code for low-frequency sound localization in mammals. Nat Neurosci 4:396–401

Schiano JL, Trahiotis C, Bernstein LR (1986) Lateralization of low-frequency tones and narrow bands of noise. J Acoust Soc Am 79:1563–1570

Stern RM, Colburn HS (1978) Theory of binaural interaction based on auditory-nerve data IV. A model for subjective lateral position. J Acoust Soc Am 64:127–140

Stern RM, Shear GD (1996) Lateralization and detection of low-frequency binaural stimuli: effects of distribution of internal delay. J Acoust Soc Am 100:2278–2288

Yin TCT, Chan JCK (1990) Interaural time sensitivity in medial superior olive of cat. J Neurophysiol 64:465–488

Zwislocki J, Feldman RS (1956) Just noticeable differences in dichotic phase. J Acoust Soc Am 28:860–864

Chapter 28
Interaural Time Processing When Stimulus Bandwidth Differs at the Two Ears

Christopher A. Brown and William A. Yost

Abstract Advances in the design of cochlear implants (CIs), as well as improved CI surgical techniques, have led to an increase in the number of patients who retain some residual low-frequency acoustic hearing in the implanted ear. Many of these patients also possess some hearing in the unimplanted ear. Although their low-frequency audiometric configurations will likely be asymmetrical across ears, they may nevertheless be able to process interaural time differences (ITDs) which might aid them in localizing sound sources and achieving a spatial release from masking. We recently published research (Brown and Yost 2011) showing how sensitivity to ITD differences was affected when the stimulus bandwidths were varied between the ears, to simulate asymmetrical hearing loss in the low-frequency region. We showed that ITD discrimination thresholds decreased as the bandwidth of the noise presented to one ear increased beyond that presented to the other ear. In the current experiment, we expand upon those conditions to further explore ITD processing in the presence of interaural spectral differences. ITD sensitivity was measured when a fixed band of noise was presented to one ear and the center frequency of a spectral band of the same width was moved upward in frequency in the other ear. The data suggest that listeners have difficulty attending to ITD differences in one spectral region when there are other spectral regions that contain conflicting or inconsistent spatial information, which is likely to be the case for many CI patients who possess bilateral residual hearing.

C.A. Brown, Ph.D.
Department of Communication Sciences and Disorders,
University of Pittsburgh, 4033 Forbes Tower, Pittsburgh, PA 15260, USA

W.A. Yost (✉)
Department of Speech and Hearing Science, Arizona State University,
870102, Tempe, AZ 85287–0102, USA
e-mail: william.yost@asu.edu

B.C.J. Moore et al. (eds.), *Basic Aspects of Hearing*,
Advances in Experimental Medicine and Biology,
DOI 10.1007/978-1-4614-1590-9_28, © Springer Science+Business Media New York 2013

1 Introduction

Implant patients who retain some residual low-frequency hearing often show improved intelligibility in a competing background due to the ability to combine their electric stimulation with low-frequency acoustic stimulation (Electric-Acoustic Stimulation, EAS, see Gifford et al. 2010). In some cases the patient may have residual low-frequency hearing in both ears, and as such one might consider stimulating both ears acoustically to take advantage of low-frequency binaural acoustic processing, especially ITD processing. This chapter is concerned with interaural time difference (ITD) processing for low-frequency acoustic stimulation when the spectra at the two ears differ, as it would if patients had different amounts of low-frequency residual hearing in each ear.

We investigated ITD discrimination thresholds as a function of the spectral overlap of low-frequency noise between the two ears. This chapter is a follow-up study to Brown and Yost (2011) in which we presented a band-pass noise stimulus to each ear and then increased the bandwidth of the noise at the right ear relative to that at the left ear (see Fig. 28.1). ITD discrimination thresholds increased as the bandwidth of the noise at the right ear increased relative to that at the left ear. In this case the spectrum level of the noise did not change at either ear, but the total noise power at the right ear increased relative to that at the left ear. Although we argued that the difference in overall level at the two ears probably did not pay a large role in the outcome of the experiment, we could not conclusively rule it out. Thus, part of the motivation for the current study was to present stimuli similar to those in Brown and Yost (2011) but when the total power was the same at both ears.

In this chapter, we used band-pass filtered noises at each ear as shown in the right panel of Fig. 28.1. The bandwidth of the noise was kept constant at both ears, and the center frequency of the noise band at the right ear was shifted upward relative to that at the left ear, reducing the area of interaural spectral overlap (ISO) and increasing the spectral regions of low-frequency (in the left ear) and high-frequency

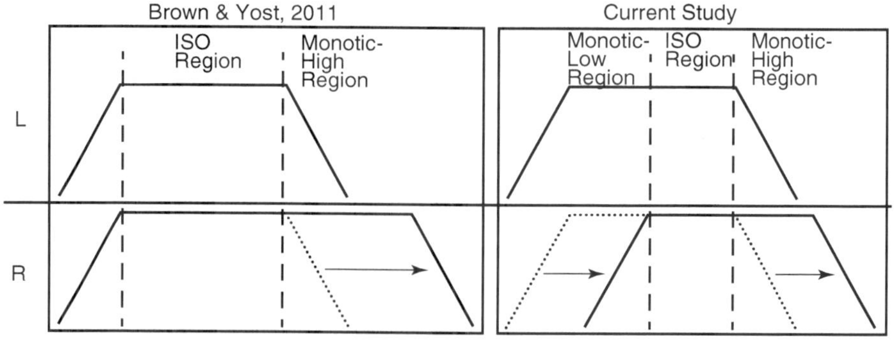

Fig. 28.1 The relationship of the noise spectra at each ear (*L* and *R*) used in Brown and Yost (2011; *left panel*) and the current study (*right panel*). In Brown and Yost (2011), the upper cutoff of the band-pass filter was increased at the right ear relative to the left ear. In the current study, the center frequency (CF) of the band of noise in the right ear was shifted upward in frequency relative to the CF of the noise band in the left ear

(in the right ear) monotic information. In these conditions the total power and spectrum level is equivalent at both ears in all conditions. These conditions are in contrast to those used in Brown and Yost (2011). There, the overall level in the right ear increased relative to that in the left ear as the upper cutoff frequency of the bandpass noise in the right ear increased, while the ISO region remained constant.

2 Methods

2.1 *Listeners*

Five listeners with normal hearing participated in the experiment. They wore Sennheiser HD250 headphones while seated in a double-walled sound proof room. Three of the subjects (L1–L3) were the same as in Brown and Yost (2011) and subject L1 was the coauthor, CB. All procedures were approved by the ASU IRB.

2.2 *Stimuli*

A 200-ms band of noise shaped with a 10-ms raised cosine rise-decay time and centered at 250 Hz was presented to the left ear with a 1/3rd-, 2/3rd-, or 2-octave bandwidth (same center frequency and bandwidths as some of the conditions of Brown and Yost 2011). The overall level of the noise was randomly varied (same at both ears) from trial to trial between 86 and 90 dB SPL. The same noise with the same bandwidth but with different center frequencies was presented to the right ear, and the center frequency was increased in 1/6 octave steps relative to the 250-Hz center-frequency noise in the left ear. Table 28.1 indicates the actual filter cutoffs of the noise bands. The ITDs were whole waveform ITDs, and a different noise was generated for each interval.

2.3 *Procedure*

A two-interval, force-choice (2AFC) task was employed. In one interval, chosen randomly, the noise contained an ITD favoring the left ear of one-half the nominal ITD, and the other interval contained an ITD of the same size, favoring the right ear. The listeners indicated which interval was perceived to be the more left. A two-up, one-down adaptive tracking procedure (tracking the 70.7 % correct point on the psychometric function) was used to estimate ITD thresholds. The initial ITD was 500 μs, and step sizes were 50 μs for the first two reversals and 20 μs for the last six reversals (thus, tracks were eight-reversals long). Thresholds

Table 28.1 The 27 filter conditions are shown

				Band-pass cutoffs (Hz)			
			Center	Left ear		Right ear	
Condition	BW (Oct)	ΔCF (Oct)	width (Oct)	HP	LP	HP	LP
1	1/3	0	2/6	223	281	223	281
2	1/3	1/6	1/6	223	281	250	315
3	1/3	2/6	0	223	281	281	354
4	1/3	3/6	−1/6	223	281	315	397
5	1/3	4/6	−2/6	223	281	354	446
6	1/3	5/6	−3/6	223	281	396	500
7	2/3	0	4/6	198	315	198	315
8	2/3	1/6	3/6	198	315	223	354
9	2/3	2/6	2/6	198	315	250	397
10	2/3	3/6	1/6	198	315	281	446
11	2/3	4/6	0	198	315	315	500
12	2/3	5/6	−1/6	198	315	353	561
13	2/3	6/6	−3/6	198	315	397	630
14	2/3	8/6	−4/6	198	315	500	794
15	2	0	12/6	125	500	125	500
16	2	1/6	11/6	125	500	140	562
17	2	2/6	10/6	125	500	158	630
18	2	3/6	9/6	125	500	177	708
19	2	4/6	8/6	125	500	198	794
20	2	5/6	7/6	125	500	222	890
21	2	6/6	1	125	500	250	1,000
22	2	8/6	4/6	125	500	315	1,260
23	2	10/6	2/6	125	500	397	1,588
24	2	12/6	0	125	500	500	2,000
25	2	14/6	−2/6	125	500	630	2,520
26	2	16/6	−4/6	125	500	794	3,174
27	2	18/6	−1	125	500	1,000	4,000

The cutoff frequencies of the band-pass filters used for the left and right ears are shown as a function of the Filter Bandwidth (BW in 1/6th octaves) and as a function of the shift in the right ear's filter center frequency (ΔCF in 1/6th octaves). The spectral width of the center band containing the ITD is shown as Center Width (in 1/6th octaves); negative values represent conditions in which there is spectral separation between the cutoffs of the filtered noises at the ears (see Fig. 28.1)

were based on the average threshold ITD over the last six reversals with the average of at least three such thresholds used to estimate the final ITD threshold.

3 Results

Figure 28.2 shows the mean ITD thresholds in μs for each listener as a function of the difference in the center frequencies of the noise bands at each ear (ΔCF), expressed in units of 1/6th octaves. The frequency characteristics of the noise band at the left ear were identical within a panel in Fig. 28.2, and the bandwidths of the noises at both

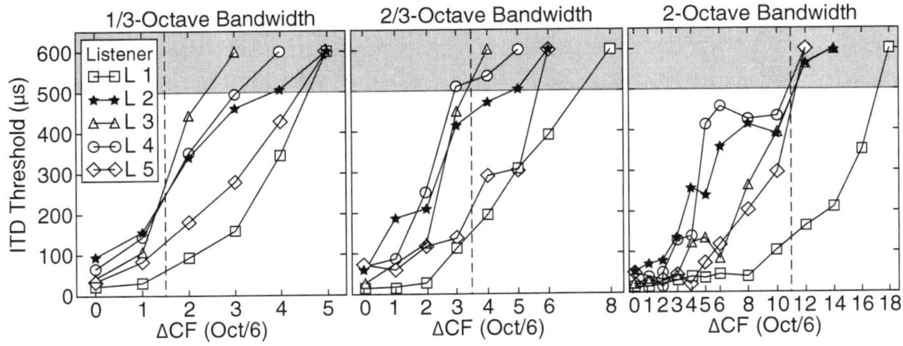

Fig. 28.2 Data for individual subjects are shown for the three bandwidth (BW) conditions (three panels). The figures display the relationship between ITD thresholds and the increase in the center frequency of the noise at the right ear (ΔCF). The *vertical dotted line* indicates the center-frequency difference when the passbands of the noises at each ear no longer overlapped (i.e., bands only overlapped in the filter skirts). *Shaded* regions indicate unreliable threshold estimates

ears were always the same. The right panel shows the data for the 1/3-octave bandwidth conditions, the middle panel the 2/3-octave bandwidth conditions, and the right panel the 2-octave bandwidth conditions. The vertical line indicates when the passbands of the two noises in each ear no longer spectrally overlap. Except for Listener 1 (author CB), listeners were usually not able to use ITD cues for discrimination when the spectra at the two ears did not overlap. When the spectra did overlap, ITD thresholds were positively correlated with the proportion of the overlap ($r^2 = 0.87$).

For all five listeners, ITD thresholds increased steeply when the proportion of the spectral overlap at the two ears was less than a particular value (approximately 30 % overlap for Listeners 2–5, and approximately 15 % for Listener 1). That is, when 70–85 % of the spectrum in the right ear was different from the spectrum in the left ear, ITD thresholds increased noticeably. This trend was the same for all three overall bandwidths (three panels of Fig. 28.2) used in the study.

4 Discussion

In Brown and Yost (2011), we found that as the bandwidth of the noise at the right ear increased beyond that presented to the left ear, ITD thresholds increased noticeably, especially for four of the five listeners (there was a more gradual change in ITD thresholds as a function of widening the bandwidth at the right ear for the coauthor, CB). In these conditions there was a low-pass band of interaural noise that could contain the ITD and a monotic band of high-frequency noise contiguous with the center band. We suggested several factors that may have contributed to the poorer ITD thresholds when the noise bandwidth was widened at the right ear: (1) an overall interaural level difference, (2) binaural interference of ITD processing caused by the high-frequency monotic band of noise at the right ear, and (3) an inability of the listeners to attend to subtle ITD changes in a dichotic image (ISO region near the middle of the head) in the presence of a fully lateralized image (monotic band at the right ear). We argued that a difficulty in attention was the most likely explanation of the data in Brown and Yost (2011).

In the present study, the results mirror those of Brown and Yost (2011) showing an increase in ITD thresholds as the spectral area of monotic noise relative to the spectral area of interaural noise increased. The ITD thresholds were in general greater for listeners L2–L5 than for L1 (coauthor CB), as was found in Brown and Yost (2011). Since there was no difference in overall level between the two ears in the present study, overall level differences are unlikely to be the basis for the poorer ITD discrimination performance as the proportion of interaural spectral overlap to monotic stimulation decreases.

While the width of the interaural spectral overlap decreases as the band of noise in the right ear is moved up in frequency, ITD thresholds do not appear to be very sensitive to different noise bandwidths when the noise is presented the same to both ears, at least over 1/3–2 octave range. That is, in Fig. 28.2 the 0 ΔCF condition represents a configuration in which the band-pass filters are the same for both ears, and the data indicate that there is a very small change in ITD threshold from a 1/3-octave wide band of noise to a 2-octave wide band. Thus, the decrease in the spectral width of the center band of interaural noise as the band in the right ear is moved upward in frequency is probably not the only reason ITD thresholds increased. This can be seen, for example, by comparing performance in conditions 1 and 23. Both conditions have a region of interaural spectral overlap that is 2/3-octaves wide, but they produce very different ITD sensitivity. Thus, the decrease in ITD threshold probably reflects a change in the proportion of the spectrum that is contained in the center interaural band relative to that in the monotic flanking bands.

The stimulus conditions and some of the results described in this chapter are similar to studies of binaural interference (see Heller and Richards 2010, for a review). In most binaural interference tasks, ITD thresholds are measured for a narrow band of noise (or a tone) in one frequency region (target band), while another narrow-band noise (or tone) is presented diotically or dichotically in a frequency region (interfering band) spectrally remote from the target. And usually only one interfering or flanking band is used. The typical result is that the presence of the interfering bands increases the ITD threshold of the target band, most often when the target band is higher in frequency than the interfering band. In the current study, if the monotic flanking bands are considered as interfering stimuli for the interaural center (target) band, then the increase in ITD thresholds may be the result of a kind of binaural interference. However, the conditions in this study differ in several ways from the past work done on binaural interference. First, the "interferers" are monotic rather than diotic or dichotic. Second, the interferers are spectrally contiguous with the target rather than spectrally separated. Finally, there are two flanking or interfering bands. Thus, if the increased ITD thresholds as a function of moving the band in the right ear up in frequency are due to binaural interference, then this study suggests that the conditions that lead to binaural interference can be expanded to include interferers that are multiple in number, monotic, and spectrally contiguous with the target band.

Models (see Buell and Hafter 1991; Heller and Trahiotis 1995) of binaural interference assume broadband processing of a weighted combination of the interaural differences of the target and interferer. In these models the ITD of the interferer "dilutes" the contribution to the weighted sum provided by the ITD of the target, forcing the target ITD to increase for threshold discrimination. One cause of the dilution may be attention. All of the subjects in this experiment and those in Brown and Yost (2011) reported on the difficulty in attending to the change in the position of the lateral image in the interaural center band caused by an ITD change, while there were lateral images at one or both ears due to the presence of the monotic flanking band(s). That is, the low- and high-frequency lateral images at the left and right ears drew the subject's attention away for the more subtle changes in the lateral image in the middle of the head caused by the changing ITD. In this experiment as the band of noise in the right ear was moved upward in frequency, the monotic lateral images appeared to become more salient making it even more difficult for subjects to attend to the changes in the lateral image caused by ITD changes in the interaural center band. As often happens when attention is a variable, some subjects attend better (e.g., the coauthor, L1, CB) than other subjects. Thus, the increases in ITD thresholds observed in this study and in the Brown and Yost (2011) study may be a result of attention being diverted to the monotic flanking bands away for the interaural center band.

This study was partially motivated by the idea that EAS patients with residual hearing in both ears might be able to process ITD differences even when the spectral regions of residual hearing are not the same at the two ears (a similar motivation was discussed in Francart and Wouters (2007)). These studies suggest that as long as there is an interaural spectral region of overlap, the EAS patients might be able to process interaural differences as long as the spectral regions of residual hearing at each ear do not vary by very much. The inability of such EAS patients to use interaural differences may not be due solely to processing ILD and ITD and cues in the region of spectral overlap. The ability to use ITD and ILD cues may also be dependent on an ability to attend to cues in the region of interaural spectral overlap (where there is similar residual hearing in both ears) when there is also stimulation occurring at only one ear.

Acknowledgments Support for this research was provided by NIDCD grants to C.A.B (Grant No. R01 DC008329) and to W.A.Y. (Grant No. R01 DC006250). We wish to thank Farris Walling, who provided valuable assistance.

References

Brown CA, Yost WA (2011) Interaural spectral asymmetry and sensitivity to interaural time differences. J Acoust Soc Am 130:EL364–EL368

Buell TN, Hafter ER (1991) Combination of binaural information across frequency bands. J Acoust Soc Am 90:1894–1900

Francart T, Wouters J (2007) Perception of across-frequency interaural level differences. J Acoust Soc Am 122:2826–2831

Gifford RH, Dorman MF, Brown CA (2010) Psychophysical properties of low-frequency hearing: implications for perceiving speech and music via electric and acoustic stimulation. Adv Otorhinolaryngol 67:51–60

Heller LM, Richards VM (2010) Binaural interference in lateralization thresholds for interaural time and level differences. J Acoust Soc Am 128:310–319

Heller LM, Trahiotis C (1995) Interference in detection of interaural delay in a sinusoidally amplitude-modulated tone produced by a second, spectrally remote sinusoidally amplitude-modulated tone. J Acoust Soc Am 97:1808–1816

Chapter 29
Neural Correlates of the Perception of Sound Source Separation

Mitchell L. Day and Bertrand Delgutte

Abstract As two sound sources become spatially separated in the horizontal plane, the binaural cues used for sound localization become distorted from their values for each sound in isolation. Because firing rates of most neurons in the inferior colliculus (IC) are sensitive to these binaural cues, we hypothesized that these neurons would be sensitive to source separation. We examined changes in the target azimuth tuning functions of IC neurons in unanesthetized rabbits caused by the concurrent presentation of an interferer at a fixed spatial location. Both target and interferer were broadband noise bursts, uncorrelated with each other. Signal detection analysis of firing rates of individual IC neurons shows that responses are correlated with psychophysical performance on segregation of spatially separated sources. The analysis also highlights the role of neural sensitivity to interaural time differences of cochlea-induced envelopes in performing this task. Psychophysical performance on source segregation was also compared to the performance of two contrasting maximum-likelihood classifiers operating on the firing rates of the population of IC neurons. The "population-pattern" classifier had access to the firing rates of every neuron in the population, while the "two-channel" classifier operated on the summed firing rates from each side of the brain. Unlike the two-channel classifier, the

M.L. Day, PhD (✉)
Eaton-Peabody Laboratories, Massachusetts Eye and Ear Infirmary,
243 Charles St., Boston, MA 02114, USA

Department of Otology and Laryngology, Harvard Medical School,
Boston, MA 02115, USA
e-mail: day@meei.harvard.edu

B. Delgutte, PhD
Eaton-Peabody Laboratories, Massachusetts Eye and Ear Infirmary,
243 Charles St., Boston, MA 02114, USA

Department of Otology and Laryngology, Harvard Medical School,
Boston, MA 02115, USA

Research Laboratory of Electronics, Massachusetts Institute of Technology,
Cambridge, MA 02139, USA

B.C.J. Moore et al. (eds.), *Basic Aspects of Hearing*,
Advances in Experimental Medicine and Biology,
DOI 10.1007/978-1-4614-1590-9_29,

population-pattern classifier could segregate the sources accurately, suggesting that some of the information contained in the heterogeneity of azimuth tuning functions across IC neurons is used to segregate sources.

1 Introduction

Sound reaching the two ears exhibits both an interaural time difference (ITD) and an interaural level difference (ILD) which are used to determine the lateral position of the sound source. Both binaural cues become distorted in the presence of a spatially separated interferer, and these changes are likely used by humans and other species to perceive spatially separated sources.

Best et al. (2004) investigated the ability of human listeners to perceptually segregate two broadband noise sources, uncorrelated with each other, at various spatial separations. The two sources could be distinguished from a single source with nearly perfect accuracy when separated by as little as 15° when one of the sources was fixed at 0° (straight-ahead). A greater spatial separation was required to achieve perceptual segregation when one of the sources was located more laterally.

We recorded from single neurons in the inferior colliculus (IC) of unanesthetized rabbits to search for neural correlates of human perception of source separation. A "target" broadband noise source was presented at various azimuths in the frontal horizontal plane in the presence of a concurrent, spatially separated noise source, uncorrelated with the target. We analyzed IC responses in two ways to compare to psychophysical results. First, we characterized neural signal detection based on changes in firing rate with spatial separation for each individual IC neuron. Next, we performed a classification of single sources versus two spatially separated sources based on the population neural activity using either of two neural decoding strategies that represent opposite extremes of how information may be combined across IC neurons.

2 Methods

Methods for surgery, acoustic stimulus delivery, and single-unit electrophysiology for the unanesthetized, head-fixed rabbit preparation were essentially the same as described previously (Devore and Delgutte 2010).

Sound stimuli in virtual acoustic space were presented binaurally through ear inserts and were first filtered with rabbit directional transfer functions (DTFs) associated with each azimuthal location in the horizontal plane. Two additional sets of DTFs with altered binaural cues were created for comparison to the "standard" condition in which ITD and ILD naturally covary with azimuth. For the "ITD-only" condition, all magnitude spectra were fixed to the spectrum at 0°, allowing ITD to vary naturally with azimuth, while ILD and monaural levels remained fixed. Similarly, for the "fixed-ITD" condition, all phase spectra were fixed to the spectrum at 0°, allowing ILD and monaural levels to vary naturally with azimuth, while ITD remained fixed.

Target and interferer stimuli were both frozen, broadband (0.1–18 kHz), 300-ms noise bursts, uncorrelated with each other, presented concurrently every 500 ms. Target and interferer were each presented at a sound level 20 dB above the neural threshold at 0°. For each IC neuron, target azimuth tuning functions (average firing rate over the stimulus duration vs. target azimuth) were collected in 15° steps with the interferer either colocated with the target (equivalent to a "single-source" condition) or fixed at 0° ("central interferer"). Azimuth tuning functions were measured under standard, ITD-only, and fixed-ITD conditions. An additional set of responses was collected under the standard condition with both target and interferer azimuths independently varied in 30° steps covering every possible spatial combination of target and interferer in the frontal hemifield. Eight stimulus trials were collected for each spatial combination.

We investigated the ability of the rate responses of individual neurons to signal the separation of the target from a central interferer (i.e., neural signal detection). For each spatial separation between target and interferer, we calculated a neural sensitivity index d' to characterize the change in neural firing rate between colocated and separated source conditions. The neural detection threshold was defined as the (interpolated) target azimuth closest to 0° at which $d'=1$ (e.g., Fig. 29.2b). Neural detection thresholds were calculated from responses collected under standard, ITD-only, and fixed-ITD conditions.

We also investigated the ability of two contrasting neural decoding strategies based on population IC activity to detect the separation of target and interferer anywhere in the frontal hemifield. For both classifiers, a conditional probability density of the population activity was estimated for each target/interferer combination, assuming multivariate Gaussian densities. "Left" and "right" populations of IC neurons were created by making a mirror twin of every neuron. The "population-pattern" classifier operated on a $2N$-dimensional vector corresponding to the firing rates of the N neurons in our sample and their mirror twins. The "two-channel" classifier operated on the 2-dimensional vector formed by the summed firing rates across the left and right IC populations. For 500 iterations of a Monte Carlo procedure, a random stimulus trial was removed from each neuron before the mean and variance of each dimension were calculated. The removed trial was then used to test the classifier performance in that iteration by classifying it to the spatial combination having the maximum likelihood. Classifier performance was assessed as the fraction of test trials in which the classifier correctly distinguished between a single source and spatially separated sources, regardless of localization accuracy.

3 Results

To investigate the physiological correlates of perception of source separation, we measured the responses to concurrent target and interferer stimuli from 99 neurons in the right IC of two female, Dutch-belted rabbits. Best frequencies (BFs) ranged from 0.19 to 17 kHz. A characteristic feature observed in many neurons was a pronounced peak or notch at 0° in the target azimuth tuning function with a central

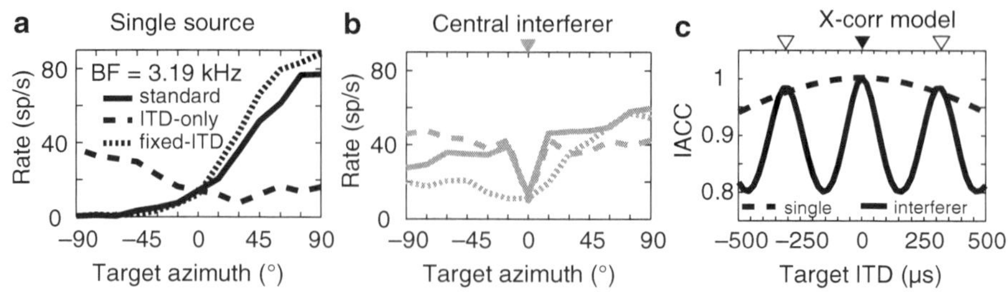

Fig. 29.1 Target azimuth tuning functions of one IC neuron under altered-cue and standard conditions for a single source (**a**) and with a central interferer (**b**). (**c**) Target ITD tuning functions from a cross-correlation model using a cochlear filter with center frequency, f_{center} = 3.19 kHz. Interferer location (*solid triangle*); ITD equal to one period of f_{center} (*open triangles*)

interferer (Fig. 29.1b, solid line, "notch"), while there was no such feature in the single-source tuning function (Fig. 29.1a, solid line). In other words, the firing rate was either sharply suppressed or enhanced as the target was separated by just 15° in either direction from the central interferer. Rate suppression or enhancement was only found in neurons with high BFs (>1.5 kHz) and occurred in a majority (72 %) of these neurons (Day et al. 2012).

The restriction of rate suppression and enhancement to high-BF neurons suggests the involvement of ILDs or envelope ITDs, as these are the binaural cues to which high-BF neurons are sensitive (Delgutte et al. 1995; Joris 2003). To identify the relevant cues, we compared azimuth tuning functions between altered-cue and standard conditions. For the high-BF neuron in Fig. 29.1a, the single-source tuning function in the fixed-ITD condition resembles that in the standard condition, while the tuning function in the ITD-only condition is nearly flat, suggesting that tuning is primarily determined by ILD. In contrast, in the presence of a central interferer (Fig. 29.1b), the fixed-ITD tuning function no longer resembles the standard tuning function, while the ITD-only tuning function shows rate enhancement similar to that observed in the standard condition. Therefore, the rate enhancement with target separation from a central interferer was dependent on sensitivity to ITD, although the single-source tuning function was primarily dependent on ILD. This neuron was representative in that rate suppression or enhancement was consistently observed in the ITD-only condition but not the fixed-ITD condition.

Rate suppression or enhancement can be explained by a cross-correlation model operating on cochlea-induced envelopes of the ear input signals. In this model, left and right sound stimuli were band-pass filtered (0.25-oct bandwidth) to mimic cochlear processing, and the normalized correlation (Trahiotis et al. 2005) was computed between the Hilbert envelopes of the left and right band-pass-filtered waveforms. As the ITD of a broadband source is varied, the interaural cross-correlation coefficient (IACC) decreases *slowly* (Fig. 29.1c, dashed line) because the envelopes being cross-correlated only contain low frequencies. However, when a central interferer is introduced (Fig. 29.1c, solid line), the IACC changes more *rapidly* with target ITD, oscillating at the period of the filter center frequency (3.19 kHz in Fig. 29.1c). While the envelopes in this case are still limited to low frequencies, the

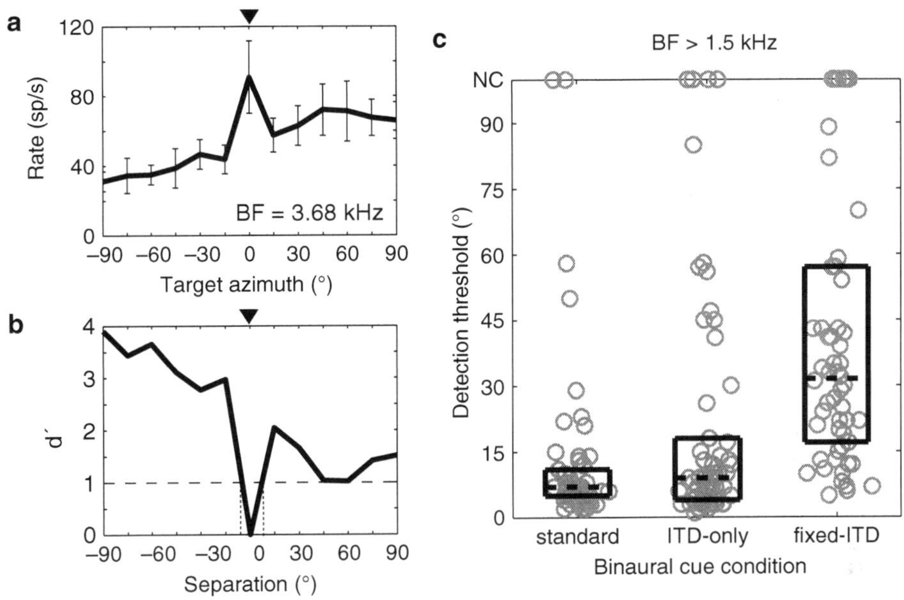

Fig. 29.2 Neural signal detection. (**a**) Target azimuth tuning function in the presence of a central interferer for a single IC neuron (bars: 1 SD). Triangle indicates interferer location. (**b**) d' calculated from the data in (a) showing the detection thresholds (*dotted lines*) where $d'=1$ (*dashed line*). (**c**) Distribution of detection thresholds across high-BF neurons ($N=58$) showing medians (*dashed lines*) and interquartile ranges (*boxes*)

rapid oscillation occurs because the waveform interactions between the target and interferer that determine the mixture envelope are dependent on the temporal fine structure after cochlear filtering. The behavior of IC neurons that are sensitive to envelope ITD is consistent with model predictions. While the firing rates of IC neurons typically change little when a single source moves by 15° (Fig. 29.1a, dashed line), the azimuth tuning functions with a central interferer show sharp changes in firing rate as the cochlea-induced envelopes become strongly decorrelated when the target is separated by 15° from the interferer.

Best et al. (2004) showed that the perceptual segregation performance for two broadband noise sources (with one fixed at 0°) remains good when the stimuli are high-pass filtered but is degraded in the fixed-ITD condition, suggesting the use of envelope ITD for segregation. The sharp rate suppression or enhancement seen in high-BF IC neurons is a possible neural correlate of this ability. To test this idea, we calculated a neural detection threshold for the spatial separation of the target from the central interferer for each high-BF neuron in our sample (Fig. 29.2a, b). The median detection threshold across the population of high-BF neurons was small (<8°) and nearly the same for ITD-only and standard conditions but was 4 times larger for the fixed-ITD condition (Fig. 29.2c). Thus, the sensitivity of IC neurons to envelope decorrelation plays a key role in the neural detection of source separation for high-pass sounds.

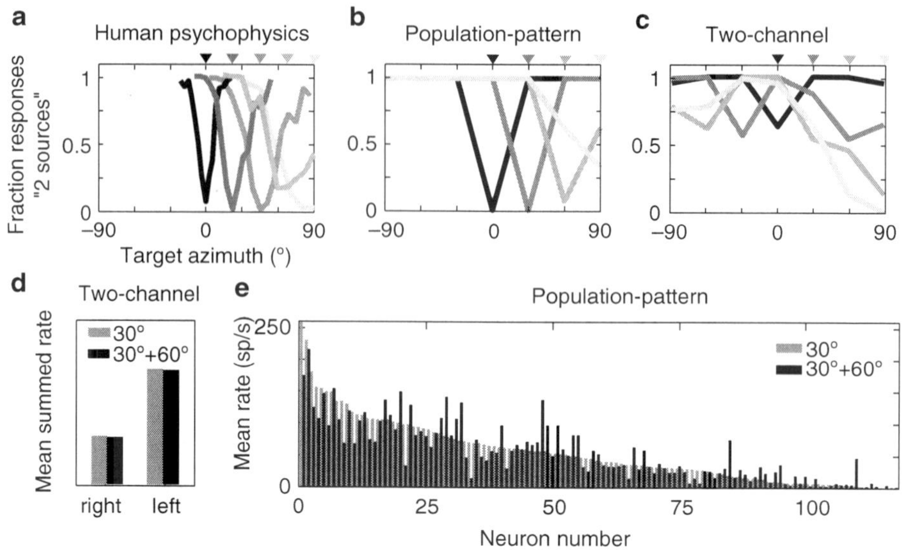

Fig. 29.3 Neural decoding. (**a**) Listener performance on the perception of separate sources as a target was separated from an interferer (Modified from Best et al. (2004), Fig. 3e). *Each line* indicates performance for an interferer fixed at the location marked by a *triangle* with matching grayscale. Performance of the population-pattern (**b**) and two-channel (**c**) classifiers on the same task. Mean rates of the two-channel (**d**) and population-pattern (**e**) classifiers for two spatial combinations (single source at 30° and two sources at 30° and 60°)

Best et al. (2004) further showed that the amount of spatial separation between target and interferer necessary for perceptual segregation increases when the interferer is placed more laterally (Fig. 29.3a). We tested the ability of two neural decoding strategies to segregate two sources anywhere in the frontal hemifield based on the firing rates of the population of IC neurons. The "population-pattern" classifier takes as input the vector of firing rates for every neuron in the population and classifies the population firing pattern into either "single source" or "two sources." This strategy represents the best performance that can be achieved from the firing rates of IC neurons under the classifier assumptions. In contrast the "two-channel" classifier operates only on the summed rates across IC neurons for each side of the brain. Such a two-channel model of sound localization has been shown to be consistent with some psychophysical results for single sources (Stecker et al. 2005; Devore et al. 2009; Lesica et al. 2010).

The population-pattern classifier was able to perfectly segregate sources with 30° separation when the interferer was fixed at 0° or 30° but required a wider separation at 60° and 90° to attain perfect segregation (Fig. 29.3b), similar to human perception. The two-channel classifier, however, failed to correctly identify single sources at 0°, 30°, and 60° (Fig. 29.3c). For example, the two-channel classifier often mistook a single source at 30° for two sources at 30° and 60°, respectively. The classifier made this error because the left and right summed rates for a single source at 30° are very similar to those for 30° and 60° (Fig. 29.3d).

The population-pattern classifier, on the other hand, has much more information available in its inputs. In Fig. 29.3e, the firing rates of each neuron in our sample in response to a single source at 30° are shown in decreasing order (gray bars). Maintaining the same order, the pattern of firing rates in response to sources at 30° and 60° (black bars) differed greatly from the pattern for 30°. The population-pattern classifier can use information contained in the heterogeneity of azimuth tuning functions across neurons to accurately segregate sources, while this information is averaged away with the two-channel classifier.

4 Discussion

The majority of high-BF neurons in the IC show a sharp rate suppression or enhancement when one broadband source is separated by just 15° from another located at 0°. These sharp rate changes are due to neural sensitivity to the decorrelation of cochlea-induced envelopes that occur with source separation. We showed that this rate suppression or enhancement is a neural correlate of the perceptual segregation of concurrent, spatially separated sources. Further, we showed that a maximum-likelihood classifier operating on the pattern of firing rates across the population of IC neurons can account for psychophysical data on perceptual segregation anywhere in the frontal hemifield by making use of the information contained in the heterogeneity of azimuth tuning functions across neurons.

The success of the population-pattern classifier on source segregation does not prove that this decoding strategy is used in the brain. This classifier assumes that some higher auditory center keeps track of the response of every IC neuron. At minimum, our result is a proof of principle that the information contained in rabbit IC firing rates is sufficient to approximate human performance on source segregation. However, the failure of the two-channel classifier does suggest that this particular neural decoding strategy cannot be used to segregate sources, even though it works adequately for the localization of a single source. Our results show that neurons that respond similarly to a single source at a given location can respond in dramatically different ways in the presence of a spatially separated interferer (Fig. 29.3e). It is likely that the neural decoding strategy used at higher auditory centers takes into account at least some of this heterogeneous tuning across neurons.

The segregation study of Best et al. (2004) required subjects to respond either "one source" or "two sources." However, a similar study of the localization of a target broadband noise in the presence of an interferer found that human listeners usually perceived a single auditory event and less often two spatially diffuse events when there were no onset or offset differences between the two stimuli (Braasch 2002). As such, the subjects' responses in the Best et al. study may be better interpreted as "single source" and "diffuse source," and similarly the population-pattern classifier may be detecting the diffuseness of the sound percept rather than two distinct source locations. The neural mechanisms involved in source separation are relevant to sound localization in reverberation, which can be interpreted as a superposition of many attenuated, separated (but coherent) sources. The effects of envelope decorrelation have also been observed in the responses of IC neurons under reverberant conditions (Devore and Delgutte 2010). Therefore, it is likely that the neural mechanisms for the perception of source separation are more generally involved in the coding of spatial properties of the surrounding environment.

Acknowledgment This work was supported by NIDCD (US) grants R01 DC002258 and P30 DC005209.

References

Best V, van Schaik A, Carlile S (2004) Separation of concurrent broadband sound sources by human listeners. J Acoust Soc Am 115:324–336

Braasch J (2002) Localization in the presence of a distracter and reverberation in the frontal horizontal plane: II. Model algorithms. Acta Acust United Ac 88:956–969

Day ML, Koka K, Delgutte B (2012) Neural encoding of sound source location in the presence of a concurrent, spatially-separated source. J Neurophysiol 108:2612–2628

Delgutte B, Joris PX, Litovsky RY, Yin TC (1995) Relative importance of different acoustic cues to the directional sensitivity of inferior-colliculus neurons. In: Manley GA, Klump GM, Koeppl C, Fastl H, Oeckinghaus H (eds) Advances in hearing research. World Scientific Publishing, Singapore, pp 288–299

Devore S, Delgutte B (2010) Effects of reverberation on the directional sensitivity of auditory neurons across the tonotopic axis: influences of interaural time and level differences. J Neurosci 30:7826–7837

Devore S, Ihlefeld A, Hancock K, Shinn-Cunningham B, Delgutte B (2009) Accurate sound localization in reverberant environments is mediated by robust encoding of spatial cues in the auditory midbrain. Neuron 62:123–134

Joris PX (2003) Interaural time sensitivity dominated by cochlea-induced envelope patterns. J Neurosci 23:6345–6350

Lesica NA, Lingner A, Grothe B (2010) Population coding of interaural time differences in gerbils and barn owls. J Neurosci 30:11696–11702

Stecker GC, Harrington IA, Middlebrooks JC (2005) Location coding by opponent neural populations in the auditory cortex. PLoS Biol 3:e78

Trahiotis C, Bernstein LR, Stern RM, Buell TN (2005) Interaural correlation as the basis of a working model of binaural processing: an introduction. In: Popper AN, Fay RR (eds) Sound source localization. Springer, New York

Chapter 30
When and How Envelope "Rate-Limitations" Affect Processing of Interaural Temporal Disparities Conveyed by High-Frequency Stimuli

Leslie R. Bernstein and Constantine Trahiotis

Abstract The purpose of this chapter is to bring together historical and current findings that reveal the presence, influence, and operation of a type of envelope "rate-limitation." The rate-limitation has been revealed in both monaural and binaural experiments. Specifically, there appears to be a low-pass envelope-filtering process that (1) functionally attenuates fluctuations of the envelope above about 150 Hz and (2) is not attributable to peripheral band-pass filtering. We show a variety of empirical outcomes and theoretical analyses that converge to demonstrate and to describe how this type of filtering constrains the processing of interaural temporal disparities (ITDs) conveyed by the envelopes of high-frequency stimuli in experiments concerning binaural detection. Included are recent behavioral and neurophysiological findings regarding how such filtering may vary with the center frequency of the stimulus.

1 Introduction

It is now well established that interaural temporal disparities (ITDs) imposed on high-frequency complex stimuli are conveyed by their envelopes. Beginning with McFadden and Pasanen (1976), who employed two-tone complexes, and Nuetzel and Hafter (1981), who employed sinusoidally amplitude-modulated (SAM) tones, it became apparent that the efficiency of processing of changes in ITD is dependent on the rate of fluctuation of the envelopes. Particularly relevant for this chapter is their finding that listeners' ability to resolve ITDs was greatly degraded when the rate of fluctuation of the envelope exceeded approximately 250 Hz.

L.R. Bernstein, PhD (✉) • C. Trahiotis
Department of Neuroscience and Surgery (Otolaryngology),
University of Connecticut Health Center,
MC3401, 263 Farmington Avenue, Farmington, CT 06030, USA
e-mail: les@neuron.uchc.edu

B.C.J. Moore et al. (eds.), *Basic Aspects of Hearing*,
Advances in Experimental Medicine and Biology,
DOI 10.1007/978-1-4614-1590-9_30,

The explanation for this type of outcome that was favored by both pairs of authors was based on peripheral auditory filtering. In order to understand their reasoning, assume the stimuli are processed via an auditory filter at the center frequency of the stimulus. It follows that increasing the rate of modulation would, necessarily, result in either decreases in the absolute level of the modulated stimulus and/or decreases in the depth of modulation of the stimulus passing through that filter. In either case, there would be a concomitant reduction in the fluctuations of internal neural events that are both synchronized to the envelope of the external stimulus and responsible for conveying the ITDs.

It is interesting, historically, that Nuetzel and Hafter (1981), who favored an explanation based solely on peripheral filtering, also explicitly noted the logical possibility that the outcomes they observed could have resulted from a degradation of the precision with which high rates of envelope modulation are encoded neurally. This type of limitation could occur even though all of the spectral components of the stimulus pass, essentially without attenuation, through the peripheral auditory filter.

2 Psychophysical Evidence for an Envelope Rate-Limitation

An experiment reported by Bernstein and Trahiotis (1994) both replicated the results of McFadden and Pasanen (1976) and Nuetzel and Hafter (1981) and supported the notion that a rate-limitation, per se, rather than peripheral auditory filtering, was primarily responsible for the degradation in the processing of ITDs at high rates of envelope fluctuation. Among their stimulus conditions, Bernstein and Trahiotis measured sensitivity to ITD conveyed by SAM tones and by two-tone complexes as a function of rate of modulation at carrier frequencies of 4 and 8 kHz. The use of SAM tones and two-tone complexes in this manner allows one to dissociate spectral separation from the rate of fluctuation of the envelope. They also measured sensitivity to changes of ITD with bands of Gaussian noise.

The top and bottom panels of Fig. 30.1 display the data obtained at center frequencies of 4 and 8 kHz, respectively. Threshold ITDs ($d'=1$) represent the average across three listeners and the error bars represent ± one standard error of the mean. The Xs represent conditions for which valid thresholds could not be obtained. The upper and lower abscissas have been aligned to reflect the fact that the expected number of envelope maxima per second ("envelope rate") for a rectangular band of noise is about 64 % of the bandwidth in Hz (Rice 1953).

In general, the results replicate the findings of the earlier studies in that resolution of ITDs is greatly degraded and/or rendered impossible (data plotted as Xs) for rates of modulation of about 250 Hz and higher. This occurs similarly for SAM tones and two-tone complexes even though, for a given rate of modulation, the spectral separation of the "sidebands" of the two-tone complexes is half that of the SAM tones. Of particular importance here is the finding that the rate of modulation above which threshold ITDs increase is substantially *lower* for stimuli centered at 8 kHz than for stimuli centered at 4 kHz. This result is the opposite of what one would expect if attenuation of spectral components via peripheral auditory filtering were

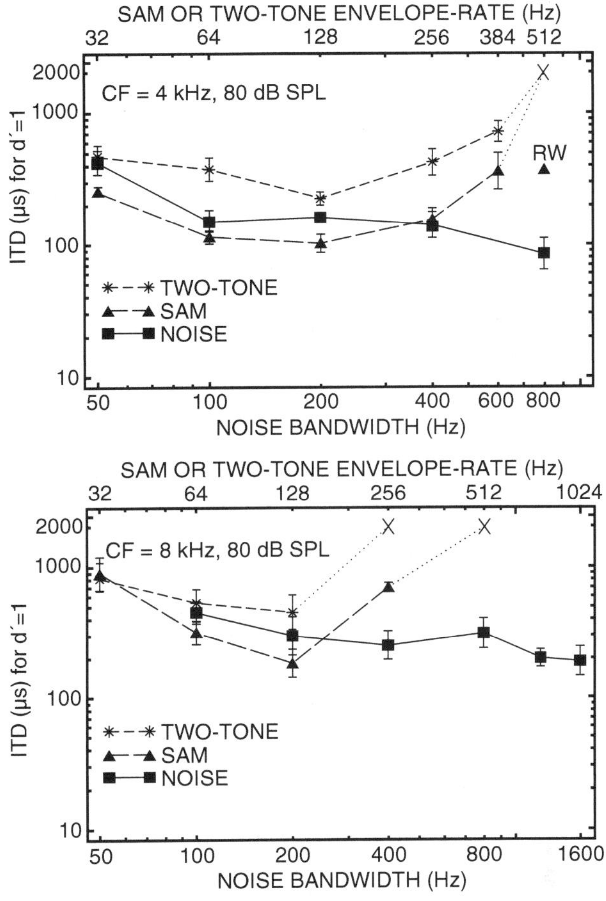

Fig. 30.1 Threshold ITD as a function of envelope-rate or noise bandwidth for stimuli centered at 4 kHz (*upper panel*) or 8 kHz (*lower panel*) (Adapted with permission from Bernstein and Trahiotis (1994). Copyright 1994, Acoustical Society of America)

responsible for the rate-limitation. This is so because the auditory filter centered at 8 kHz is roughly twice as wide as the one centered at 4 kHz. Note that a rate-limitation was not evident for ITDs conveyed by the envelopes of bands of noise. This is expected because, unlike the case for the spectrally discrete stimuli, increasing the bandwidth of noise beyond one auditory filter would not be expected to alter the internal representation of the envelope of the continuous-spectrum noise passed by that filter. Thus, threshold ITDs would not be expected to increase as the bandwidth of a noise is increased. On the contrary, threshold ITDs would be expected to, if anything, become smaller to the degree that independent information concerning ITD could be accrued across separate filters. The data obtained at both center frequencies do, in fact, exhibit such a decline.

Three subsequent studies, Kohlrausch et al. (2000), Ewert and Dau (2000), and Moore and Glasberg (2001), showed that an envelope rate-limitation also affects processing of envelope fluctuations at high spectral frequencies when the cues are monaural rather than binaural. In those studies, temporal modulation transfer functions (TMTFs) were measured at various center frequencies. The patterning of the

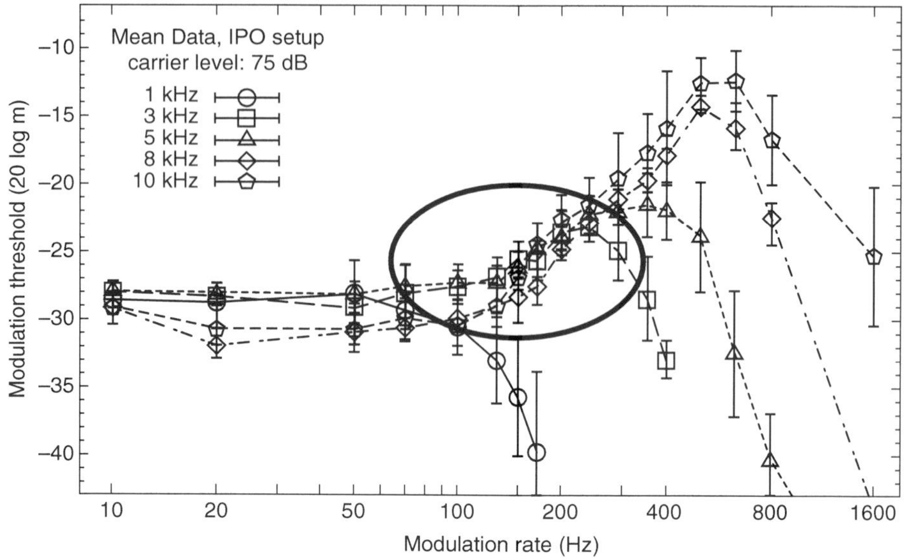

Fig. 30.2 TMTFs as a function of center frequency (Reprinted with permission from Kohlrausch et al. (2000). Copyright 2000, Acoustical Society of America)

data and their quantitative analyses led Kohlrausch et al. (2000) and Ewert and Dau (2000) to include in their modeling a low-pass filter that attenuates, independent of the center frequency of the stimulus and, correlatively, independent of the width of the auditory filter, fluctuations of the envelope that are more rapid than 150 Hz. Representative TMTF data taken from Kolhraush et al. are shown in Fig. 30.2. The superimposed oval highlights the region of the data demonstrating the rise of the thresholds associated with increases in the rate of modulation.

It remained to be seen whether the incorporation of such an envelope low-pass filter into a comprehensive cross-correlation-based model of binaural hearing would provide a quantitative accounting of the rise in threshold ITDs observed for high rates of modulation. To that end, Bernstein and Trahiotis (2002) measured threshold ITDs for SAM tones and transposed tones centered at 4, 6, or 10 kHz while varying rate of modulation. Their results, averaged across four listeners, are shown in Fig. 30.3 as solid symbols. Note that, while listeners were consistently more sensitive to changes in ITD conveyed by the transposed tones, the rates of modulation beyond which thresholds ITDs increased were essentially the same for the two types of stimuli at each center frequency. That outcome notwithstanding, as indicated by the threshold ITDs obtained at a rate of modulation of 256 Hz, the "limiting" rate of modulation appears to have *decreased* as center frequency increased, in line with the results of Bernstein and Trahiotis (1994). In fact, at that rate of modulation, the listeners were unable to perform the task when the center frequency was 10 kHz even for ITDs as large as 1 ms.

The dotted lines in each panel of Fig. 30.3 represent predictions of a general cross-correlation-based model (e.g., Bernstein and Trahiotis 2002) that does not incorporate the envelope low-pass filter of interest. Those predictions can be used to

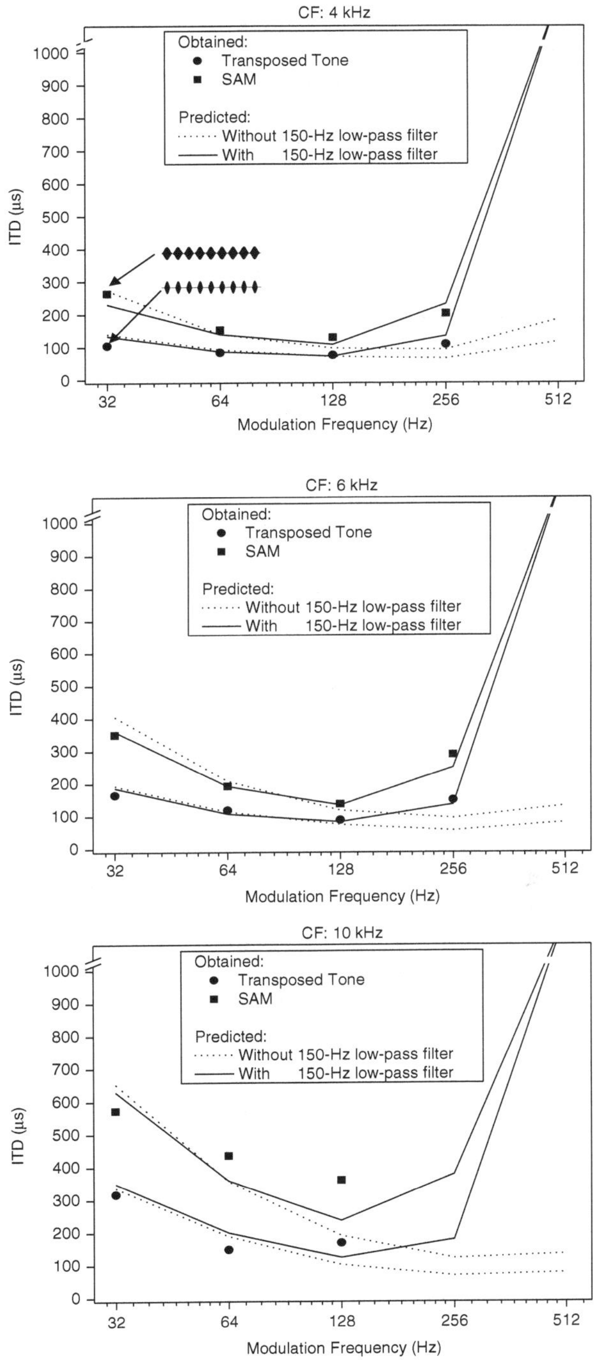

Fig. 30.3 Threshold ITDs as a function of modulation frequency for SAM and transposed tones centered at 4, 6, or 10 kHz (Adapted with permission from Bernstein and Trahiotis (2002). Copyright 2002, Acoustical Society of America)

evaluate Nuetzel and Hafter's (1981) notion that peripheral auditory filtering degrades resolution of ITDs at high rates of modulation by attenuation of the "sidebands" of the stimulus. Consistent with that view is the small rise in predicted threshold ITDs at rates of modulation ≥256 Hz for stimuli centered at 4 kHz and progressively smaller rises in predicted threshold ITDs at 6 and 10 kHz, respectively. Although such a model does a fairly good job of predicting threshold ITDs for both types of stimuli at low and moderate rates of modulation, it fails to predict the magnitude of the increases in threshold ITD at high rates of modulation at any of the center frequencies tested.

The solid lines represent predicted threshold ITDs derived from the cross-correlation-based model after it was augmented to include a 150-Hz low-pass envelope filter as suggested by Kohlrausch et al. (2000). As is evident from the figure, including the low-pass filter resulted in predictions that are much more accurate and even successfully predict when the task could not be performed at center frequencies of 4 and 6 kHz. A failure of the model is its prediction that the task could have been performed at a rate of modulation of 256 Hz for the 10-kHz-centered stimuli. As indicated by the absence of data points for this condition, the task could not be performed. The model's failure is a consequence of the underlying (apparently false) assumption that the cutoff frequency of the envelope low-pass filter is independent of center frequency in tasks measuring threshold ITDs. That assumption runs counter to the patterning of the data in Fig. 30.3 and to the patterning of the data in Fig. 30.1.

Recently, Majdak and Laback (2009) reported the results of an experiment designed explicitly to measure effects of center frequency and rate of fluctuation of the envelope on threshold ITDs. The stimuli they employed were 1,500, 2,121, and 3,000-Hz-wide band-pass filtered click trains centered near 4,600, 6,500, or 9,200 Hz, respectively. Such stimuli appear to be advantageous because the temporal signatures of their filtered envelopes have especially steep slopes that would be expected to facilitate sensitivity to ITD and do so in a manner that might help to reveal differences across center frequency by increasing the "dynamic range" of the data. Based on their results, Majdak and Laback concluded that while, overall, there was a decrease in sensitivity to changes in ITD with increasing center frequency, there was no change in the nature of envelope low-pass filtering or envelope rate-limitation *across* center frequency. We were puzzled by this outcome, given the unequivocal patterning of the data in Figs. 30.1 and 30.3.

Close scrutiny of Majdak and Laback's (2009) experimental procedures revealed potentially important differences as compared to the experimental procedures routinely employed in our laboratory. Perhaps the most salient concerned the nature of the "background noise." While they employed a continuous, interaurally uncorrelated broadband noise (50–20 kHz) presented at a spectrum level of about 9 dB, we employed a continuous, diotic noise low passed at 1,300 Hz and presented at a spectrum level of 30 dB. It seems possible that the nature of Majdak and Laback's background noise may have affected the processing of ITDs by their listeners in two distinct manners. First, the spectral extent of their interaurally uncorrelated background noise overlapped with the spectral regions of the "target" stimuli that conveyed the ITD, while ours did not. Second, the substantially lower spectrum level of their background noise may not have precluded their listeners' use of ITDs conveyed by low-level, low-frequency, fine-structure-based information. This could explain the very low threshold ITDs (below 50 μs) obtained from their listeners NH2, NH7, and NH8 at a pulse rate of 200 Hz at three center frequencies.

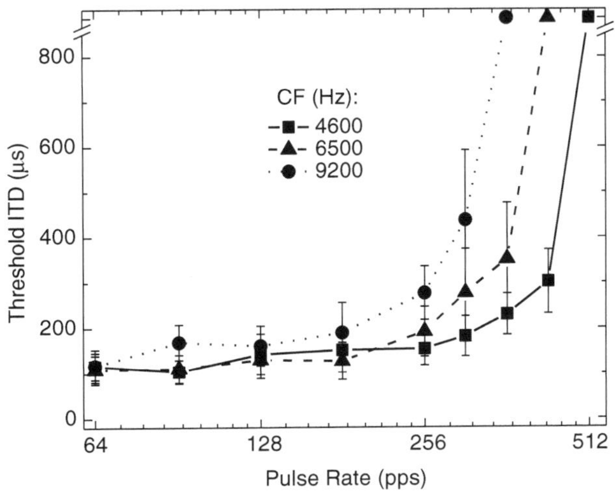

Fig. 30.4 Threshold ITD as a function of pulse rate. Error bars represent ± one standard error of the mean. Points plotted above the break in the ordinate represent conditions for which the task was essentially impossible

In light of these differences, we decided to measure threshold ITDs using Majdack and Laback's (2009) target stimuli in our laboratory while employing our psychophysical procedures and background noise. Figure 30.4 displays the results averaged across four listeners. Note that the patterning of the data reveals that threshold ITDs begin to increase at progressively lower pulse (envelope) rates as center frequency is increased. Using the cross-correlation-based model discussed earlier, we sought, for each center frequency, the cutoff of the second-order envelope low-pass filter that best fit the data. Those cutoffs were 330, 240, and 170 Hz for center frequencies of 4,600, 6,500, and 9,200 Hz, respectively. The predictions of the model accounted for 90% of the variance in the behavioral data. These empirical and theoretical outcomes are consistent with the findings obtained previously in our laboratory (Figs. 30.1 and 30.3) suggesting that the cutoff frequency of the envelope low-pass filter decreases with increases in center frequency. They are not, however, consistent in that the apparent and derived cutoff frequencies of the envelope low-pass filter centered near 4 kHz is about twice the value obtained earlier with different stimuli. At this point, we have no satisfactory explanation for the difference but are planning the type of large-scale multifactor, parametric study that we believe is required in order to understand and to resolve the differences. Ultimately, the challenge will entail gaining a much deeper understanding of the mechanisms underlying the low-pass nature of the envelope rat limitation.

3 Neurophysiological Evidence for an Envelope Rate-Limitation

The results of two recent neurophysiological investigations are consistent with the behavioral and theoretical results described above. Both Rodríguez et al. (2010) and Middlebrooks and Snyder (2010) measured responses of neural units within the

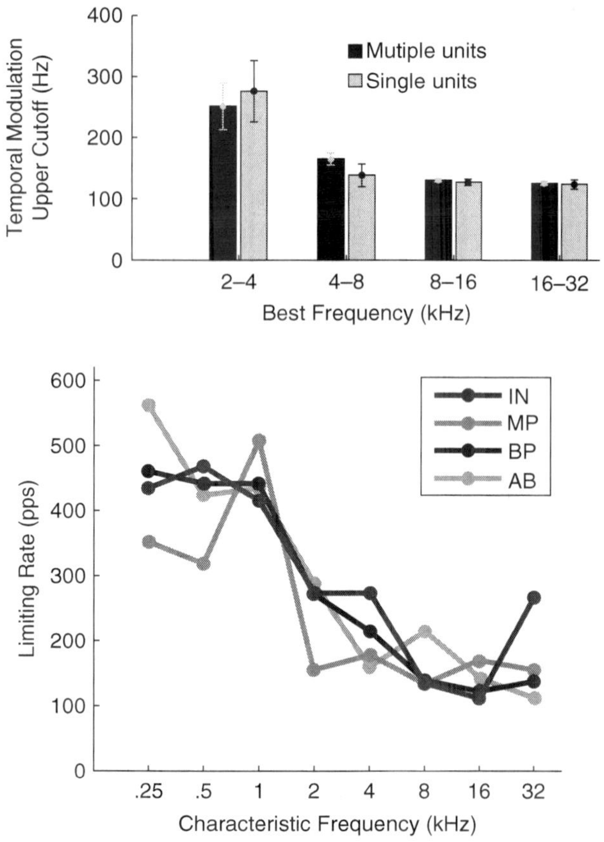

Fig. 30.5 Neurophysiological data obtained from the inferior colliculi of cats with four different electrode types. *Upper panel*: Upper cutoff of temporal modulation rate as a function of center frequency (Adapted from Rodríguez et al. (2010) with permission from APS). *Lower panel*: Limiting rate of phase-locking to pulsatile stimuli as a function of center frequency (Adapted from Middlebrooks and Snyder (2010) with permission from SFN)

inferior colliculus (IC) of cats. Rodríguez et al. stimulated acoustically with spectrally and temporally dynamically changing "ripple" stimuli, while Middlebrooks and Snyder (2010) stimulated the auditory nerve electrically using pulsatile stimuli. Representative data from each study are shown in Fig. 30.5. The results of both studies are in agreement in that the rates of fluctuation beyond which envelope coding degrades systematically decrease with increases in the spectral frequency to which the unit is best tuned. This can be seen in the responses obtained from units having "best" or "characteristic" frequencies of 2 kHz and higher, for which neural responses are synchronized to the envelope, rather than to the fine-structure of the stimulus. In both studies, those rates (between about 100 and 250 Hz) are remarkably similar to the ones derived behaviorally (Fig. 30.4).

4 Summary

In toto, both behavioral and neurophysiological data and their respective quantitative analyses strongly suggest that both monaural and binaural temporal processing are constrained by a process or processes that can be described as a low-pass filtering of the envelope of stimuli centered at high frequencies. Furthermore, the ability to encode precisely rapid fluctuations of such envelopes appears to decline as the center frequency of the stimulus and, thus, the center frequency of the cochlear tonotopic region stimulated increases.

Acknowledgment This research was supported by research grant NIH DC-04147 from the National Institute on Deafness and Other Communication Disorders, National Institutes of Health.

References

Bernstein LR, Trahiotis C (1994) Detection of interaural delay in high-frequency SAM tones, two-tone complexes, and bands of noise. J Acoust Soc Am 95:3561–3567

Bernstein LR, Trahiotis C (2002) Enhancing sensitivity to interaural delays at high frequencies by using "transposed stimuli". J Acoust Soc Am 112:1026–1036

Ewert SD, Dau T (2000) Characterizing frequency selectivity for envelope fluctuations. J Acoust Soc Am 108:1181–1196

Kohlrausch A, Fassel R, Dau T (2000) The influence of carrier level and frequency on modulation and beat-detection thresholds for sinusoidal carriers. J Acoust Soc Am 108:723–734

Majdak P, Laback B (2009) Effects of center frequency and rate on the sensitivity to interaural delay in high-frequency click trains. J Acoust Soc Am 125:3903–3913

McFadden D, Pasanen EG (1976) Lateralization at high frequencies based on interaural time differences. J Acoust Soc Am 59:634–639

Middlebrooks JC, Snyder RL (2010) Selective electrical stimulation of the auditory nerve activates a pathway specialized for high temporal acuity. J Neurosci 30:1937–1946

Moore BCJ, Glasberg BR (2001) Temporal modulation transfer functions obtained using sinusoidal carriers with normally hearing and hearing-impaired listeners. J Acoust Soc Am 110:1067–1073

Nuetzel JM, Hafter ER (1981) Discrimination of interaural delays in complex waveforms: spectral effects. J Acoust Soc Am 69:1112–1118

Rice SO (1953) Mathematical analysis of random noise. In: Wax N (ed) Selected papers on noise and stochastic processes. Dover, New York, pp 133–294

Rodríguez FA, Read HL, Escabí MA (2010) Spectral and temporal modulation tradeoff in the inferior colliculus. J Neurophysiol 103:887–903

Chapter 31
The Sound Source Distance Dependence of the Acoustical Cues to Location and Their Encoding by Neurons in the Inferior Colliculus: Implications for the Duplex Theory

Heath G. Jones, Kanthaiah Koka, Jennifer Thornton, and Daniel J. Tollin

Abstract For over a century, the Duplex theory has posited that low- and high-frequency sounds are localized using two different acoustical cues, interaural time (ITDs) and level (ILDs) differences, respectively. Psychophysical data have generally supported the theory for pure tones. Anatomically, ITDs and ILDs are separately encoded in two parallel brainstem pathways. Acoustically ILDs are a function of location and frequency such that lower and higher frequencies exhibit smaller and larger ILDs, respectively. It is well established that neurons throughout the auditory neuraxis encode high-frequency ILDs. Acoustically, low-frequency ILDs are negligible (~1–2 dB); however, humans are still sensitive to them and physiological studies often report low-frequency ILD-sensitive neurons. These latter findings are at odds with the Duplex theory. We suggest that these discrepancies arise from an inadequate characterization of the acoustical environment. We hypothesize that low-frequency ILDs become large and useful when sources are located near the head. We tested this hypothesis by making measurements of the ILDs in chinchillas as a function of source distance and the sensitivity to ILDs in 103 neurons in the inferior colliculus (IC). The ILD sensitivity of IC neurons was found to be frequency *independent* even though far-field acoustical ILDs were frequency *dependent*. However, as source distance was decreased, the magnitudes of low-frequency ILDs increased. Using information theoretic methods, we demonstrate that a population of IC neurons can encode the full range of acoustic ILDs across frequency that would be experienced as a joint function of source location and distance.

H.G. Jones • K. Koka • J. Thornton • D.J. Tollin (✉)
Department of Physiology and Biophysics,
University of Colorado School of Medicine,
RC1-N, Stop 8307, 12800 E. 19th Avenue,
Aurora, CO 80045, USA
e-mail: daniel.tollin@ucdenver.edu

B.C.J. Moore et al. (eds.), *Basic Aspects of Hearing*,
Advances in Experimental Medicine and Biology,
DOI 10.1007/978-1-4614-1590-9_31,

1 Introduction

The Duplex theory states that interaural time differences (ITDs) are used to localize low-frequency sounds, whereas interaural level differences (ILDs) are used to localize high-frequency sounds (Rayleigh 1907). Over a century of anatomical, physiological, and behavioral research has generally supported the Duplex theory for pure tone stimuli in both humans and other animal models (Hafter 1984). While there have been some modifications to the Duplex theory (Hafter 1984), for absolute localization of free-field sounds, the theory continues to hold up (Macpherson and Middlebrooks 2002). A caveat, however, is that humans can discriminate ~1 dB differences in ILDs for frequencies as low as 200 Hz (Hafter 1984). Additionally, physiological studies have consistently observed low-frequency tuned (<~2–4 kHz) ILD-sensitive neurons in brainstem nuclei (Sanes and Rubel 1988; Semple and Kitzes 1987) and cortex (Benson and Teas 1976). Given that ILDs for low frequencies are negligible, the joint presence of neurons with the ability to encode large low-frequency ILDs and behavioral sensitivity is peculiar.

According to ecological and efficiency principles, it is hypothesized that neural systems have evolved strategies to faithfully represent the full spectrum of sensory signals experienced by an organism in its natural habitat (Barlow 1961). That is, through experience neurons match their limited operating ranges to the range of signals they must encode. On these grounds, low-frequency ILD-sensitive neurons should not exist nor should behavioral sensitivity because there is no such physical cue. We suggest that the past studies have not considered the full set of acoustical cues that are actually experienced in real-world environments.

2 Methods

2.1 *Electrophysiological Methods*

Extracellular responses of well-isolated neurons were recorded in the ICC of ketamine-anesthetized chinchillas. Neuronal ILD sensitivity was examined using 50-ms duration characteristic frequency (CF) tones by holding the signal level to the contralateral ear (~20 dB re: threshold) constant and varying the level at the ipsilateral ear from 30 dB below to 30 dB above ipsilateral threshold. The rate vs ILD for each neuron was fitted with a 4-parameter sigmoid, $rate(ILD) = y_o + \alpha/(1+exp(-(ILD - ILD_o)/\beta))$; before fitting, the data were normalized to the maximum rate (Fig. 31.1a). The fits accurately described the rate-ILD data ($R>0.9$ in 96/103 neurons). The parameters of the functions (Fig. 31.1) were used for analysis. Half-max ILD is the ILD at 50 % of the maximal rate, rate vs ILD slope (spikes/s/dB, not

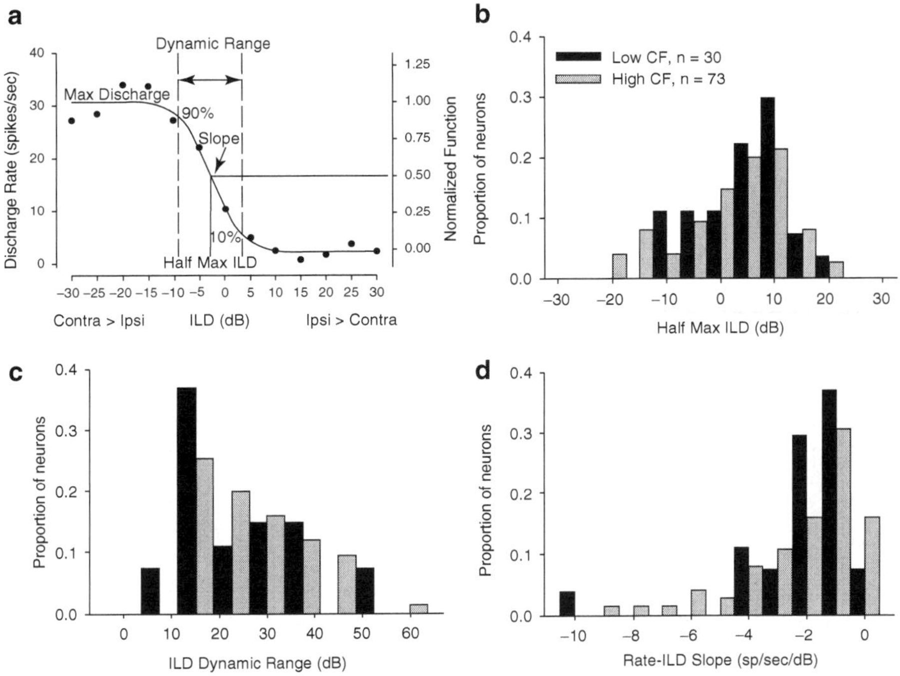

Fig. 31.1 (**a**) Example rate vs ILD function. Distributions of half-max ILD (**b**), ILD dynamic range (**c**), and rate-ILD slope (**d**) for low (<4 kHz)- and high (>4 kHz)-frequency ICC neurons

normalized) was computed at half-max ILD, and ILD dynamic range was defined between 90 and 10 % of maximal rate.

2.2 *Acoustical Measurements of ILD as a Function of Source Distance*

ILDs were calculated from acoustical measurements made at different sound source-to-observer distances (10, 20, and 100 cm). Recordings took place in a double-walled, acoustic chamber (see Koka et al. 2011 for details). A cone-shaped speaker (MF1-M, TDT, Alachua, FL), creating an acoustic point source, was attached to an industrial sliding camera tripod so that the source could be placed precisely at the required distances from the animal. The animal was placed on a platform such that it could be rotated around its interaural axis through ±90° in 10° steps. The free-field-to-eardrum acoustic impulse response for each ear and each location was measured from which head-related transfer functions (HRTFs) were computed. The acoustic ILDs at each position were calculated by subtracting the left ear HRTFs from right ear.

3 Results

3.1 Low-Frequency ICC Neurons Code for ILDs Outside the Acoustic Physiological Range

Physiological data were based on 103 ILD-sensitive ICC neurons. Figure 31.2 plots the neuronal half-max ILD (symbol) plus the ILD dynamic range (error bars) for each neuron as a function of CF. The dynamic range indicates the ILDs over which the neural response was modulated, while the half-max ILD gives the midpoint ILD about which the response is modulated maximally (Fig. 31.1a). The black line in Fig. 31.2 indicates the maximum acoustical ILD measured at a 1 m (100 cm) distance as a function of frequency computed across all locations in the frontal hemisphere. Consistent with the Duplex theory, ILDs are small (<~1–3 dB) until the frequency exceeds 4 kHz.

Consistent with the acoustics, Heffner et al. (1994) demonstrated that chinchillas use ILDs to localize free-field sounds above 4 kHz and likely use ITDs below 4 kHz. Based on these empirical behavioral and acoustics data, our neurons were separated into low (<4 kHz) and high (>4 kHz) CF groups (vertical dashed line, Fig. 31.2). Two points are observed from Fig. 31.2. First, the distributions of the ILD sensitivity metrics (e.g., Fig. 31.1a) for neurons with high CFs fall within (half-max ILD) and generally span (ILD dynamic range) the range of the empirically measured ILDs available for far-field sources (100 cm; solid black lines, Fig. 31.2). The majority of the half-max ILD values (95 %) for high-CF neurons (69/73) fell within the range of maximum ILDs.

On the other hand, the half-max ILDs of the majority (~60 %, 18/30) of low CF neurons (<4 kHz) fell at the edge (within 1 dB) or outside of the acoustic range of ILDs. That is, the neural sensitivities of low-frequency ICC neurons were not constrained to the physiological range of ILDs. To test the hypothesis that low-frequency neurons might have different ILD coding capabilities than high-frequency neurons, the distributions of half-max ILD, ILD dynamic range, and the rate-ILD slope

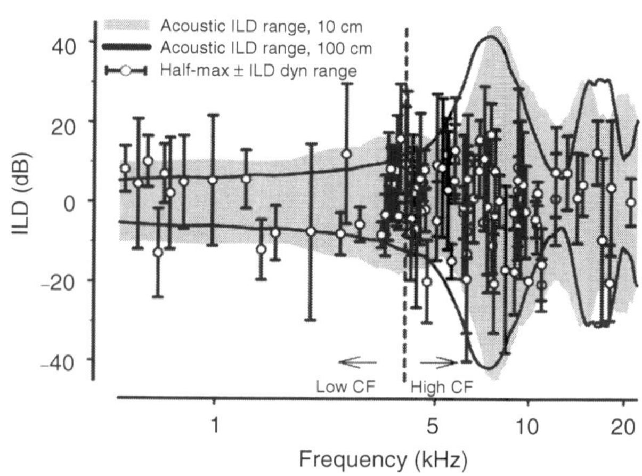

Fig. 31.2 Maximum acoustic ILDs (100 cm, *black line*) and ICC neuron coding as a function of frequency. Half-max ILD (○) and dynamic range (*error bars*) are shown for each ICC neuron. *Shaded* area indicates max acoustic ILDs for source distances 10 cm from the head

(Figs. 31.1b–d) were examined. No significant frequency-dependent differences between the means were observed for any of the ILD coding parameters [half max: $t_{101} = 1.5$, $p = 0.14$; dynamic range: $t_{101} = -0.42$, $p = 0.68$; slope: $t_{101} = 0.21$, $p = 0.83$]. These data suggest that the coding of ILDs by ICC neurons is frequency *independent* and that the low-frequency neurons share essentially identical sensitivities to the same overall ranges of ILDs as high-frequency neurons.

Given a strict interpretation of the Duplex theory, the discrepancy between the frequency *independence* of neural ILD coding (Fig. 31.1b–d) and frequency *dependence* of the acoustic ILDs is unexpected as it suggests that there are neurons that can encode sound features that may not be experienced in the natural environment. Alternatively, we suggest that this discrepancy arises from an inadequate characterization of the acoustical environment. Based on prior studies (Duda and Martens 1998; Brungart and Rabinowitz 1999; Kim et al. 2010), we hypothesize that low-frequency ILD-sensitive neurons are necessary because acoustically low-frequency ILDs become large and potentially useful when sound sources are located near the head.

3.2 The Effects of Sound Source Distance on the ILD Cues to Location

Acoustic ILDs were measured as a function of azimuth and three source distances (10, 20, and 100 cm) in three chinchillas. These findings are summarized in Fig. 31.3 where the ILDs are plotted at one azimuth, 75°, as a function of frequency for the three different source distances in each animal – the trends at this source location are representative of other locations. ILD magnitudes generally increased with (1) increasing frequency, (2) with more lateral azimuths away from midline, and (3) with decreasing source distance. At 100 cm (far field), maximum ILDs were very small for low frequencies but become systematically larger for frequencies above ~4 kHz. From ~6 to 18 kHz, the spectral notches created by the contralateral pinna (farthest from the source) produce large ILDs. For sources very close (10 cm) to the head, there was a substantial increase (10 dB or more) in ILD magnitude particularly for lower frequencies (<4 kHz). These data demonstrate that ILDs are not only a joint function of azimuth and frequency but also sound source distance.

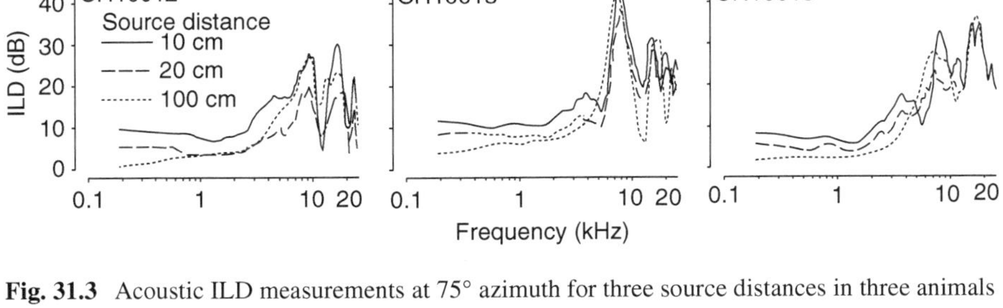

Fig. 31.3 Acoustic ILD measurements at 75° azimuth for three source distances in three animals

3.3 ILD Sensitivity of ICC Neurons Spans the Range of Physically Available Acoustic ILDs when Source Distance Is Considered

The neuronal half-max ILD (symbol) plus the associated ILD dynamic range (error bars) for each neuron were plotted again as a function of the neuron's CF and then compared to the maximum ILDs that were recorded at distances of 10 cm (gray-shaded area, Fig. 31.2). After taking into consideration the ecologically relevant effect of source distance on ILDs, the majority (93 %) of low-frequency ILD-sensitive neurons (28/30, see Fig. 2 below the 4 kHz dashed line) now have half-max ILD values that fall *within* the range of available ILDs. As seen in Fig. 31.2, ILDs systematically increase with decreases in the distance between a sound source and observer, particularly for frequencies <~4–5 kHz. There were still some low-frequency half-max values (2/30) that fell just outside the acoustic ILDs; however, it is expected that for sound source distances <10 cm (the closest distance used here), the maximum ILDs available would be considerably larger, up to 20 dB (Duda and Martens 1998). Thus, the half-max ILDs of these outliers would then fall into the acoustical range of ILDs experienced for sources occupying space close to the head.

3.4 Fisher Information Reveals Similar ILD Coding Ranges in Low- and High-Frequency ICC Neurons

We computed the Fisher information (FI) separately in the high- and low-frequency populations of ICC neurons. The FI is a measure of the amount of information about a particular stimulus value, such as a single ILD, and thus quantifies the accuracy with which a stimulus is encoded via spikes (Dayan and Abbott 2001). Higher FI implies higher coding accuracy. Practically, this translates to a higher capacity for a neuron to discriminate, for example, two nearby values of ILD. FI is computed from the probability of producing r spikes for a given stimulus S. Here we used an approximation of FI (Dayan and Abbott 2001) given by $FI(ILD) = y'(ILD)^2/\sigma(ILD)^2$ where y'(ILD) is the derivative of the spike count with respect to the *ILD* (count-ILD tuning curve) and σ(ILD)^2 is the count variance. Figure 31.4 shows the computation of the population Fisher information. Figure 31.4a shows the mean discharge rate for the high- and low-frequency ICC neurons. The assumption was made from the symmetry of the midbrain that there were identical populations of ILD-sensitive neurons in the ICC on each side; this assumption was implemented by simply rotating the rate-ILD curves for each neuron about the 0 ILD point. This results at the population level in two broadly tuned channels for ILD, with similar sensitivities for low- and high-frequency neurons. Figure 31.4b and c shows the Fisher information for individual neurons in both the low- and high-frequency neurons, respectively. Finally, Fig. 31.4d shows the population FI constructed from the sum of the FI for 30 neurons drawn randomly with replacement from the low- and high-frequency

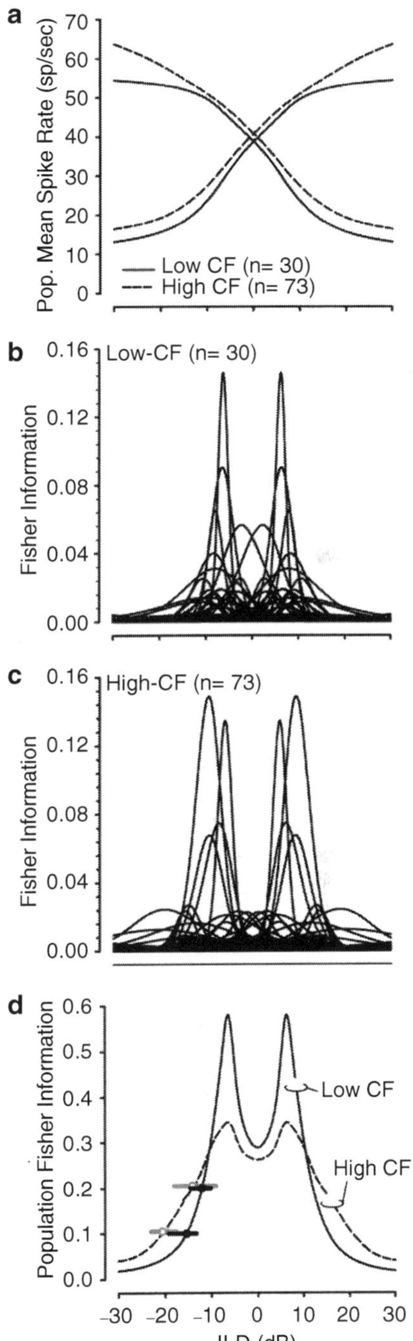

Fig. 31.4 Computation of Fisher information. Mean population rate-ILD functions (**a**) and individual neuron Fisher information (**b**, **c**) for low (<4 kHz)- and high (>4 kHz)-frequency ICC neurons. (**d**) Population Fisher information. *Error bars*, 95 % confidence interval

neurons (Fig. 31.4b,c); the data in Fig. 31.4d results from the mean of 1,000 bootstrapped computations of population FI. The low- and high-frequency neural populations exhibited good coding accuracy over similar ranges of ILD. At an ILD of ~15 dB, coding accuracy was not significantly different (overlapping 95 % confidence intervals), but by 20 dB, the high-frequency population exhibited significantly better coding accuracy than the low.

4 Discussion

As far as we are aware, this is the first study to systematically compare the ILD sensitivity of neurons at any level of the auditory system to the physiological range of ILDs acoustically available to an animal species. Although there have been prior studies on the ILD sensitivity of neurons and a growing amount of research on the effect of sound source distance on the cues to location (Duda and Martens 1998; Brungart and Rabinowitz 1999; Zahorik 2002; Kim et al. 2010), this study demonstrated that ILD-sensitive ICC neurons are sufficient to encompass (Fig. 31.2) and accurately encode (Fig. 31.4) the physiological range of ILDs.

Sound localization has long been thought to be based on the Duplex theory (Rayleigh 1907). Additionally, based on ecological and efficiency principles, it is hypothesized that neural systems have evolved strategies to faithfully represent the full spectrum of sensory signals experienced by an organism in its natural habitat (Barlow 1961). However, if the mechanisms responsible for the Duplex theory were implemented at the ICC and if the auditory system evolved strategies to represent the full spectrum of signals, then it would not be unreasonable to expect that there would either be no (or very few) low-frequency ILD-sensitive neurons or that they would have sensitivities that were constrained within the small range of acoustic ILDs (e.g., Fig. 31.2). The initial analysis of the data did not support these hypotheses, as the distributions of neural ILD coding parameters (Fig. 31.1) were virtually identical for low- and high-CF neurons. The data suggested that the notions of experience driven neural coding (i.e., plasticity) and/or ecological and efficiency neural coding principles may not be supported as properties of ILD-sensitive neurons were found to be *frequency independent* (counter to what would be expected given the posits of the Duplex theory) even though the acoustical ILD cues themselves are highly frequency *dependent*.

However, these classic hypotheses were based on the distributions of acoustic cues measured in the far field (1 m or more) and in unnatural, noise, and reflection-free environments. It has been shown acoustically that the ILD cue to location is greatly affected by source distance; in particular, low-frequency ILDs increase substantially as the source-to-observer distance decreases (Duda and Martens 1998; Brungart and Rabinowitz 1999; Kim et al. 2010). Thus, for a given source location, the ILD cue is not invariant with distance. Here, after considering the physiologically relevant acoustical ILDs experienced by the chinchilla at more proximal sound source distances (<1 m), it was shown that the area of greatest change in the rate-ILD functions of IC neurons (Figs. 31.2 and 31.4a) covers the range of ILDs experienced by the chinchilla for low-frequency sounds.

Although many studies have demonstrated low-frequency ILD-sensitive neurons, few have suggested a functional significance for these neurons. The presence of these neurons may offer a possible mechanism for which information about sound source distance may be neurally represented. Although these neural responses may not be encoding source distance per se, the comparison of an ILD cue with an ITD cue could provide information about the proximity of the source. Given the fact that ITDs are not greatly affected by sound source distance (Duda and Martens 1998), source locations along the same angular plane at different distances from the observer would produce similar ITDs but different ILDs. For example, a low-frequency sound generating a particular ITD and an extremely large ILD might indicate that the source is closer than if the sound produced the same ITD but with a much smaller ILD. Zahorik (2002) suggests that the perceptual processes subserving distance localization likely combine and weigh multiple cues in order to produce stable estimates of source distance. Thus, the weighing and combining of neural representations of binaural localization cues from different distances may provide additional information in conjunction with other traditional distance cues such as intensity and direct-to-reverberant energy ratio. This study suggests that low-frequency tuned neurons exhibiting ILD sensitivity may provide the neural circuitry needed to encode one aspect of source distance, namely, the large low-frequency ILDs produced by nearby sources.

Acknowledgement This research was supported by NIH R01-DC011555 (DJT).

References

Barlow H (1961) Possible principles underlying the transformation of sensory messages. In: Rosenblith, W (ed) Sensory Communication. Wiley, New York

Benson DA, Teas DC (1976) Single unit study of binaural interaction in the auditory cortex of the chinchilla. Brain Res 103:313–338

Brungart DS, Rabinowitz WM (1999) Auditory localization of nearby sources. Head-related transfer functions. J Acoust Soc Am 106:1465–1479

Dayan P, Abbott LF (2001) Theoretical neuroscience: computational and mathematical modeling of neural systems. MIT Press, Cambridge, MA

Duda RO, Martens WL (1998) Range dependence of the response of a spherical head model. J Acoust Soc Am 104:3048–3058

Hafter ER (1984) Spatial hearing and the Duplex theory: how viable is the model. In: Edelman GM, Gall WE, Cowan WM (eds) Dynamic aspects of neocortical function. Wiley, New York, pp 425–448

Heffner RS, Heffner HE, Kearns D, Vogel J, Koay G (1994) Sound localization in chinchillas. I: left/right discriminations. Hear Res 80:247–257

Kim DO, Bishop B, Kuwada S (2010) Acoustic cues for sound source distance and azimuth in rabbits, a racquetball and a rigid sphere. J Assoc Res Otolaryngol 11:541–557

Koka K, Jones HG, Thornton JL, Lupo JE, Tollin DJ (2011) Sound pressure transformations by the head and pinnae of the adult Chinchilla (Chinchilla lanigera). Hear Res 272:135–147

Macpherson EA, Middlebrooks JC (2002) Listener weighting of cues for lateral angle: the Duplex theory of sound localization revisited. J Acoust Soc Am 111:2219–2236
Rayleigh L (1907) On our perception of the direction of a source of sound. Philos Mag 13:214–232
Sanes DH, Rubel EW (1988) The ontogeny of inhibition and excitation in the gerbil lateral superior olive. J Neurosci 8:682–700
Semple MN, Kitzes LM (1987) Binaural processing of sound pressure level in the inferior colliculus. J Neurophysiol 57:1130–1147
Zahorik P (2002) Assessing auditory distance perception using virtual acoustics. J Acoust Soc Am 111:1832–1846

Chapter 32
Cochlear Contributions to the Precedence Effect

Sarah Verhulst, Federica Bianchi, and Torsten Dau

Abstract Normal-hearing individuals have sharply tuned auditory filters, and consequently their basilar-membrane (BM) impulse responses (IRs) have durations of several ms at frequencies in the range from 0 to 5 kHz. When presenting clicks that are several ms apart, the BM IRs to the individual clicks will overlap in time, giving rise to complex interactions that have not been fully understood in the human cochlea. The perceptual consequences of these BM IR interactions are of interest as lead-lag click pairs are often used to study localization and the precedence effect. The present study aimed at characterizing perceptual consequences of BM IR interactions in individual listeners based on click-evoked otoacoustic emissions (CEOAEs) and auditory brainstem responses (ABRs). Lag suppression, denoting the level difference between the CEOAE or wave-V response amplitude evoked by the first and the second clicks, was observed for inter-click intervals (ICIs) between 1 and 4 ms. Behavioral correlates of lag suppression were obtained for the same individuals by investigating the percept of the lead-lag click pairs presented either monaurally or binaurally. The click pairs were shown to give rise to fusion (i.e., the inability to hear out the second click in a lead-lag click pair), regardless of monaural or binaural presentation. In both cases, the ICI range where the percept was a fused image correlated well with the ICI range for which monaural lag suppression occurred in the CEOAE and ABR (i.e., for ICIs below 4.3 ms). Furthermore, the

S. Verhulst (✉)
Department of Biomedical Engineering,
Center for Computational Neuroscience and Neural Technology,
Boston University, 677 Beacon St., Boston, MA 02215, USA

Department of Electrical Engineering,
Center for Applied Hearing Research, Technical University of Denmark,
Oersted plads Bld 352, 2800 Kgs., Lyngby, Denmark
e-mail: save@bu.edu

F. Bianchi • T. Dau
Department of Electrical Engineering,
Center for Applied Hearing Research, Technical University of Denmark,
Oersted plads Bld 352, 2800 Kgs., Lyngby, Denmark

B.C.J. Moore et al. (eds.), *Basic Aspects of Hearing*,
Advances in Experimental Medicine and Biology,
DOI 10.1007/978-1-4614-1590-9_32,

lag suppression observed in the wave-V amplitudes to binaural stimulation did not show additional contributions to the lag suppression obtained monaurally, suggesting that peripheral lag suppression up to the level of the brainstem is dominant in the perception of the precedence effect.

1 Introduction

The tuning of human auditory filters (Q_{ERB}), derived from behavioral (tone-on-tone forward masking) and objective methods (otoacoustic emission phase gradient), has been estimated to 12.7, 15.63, and 19.24 for center frequencies of 1, 2, and 4 kHz, respectively (Oxenham and Shera 2003; Shera et al. 2010). Applying a suitable model such as the gammatone filter (Irino and Patterson 1997),

$$\text{BM IR}_{GT}(t) = at^{n-1}\exp(-2\pi bf_c Q_{ERB}^{-1}t)\cos(2\pi f_c t+\phi) \quad (t>0), \qquad (32.1)$$

with parameters $n=4$, $\phi=\pi/2$ and $a=1$, $b=1.018$, leads to a basilar-membrane impulse response (BM IR) description for the human auditory filter. The BM IR durations (i.e., the time until the amplitude reduced by 95 %) calculated from Eq. 32.1 are 18.7, 11.6, and 7.1 ms for frequencies of 1, 2, and 4 kHz, respectively. It is thus expected that BM IRs to clicks that are separated by only several ms will overlap in time. Indeed, already in 1969, Goblick and Pfeiffer observed BM IR interactions in the firing pattern of cochlear nerve fibers in cat in response to temporally spaced acoustical clicks. They described their observations in terms of two systems: an instantaneous compression mechanism and an unknown dynamic compression process.

The present study aimed at characterizing human BM IR interactions noninvasively by using click-evoked otoacoustic emissions (CEOAEs) and auditory brainstem responses (ABRs). Individual correlations between the objective measures and the perception of *monaural* click pairs were performed to determine the contribution of BM IR interactions to the percept of fusion, i.e., the inability to hear out the second click in a click pair (Litovsky et al. 1999). Secondly, the influence of monaural BM IR interactions on the perception of *binaural* click pairs, known to lead to the perception of the precedence effect (e.g., Wallach et al. 1949; Litovsky et al. 1999), was investigated.

2 Materials and Methods

Six normal-hearing subjects with audibility thresholds below 20 dB HL (3 females and 3 males), aged between 24 and 34 yrs, participated in the experiments. All experiments were performed in a double-walled soundproof booth. The 83-µs-long clicks were generated digitally in Matlab with a sampling rate of 48 kHz and were presented over

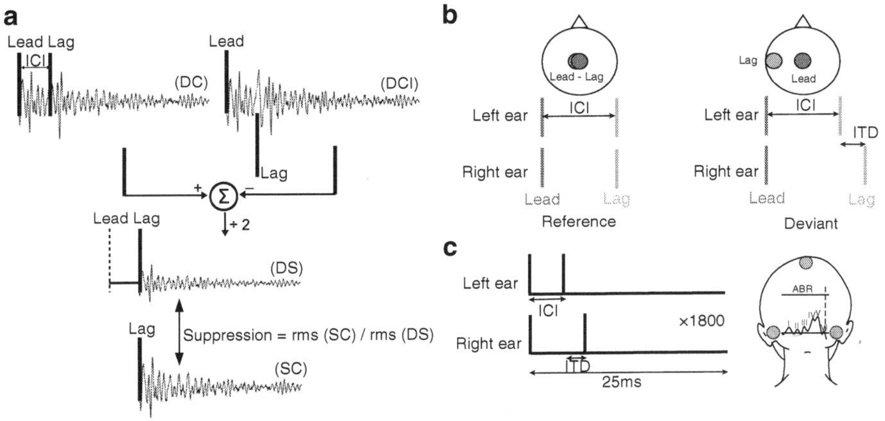

Fig. 32.1 (**a**) Method for obtaining lag suppression in the CEOAE recordings. The stimuli were presented using an interleaved procedure: for each ICI and ITD condition, 1,800 repetitions of the following three stimuli were presented – single click (SC), double click (DC; two condensation clicks), and double click inverted (DCI; one condensation and one rarefaction click). The CEOAE recorded to a DC stimulus contains a CEOAE to the leading click, a CEOAE to the lagging click, and a nonlinear component due to the ICI. To remove the CEOAE to the leading click while maintaining the CEOAE to the lagging click, a derived-suppressed response (DS) was obtained by subtracting the DCI response from the DC response and halving the result (Fig. 1A; Kemp and Chum 1980; Kapadia and Lutman, 2000a, b; Verhulst et al. 2011a). (**b**) Lead-lag click pairs used for the psychoacoustical and objective experiments. For ICIs above the echo-threshold, the reference stimulus leads to a centered percept of two separate clicks, whereas the deviant stimulus leads to a laterized percept to the left for the second click. (**c**) Stimulus configuration for the ABR recordings and visualization of the placement of the Cz, M1, and M2 electrodes

ER-2 earphones (CEOAE and ABR) or Sennheiser HD580 headphones (psychoacoustics). In the left ear, click-pair stimuli were presented at 65 dB peSPL (CEOAE) or 75 dB peSPL (ABR, psychoacoustics) for 7 different inter-click intervals (ICI): 0, 1, 2, 3, 4, 5, and 8 ms. In the right ear, the delay between the clicks corresponded to the ICI in the left ear plus an interaural time difference (ITD) of 300 μs (Fig. 32.1c). The ear-canal recordings were performed binaurally for the CEOAE recordings and both monaurally (L,R) and binaurally for the ABR and psychoacoustical experiments.

Stimulus presentation and data acquisition for the CEOAE recordings were described in Verhulst et al. (2011a). CEOAE lag suppression was calculated as the rms level difference between the derived-suppressed (DS) and single-click (SC) responses in a time frame of 6–18 ms after the onset of the lagging click (see Fig. 32.1b for procedure). The ABR wave-Vs were recorded as described in Bianchi et al. (2013) (Fig. 32.1c). ABR lag suppression was calculated as the wave-V amplitude difference (in dB) in response to the second and the first clicks.

The psychoacoustical echo-threshold (i.e., the smallest ICI at which two separate clicks were perceived) was determined using an adaptive one-interval, two-alternative forced choice (2AFC) procedure. Each trial consisted of a reference and a deviant (Fig. 32.1b) and the subjects indicated whether they perceived one single click (fused image) or two separate clicks (lead and lag). The starting value of the

ICI was 1 ms. After each "single-click response," the ICI was increased, and after each "two-click response," the ICI was decreased. The initial step size was 1 ms and was reduced to 0.5 and 0.3 ms as the threshold was approached. The echo-threshold was obtained after six reversals and corresponded to the 70.7 % point on the psychometric function. Thresholds were obtained as the average of three repeated experimental runs. Additional experiments investigating the laterization of the click pairs used here were performed in Bianchi et al. (2013) for the same subjects.

3 Results

3.1 Monaural BM IR Interactions

Monaural lag suppression, representing the level difference between the CEOAE to a single click (SC) and the lag CEOAE derived from the CEOAE to a lead-lag click pair (DS; see Fig. 32.1a), is shown in Fig. 32.2a as the mean over six subjects. Lag suppression was observed for ICIs up to 8 ms. Maximal suppression with individual levels up to 10 dB was found for ICIs of 1–2 ms, in agreement with earlier studies (Kapadia and Lutman 2000b; Verhulst et al. 2011a). These results demonstrate that the CEOAE to the lagging click in a double-click pair is suppressed with respect to the leading click if the ICI between the lead-lag pair is below 8 ms. Individual results for subject KE were also indicated in panel A, as for this subject, a detailed comparison across methods is shown in panels b and c.

The forward traveling wave to a click excites the whole BM, yet the broadband CEOAE only contains information about local BM processing at those locations where reflections were generated. These locations are observed as components in the CEOAE spectrum, with the strength, number, and frequencies being subject dependent and resulting from the individual BM irregularity pattern (Sisto and Moleti 2008; Shera and Guinan 2007). Figure 32.2b shows a CEOAE spectrum to a single click (i.e., the gray area under SC in all panels) for subject KE, reflecting the broadband nature of the emission. Overlaid in each panel are the spectra of the CEOAE to the lagging click (i.e., the white area under DS) for several ICI conditions. For ICIs up to 2 ms, lag suppression affects all components in the CEOAE spectrum equally, but for larger ICIs, a release from suppression is observed. This release from suppression affects the higher frequencies first as the ICI increases. For an ICI of 5 ms, low-frequency components in the CEOAE are still suppressed, whereas the higher-frequency components show a release from suppression. The frequency dependence of the release from suppression provides a link between CEOAE lag suppression and the local BM IR duration (Verhulst et al. 2011b). Even though it is unclear which exact local BM mechanism is responsible for lag suppression exceeding 6 dB at ICIs larger than zero, BM models based on instantaneous compression can account for the ICI range and frequency dependence of the lag suppression (Kapadia and Lutman 2000a; Harte et al. 2005; Verhulst et al. 2011b).

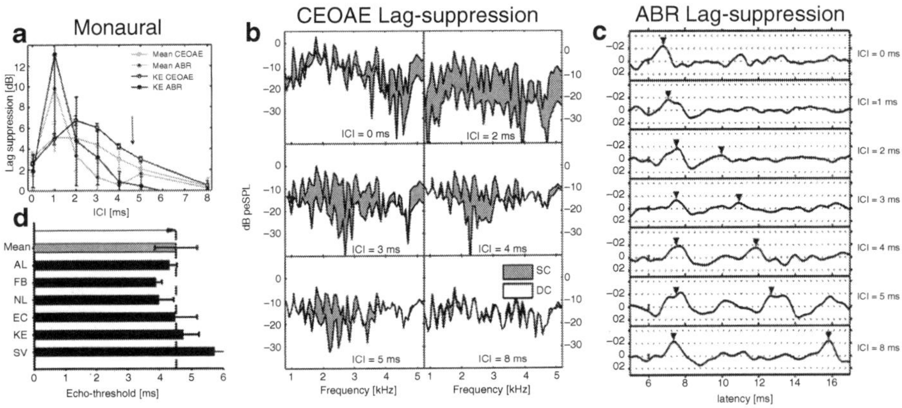

Fig. 32.2 (**a**) Mean and individual levels of monaural lag suppression obtained from CEOAE and ABR recordings. (**b**) Single-click CEOAE (6–20 ms window) spectra for subject KE, overlaid with derived-suppressed CEOAE spectra for different ICI conditions. The noise floor on these recordings is situated around –30 dB peSPL. (**c**) Recorded ABR waveforms to monaural click pairs with varying ICI for subject KE. Wave-Vs are indicated with *triangles*. (**d**) Monaural echo-threshold obtained from monaural stimulation with the reference click pairs in Fig. 32.1b. For ICIs below the threshold, a fused single-click percept was reported for all subjects

The CEOAE lag-suppression results in Fig. 32.2a and b have demonstrated that the amplitude of specific components in the CEOAE to the second click is reduced after the presentation of an earlier click. As CEOAEs predominantly originate from reflections of the forward traveling wave to place-fixed BM irregularities (Zweig and Shera 1995; Shera and Guinan 1999, 2007), it is inferred that these CEOAE amplitude reductions reflect gain reductions in local BM processing. Consequently, there should be correlates of this gain reduction caused by presenting two temporally spaced clicks along the ascending auditory pathway. Figure 32.2c confirms this by showing monaurally recorded ABR waveforms evoked by the reference stimulus in Fig. 32.1b as a function of increasing ICI. Two wave-V components are evoked and reflect activity from the superior olivary complex (indicated with downward pointing triangles; Picton 2011). Whereas the amplitude of the wave-V to the leading click is constant with ICI, the amplitude of the second wave-V increases as the ICI increases. The level difference between the amplitude of wave-V to the leading and the lagging clicks reflects ABR lag suppression and is shown in Fig. 32.2a for subject KE. ABR lag suppression was maximal for an ICI of 1 ms and decreased with increasing ICI. The ICI range of lag suppression was somewhat shorter than the range observed for CEOAE lag suppression, with maximal levels that were up to 5 dB higher for ABR than for CEOAE lag suppression. Even though there may be effects related to inner-hair-cell (IHC) processing and across-channel synchrony that are reflected in ABR lag suppression but not in CEOAE lag suppression, both objective measures exhibit substantial amounts of suppression for ICIs less than 4 ms.

Perceptual correlates of monaural BM IR interactions were investigated with a fusion test (Litovsky et al. 1999; Bianchi et al. 2013), where subjects were asked to report whether 1 or 2 clicks were perceived when listening to monaurally presented click pairs with varying ICIs. For the ICIs where fusion occurs, the leading and lagging clicks are perceived as a single fused event. The results in Fig. 32.1d demonstrate that subjects were unable to detect the second click in a click pair (i.e., perceptual lag suppression) when the ICI was below 4.3 ms. For subject KE, this *monaural echo-threshold* occurred for an ICI of 4.8 ms. ABR and CEOAE lag suppression were always higher below than above this threshold for all subjects tested. This result was significant for all subjects for the CEOAE measure and for 3 out of 6 subjects for the ABR measure (Bianchi et al. 2013).

3.2 Consequences of Monaural BM IR Interactions for Binaural Processing

Given the correlation between objective and perceptual lag-suppression data from Fig. 32.2, this monaural and peripheral component might also affect the perception of binaural click pairs. Binaural click pairs of the deviant type in Fig. 32.1b are known to evoke the *precedence effect* (Litovsky et al. 1999). For ICIs where fusion takes place (i.e., ICIs less than 4.3 ms in this study), a single click at the center of the head was perceived, i.e., at the location of the leading click. When the ICI increased above the echo-threshold, two clicks were heard, with the second click perceived at a location corresponding to the ITD introduced (i.e., to the left for an ITD of 300 μs).

The top panel in Fig. 32.3a shows CEOAE lag suppression obtained for subject KE for the deviant stimuli in Fig. 32.2b. Levels of CEOAE lag suppression were not identical in the two ears. There are two main reasons for this: (1) the frequency content related to the underlying BM irregularity pattern in the CEOAEs was different, and (2) an ITD was introduced in the right ear whereas the left ear only contained an ICI. Nevertheless, both ears demonstrate a substantial amount of BM IR lag suppression for ICIs below 6 ms. ABR lag suppression in the bottom panel of Fig. 32.3a was largest for ICIs below 4 ms. Even though ABR lag suppression was observed for a smaller range of ICIs (0–3 ms) than CEOAE lag suppression (top panel), both objective methods showed suppression for ICIs below 4 ms. The double-click pairs used to evoke the CEOAE and ABRs (Fig. 32.3a) lead to the perception of a fused event in the center of the head for ICIs below 4.8 ms (Fig. 32.3b, KE). Above this *binaural echo-threshold*, the second click in the click pair was lateralized to the left. Individual monaural and binaural echo-thresholds in Figs. 32.2d and 32.3c were similar, in agreement with the fusion thresholds found in Litovsky et al. (1997) and Rakerd et al. (1997). This suggests that lag suppression hinders the perception of the second click, regardless of whether it is presented monaurally or binaurally. The ABR results in Fig. 32.3a furthermore showed *binaural* lag-suppression levels in between the lag-suppression levels obtained for each ear individually

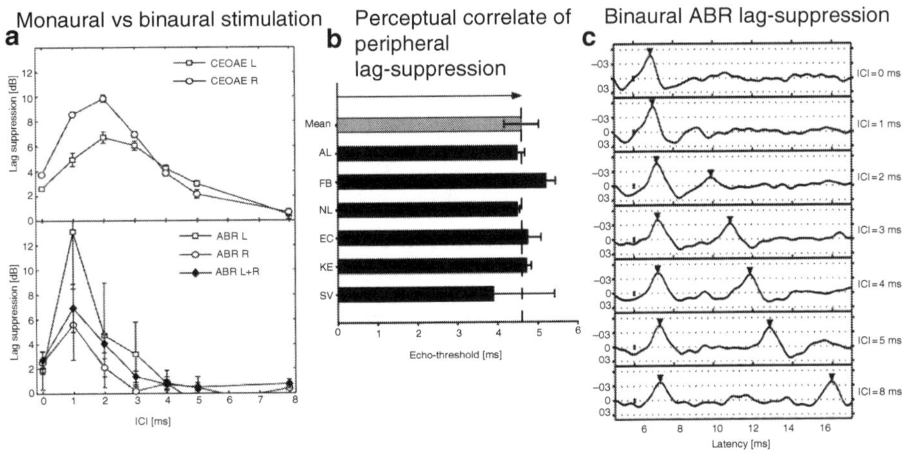

Fig. 32.3 (**a**) Lag suppression obtained from CEOAE and ABR recordings to monaural and binaural stimulation with the click pairs of the deviant type in Fig. 32.1b for subject KE. (**b**) Binaural echo-thresholds for a binaural click pair with ITD of 300 μs. Below the threshold, a single-fused click was perceived, and above the threshold, the second click in the pair was lateralized to the left. (**c**) Recorded ABR waveforms in response to binaural click pairs of the deviant type in Fig. 32.1b for subject KE. Wave-Vs are indicated with *triangles*

(Bianchi et al. 2013). Binaural ABR lag suppression thus seems to reflect monaural lag suppression, as demonstrated in Fig. 32.3c for ABR waveforms recorded to binaural lead-lag click pairs.

4 Discussion

Perceptual consequences of BM IR interactions were investigated by comparing CEOAE, ABR, and perceived lag suppression for lead-lag click pairs in individual subjects. The objective CEOAE and ABR methods showed lag suppression for ICIs below 4 ms. They did not, however, yield identical patterns across ears and methods, which is a consequence of the nature of the signals. Whereas ABR wave-V reflects neural activity across many fibers of cochlear and brainstem sites (Junius and Dau 2005), the CEOAE contains information about those cochlear locations where reflections from the forward traveling wave took place (Shera and Guinan 1999). The CEOAE suppression patterns thus only contain a subset of frequencies (predominantly in the 1–2 kHz region; Puria (2003)) of the synchronously excited region on the BM to a broadband click, leading to higher across-ear-and-subject variability in the CEOAE lag-suppression patterns (Verhulst et al. 2011a). Since release from CEOAE lag suppression is frequency dependent, as observed in Fig. 32.2b, the frequency components in the CEOAE will determine the ICI for which the release of lag suppression is obtained. The higher frequencies contributing to ABR wave-V versus the mid-frequency content in the CEOAE may explain why ABR suppression patterns are generally restricted to shorter ICIs than the CEOAE patterns.

The importance of peripheral processing for the perception of the precedence effect was earlier emphasized by the study of Hartung and Trahiotis (2001), who showed how model predictions based on auditory filtering (gammatone filter bank; e.g., Patterson and Allerhand 1995) and auditory hair cells (Meddis 1986) could account for binaural ITD processing. The results in the present study provide experimental evidence for the proposed bottom-up approach in Hartung and Trahiotis (2001), by showing correlations between CEOAE/ABR lag suppression and the precedence effect. Although it is unclear which cortical or cognitive processes add to the processing of binaural click pairs, the results presented here and in Bianchi et al. (2013) provide evidence for a monaural peripheral contribution to the perception of lead-lag pairs up to the level of the brainstem.

Acknowledgements Work supported by Technical University of Denmark and Oticon Foundation.

References

Bianchi F, Verhulst, S, Dau T (2013) Experimental evidence for a cochlear source of the precedence effect. Journal of the Association for Research in Otolaryngology (in press)

Goblick T Jr, Pfeiffer R (1969) Time-domain measurements of cochlear nonlinearities using combination click stimuli. J Acoust Soc Am 46(4):924–938

Harte JM, Elliott SJ, Kapadia S, Lutman ME (2005) Dynamic nonlinear cochlear model predictions of click-evoked otoacoustic emission suppression. Hear Res 207(1–2):99–109

Hartung K, Trahiotis C (2001) Peripheral auditory processing and investigations of the "precedence effect" which utilize successive transient stimuli. J Acoust Soc Am 110(3):1505–1513

Irino T, Patterson RD (1997) A time-domain, level-dependent auditory filter: the gammachirp. J Acoust Soc Am 101(1):412–419

Junius D, Dau T (2005) Influence of cochlear traveling wave and neural adaptation on auditory brainstem responses. Hear Res 205(1–2):53–67

Kapadia S, Lutman ME (2000a) Nonlinear temporal interactions in click-evoked otoacoustic emissions. I. Assumed model and polarity-symmetry. Hear Res 146(1–2):89–100

Kapadia S, Lutman ME (2000b) Nonlinear temporal interactions in click-evoked otoacoustic emissions. II. Experimental data. Hear Res 146(1–2):101–120

Kemp D, Chum R (1980) Properties of the generator of stimulated acoustic emissions. Hear Res 2:213–232

Litovsky RY, Rakerd B, Yin TC, Hartmann WM (1997) Psychophysical and physiological evidence for a precedence effect in the median sagittal plane. J Neurophysiol 77(4):2223–2226

Litovsky RY, Colburn HS, Yost W, Guzman SJ (1999) The precedence effect. J Acoust Soc Am 106(4):1633–1654

Meddis R (1986) Simulation of mechanical to neural transduction in the auditory receptor. J Acoust Soc Am 79(3):702–711

Oxenham AJ, Shera CA (2003) Estimates of human cochlear tuning at low levels using forward and simultaneous masking. J Assoc Res Otolaryngol 4(4):541–554

Patterson RD, Allerhand MH (1995) Time-domain modeling of peripheral auditory processing: a modular architecture and a software platform. J Acoust Soc Am 98(4):1890–1894

Picton TW (2011) Human auditory evoked potentials. Plural, San Diego

Puria S (2003) Measurements of human middle ear forward and reverse acoustics: implications for otoacoustic emissions. J Acoust Soc Am 113(5):2773–2789

Rakerd B, Hsu J, Hartmann WM (1997) The Haas effect with and without binaural differences. J Acoust Soc Am 101:3083

Shera CA, Guinan JJ (1999) Evoked otoacoustic emissions arise by two fundamentally different mechanisms: a taxonomy for mammalian OAEs. J Acoust Soc Am 105(2):782–798

Shera CA, Guinan JJ (2007) Mechanisms of mammalian otoacoustic emission. In: Manley A, Fay FF, Popper AN (eds) Active processes and otoacoustic emissions in hearing. Springer, New York, pp 306–342

Shera CA, Guinan JJ, Oxenham AJ (2010) Otoacoustic estimation of cochlear tuning: validation in the chinchilla. J Assoc Res Otolaryngol 11(3):343–365

Sisto R, Moleti A (2008) Transient evoked otoacoustic emission input/output function and cochlear reflectivity: experiment and model. J Acoust Soc Am 124(5):2995–3008

Verhulst S, Harte JM, Dau T (2011a) Temporal suppression of the click-evoked otoacoustic emission level-curve. J Acoust Soc Am 129(3):1452–1463

Verhulst S, Shera CA, Harte JM, Dau T (2011b) Can a static nonlinearity account for the dynamics of otoacoustic emission suppression? In: Shera CA, Olsen E (eds) What fire is in mine ears: progress in auditory biomechanics. AIP, Melville, pp 257–263

Wallach H, Newman E, Rozenzweig R (1949) The precedence effect in sound localization. Am J Psychol 42(3):315–336

Zweig G, Shera CA (1995) The origin of periodicity in the spectrum of evoked otoacoustic emissions. J Acoust Soc Am 98(4):2018–2047

Chapter 33
Off-Frequency BMLD: The Role of Monaural Processing

Steven van de Par, Bjoern Luebken, Jesko L. Verhey, and Armin Kohlrausch

Abstract Large binaural masking-level differences (BMLDs) can be observed when a tonal signal with an interaural phase difference of π is presented against a diotic masker. The BMLD is large when the signal is spectrally centered in the masker and decreases strongly for off-frequency signals. No such reduction in BMLD would be expected, if monaural detection were governed by energy cues and binaural detection by changes in interaural cross-correlation. The reduction in BMLD thus suggests either that binaural processing is impaired or, alternatively, that additional monaural cues are available in off-frequency conditions. In this study, a stimulus paradigm is used that is expected to impair the processing of additional monaural cues. In the base experiment, a 25-Hz-wide band of diotic noise centered at 700 Hz served as masker. A target tone was presented at 0, 30, 60, and 100 Hz above the masker center frequency, either interaurally in phase (S_0) or out of phase (S_π). In the extended experiment, an additional interference tone was always presented spectrally below the masker at the same frequency distance as the target tone was positioned above the masker. The interferer level was 6 dB below the level of the 65 dB masker. By presenting the interferer, a strong modulation is introduced,

S. van de Par (✉) • B. Luebken
Acoustics Group, Carl von Ossietzky University,
Carl-von-Ossietzky Strasse 9-11, 26129 Oldenburg, Germany
e-mail: steven.van.de.par@uni-oldenburg.de

J.L. Verhey
Department of Experimental Audiology, University of Magdeburg, Leipziger Str. 44, 39120 Magdeburg, Germany

Forschungszentrum Neurosensorik, Carl von Ossietzky University, 26111 Oldenburg, Germany

A. Kohlrausch
Philips Group Innovation, Smart Sensing & Analysis,
High Tech Campus 36, NL-5656 AE Eindhoven, The Netherlands

Eindhoven University of Technology,
513, NL-5600 MB Eindhoven, The Netherlands

B.C.J. Moore et al. (eds.), *Basic Aspects of Hearing*,
Advances in Experimental Medicine and Biology,
DOI 10.1007/978-1-4614-1590-9_33,

which should impair the detectability of the target tone based on the beating of masker and target. Results show a small off-frequency BMLD in the base experiment in line with literature. Adding the interference tone produced an increase in both N_0S_0 and the N_0S_π thresholds, suggesting that monaural modulation cues were indeed used, but also suggesting that detection performance in the N_0S_π condition was dominated by monaural processing. Additional conditions with modulated interference tones at 500 Hz further supported our hypothesis that monaural modulation cues contributed to reduced off-frequency BMLDs.

1 Introduction

The binaural masking-level difference (BMLD) refers to a reduction in masked threshold when a target has a different interaural configuration than the masker, for example, when the target is interaurally out of phase and the masker is interaurally in phase. BMLDs have been predominantly studied for conditions where a tonal target is spectrally centered in a noise masker.

Some studies have also investigated BMLDs in conditions where the target lies outside the masker spectrum (e.g., Zwicker and Henning 1984; Henning et al. 2007; Buss and Hall 2010; Nitschmann and Verhey 2012). These studies consistently show that the BMLD is strongly reduced in off-frequency conditions.

The BMLD measures a difference between a monaural and binaural threshold. Thus a constant BMLD which is independent of frequency separation between masker and target would be in line with the idea that the only significant effect that occurs due to the frequency separation is peripheral filtering. This would then reduce monaural and binaural thresholds equally provided that monaural detection is purely based on energy cues, and binaural detection is only based on changes in the interaural cross-correlation.

Binaural thresholds may also be affected by an effectively larger binaural critical bandwidth (cf. Hall et al. 1983; Nitschmann et al. 2010). This would result in a smaller reduction of binaural thresholds as compared to monaural thresholds in off-frequency conditions, lowering the off-frequency BMLD. In addition, in off-frequency conditions, the rate of change in binaural ITD and ILD cues increases, and these interaural differences might not be processed as effectively as the slower fluctuation cues in the on-frequency conditions; this would also reduce the BMLD. The detrimental effect of the rate of fluctuation has been discussed by Zurek and Durlach (1987) in connection to masker bandwidth effects in on-frequency BMLDs.

An alternative reason for a reduction of the off-frequency BMLD could be that in off-frequency conditions, monaural cues are available that allow for a more effective monaural detection than possible in on-frequency conditions. In off-frequency conditions, the presence of the target within the noise masker causes a distinct modulation of the stimulus envelope that could be detected effectively. These modulation cues are not available in the on-frequency condition and therefore modulation cues could lead to a reduction in off-frequency BMLDs (cf. van de Par and Kohlrausch 2005; Nitschmann and Verhey 2012). Distortion products causing a cubic difference tone can also facilitate the monaural detection of tones in an off-frequency condition, as argued by Greenwood (1971). This could also be a reason for the small off-frequency BMLDs, provided that the distortion products in the dichotic situation lead to no binaural advantage over the diotic conditions.

In a study investigating off-frequency BMLDs, van der Heijden et al. (1997) studied masked thresholds for $N_\tau S_0$ conditions with off-frequency targets that were placed below a masker with center frequency of 850 Hz and a bandwidth of 500 Hz. If masking was caused by the lowest components in the masker, the dependence of thresholds on interaural masker delay (τ) would match the periodicity of the lowest masker components. It was, however, found that thresholds were indicative of on-frequency masking in line with on-frequency distortion products being generated that masked the target.

The experiments that are presented in this study aim to reduce the effectiveness of the extra monaural cues that are available in off-frequency conditions. In this way, we want to investigate the role of monaural cues in reducing the BMLD. For this purpose, we will use a standard off-frequency BMLD paradigm, with the target placed above the masker with a certain frequency offset. By then adding an interferer tone with the same frequency offset below the masker in all presentation intervals, we will introduce envelope modulations in the masker which will make it more difficult to use these cues for the detection of the target tone. In addition, distortion products caused by the interaction of masker and target will be effectively masked by the interferer tone. In this manner, the additional off-frequency monaural cues will be made less effective and should lead to larger BMLDs.

2 Experiment I: Target Tone Placed at a Frequency Above the Masker

2.1 Stimuli and Methods

N_0S_0 and N_0S_π thresholds were measured in two conditions. In the baseline condition, the masker was a bandpass-filtered noise centered at 700 Hz with a bandwidth of 25 Hz and a sound pressure level of 65 dB. The tonal target was placed above the masker with a frequency offset of 0, 30, 60, and 100 Hz relative to the masker center frequency. In the interference condition, an additional tonal interferer with a level of 59 dB SPL was placed below the masker with the same frequency offset as the target. The interferer was present in all three presentation intervals. The masker and interferer had durations of 500 ms, the target of 400 ms, and all were gated with a 50 ms raised-cosine window.

Thresholds were measured using a 3-interval forced-choice 2-down 1-up adaptive staircase method that started with a step size of 8 dB which was halved after every second reversal of the track until a step size of 1 dB was reached. With this step size, another eight reversals were measured which served to determine the mean threshold value. For each condition, three threshold values were obtained of which the mean value is reported in this study. Five normal-hearing subjects participated in this first experiment. Stimuli were presented over an RME Fireface UC digital-to-analogue converter using a Tucker-Davis HD7 headphone attenuator and Sennheiser HD 650 headphones.

2.2 Results and Discussion

Figure 33.1 shows the results of the first experiment. The filled symbols represent the condition without an interferer. In the on-frequency condition ($\Delta f = 0$), a BMLD of about 18 dB is observed, while for off-frequency conditions, the BMLD reduces to almost 0 dB in line with the findings of previous literature studies.

In the conditions with interferer (open symbols), both N_0S_0 (circles) and N_0S_π (squares) thresholds increase relative to the condition without interferer. The increase in N_0S_0 thresholds was expected since the interferer should reduce the processing of monaural modulation cues and/or distortion products. Due to the presentation level of the interferer that was 6 dB lower than the masker, it is not expected that additional energetic masking will occur. The increase in N_0S_π thresholds was not expected because the interferer should only affect monaural processing. Such an increase is, however, consistent with the idea that in the off-frequency condition, N_0S_π detection thresholds were predominantly based on monaural

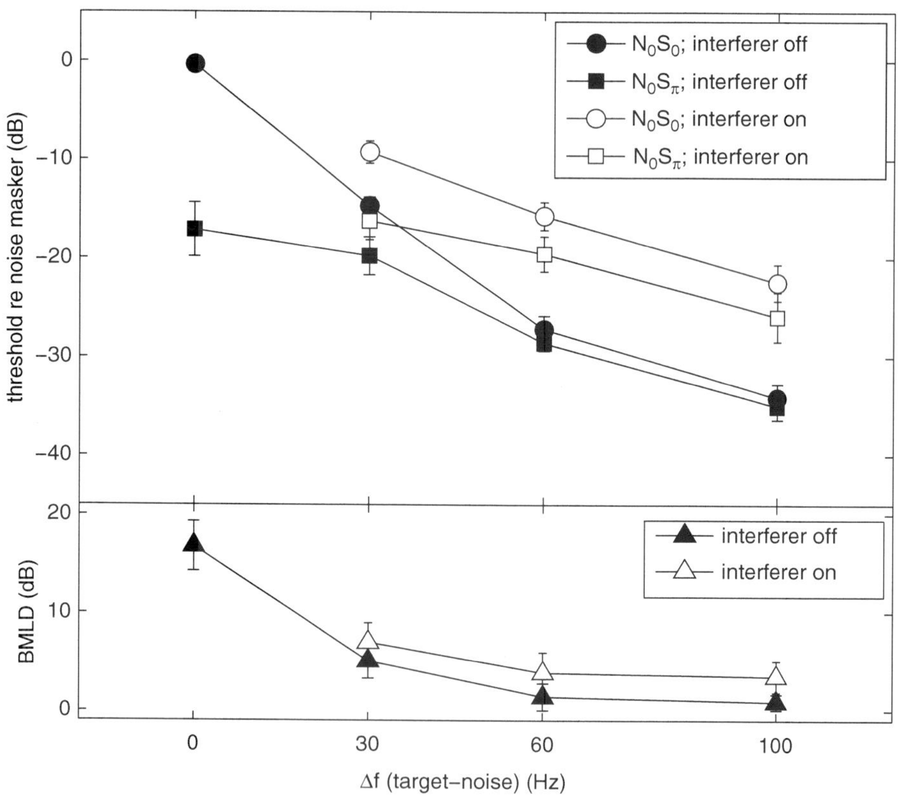

Fig. 33.1 The *top panel* shows N_0S_0 (*circles*) and N_0S_π (*squares*) thresholds as a function of the frequency separation between target and noise masker for conditions without interferer (*filled symbols*) and with interferer (*open symbols*). The *lower panel* shows corresponding BMLD values

processing, that is, that also in N_0S_π conditions, monaural cues could be detected at lower target-to-masker ratios than binaural cues.

3 Experiment II: Target Tone Placed at a Frequency Below the Masker

The second experiment extends the previous one by presenting target below the masker and the interferer above the masker. This rearrangement allows us to estimate the level that distortion products could have on the premise that they are the main cue for target detection in the off-frequency N_0S_0 conditions without interferer.

3.1 *Stimulus and Method*

Besides the exchange in spectral position of target and interferer, all other experimental details are the same as in the previous experiment. Masker-target separations of 100 Hz were not measured.

3.2 *Results and Discussion*

The results of the second experiment are shown in Fig. 33.2 in the same manner as in Fig. 33.1. In off-frequency conditions without interferer, there exists a somewhat larger BMLD than in the previous experiment. The presence of the interferer leads to an increase in the N_0S_0 and the N_0S_π thresholds, but this increase is now smaller, in particular for the N_0S_π thresholds. As in the first experiment, these results are in line with the hypothesis that the usage of modulation cues is reduced by the presence of the interferer.

The data obtained in this experiment allows us to make an estimate for the maximum level of the distortion products in the configuration of Experiment 1. For this purpose, the distortion products are first analyzed for Experiment 2, where they are generated by the interaction between masker and interferer. The resulting distortion products would spectrally overlap with the target and could thus serve as an additional on-frequency masker. If the *observed* masked thresholds indeed reflect such a process, the levels of the distortion products can be estimated by assuming that they should be about equal to the masked threshold (SNR 0 dB at threshold). In Fig. 33.2, these masked thresholds (open circles) amount to 45 dB SPL for $\Delta f=-60$ Hz and 55 dB SPL for $\Delta f=-30$ Hz.

In the first experiment, these distortion products are not generated by the 59 dB SPL interferer but by the target tones. These target tones had N_0S_0 thresholds of 36

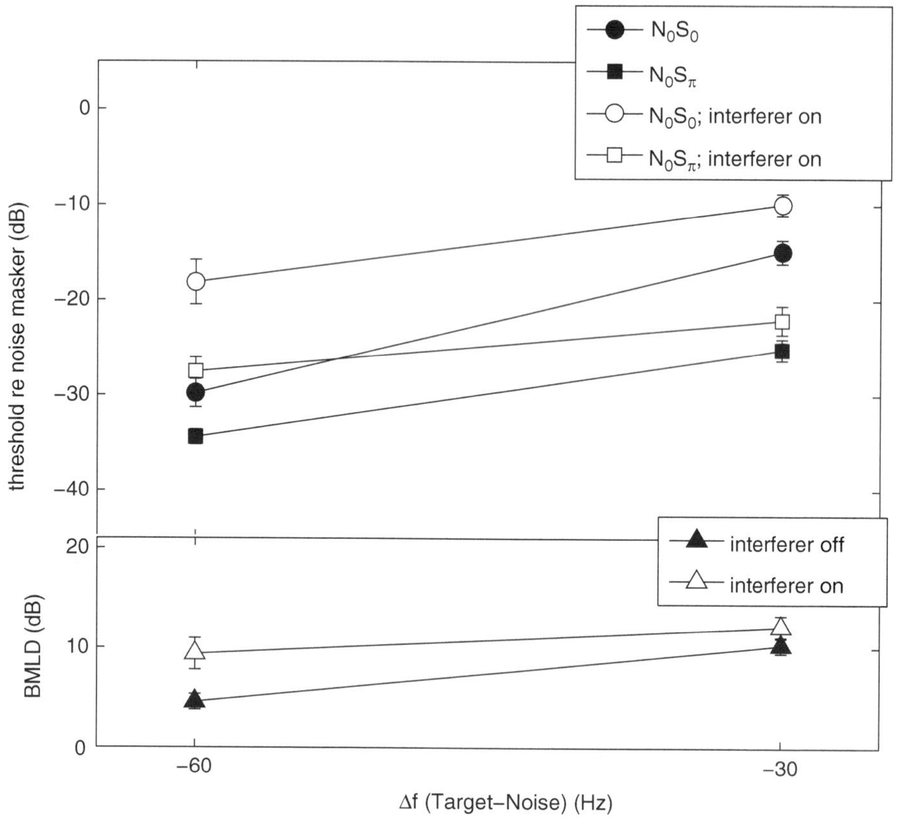

Fig. 33.2 Similar to Fig. 33.1, now with the interferer placed above the masker and the target tone below the masker

and 50 dB SPL for Δf of 60 and 30 Hz, respectively. Assuming that the level of the distortion products scales linearly with the level of the tonal component above the masker, the levels of the distortion components in Experiment I should be 22 dB SPL for $\Delta f = 60$ Hz and 46 dB SPL for $\Delta f = 30$ Hz in the N_0S_0 condition without interferer.

4 Experiment III: Two-Tone Interferer at 500 Hz

4.1 Stimulus and Method

Although the addition of the interferer in Experiments 1 and 2 shows that extra monaural processing cues are available in off-frequency conditions, it is not clear what the nature of these monaural cues is. They may be modulation cues but could also consist of cues due to the generation of distortion products. For this purpose, in this third experiment, the interferer is not placed symmetrically below the masker. Instead, two interferer tones at 59 dB SPL each with a frequency separation that was equal to the masker-target separation are placed at a lower frequency such that the highest of the interferer frequencies was 500 Hz. These interferer tones should, however, still be effective in causing modulation detection interference (Yost et al. 1989) in the detection of the target tone.

4.2 Results and Discussion

In Fig. 33.3, the results of the new experiment are shown together with the results of Experiment 1. For the N_0S_0 conditions, we get very similar results for both types of interferers used here and in Experiment 1. This suggests an equal ability of both interferers to remove the extra monaural cues that are available in the off-frequency conditions.

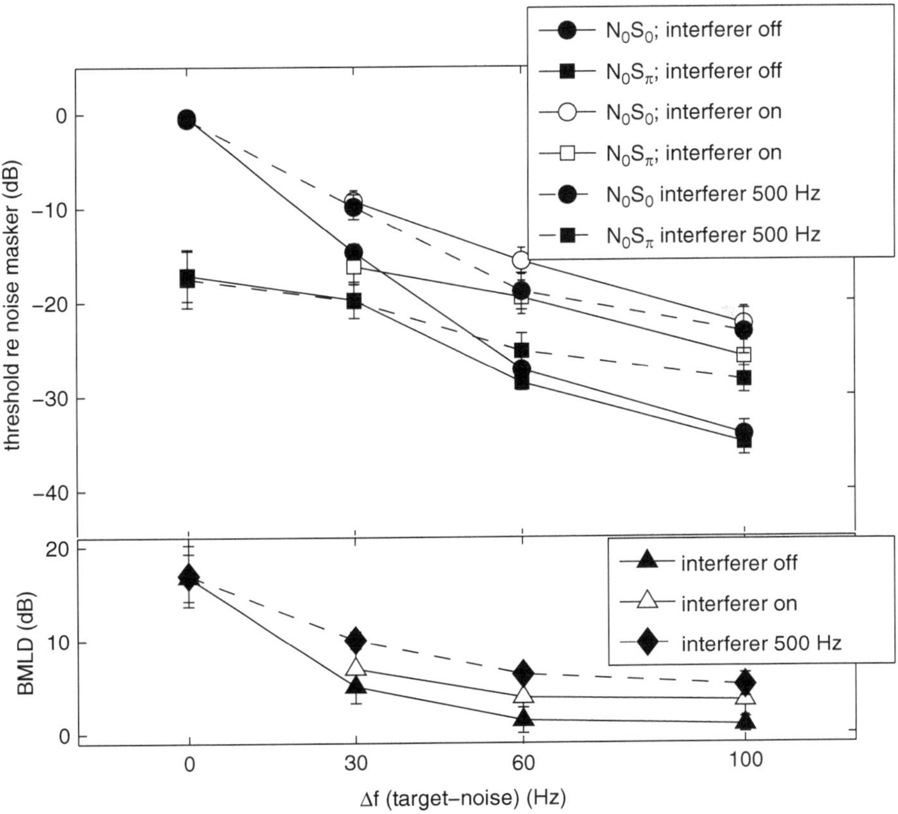

Fig. 33.3 Similar to Fig. 33.1, only now data are also shown for the new interferer that is placed at 500 Hz (*dashed lines*)

The intended role of the 500-Hz interferer is to impair the use of modulation cues in the detection task. Alternatively, the effect of the interferer could be the direct masking of distortion products. For the parameter setting used in Experiment III, such an effect is expected to be strongest for $\Delta f = 60$ Hz. The level analysis at the end of the previous experiment will now be used to decide how likely such a masking process is.

For the condition of $\Delta f = 60$ Hz, the level of the distortion products centered at 640 Hz is 22 dB SPL. This level is 37 dB lower than the level of the 500-Hz interferer. We need to relate this to the decrease in masked thresholds above a narrowband masker. In Experiment I, we have seen that for a target 100 Hz above the masker, thresholds were 35 dB below the masker level. We can therefore safely assume that for a separation of 140 Hz, thresholds will be at least 45 dB lower than the masker level, that is, it is unlikely that the interferer masked the distortion products. Thus, in summary, the increase in thresholds caused by the addition of the 500-Hz interferer is difficult to reconcile with masking of distortion products by the interferer.

5 Discussion

In this study, off-frequency BMLD experiments were conducted with and without additional interferers. The goal was to investigate whether in N_0S_0 conditions, modulation cues provided extra cues that could only be used in the off-frequency condition. The usage of such cues would help to explain why the BMLD is small in off-frequency conditions. The interferer was designed to cause similar modulations independent of whether the target was present or not, making these cues less useful for detecting the presence of the masker.

Results were in line with this assumption, Experiments 1 and 3 showed an increase in N_0S_0 threshold in the presence of the interferer. Interestingly, also N_0S_π thresholds increased which could indicate that also these thresholds are mediated by monaural modulation detection cues. The addition of the interferer did lead to an increase of the BMLD; it was, however, not restored to a level comparable to those found for on-frequency masking. This may indicate that there are additional factors in play that reduce the effectiveness of binaural processing in off-frequency conditions. An alternative explanation may be that the interferers did not fully remove the ability to process modulation cues.

In another explanation for the increase in N_0S_0 thresholds due to the addition of the interferers, distortion products were masked which were responsible for an improved detection in off-frequency N_0S_0 conditions without interferer. This assumption, however, would imply distortion product levels that were too high to explain the increase in N_0S_0 thresholds when the 500-Hz interferer was used.

In an earlier study by van der Heijden et al. (1997), the role of distortion products was demonstrated in the periodicity of off-frequency $N_\tau S_0$ thresholds that were in line with the presence of on-frequency distortion products. In that study, however, the distortion products were generated by a wider bandwidth masker and resulted effectively in increased thresholds. In our experiment, we have narrowband maskers where distortion products could have potentially been responsible for lowering thresholds.

In summary, our experiments support the hypothesis that at least part of the reduction of off-frequency BMLD is due to a rather effective monaural processing based on the use of monaural modulation cues.

References

Buss E, Hall JW III (2010) The role of off-frequency masking in binaural hearing. J Acoust Soc Am 127:3666–3677

Greenwood DD (1971) Aural combination tones and auditory masking. J Acoust Soc Am 50:502–543

Hall JW, Tyler R, Fernandes MA (1983) Monaural and binaural auditory frequency resolution measured using bandlimited noise and notched-noise masking. J Acoust Soc Am 73:894–898

Henning GB, Yasin I, Witton C (2007) Remote masking and the binaural masking-level difference. In: Kollmeier B, Klump G, Hohmann V, Langemann U, Mauermann M, Uppenkamp S, Verhey J (eds) Hearing – from basic research to applications. Springer, Heidelberg

Nitschmann M, Verhey JL (2012) Modulation cues influence binaural masking-level difference in masking-pattern experiments. J Acoust Soc Am Express Lett 131:EL223–EL228

Nitschmann M, Verhey JL, Kollmeier B (2010) Monaural and binaural frequency selectivity in hearing-impaired subjects. Int J Audiol 49:357–367

van de Par S, Kohlrausch A (2005) The role of intrinsic masker fluctuations on the spread of masking. In: Proceedings of the Forum Acusticum, Budapest, 2005

van der Heijden M, Trahiotis C, Kohlrausch A, van de Par S (1997) Binaural detection with spectrally nonoverlapping signals and maskers: evidence for masking by aural distortion products. J Acoust Soc Am 102:2966–2972

Yost WA, Sheft S, Opie J (1989) Modulation interference in detection and discrimination of amplitude modulation. J Acoust Soc Am 86:2138–2147

Zurek PM, Durlach NI (1987) Masker-bandwidth dependence in homophasic and antiphasic tone detection. J Acoust Soc Am 81:459–464

Zwicker E, Henning GB (1984) The four factors leading to binaural masking-level differences. Hear Res 47:19–29

Chapter 34
Measuring the Apparent Width of Auditory Sources in Normal and Impaired Hearing

William M. Whitmer, Bernhard U. Seeber, and Michael A. Akeroyd

Abstract It is often assumed that single sources of sound are perceived as being punctate, but this cannot be guaranteed, especially for hearing-impaired listeners. Any impairment that gives a reduction at the periphery in the accuracy of coding fine-scale temporal information must give a slight interaural jitter in the temporal information passed to higher centres, and so would be expected to lead to an effective reduction in the interaural coherence (IC) of any stimulus. This would lead to deficits in locating sounds, but deficits of imprecision, not inaccuracy. In turn, this implies that older hearing-impaired individuals should have a diminished perception of auditory space, affecting their abilities to perceive clear, concise, punctate spatial impressions or to separate sounds by location. The current work tested this hypothesis by using two separate visual-analogy methods to measure auditory source width for broadband sounds. In one method, the listener sketched the auditory image, a visual-description task, and for the other, the listener selected the closest one of a set of pre-drawn visual sketches (note that the first is an open-set experiment, whereas the second is a closed-set experiment). We found that older hearing-impaired listeners had increased difficulty in judging changes in interaural coherence, showing a corresponding insensitivity to auditory source width in the visual-analogy tasks.

W.M. Whitmer, PhD (✉) • M.A. Akeroyd
MRC Institute of Hearing Research (Scottish Section),
Glasgow Royal Infirmary, G31 2ER, Glasgow UK
e-mail: bill@ihr.gla.ac.uk

B.U. Seeber
Audio Information Processing,
Technische Universität München,
80333 München, Germany

B.C.J. Moore et al. (eds.), *Basic Aspects of Hearing*,
Advances in Experimental Medicine and Biology,
DOI 10.1007/978-1-4614-1590-9_34,

1 Introduction

When a sound is coherent, the acoustic information arriving at the two ears is the same. Architectural acoustics studies have shown that these sounds are perceived as punctate images by normal-hearing (NH) listeners. The reflections encountered in most spaces create interaural differences, decreasing the similarity between the sounds reaching the two ears. This can be measured by *interaural coherence* (IC), the normalised peak in the interaural cross-correlation function. The NH listener perceives these less interaurally coherent sounds as broader, more diffuse sounds.[1] While it has been shown across numerous methods that IC affects the width of sounds for normal-hearing listeners (e.g. Keet 1968), less is known about how it affects the perception of width by older, hearing-impaired (HI) listeners.

There is indirect psychophysical and physiological evidence to suggest that older HI individuals do perceive sound sources differently. In comparison with the azimuthal localisation of broadband sounds, older HI listeners have shown roughly threefold increases in imprecision – the intrasubject variability or scatter – relative to younger NH groups, though without showing significant changes in location bias (e.g. Lorenzi et al. 1999a, b; Dobreva et al. 2011). Furthermore, the ability to discriminate dichotic from diotic stimuli using supra-threshold interaural time and phase information has been shown to decrease with age (e.g. Herman et al. 1977; Grose and Mamo 2010). Physiological examinations of the coding of sound-source location have shown broader tuning and decreased sensitivity along the aged auditory pathway (cf. Ison et al. 2010). This evidence, however, can only *suggest* that there are age- and impairment-related differences in the spatial percept of sound sources, not *show* these differences.

A recent study by Boyd et al. (2012) has shown that hearing loss can contract the externalisation of sounds. Impulse responses from loudspeakers in the front hemifield separated by 30° at a 3-m radius were recorded with and without the listener present. Mixes between the head and no-head impulse responses were convolved with speech, and then the resulting stimuli were presented to younger NH and older HI listeners who were asked to rate the perceived depth on a continuous, semantically anchored egocentric scale. While NH listeners rated sounds to shift from *inside their head* to *at the loudspeakers* with increasing mixes of their individualised impulse response, HI listeners rated sounds to only move from *at their ear* to *in the room*. That is, the older HI listeners' results suggest a reduced perceptual range along the radial dimension.

How these age and impairment changes affect spatial perception can be most easily demonstrated using visual analogies to acoustic space. Licklider and Dzendolet (1948) first reported a visual analogy to the diffuseness of sounds based on IC by mixing three independent noises to the *x*- and *y*-axis inputs of an oscilloscope. Pollack (1960) used this method to visually replicate IC discrimination thresholds (Pollack and Trittipoe 1959). Plenge (1972) used a visual-description method to demonstrate the varying percept of sound-source size. The stimuli were two independent narrowband noises (400-Hz centre frequency, 300-Hz bandwidth) that were mixed at varying ratios into two channels to vary the IC. The two channels were presented through two loudspeakers in an anechoic chamber to 12 NH listeners who sketched the size of the stimuli onto paper showing the placement of the loudspeakers, themselves and the walls (i.e. a floor-plan view). Near-diotic stimuli – where the second noise was attenuated more than 25 dB – were drawn as small shapes between the loudspeakers, but when the second noise was attenuated by only 4 dB, the widths of the responses covered the separation of the loudspeakers. Blauert and Lindemann (1986) further explored changes to the size of sketched responses based on the IC. They used pink and bandpass noises presented over headphones. The total proportion of area within the head taken by sketches was measured and analysed: fully coherent stimuli were drawn significantly smaller than partially coherent stimuli, but partially coherent stimuli, with ICs of 0.25–0.75, did not have significantly different sizes. Partially coherent stimuli were often drawn as two events (just inside the ears) or three events (centre and ears). Using a similar sketching template to Plenge, Martens (1999) demonstrated that the width and envelopment of percussive sounds increased for NH listeners when presented through a two vs. one subwoofer system. Merimaa and Hess (2004) used a computerised visual-mapping system for NH listeners to describe the width and envelopment of sounds recorded in anechoic and reverberant rooms. Listeners were instructed to visualise the sound as a circular arc for width judgments and were able to adjust the radius, angle (extent) and midpoint of the arc on a GUI. The results, given as angular width of the responses, showed significant differences between stimuli, acoustic spaces and listeners.

Nevertheless, a primary issue with demonstrating spatial percepts through visual analogy is that intersubject variability can overwhelm potential perceptual differences. A balance must be struck between having a method that allows the listener to describe their percept of the sound source(s) and a method that constrains extraneous variability to detect changes in that percept. While it is possible to constrain variability through specific instruction, the individual responses can still vary widely (cf. Merimaa and Hess 2004). A closed-set task, such as establishing similarity through pairwise comparison (Martens 1999), may also constrain variability, but may not describe the way in which the individual perceives their environment. To illustrate this point, Fig. 34.1 shows numerous but not exhaustive different potential percepts of a given acoustic scenario: a point source directly in front of the listener. Despite being only two-dimensional representations – and schematic representations at that – the possible combinations are overwhelming for any pairwise fixed-level testing design.

To experimentally determine if there were differences in the percept of sound sources between NH and HI listeners, the current study employed two methods of visual analogy: a drawing task and an identification task. In order to reduce variability, emphasis was placed on the apparent width of the stimulus presented over headphones, not the stimulus location. In the drawing or visual-description task, listeners drew their representation of sound sources with different interaural coherences and simulated positions onto a pre-drawn schematic of a mannequin head. In the identification task, listeners chose the closest match to their perception of the

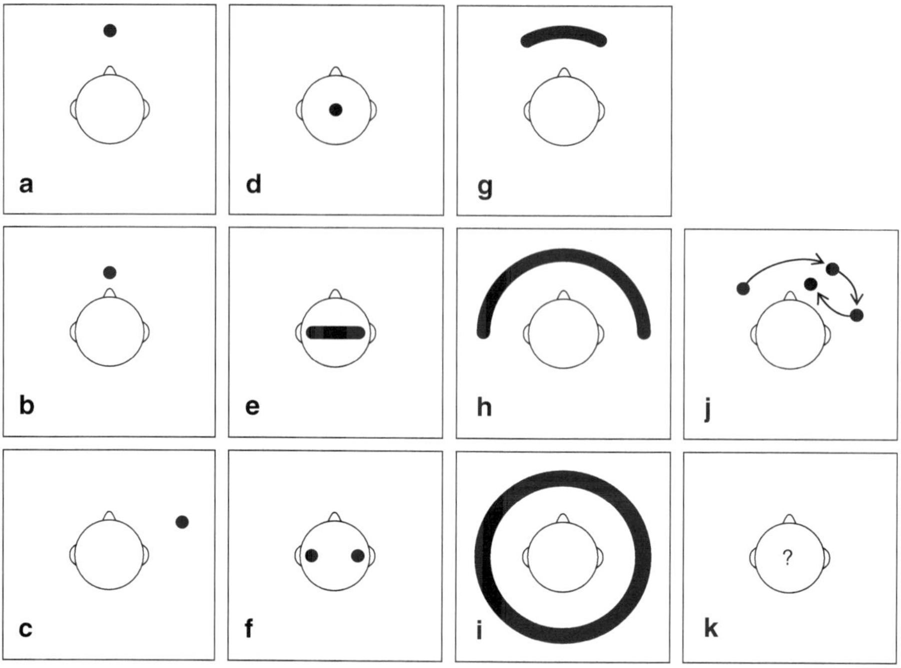

Fig. 34.1 Top-view schematic examples of the potential spatial percepts of an acoustic scene with a point source presented in front of the listener. Panel **a** shows a punctate percept accurately representing the direction and distance of the acoustic event. Panels **b** and **c** show the punctate source perceived at different radial and angular positions (i.e. a localisation bias). Panels **d–f** show intracranial images: a punctate image (d), a broad image (e) and dual images at the two ears (f); headphone presentation without applying proper ear and headphone equalisation would be expected to result in these types of images. Panels **g–i** show broadened distal images, from slightly broader (g) to stretching across the azimuth (h) to being perceived as all around the listener or fully diffuse (i). Images resembling these could be expected from presentation in a reverberant or multiple-source environment. Panel **j** shows an unstable spatial percept, with the source location moving, and panel **k** indicates an indescribable percept

source from a predefined set of 15 images representing hypothetically narrow to wide images presented to the left, centre and right.

2 Methods

A group of 21 HI adults matched for hearing loss with ages ranging from 48 to 77 years were recruited by post. Better ear pure-tone threshold averages ranged from 33- to 43-dB HL with asymmetries of 0–10-dB HL. Four NH listeners (2 female), aged 28–41 years with normal hearing based on pure-tone audiometric thresholds between 500 and 4,000 Hz less than 20-dB HL, were also recruited from employees and students of the Institute of Hearing Research. All listeners had participated in an auditory-source-width discrimination task prior to testing, giving them some familiarity with the concept of source width and the stimuli being used in the current tasks.[2]

The stimuli were composed of 500-ms third-octave narrowband noises with octave-spaced centre frequencies from 500 to 4,000 Hz. Each narrowband component was generated using Plenge's symmetric generator method (1972) to reduce variability within each band: two independent noises were independently attenuated, then added and subtracted, respectively, from each other in the left and right channel for the desired IC. In the drawing task, three IC values were tested: 0.6, 0.8 and 1. In the identification task, five IC values were tested: 0.6–1 in 0.1 increments. The broadband stimuli were presented at 75 dB(A) over circumaural headphones. To examine the effect of location and maintain interest throughout the experimental session, the stimuli were presented from three simulated positions (−30°, 0° and +30°) and monaurally left or right, randomly chosen on each trial. The ±30° positions were created by applying global ITDs and ILDs of 229 μs and 4.8 dB (being the average values across multiple HRTF databases) to the stimuli.

In the visual-description procedure, HI and NH listeners were required to sketch, using a touch screen, the perceived size of the sound they heard on each trial onto a 450-pixel square image of a mannequin head with an ear-to-ear distance of 360 pixels. Listeners' responses were sometimes incomplete, necessitating a recursive sliding or boxcar average to complete the shapes. The only additional instruction was that, as the experiment was concerned with the size of the sound the listener heard, they were to project any sounds heard to the rear to the front for their response. Each listener was given fully coherent (IC = 1) stimuli with the five locations (L, −30°, 0°, +30° and R) as practice. All listeners then commenced with the experiment, sketching their percept of ten presentations of each of 11 combinations of IC and position (3 IC × 3 positions plus monaural stimuli). Stimuli were presented in random order.

A problem was evident in the visual-description responses for the ±30° stimuli: sketches were constrained by the image of the head. Listeners did not draw images beyond the two ears, so that the area and width were confounded by the centre of the response. As responses for ±30° stimuli were confounded by the visual anchor of the mannequin head, those stimuli were not included in further analysis of the measurement technique. Two of the older HI listeners did not draw shapes at all, only drawing dots to indicate positions (e.g. panel d in Fig. 34.1 above), and three older HI listeners placed left- and right-positioned stimuli in the opposite hemifield, indicating a possible momentary lapse in understanding the mirror-image aspect of the task. The results of these five listeners were not included in the analysis.

Immediately following the drawing task, HI and NH listeners completed the visual identification task. They were presented with stimuli with ICs of 0.6–1 and simulated positions of −30°, 0° and 30° (no monaural stimuli were used). After each stimulus presentation, a 5 × 3 matrix of images was displayed with 20-pixel high grey bars made of visual noise. The widths of the pre-drawn bars ranged across the five columns from 20 to 100 pixels in 20-pixel increments based on previously established near-linear relationships between IC and width (cf. Keet 1968). The positions of the bars across the three rows were at approximately −30°, 0° and +30°. The 15 conditions were presented in a randomised order for each block; after two blocks of practice, listeners completed four test blocks (i.e. a total of 90 trials).

3 Results and Discussion

For the drawing task, the raw results of the 16 HI listeners and four NH listeners are shown in Fig. 34.2; individual sketches were collated into density plots for the 0° stimuli with ICs of 0.6. 0.8 and 1 and the monaural stimuli. Three basic results are clear: (1) changes in IC did not produce noticeably different sized responses for HI listeners, though they did for NH listeners, (2) diotic (IC = 1) stimuli were drawn smaller and narrower by NH listeners relative to HI listeners, and (3) drawing methods varied greatly, especially among HI listeners.

In the identification task, each image of expanding width was assigned a descending IC (1–0.6) based on pilot results. The last four response ICs for each position (−30°, 0° and +30°) were averaged for each listener. The raw results for the identification task are shown in Fig. 34.3. The NH results show a clear change in response selection with increasing IC and relatively good agreement among listeners. The HI listeners show a clear insensitivity to changes in stimulus IC, with linear-regression slopes near zero. Furthermore, the variability in the mean response for HI listeners varied widely; the average response IC ranged from 0.63 to 0.88, with only a slight tendency towards hypothetically broader (lower IC) images. The mean data in the identification task mirror the mean data in the drawing task: NH listeners judged the least coherent stimuli as significantly wider and the fully coherent stimuli as significantly narrower than HI listeners.

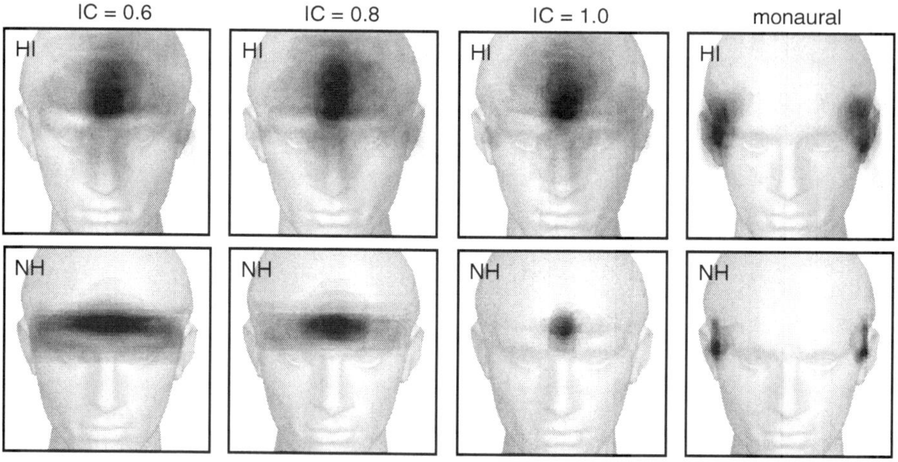

Fig. 34.2 Density plot of drawing experiment responses formed by overlaying all responses for HI ($n = 24$) and NH ($n = 7$) participants (rows) and IC (columns) for stimuli presented over headphones with simulated lateral position of 0°. *Greyscale* indicates frequency, with *black* being most frequent. The *rightmost* column shows responses for the monaural stimuli. The mannequin-head image was inverted for contrast during testing. Responses were less affected by IC for HI relative to NH participants

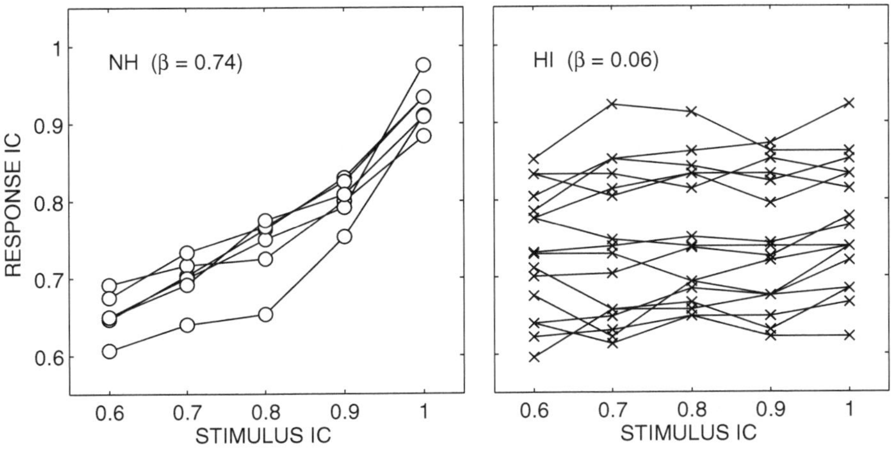

Fig. 34.3 Mean response interaural coherence (IC) of the chosen visual stimulus as a function of stimulus IC for all NH (*left panel*) and HI (*right panel*) listeners. The average linear-regression slopes (β) are given for both groups

4 Summary

Based on their insensitivity to changes in IC, and the significantly broader width with which they drew diotic (IC = 1) stimuli, older HI individuals do not appear to hear punctate images for binaural sounds. The increased temporal jitter found in the aging auditory pathway (Pichora-Fuller and Schenider 1991) is manifest in the current results as a reduced sensitivity to changes in interaural coherence; that is, all sound sources are perceived as diffuse images with weaker neural representations of location (Ison et al. 2010). Like many previous studies, these visual-analogy experiments were conducted over headphones for precise control of IC. If these deficits persist in the free field, they could undermine the benefit of source-separation strategies in hearing prostheses, as well as impact our understanding of how the early reflections that change IC can affect HI individuals.

5 Endnotes

1. The focus here is on the interaural disparities that cause changes in source width, which distinguishes the study of auditory source width from the early twentieth-century behaviourist studies of "tonal volume" that examined pure-tone frequency difference limens by means of the assumption that higher frequency tones sound more punctate than lower frequency tones (e.g. Rich 1916).
2. Further details on the methods reported here can be found in Whitmer et al. (2012).

Acknowledgments The Scottish Section of IHR is supported by intramural funding from the Medical Research Council (grant number U135097131) and the Chief Scientist Office of the Scottish Government.

References

Blauert J, Lindemann W (1986) Spatial mapping of intracranial auditory events for various degrees of interaural coherence. J Acoust Soc Am 79:806–813

Boyd A, Whitmer W, Soraghan J, Akeroyd M (2012) Auditory externalization in hearing-impaired listeners: the effect of pinna cues and number of talkers. J Acoust Soc Am 131:EL268–EL274

de villiers Keet W (1968) The influence of early lateral reflections on the spatial impression. In: Proceedings of the sixth international congress on Acoustics, Tokyo, 1968, pp E2–E4

Dobreva M, O'Neill W, Paige G (2011) The influence of aging on human sound localization. J Neurophysiol 105:2471–2486

Grose J, Mamo S (2010) Processing of temporal fine structure as a function of age. Ear Hear 31:755–760

Herman G, Warren L, Wagener J (1977) Auditory lateralization: age differences in sensitivity to dichotic time and amplitude cues. J Gerontol 32:187–191

Ison J, Tremblay K, Allen P (2010) Closing the gap between neurobiology and human presbycusis: behavioral and evoked potential studies of age-related hearing loss in animal models and in humans. In: Gordon-Salant S, Frisina R, Fay R, Popper A (eds) The aging auditory system. Springer, New York, pp 75–110

Licklider J, Dzendolet E (1948) Oscillographic scatterplots illustrating various degrees of correlation. Science 107:121–124

Lorenzi C, Gatehouse S, Lever C (1999a) Sound localization in noise in normal-hearing listeners. J Acoust Soc Am 105:1810–1820

Lorenzi C, Gatehouse S, Lever C (1999b) Sound localization in noise in hearing-impaired listeners. J Acoust Soc Am 105:3454–3463

Martens W (1999) The impact of decorrelated low-frequency reproduction on auditory spatial imagery: are two subwoofers better than one? In: Proceedings of the Audio Engineering Society 16th international conference of spatial sound reproduction, Rovaniemi, 1999, pp 67–77

Merimaa J, Hess W (2004) Training of listeners for evaluation of spatial attributes of sound. In: Proceedings of the 117th Convention of Audio Engineering Society, San Francisco, 2004. Preprint 6237

Pichora-Fuller K, Schenider B (1991) Masking level differences in the elderly: a comparison of antiphasic and time-delay dichotic conditions. J Speech Hear Res 34:1410–1422

Plenge G (1972) Über das Problem der Im-Kopf-Lokalisation. Acustica 26:241–252

Pollack I (1960) Identification of visual correlational analysis. J Exp Psychol 59:351–360

Pollack I, Trittipoe W (1959) Binaural listening and interaural noise correlation. J Acoust Soc Am 31:1250–1252

Rich G (1916) A preliminary study of tonal volume. J Exp Psychol 1:13–22

Whitmer W, Seeber B, Akeroyd M (2012) Apparent auditory source width insensitivity in older hearing-impaired individuals. J Acoust Soc Am 132:369–379

Chapter 35
Psychophysics of Human Echolocation

Sven Schörnich, Ludwig Wallmeier, Nikodemus Gessele, Andreas Nagy, Michael Schranner, Daniel Kish, and Lutz Wiegrebe

Abstract The skills of some blind humans orienting in their environment through the auditory analysis of reflections from self-generated sounds have received only little scientific attention to date. Here we present data from a series of formal psychophysical experiments with sighted subjects trained to evaluate features of a virtual echo-acoustic space, allowing for rigid and fine-grain control of the stimulus parameters. The data show how subjects shape both their vocalisations and auditory analysis of the echoes to serve specific echo-acoustic tasks. First, we show that humans can echo-acoustically discriminate target distances with a resolution of less than 1 m for reference distances above 3.4 m. For a reference distance of 1.7 m, corresponding to an echo delay of only 10 ms, distance JNDs were typically around 0.5 m. Second, we explore the interplay between the precedence effect and echolocation. We show that the strong perceptual asymmetry between lead and lag is weakened during echolocation. Finally, we show that through the auditory analysis of self-generated sounds, subjects discriminate room-size changes as small as 10%.

In summary, the current data confirm the practical efficacy of human echolocation, and they provide a rigid psychophysical basis for addressing its neural foundations.

S. Schörnich • L. Wallmeier • N. Gessele • A. Nagy • M. Schranner • L. Wiegrebe (✉)
Division of Neurobiology, Department of Biology II,
University of Munich, Grosshaderner Str 2, Martinsried 82152, Germany
e-mail: lutzw@lmu.de

D. Kish
World Access for the Blind, Huntington Beach, CA, USA
e-mail: daniel.kish@worldaccessfortheblind.org

B.C.J. Moore et al. (eds.), *Basic Aspects of Hearing*,
Advances in Experimental Medicine and Biology,
DOI 10.1007/978-1-4614-1590-9_35,

1 Introduction

Echolocation, defined as imaging of the environment through the auditory analysis of acoustic reflections elicited by self-generated sounds, allows gaining information about one's surroundings even in complete darkness. Therefore, this ability is found in mammals whose habitat or way of life renders the use of vision difficult or impossible, like toothed whales or bats.

Several studies have shown that blind or blindfolded human subjects can echo-acoustically detect and discriminate objects of different shape or texture (Kellogg 1962; Rice and Feinstein 1965; Rice 1967; Schenkman and Nilsson 2010). Also object localisation has been shown to be quite precise in blind human echolocation experts (Teng et al. 2011).

This chapter summarises results from three formal psychophysical studies addressing the efficacy of human echolocation in fully controlled virtual echo-acoustic space.

The principal outline of an echo-acoustic experiment is fully described in terms of a linear system. The outgoing sound is reflected by an ensonified object and perceived through the subjects' ears. The spatial characteristics of the outgoing sound, the way it is reflected by an object, and the path the reflection takes from the object to the subject's ear drums can be described by acoustic impulse responses (IRs). To transfer such an experiment into virtual echo-acoustic space (VEAS), these IRs have to be known and applied in real time to the sounds generated by the subject.

VEAS was created by picking up the subjects' vocalisation from a headset microphone and feeding them back to earphones. The latter block external sound quite effectively and thus the subjects would not perceive their own vocalisations as they would in the free sound field. Therefore, direct sound and echoes were presented separately via two paths: the first path was a direct, level-adjusted path from the microphone to the earphones. The level of the direct path was set such that the subject's percept of their own voice in the anechoic chamber was most similar to the percept of their voice with the earphones removed from the ear canals. The second (echo) path incorporated the IRs including the vocal IR (which describes the azimuth-dependent spread of sound from the mouth), the acoustic IR from the ensonified object, and the head-related IR, which describes the azimuth-dependent path from the object to the two ears.

2 Target Ranging in VEAS

In contrast to vision, where distance information is relatively difficult to infer (Palmer 1999), the distance to a sound-reflecting surface can be echo-acoustically determined by estimating the time delay between emission and echo reception (Simmons 1973; Denzinger and Schnitzler 1998; Goerlitz et al. 2010). A formal quantification of human echo-acoustic sensitivity to target range is not available to date.

2.1 Stimuli

The echo delay corresponding to the required range was applied by preceding the IR for the second path with so many zeros that, together with the digital IO delay of the hardware and convolution software, the delay corresponding to the required reflector range was generated. Next, the IR amplitude was scaled to match the range-dependent geometric attenuation of a virtual echo. For each reference range, the geometric attenuation was globally set, i.e., compared to the reference range of 1.7 m, the IR amplitude was decreased by 6 and 12 dB for ranges of 3.4 and 6.8 m, respectively.

2.2 Procedure

To solve this task, the subjects had to produce tongue clicks and analyse the real-time-generated virtual echoes from these clicks. In an adaptive two-alternative, forced-choice paradigm, subjects were trained to find the interval in which the reflective surface was further away from the subject; i.e., in which the delay between the emission and echo was longer than the reference delay. The latter was roved across trials by ±5 %. Each interval began with a 50 ms, 1 kHz tone pip. Directly after the tone pip, both the direct path and the echo path were activated for 5 s, such that when the subject produced a sound, the direct path would feed directly into their ears, while the echo path provided a real-time-generated echo of the sound with the appropriate range, spectral content, and binaural characteristics. The end of the interval was signalled by another tone pip (50 ms, 2 kHz) which was presented directly after the direct and echo path had been switched off. Subjects were given audio feedback consisting of a 250 ms frequency chirp which was upward modulated for positive feedback and downward modulated for negative feedback.

2.3 Results

At a reference range of 1.7 m (reference-echo delay = 10 ms, Fig. 35.1, top), most subjects could detect a change in target range of only 30–40 cm; only one subject performed significantly worse, with a range JND of 93 cm. However, while for this subject the range JND stayed approximately constant, even when the reference range was increased to 3.4 or 6.8 m, two other subjects (red and white) showed increasing JNDs with increasing reference range.

To solve this task, subjects produced relatively short vocalisations (tongue clicks) with durations between 3 and 12 ms. The click SPLs, measured at the headset microphone, were more variable, ranging from about 60–105 dB. Also the number

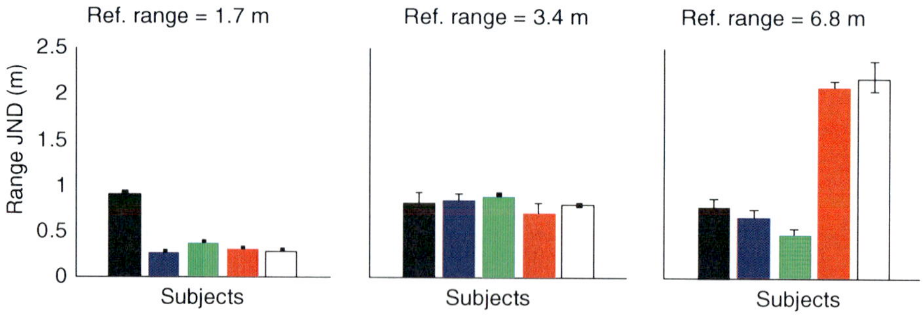

Fig. 35.1 Echo-acoustic sensitivity to target range. *Colour bars* represent performances of individual subjects

of clicks produced to evaluate the reflection properties in a 5 s interval of the 2AFC task varied across the subjects with individual averages between only 6 and up to 23 clicks. Most subjects produced relatively high-frequency clicks with peak frequencies around 6–7 kHz and −15 dB bandwidths from 3 to 10 kHz or 15 kHz.

3 Echolocation vs. Echo Suppression in Humans

The precedence effect predicts a conflict of echolocation and echo suppression: when localising sound sources, the human auditory system suppresses spatial information about echoes, but just this information underlies effective echolocation. A common approach to investigate the precedence effect is the arrangement of two sound sources that present a direct sound (lead) and a delayed copy (lag). Several experiments on lag-discrimination suppression have quantified the deterioration of spatial information about the lag produced by the lead. Here, minimum audible angles (MAAs) were measured in VEAS. In the 'listening' version, the subjects had to discriminate between positions of a single sound source, the leading or the lagging of two sources. In the 'echolocation' version, the sound sources were replaced by sound reflectors. Here, the subjects evaluated the echoes generated in real time from self-produced tongue clicks and thereby discriminated between positions of a single reflector, the leading or the lagging of two reflectors.

3.1 Stimuli

Individual HRIRs and VIRs, measured at 10° azimuthal resolution, were interpolated to 0.2° to construct the stimuli. In the 'listening' version, the stimuli were acoustic impulses at a repetition rate of 2.5 Hz, convolved with the required HRIRs. The lead-lag delay was 2 ms. In the 'echolocation' version, stimuli were generated

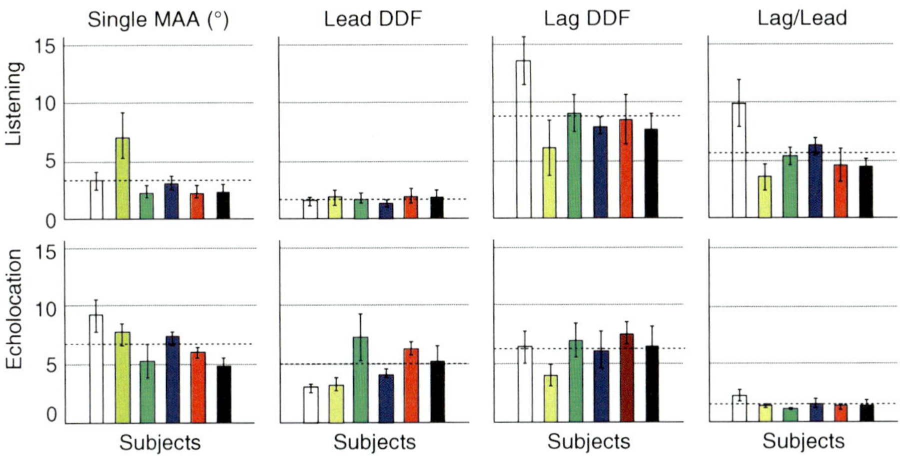

Fig. 35.2 Precedence in a listening and an echolocation task. *Colour bars* represent individual data

by the listeners (typically tongue clicks), and, in addition to the direct path described above, they were convolved in real time with first the VIR and then the HRIR and played back with a delay of 10 ms (lead) or 12 ms (lag). Thus the lead-lag delay was fixed at 2 ms, the distance to the leading virtual reflector was 1.7 m, and the distance to the lagging reflector was 2.04 m. Note that in the echolocation version, the perception of the echoes from the leading and lagging reflections was always preceded by the percept of the outgoing sound.

3.2 Procedure

In an adaptive 2AFC paradigm with audio feedback, MAAs were measured following a three-down, one-up rule. The beginning and end of each 5 s interval were marked by 50 ms tone pips. Intervals were long enough to allow listeners to explore the spatial layout of the reflectors in the 'echolocation' version with several tongue clicks. Listeners were extensively trained until they reached stable performance.

3.3 Results

MAA measurements for single objects presented in virtual space are shown in the top row of Fig. 35.2. Objects are either a sound source emitting impulses (listening) or a surface reflecting the subjects' tongue clicks (echolocation).

The data show that sighted subjects can be trained to discriminate reflective surfaces by echolocation with an accuracy comparable to sound-source localisation. Data from six subjects reveal a mean MAA of 3.4° in the listening version and 6.7° in the echolocation version; individual MAAs are shown in the first row of Fig. 35.2.

An appropriate method for comparing results across versions is the calculation of the discrimination deterioration factor (DDF), defined as ratio of lead/single or lag/single thresholds (Litovsky 1997; Tollin and Henning 1998). Individual DDFs for lead and lag discrimination are shown in the second and third rows of Fig. 35.2, respectively.

In the listening version, the presence of a lagging source impaired lead discrimination only slightly by a factor of 1.6 (second row, left), while a leading source impaired lag discrimination considerably by a factor of 8.8 (third row, left). This asymmetry between lead and lag discrimination is consistent with previous studies on discrimination suppression (Litovsky et al. 1999).

In the echolocation version, however, this asymmetry was significantly weaker: lead and lag discrimination deteriorated by factors of 4.8 and 6.2, respectively. To quantify the asymmetry between lead and lag discrimination as the defining measure of the influence of the precedence effect, the ratio of lag/lead thresholds was calculated and depicted in the bottom row of Fig. 35.2. The strength of precedence was significantly higher in the listening experiment.

These data indicate that the precedence effect is weakened in an echolocation context.

4 Human Sonar Sensitivity to Room Size

Echolocation does not only allow determination of the distance and position of objects in space but also assessment of the dimensions of an enclosed space itself. From architectural acoustics, it is known that the reverberant qualities of an enclosure can be quantified in terms of a binaural room impulse response (BRIR): the BRIR incorporates the spatial and temporal distribution of reflections which a sound undergoes on its way from a specific source to a human, binaural receiver. In the current experiment, we aimed to formally quantify the ability of sighted human subjects to detect changes in the size of an enclosed space by evaluating echoes of the subjects' own vocalisations.

4.1 Stimuli

We recorded the BRIR of a real room (a chapel with a maximum width of 7.18 m, a maximum length of 17.15 m, and a maximum height of 5.54 m; reverberation time ~ 1.8 s) and compressed the BRIR along the time axis to simulate decreases in room size (Blauert and Xiang 1993). As reference BRIRs the recorded BRIR was compressed with compression factors of 0.2, 0.5, and 0.7; a compression factor of 0.2 produced the smallest perceived room.

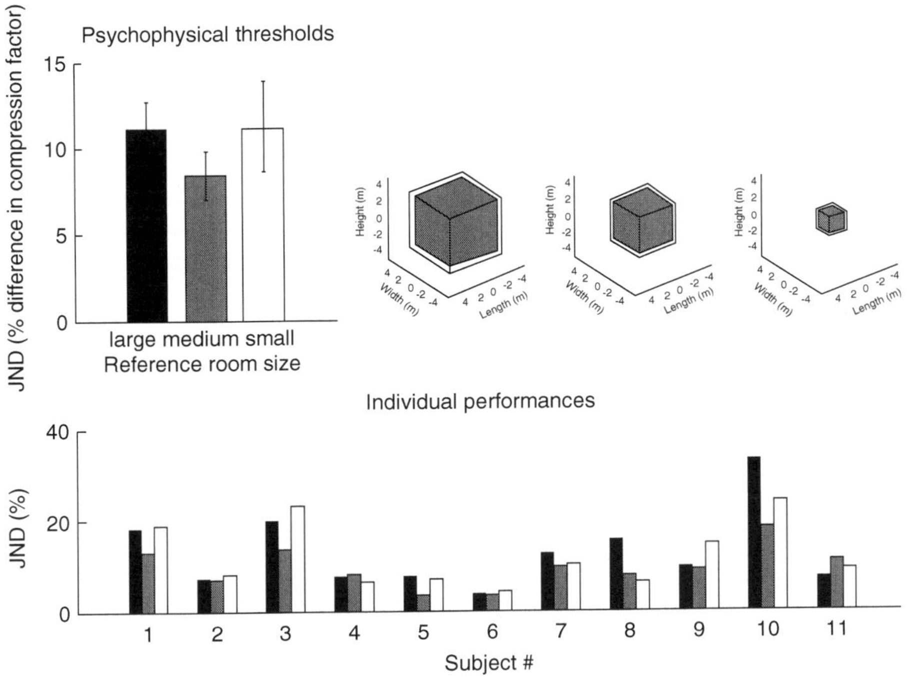

Fig. 35.3 Echo-acoustic JNDs for room size. The *upper left panel* shows average JNDs as a function of the reference room size; these data are geometrically illustrated on the *right*. Individual data are shown in the *bottom row*

4.2 Procedure

Similar to the previous experiments, we used a 2AFC paradigm and VEAS to quantify the just-noticeable differences (JNDs) in acoustic room size. Room-size JNDs specify by which percentage each side of the virtual room must be increased such that the corresponding changes in the BRIR can be perceived via echolocation. In each interval, a virtual room was presented for 5 s, delimited by 50 ms tone pips. The subjects produced tongue clicks in each interval and were asked to indicate which of the two rooms was the smaller one after the second interval. Again, listeners were given audio feedback.

4.3 Results

Figure 35.3 shows that listeners are quite sensitive to changes in the echo-acoustically perceived room size; their JNDs are on the order of 10 %. A graphic illustration of this echo-acoustic sensitivity is shown in the upper right panel of Fig. 35.3. For each of the reference room sizes, the psychophysical performance quantified here is sufficient to discriminate the grey-filled room from the transparent room surrounding it.

5 Discussion and Conclusions

This chapter summarises data from a series of formal psychophysical experiments on the efficacy of human echolocation. Placing the experiments in virtual echo-acoustic space allows for unprecedented experimental control of stimulus parameters and detailed documentation of the sensory-motor interactions underlying echolocation in humans. The data show that sighted subjects can be successfully trained to echo-acoustically detect changes in the range of a reflector positioned in virtual echo-acoustic space. The subjects accomplish this task by vocally emitting short broadband sounds and evaluating the echoes generated by the reflector or reflectors. The data show that range JNDs were typically below 1 m and, for a reference range of 1.7 m, they were typically below 0.5 m.

The fact that listeners used clicks with a high-spectral centre of gravity and short durations argues for a temporal auditory analysis as opposed to a pitch-based spectral analysis which would benefit from long-lasting, low-frequency stimulation to produce spectrally resolved harmonics.

The second set of experiments sheds some light on the interplay between echo suppression (the precedence effect) and echolocation. Using a lag-discrimination-suppression paradigm, both in a 'listening' and an 'echolocation' version, the data show that sighted subjects can be trained to discriminate locations of single reflective surfaces almost as well as locations of external sound sources. Second, the data show that in the listening version, the presence of a leading sound source impaired lag discrimination much more than vice versa. This strong asymmetry between lead and lag discrimination was not observed in the echolocation version. These data indicate that the precedence effect, which facilitates the localisation of a leading sound source at the expense of a lagging source, is weakened in an echolocation context. While this decreases presumed sonar localisation accuracy of the closer reflective surface, it allows for a more balanced assessment of complex spatial layouts through echolocation.

The final experiments investigated human sonar sensitivity to the size of an enclosed space. Again, using a formal 2AFC procedure with virtual echo-acoustic space stimuli, the data show that sighted human subjects can discriminate the size of ensonified spaces with JNDs around 10 %. Subjects were very little affected by the sound level of the reverberations but evaluated the temporal decay of the reverberations to solve this task.

The current data were all gathered from highly trained but sighted listeners. Research has shown that blind human echolocation experts often perform significantly better even in such formalised tasks (e.g. Teng et al. 2011). These findings stress the need to extend the current techniques and paradigms to those subjects who rely on echo-acoustic information every day.

Acknowledgements This work was supported by the 'Deutsche Forschungsgemeinschaft' (Wi 1518/9 to Lutz Wiegrebe), the 'Studienstiftung des Deutschen Volkes' (Stipend to Ludwig Wallmeier), and the Danish Research Foundation.

References

Blauert J, Xiang N (1993) Binaural scale modelling for auralization and prediction of acoustics in auditoria. J Appl Acoust 38:267–290

Denzinger A, Schnitzler HU (1998) Echo SPL, training experience, and experimental procedure influence the ranging performance in the big brown bat, Eptesicus fuscus. J Comp Physiol A 183:213–224

Goerlitz HR, Geberl C, Wiegrebe L (2010) Sonar detection of jittering real targets in a free-flying bat. J Acoust Soc Am 128:1467–1475

Kellogg WN (1962) Sonar system of the blind. Science 137:399–404

Litovsky RY (1997) Developmental changes in the precedence effect: estimates of minimum audible angle. J Acoust Soc Am 102:1739–1745

Litovsky RY et al (1999) The precedence effect. J Acoust Soc Am 106:1633–1654

Palmer SE (1999) Vision Science: Photons to Phenomenology. MIT Press, Cambridge

Rice CE (1967) Human echo perception. Science 155:656–664

Rice CE, Feinstein SH (1965) Sonar system of the blind: size discrimination. Science 148: 1107–1108

Schenkman BN, Nilsson ME (2010) Human echolocation: blind and sighted persons' ability to detect sounds recorded in the presence of a reflecting object. Perception 39:483–501

Simmons JA (1973) The resolution of target range by echolocating bats. J Acoust Soc Am 54: 157–173

Teng S, Puri A, Whitney D (2011) Ultrafine spatial acuity of blind expert human echolocators. Exp Brain Res 216:483–488

Tollin DJ, Henning GB (1998) Some aspects of the lateralization of echoed sound in man. I. The classical interaural-delay based precedence effect. J Acoust Soc Am 104:3030–3038

Part V
Speech and Temporal Processing

Chapter 36
Formant-Frequency Variation and Its Effects on Across-Formant Grouping in Speech Perception

Brian Roberts, Robert J. Summers, and Peter J. Bailey

Abstract How speech is separated perceptually from other speech remains poorly understood. In a series of experiments, perceptual organisation was probed by presenting three-formant (F1+F2+F3) analogues of target sentences dichotically, together with a competitor for F2 (F2C), or for F2+F3, which listeners must reject to optimise recognition. To control for energetic masking, the competitor was always presented in the opposite ear to the corresponding target formant(s). Sine-wave speech was used initially, and different versions of F2C were derived from F2 using separate manipulations of its amplitude and frequency contours. F2Cs with time-varying frequency contours were highly effective competitors, whatever their amplitude characteristics, whereas constant-frequency F2Cs were ineffective. Subsequent studies used synthetic-formant speech to explore the effects of manipulating the rate and depth of formant-frequency change in the competitor. Competitor efficacy was not tuned to the rate of formant-frequency variation in the target sentences; rather, the reduction in intelligibility increased with competitor rate relative to the rate for the target sentences. Therefore, differences in speech rate may not be a useful cue for separating the speech of concurrent talkers. Effects of competitors whose depth of formant-frequency variation was scaled by a range of factors were explored using competitors derived either by inverting the frequency contour of F2 about its geometric mean (plausibly speech-like pattern) or by using a regular and arbitrary frequency contour (triangle wave, not plausibly speech-like) matched to the average rate and depth of variation for the inverted F2C. Competitor efficacy depended on the overall depth of frequency variation, not depth relative to that for the other formants. Furthermore, the triangle-wave competitors were as effective as their more

B. Roberts (✉) • R.J. Summers
Psychology, School of Life and Health Sciences,
Aston University, Birmingham B4 7ET, UK
e-mail: b.roberts@aston.ac.uk

P.J. Bailey
Department of Psychology,
University of York, Heslington, York YO10 5DD, UK

B.C.J. Moore et al. (eds.), *Basic Aspects of Hearing*,
Advances in Experimental Medicine and Biology,
DOI 10.1007/978-1-4614-1590-9_36,

speech-like counterparts. Overall, the results suggest that formant-frequency variation is critical for the across-frequency grouping of formants but that this grouping does not depend on speech-specific constraints.

1 Introduction

Speech comprises dynamic and heterogeneous acoustic elements yet it is heard as a single perceptual stream, even when accompanied by other sounds. Spectral prominences in speech, called formants, are perceptually important because they convey articulatory information about vocal tract shape and its change over time. Hence, knowledge of formant frequencies and their change over time is of great benefit to listeners trying to understand a spoken message. When more than one talker is speaking at once, choosing and grouping together the right set of formants from the mixture is critical for intelligibility, but few studies have focussed on this across-formant grouping. The relative contributions of grouping 'primitives' (Bregman 1990) and of speech-specific factors to the perceptual coherence of speech remain unclear, and the critical acoustical correlates of the latter – if they exist at all – remain almost unknown (Darwin 2008; Mesgarani and Chang 2012).

The parametric manipulations possible with simplified speech signals make them attractive stimuli with which to explore these issues. We used sine-wave and synthetic-formant analogues of natural speech. The factors governing perceptual organisation are generally revealed only where competition operates. Therefore, the second-formant competitor (F2C) paradigm was used (Remez et al. 1994), or variants of it, in which the listener must reject an extraneous formant to optimise recognition. The main advantages of the F2C paradigm are as follows: (a) the dichotic configuration minimises energetic masking of target formants by competitor formants, (b) changes in grouping can be indexed by changes in intelligibility, and (c) no assumptions are required about whether the effect of the competitor is to displace or dilute the phonetic contribution of the target F2.

Remez et al. (1994) showed that an F2C created by time-reversing F2 was an effective competitor for sine-wave speech but that a pure tone of constant frequency and amplitude was not. These results suggest that the modulation patterns of formant contours are critical for across-formant grouping. Four experiments are reported which explored the critical aspects of this frequency and amplitude variation. Experiment 1 used separate manipulations of the frequency and amplitude contours of competitor formants to tease apart their impact on the intelligibility of sine-wave speech (Roberts et al. 2010). Experiments 2–4 all used synthetic-formant speech. Experiment 2 measured the effect on intelligibility of manipulating the rate of variation in the frequency contours of competitor formants, relative to that for the target formants (Summers et al. 2012). Experiments 3 and 4 explored the effect on intelligibility of manipulating the depth and pattern of variation in the formant-frequency contour of F2C, relative to that for the target formants (Roberts B et al. 2012, unpublished data). The former used F2Cs whose frequency contours were created by spectral inversion of F2; the latter used F2Cs with regular and arbitrary frequency contours that were not plausibly speech-like.

2 General Method

Stimuli were derived from sentences spoken by a British male talker. The frequency contours of the first three formants were estimated automatically using Praat (Boersma and Weenink 2010). Gross errors in formant-frequency estimates were hand-corrected; corresponding amplitude contours were extracted. For experiment 1, these contours were used to control three time-varying sinusoids to generate sine-wave analogues of the sentences (Bailey et al. 1977; Remez et al. 1981). For experiments 2–4, these contours controlled three parallel second-order resonators, whose outputs were summed. The resonator input was an excitation pulse modelled on the glottal waveform (Rosenberg 1971). The pitch was monotonous (F0 = 140 Hz), and the 3-dB bandwidths of F1, F2, and F3 were 50, 70, and 90 Hz. When present, competitor formants had the same bandwidths and F0 as their target counterparts.

The three target formants were always presented dichotically, such that a competitor for F2 (F2C), or for F2+F3 (F2C+F3C), could be added in the same ear as F1 but in the opposite ear to the corresponding target formant(s). The frequency and amplitude contours of the competitor formants were generated in various ways – in some cases they were derived from the corresponding contours of the target formants by simple transformations (e.g. time reversal); in others, their properties were less closely related to those of the target formants.

For each listener, the sentences were divided equally across conditions (six per condition) using an allocation counterbalanced by rotation across each set of listeners. All listeners were native English speakers with normal hearing; most were undergraduates. Stimuli were presented over headphones in random order at ~75 dB SPL in a sound-attenuating booth. Participants could listen to each stimulus up to six times before entering their transcription; no feedback was given. Beforehand, listeners completed a training session with feedback to improve recognition performance. The measure of intelligibility used was % keywords correctly identified (tight scoring).

3 Experiment 1

3.1 Method

Sine-wave analogues were derived from sentences comprising almost continuously voiced speech. For each sentence, a set of F2 competitors was created by manipulating the frequency (*f*) and amplitude (*a*) contours of F2. The frequency contour of F2C could be time-reversed (R), inverted about its spectral centroid (I), or constant at its spectral centroid (C). The amplitude contour could be time-reversed (R), normal (N), or constant at a value preserving RMS power (C). Stimuli were presented dichotically (left ear = F1+F2C+F3; right ear = F2). There were 12 conditions, five of which were controls (no F2) and one was the dichotic reference (no F2C). Six F2Cs were tested from the nine possible combinations of f and a parameters.

3.2 Results and Discussion

The results ($n = 24$) were as follows. The control conditions indicated that intelligibility was near floor for F1+F3 alone, and when any of the F2Cs was added without the target F2. Hence, these F2Cs did not in themselves support intelligibility. All F2Cs with time-varying frequency contours were highly effective competitors, regardless of their amplitude characteristics. Compared with the dichotic reference and control conditions, these competitors caused performance to fall by ~2/3. In contrast, both F2Cs with constant-frequency contours were completely ineffective, irrespective of whether the amplitude contour was identical to that of the target F2 or was constant.

The findings confirm and extend those of Remez et al. (1994). F2 competitors typically reduced intelligibility, plausibly by providing an alternative to F2 in the perceptual organisation of the sentences. The results suggest that it is the modulation patterns of the formant-frequency contours, but not those of the amplitude contours, that are critical for across-formant grouping. In principle, the impact of frequency variation on across-formant grouping might arise either from speech-specific properties or from the operation of a hitherto undescribed 'primitive' which has more general implications for the perceptual organisation of dynamic broadband signals.

4 Experiment 2

What aspects of formant-frequency variation might be important in the context of across-formant grouping? One dynamic property of speech that merits consideration is the rate of formant-frequency variation. Many properties are influenced by changes in natural speech rate, but these commonly include changes in the rate of formant-frequency variation (e.g. Weismer and Berry 2003). Hence, in principle, differences in the rate of formant-frequency variation between talkers might provide a basis for the appropriate grouping of formants. A hypothesis based on grouping by similarity predicts that the impact of a competitor formant on intelligibility will be rate tuned, such that maximum interference will occur when the rate of formant-frequency variation for the competitor is most like that for the target formants. An alternative hypothesis is that faster variations are more disruptive, such that interference is proportional to the rate of formant-frequency variation in the competitor. We evaluated these hypotheses by manipulating the rate of formant-frequency change.

4.1 Method

Synthetic-formant analogues were derived from sentences comprising ≤25 % phonemes involving closures or unvoiced frication; stimuli were normalised to the mean rate for the set (~3.7 syllables/s). Two-formant competitors (F2C+F3C) were derived from long passages of almost continuously voiced sentences spoken by the same talker, allowing competitors with rates above baseline to be generated without splicing formant tracks together. The frequency and amplitude contours of F2 and F3 were extracted from these passages and time-reversed. At baseline, the time-reversed formant contours were normalised to the mean rate for the target sentences.

Competitors were generated at various rates relative to baseline; parallel changes were made for F2C+F3C. The rates used relative to baseline were 0 (= constant frequency), 0.25, 0.5, 1, and 2. Higher rates were excluded owing to timbre changes arising from synthesis artefacts; these cues might have assisted segregation of target and competitor formants. For a relative rate of 0, the two frequency contours were set to be constant at the geometric mean frequency for the appropriate formant in the spliced segment. For a given sentence, competitor parameters were created by randomly selecting one of a set of extracted and time-reversed formant tracks and splicing out a segment corresponding (after scaling to the desired rate) to the duration of the target sentence.

To explore the grouping role of formant-amplitude contours in the context of synthetic-formant speech, the effect of varying the rate of frequency change in the competitor was measured when the amplitude contour of the competitor was time-reversed (scaled to the desired competitor rate, in parallel with the time-reversed frequency contours) and when it was constant (preserving RMS power for each formant). Stimuli were presented dichotically (F1+F2C+F3C; F2+F3). There were 11 conditions – nine experimental (rate manipulation), one control (no F2+F3), and the dichotic reference (no competitor).

4.2 Results and Discussion

The results (Fig. 36.1; $n=39$) indicate that competitor efficacy is dependent on competitor rate. The impact of F2C+F3C on intelligibility increases gradually and progressively as the rate of frequency variation in the competitor formants increases, at least for rates up to twice baseline. As intelligibility was not minimal when competitor rate = 1, it appears that competitor efficacy is not tuned to the rate of the target sentences; most probably it depends primarily on the overall rate of frequency change in the competitor formants. This finding is consistent with the hypothesis that faster variation in the frequencies of extraneous formants is more disruptive. In contrast, and consistent with the results for sine-wave speech, rate of amplitude change in the competitor had no discernible effect on intelligibility. The results suggest that differences in speech rate as such may not be a

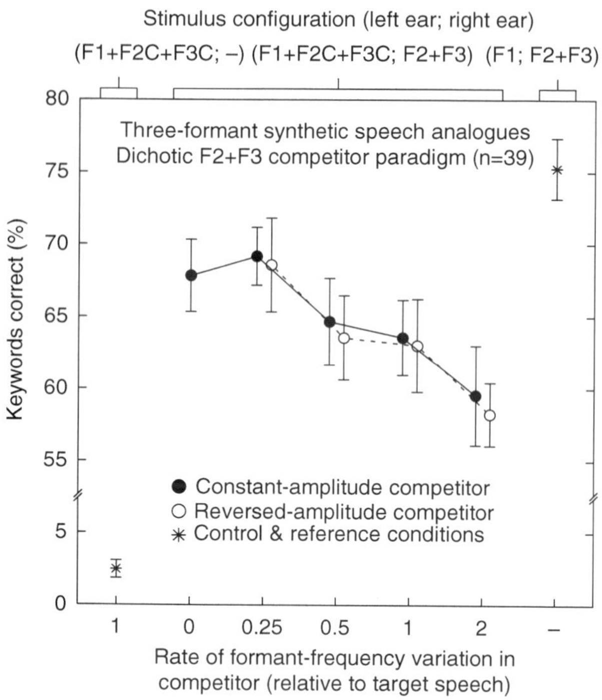

Fig. 36.1 Influence of rate of formant-frequency variation on the effect of competitors (F2C+F3C) on intelligibility (synthetic-formant sentences). Mean scores and standard errors are shown separately for the constant- and reversed-amplitude conditions. The results for the control and dichotic-reference conditions are shown on the left and right sides, respectively. The bottom axis indicates the relative rate for the competitor formants (Reproduced from Summers et al. (2012) with kind permission from Springer Science + Business Media B.V.)

significant cue for across-frequency grouping of formants when segregating the speech of concurrent talkers.

5 Experiment 3

Remez (1996) reported that reducing the frequency and amplitude variation in a competitor created by time-reversing F2 reduced its impact on the intelligibility of sine-wave speech. We extended this approach to synthetic-formant speech and refined it by manipulating the depth of variation in the frequency contour of a time-varying F2C without changing the amplitude contour.

5.1 Method

In a preliminary study, the depth of variation in the frequency contour of each target formant was scaled to a range of values about its geometric mean (100 to 0 % [i.e. constant]). The acoustical effect of this manipulation is similar to that of physically constraining the excursions made by the main articulators away from their average positions. The results indicated that scaling the frequency contours to 50 % depth had relatively little effect on diotic intelligibility. Hence, in the main experiment, we were able to use three-formant analogues of the target sentences whose formant-frequency contours were scaled to 50 % depth, but which were reasonably intelligible. This made it possible to explore the effect of scaling F2Cs to have greater, as well as smaller, variation in their formant-frequency contours than that of the target formants without exceeding the natural range. For each sentence, a set of F2Cs was created using a constant amplitude contour (matching the RMS power of F2). The frequency contour of F2C was derived from that of F2 by inversion about its geometric mean and scaling to one of five values (depth = 100 to 0 %, 25 % steps). Stimuli were presented dichotically (F1+F2C; F2+F3). There were seven conditions – five experimental (depth of F2C was varied), one control (no F2), and the dichotic reference (no F2C).

5.2 *Results and Discussion*

The results ($n = 21$) were as follows. The control condition indicated that intelligibility was near floor when F2C was added full scale without the target F2. When the target F2 was present, adding F2C reduced intelligibility; this reduction was greatest for 100 %-depth (19.6 % points) and least for 0 %-depth (constant) F2Cs (6.3 % points). The smooth and progressive decline in intelligibility as the scaling factor for the inverted F2C increased indicates that competitor efficacy depends on the overall depth of its frequency variation, not its depth relative to that of the other formants (all set to 50 % depth). This pattern is similar to that observed for the effect of differences in competitor rate. Though not conclusive, this outcome is consistent with the idea that the grouping heuristic involved is not speech specific. In contrast, Remez et al. (1994) interpreted their findings in terms of the plausibility of speech-like variation in the competitor. One way to evaluate this interpretation is through manipulation of the acoustic properties of F2C in ways that change its articulatory plausibility.

6 Experiment 4

6.1 *Method*

The importance of speech-like variation for across-formant grouping was explored using an F2C with a regular and arbitrary formant-frequency contour. A triangle wave was used, which does not constitute a plausibly speech-like frequency contour. Specifically, it differs from natural formant contours in having precise periodicity and a wave shape with sharp peaks and troughs. Sharp peaks and troughs would not be expected from a dynamical system like the vocal tract, composed of articulators having mass and, when in motion, momentum. The same dichotic configuration and procedure were used as for experiment 3. There were eight conditions – five experimental (depth of triangle-wave contour for F2C was varied from 100 to 0 % in 25 % steps), the dichotic reference (no F2C), and two controls. One control comprised F2+F3 alone to provide a measure of intelligibility when F1 does not contribute to the sentence. The other was the 100 %-depth inverted F2C condition from experiment 3, as a comparator for the 100 %-depth triangle-wave case. For each sentence, the triangle-wave frequency contour was matched to the average rate and depth of modulation for its inverted F2C counterpart, derived from F2. Modulation rate was set in relation to zero crossings at the geometric mean frequency of the target F2. Peak-to-trough depth was matched to that of F2 on a log-frequency scale and centred on the geometric mean frequency. All F2Cs were synthesised using a constant amplitude contour.

6.2 *Results and Discussion*

The results ($n = 24$) were as follows. Intelligibility more than halved, relative to the dichotic reference, when F2+F3 were presented alone. This provides a benchmark against which to assess the impact of different competitor formants. Increasing the depth of frequency modulation for the triangle-wave F2C caused intelligibility to decline progressively, relative to the dichotic reference. F2Cs with constant-frequency contours (0 % depth) were least effective and those with 100 %-depth contours were most effective (reductions = 11.4 and 28.9 % points). As for experiment 3, the results indicate that competitor efficacy depends on the overall depth of frequency variation, not depth relative to that for the other formants (all set to 50 % depth). The second control indicates that the reduction in recognition arising from adding a 100 %-depth inverted F2C (28.0 % points) is almost identical to that observed for the 100 %-depth triangle-wave F2C. Contrary to the argument that across-formant grouping depends on speech-specific constraints (Remez et al. 1994; Remez 1996), the triangle-wave competitors were as effective as their more speech-like counterparts. Also, performance in both 100 %-depth cases was not much greater than for the F2+F3 control, suggesting that F1 may in effect have been excluded from the percept of the target sentences. This offers a potential alternative account for the efficacy of F2C, in terms of the perceptual capture of the target F1. Given, however, that there were no systematic differences between the F1+F2+F3 and F2+F3 conditions in the types of phoneme selected by listeners in the transcriptions, it was not possible to evaluate this hypothesis.

7 Conclusions

The results confirm and extend those of earlier studies. Adding competitor formants typically reduces intelligibility; this effect is one of informational masking rather than energetic masking. A more specific interpretation is that competitors act by influencing the perceptual organisation of the sentences, either by displacing/

diluting the phonetic contribution of their target counterparts or by capturing the target F1. The results indicate that the overall extent of formant-frequency variation, but not formant-amplitude variation, is critical for the across-formant grouping. Competitor efficacy does not depend on the plausibility of the articulatory motions implied by F2C. Hence, we can conclude that there are at least some circumstances in which across-formant grouping does not depend on speech-specific constraints.

Acknowledgements This research was supported by Research Grant EP/F016484/1 from the Engineering and Physical Sciences Research Council (UK).

References

Bailey PJ, Summerfield Q, Dorman M (1977) On the identification of sine-wave analogues of certain speech sounds. Haskins Lab Status Rep SR-51/52:1–25

Boersma P, Weenink D (2010) Praat, a system for doing phonetics by computer. Institute of Phonetic Sciences, University of Amsterdam

Bregman AS (1990) Auditory scene analysis: the perceptual organization of sound. MIT Press, Cambridge

Darwin CJ (2008) Listening to speech in the presence of other sounds. Philos Trans R Soc Lond B Biol Sci 363:1011–1021

Mesgarani N, Chang EF (2012) Selective cortical representation of attended speaker in multi-talker speech perception. Nature 485:233–236

Remez RE (1996) Perceptual organization of speech in one and several modalities: common functions, common resources. In: ICSLP-1996, Philadelphia, p 1660–1663

Remez RE, Rubin PE, Pisoni DB, Carrell TD (1981) Speech perception without traditional speech cues. Science 212:947–950

Remez RE, Rubin PE, Berns SM, Pardo JS, Lang JM (1994) On the perceptual organization of speech. Psychol Rev 101:129–156

Roberts B, Summers RJ, Bailey PJ (2010) The perceptual organization of sine-wave speech under competitive conditions. J Acoust Soc Am 128:804–817

Rosenberg AE (1971) Effect of glottal pulse shape on the quality of natural vowels. J Acoust Soc Am 49:583–590

Summers RJ, Bailey PJ, Roberts B (2012) Effects of the rate of formant-frequency variation on the grouping of formants in speech perception. J Assoc Res Otolaryngol 13:269–280

Weismer G, Berry J (2003) Effects of speaking rate on second formant trajectories of selected vocalic nuclei. J Acoust Soc Am 113: 3362–3378

Chapter 37
Do We Need STRFs for Cocktail Parties? On the Relevance of Physiologically Motivated Features for Human Speech Perception Derived from Automatic Speech Recognition

B. Kollmeier, M.R. René Schädler, A. Meyer, J. Anemüller, and B.T. Meyer

Abstract Complex auditory features such as spectro-temporal receptive fields (STRFs) derived from the cortical auditory neurons appear to be advantageous in sound processing. However, their physiological and functional relevance is still unclear. To assess the utility of such feature processing for speech reception in noise, automatic speech recognition (ASR) performance using feature sets obtained from physiological and/or psychoacoustical data and models is compared to human performance. Time-frequency representations with a nonlinear compression are compared with standard features such as mel-scaled spectrograms. Both alternatives serve as an input to model estimators that infer spectro-temporal filters (and subsequent nonlinearity) from physiological measurements in auditory brain areas of zebra finches. Alternatively, a filter bank of 2-dimensional Gabor functions is employed, which covers a wide range of modulation frequencies in the time and frequency domain. The results indicate a clear increase in ASR robustness using complex features (modeled by Gabor functions), while the benefit from physiologically derived STRFs is limited. In all cases, the use of power-normalized spectral representations increases performance, indicating that substantial dynamic compression is advantageous for level-independent pattern recognition. The methods employed may help physiologists to look for more relevant STRFs and to better understand specific differences in estimated STRFs.

1 Introduction

Automatic speech recognition (ASR) has developed independently from hearing research. It is driven by the dream of computers to understand speech as well as human listeners. Even though this dream had to be postponed repeatedly by each new

---

B. Kollmeier (✉) • M.R.R. Schädler • A. Meyer • J. Anemüller • B.T. Meyer
Medizinische Physik,
Carl von Ossietzky University, Oldenburg D-26111, Germany
e-mail: birger.kollmeier@uni-oldenburg.de

B.C.J. Moore et al. (eds.), *Basic Aspects of Hearing*,
Advances in Experimental Medicine and Biology,
DOI 10.1007/978-1-4614-1590-9_37, © Springer Science+Business Media New York 2013

generation of ASR researchers, the technology has reached a remarkable performance level. In comparison to human speech recognition (HSR), however, a considerable gap still exists, especially when expressed in terms of speech reception thresholds (SRT), i.e., the signal-to-noise ratio required to achieve a certain level of recognition performance. In 2011, Meyer et al. have shown that two factors contribute to this gap: (1) a "representation component," i.e., the nonideal transformation from a speech signal to ASR features (in comparison to the "ideal" peripheral and central auditory system of humans); and (2) a "cognitive gap," i.e., the imperfect world knowledge of the ASR system, resulting from a relatively small number of training examples (or templates) and very limited capabilities to include context information during the recognition process. While the reduction of the cognitive gap would require enormous computational efforts, large audio databases, and complex cognitive models, the former gap is of considerable interest for hearing researchers. It bears the potential of comparing an ASR system, with a feature set tailored by the auditory scientist, to an HSR performance as a benchmark. The closer the ASR system performance matches that of humans, the better the front end should match the one of human auditory processing assuming a fair comparison (i.e., using the same speech material and task with identical noise conditions). Several studies have demonstrated the usability of this approach (Jürgens and Brand 2009; Kim and Stern 2009). Typically, the gap can be reduced if the ASR back end is given the same (restricted) information as the human listener (e.g., when listeners are restricted to low-lexical information).

In the current approach, this comparison between ASR and HSR is used to evaluate the usability of speech features derived from physiological spectro-temporal receptive fields (STRFs) measured in zebra finches (Gill et al. 2006) in comparison to a simple model for STRFs based on 2-dimensional Gabor features (Schädler et al. 2012). Each of the approaches can be combined with power normalization of the spectrum prior to further processing. The aims are (a) to find out how these feature sets compare to "classical" feature sets employed in ASR, such as mel-frequency-scaled cepstral coefficients (MFCCs) that operate on a frame-by-frame-basis (i.e., primarily spectral with very limited temporal processing); (b) to evaluate the utility of using a physiologically inspired feature set as opposed to the stylized but mathematically elegant set of Gabor time-frequency features; and (c) to demonstrate the utility of this approach (ASR and HSR comparison using special feature sets) in evaluating the results from auditory research.

2 Methods

2.1 *Time-Frequency Representations*

Spectro-temporal features are calculated by filtering time-frequency representations of the speech signal. In past work (Schädler et al. 2012), mel-spectrograms have been used that incorporate some properties of the auditory system to a limited extent but also largely ignore physiological and psychoacoustic findings (such as the asymmetric filters in the inner ear or masking properties). We compare performance of this "classical" ASR to that based on power-normalized cepstral coefficients (PNCCs), a feature type that has been shown to be quite robust against a large variety of noises (Kim and Stern 2009). Mel-spectrograms are computed from time frames every 10 ms using an overlapping analysis window of 25 ms duration. A preemphasis is applied to each segment before calculating the magnitude-squared short-time Fourier transform. Each frame is then processed by a mel-filter bank with triangular filters and compressed with the logarithm. Power-normalized spectra (PNS) are calculated in a similar way but differ in important aspects: The differences between mel-spectra and PNS are that PNS integrate the squared spectra using a gammatone filter bank, which has been shown to better approximate the place-frequency mapping of the basilar membrane. A power function nonlinearity that mimics the dependency of the input sound level and the perceived loudness is used for compression. The exponent of 0.1 was derived from the relation between auditory nerve firings and the level of a presented tone. Finally, the medium time power bias is removed, which is motivated by the fact that the auditory system is sensitive to changes of the incoming signal (in contrast to low-modulated background noises, which are largely ignored). These enhancements result in a representation that is inherently more robust than mel-spectra.

2.2 Estimation of Spectro-Temporal Receptive Fields

STRF models the stimulus features to which an auditory neuron is sensitive by indicating the optimal stimulus pattern preceding a spike in a specific time window. Here, we use a linear estimator based on ridge regression to account for second-order correlations in the stimulus ensemble. The best regularization parameter is found using cross-validation. The STRFs used in the experiments (Fig. 37.1a) are estimated using recordings from different auditory regions in male adult zebra finches for responses to conspecific vocalizations (Gill et al. 2006) (freely available at www.crcns.org). Since the temporal resolutions of the STRFs (approx. 1 ms) and speech recognition back end (approx. 10 ms) differed, the predicted responses are downsampled. Furthermore, frequency-shifted STRFs are added to the original STRF ensemble to cover the whole frequency range.

2.3 Spectro-Temporal Features

Besides the estimated STRFs (Fig. 37.1a), idealized STRFs modeled with 2D-Gabor functions are used to extract features that encode spectro-temporal modulations. In 2012, Schädler et al. proposed an approach in which a set of 2D-Gabor filters, suitable for ASR, is generated by means of a modulation filter bank (Fig. 37.1b). While the estimated STRFs show that the measured neurons are tuned to more complex patterns, the 2D-Gabor filters are simpler and only tuned to a specific

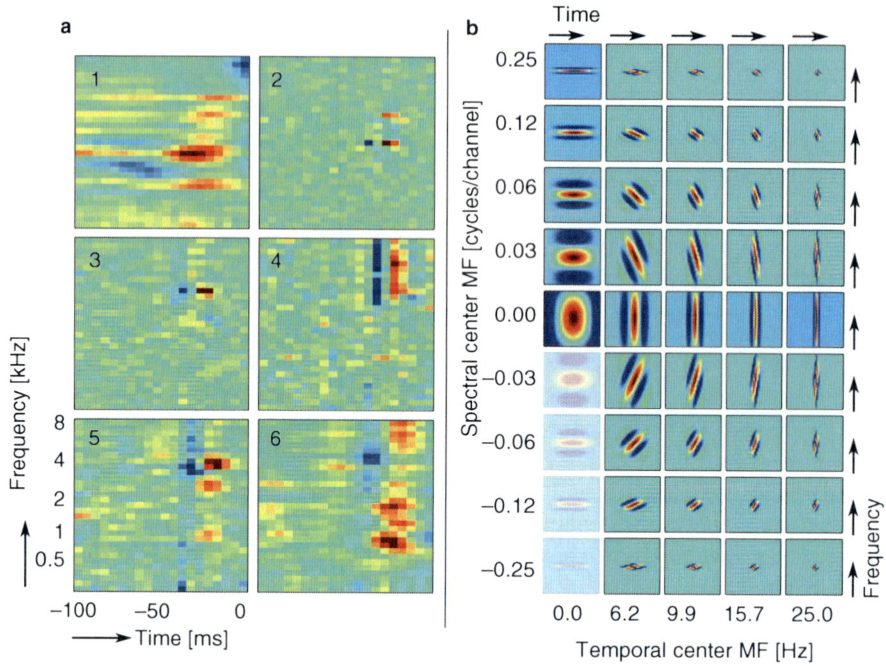

Fig. 37.1 Spectro-temporal receptive fields estimated from recordings of six example units, numbered 1-6 (**a**) and 2-dimensional Gabor functions (**b**) used to extract relevant information from time-frequency representations of speech

combination of a spectral and a temporal modulation frequency. The spectro-temporal features are extracted from the time-frequency representation by 2D convolution with the STRFs or 2D-Gabor filters. The filtered spectro-temporal representations contain the expected activation pattern of a neuron with the specific STRF at different center frequencies. The center frequencies are selected such that the STRFs or Gabor filters overlap, measured by the 2D correlation, and does not exceed a threshold. For STRFs, the threshold was set to 0.5; for the Gabor filter bank features, the values are taken from Schädler et al. (2012). The feature vector is composed of the output at the selected center frequencies. The dimensionality of the feature vectors for STRF and Gabor features is approx. 1,000 and 311, respectively.

2.4 Human Listening Tests and ASR Experiments

To establish a valid comparison of human and machine performance, HSR and ASR results were obtained with the same speech database (noisy digit sequences from the Aurora2 corpus (Hirsch and Pearce 2000)). Results reported in this work were obtained by training the ASR system with a mixture of clean and noisy speech ("multi-condition training"). Testing is performed with clean and noisy digit strings using eight noises (four of which were used during training) with SNRs ranging from −5 to 20 dB. Speech

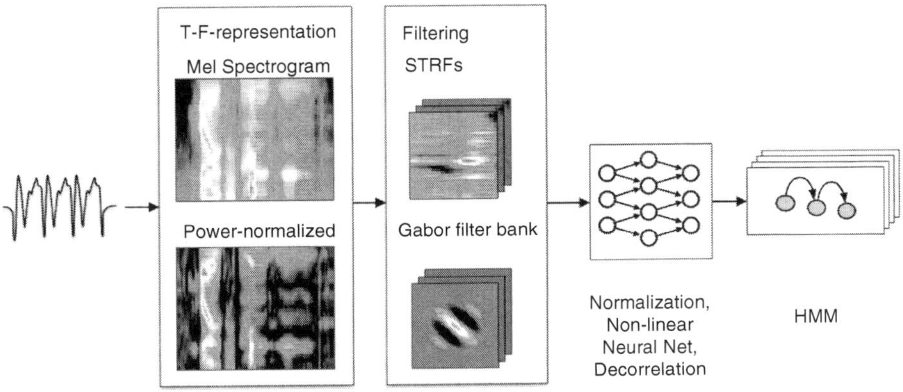

Fig. 37.2 Speech recognition setup: A time-frequency representation (a mel- or power-normalized spectrogram) is filtered with spectro-temporal receptive fields or a Gabor filter bank to capture relevant speech features. These are used as input to a Tandem ASR system

items from 214 speakers were used for either training or testing. Ten subjects aged between 25 and 39 with normal hearing listened to the audio material in a sound-insulated booth via audiological headphones. Signals were presented at a comfortable listening level. Since Aurora2 test material contains more than 70,000 digit strings, a subset was compiled that is suitable for listening tests with humans. Pilot experiments were performed to identify the SNRs at which listeners actually produce errors. With clean signals, a 0 % error rate was obtained with a list containing 650 words, and even at 5 dB SNR, the error rate for two listeners was below 1 %. Hence, the tests with 10 listeners were performed at the lowest SNRs from the Aurora2 database (0 and −5 dB).

The Aurora2 reference ASR recognizer uses 13-dimensional MFCCs with delta and double-delta features, which are computed from the speech data using the front end provided with the hidden Markov toolkit. Results for PNCCs were obtained as a second baseline, since they are related to one of the time-frequency representations investigated in this work.

The resulting 39-dimensional features are used to train and test the hidden Markov model (HMM). Spectro-temporal features are used as input to a Tandem that consists of a nonlinear neural net (multilayer perceptron, MLP) and an HMM (Fig. 37.2). The MLP maps the input features to phone posteriors, which are decorrelated with a principal component analysis and fed to the HMM.

3 Results

Recognition scores for humans, ASR baseline features, and ASR spectro-temporal features are presented in Table 37.1. The last column of Table 37.1 shows the speech reception threshold (the SNR at which 50 % of words are correctly identified). The SRTs were obtained by linear interpolation of recognition scores shown in Table 37.1 and Fig. 37.3.

Table 37.1 Accuracy for the recognition of noisy digits for humans and machines

		−5 dB	0 dB	5 dB	10 dB	15 dB	20 dB	Clean	Avg.	SRT/dB
Human listeners		80.1	95.6	99.3	–	–	–	100.0	93.8*	−10.1
ASR baseline features										
MFCC		18.4	48.9	78.0	90.3	94.7	96.5	98.0	75.0	0.2
PNCC		36.4	70.9	89.1	95.3	97.4	98.2	98.8	83.7	−3.0
Spectro-temporal features										
Mel-spec	Gabor filter bank	29.8	63.6	84.6	93.5	96.5	97.7	98.4	80.6	−2.0
Mel-spec	STRF	21.0	52.5	79.4	91.7	96.0	97.2	98.2	76.5	−0.4
Power-normalized	Gabor filter bank	42.4	73.0	88.4	94.6	96.9	97.3	97.4	84.3	−3.8
Power-normalized	STRF	39.1	71.2	89.5	95.9	97.7	98.1	98.3	84.2	−3.6

*Note that the average performance measure is based on all SNRs for the ASR systems but is biased towards lower SNRs for humans. The SRT is presented in the last column

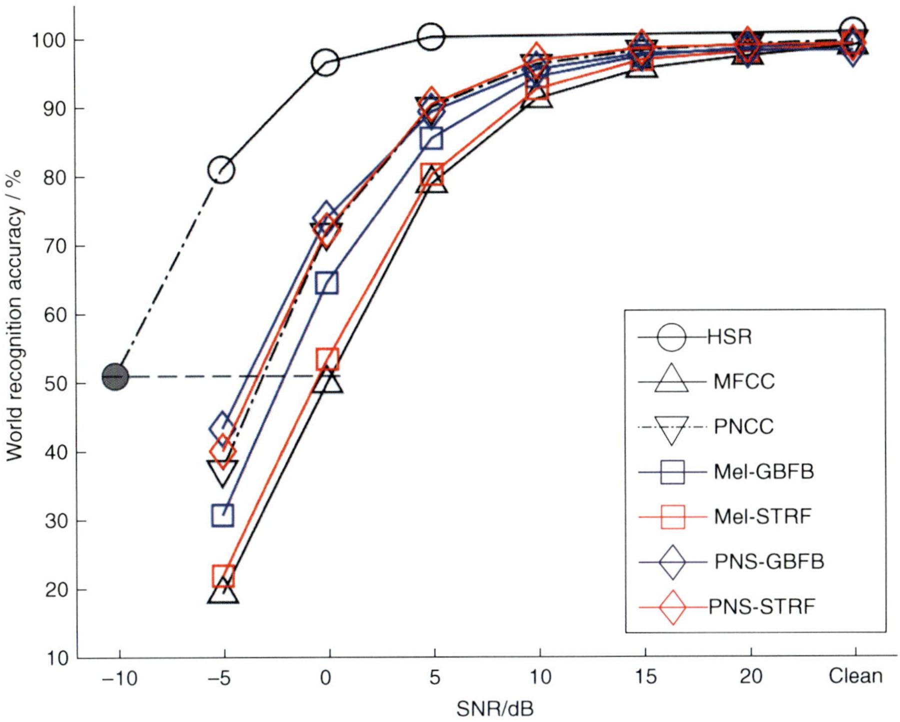

Fig. 37.3 Overview of recognition accuracies as presented in Table 37.1

All systems based on spectro-temporal processing outperform the MFCC baseline, and the already competitive PNCC baseline is further improved when using features based on power-normalized spectrograms. Feature extraction with a Gabor filter bank produces better results than feature extraction with STRFs based on zebra finches. This effect is stronger for features based on mel-spectrograms than on power-normalized spectrograms.

In speech recognition tasks, the relative increase of error rates is employed to compare results across systems. However, when the performance differences are as large as reported here, the increase of errors appears to be a rather fragile measure since high values are obtained as soon as HSR approaches 100 % recognition rates. The gap between humans and ASR systems is therefore reported in terms of the speech reception threshold (SRT), i.e., the SNR at which listeners (or the ASR system, respectively) achieve 50 % accuracy. The SRT for humans is 10.1 dB, while the SRTs of ASR systems range from −3.8 (power-normalized Gabors) to 0.2 dB (MFCCs), i.e., the human-machine gap ranges from 6.2 to 10.3 dB, depending on the feature type being used.

4 Discussion and Conclusions

- The human-machine gap for speech recognition in noise reported here (6.2–10.3 dB) is smaller but still in the same range as reported in (Meyer et al. 2011; Sroka and Braida 2005). Since the gap is a consequence of both an imperfect representation of auditory features in the ASR system ("bottom-up processing gap" produced by an imperfect front end, amounting to approximately 10 dB) and an imperfect recognition stage due to the lack of appropriate templates or world knowledge ("top-down processing gap" produced by an imperfect back end and amounting to approximately 5 dB, as estimated in (Meyer et al. 2011)), it seems the current approach primarily reduced the bottom-up processing gap to just a few dB. This indicates that the current approach captures most of the relevant information of the acoustic input signal. Hence, any findings about the relative benefit of certain processing features obtained in the ASR experiment should bear some significance for modeling human auditory processing as well.
- The auditory model-based preprocessing using either Gabor functions as stylized STRFs or physiologically measured STRFs as the major speech feature extraction stage outperforms the standard speech features routinely employed in automatic speech recognition. This indicates that a reasonable amount of processing across adjacent audio frequencies and time frames as provided by some auditory models seems to be a key factor in increasing robustness of auditory processing in noise. Auditory physiologists, psychoacousticians, and modelers should therefore avoid purely spectral or purely temporal auditory representations of the input signal.
- For both filter types (STRF data or Gabor models), the performance improves when power-normalized spectrogram representations are used in addition to the standard (logarithmic) compressive nonlinearity already employed in the speech feature extraction stage. This indicates that the robustness in sound recognition against absolute level changes encountered in the auditory system relies on an "effective" compression that is even more compressive than the logarithm. It is unclear whether such compression is physiologically implemented at a peripheral representation stage (as implemented here by a power normalization) or rather at a more central recognition stage (which could be modeled in our case as a separate ASR system trained for each input speech level or even SNR). Hence, the physiological basis of an auditory representation inspired by normalized power spectra would be an interesting research topic.

Gabor feature-based STRF consistently showed a better performance than the STRFs obtained from physiological recordings with zebra finches. Even though the observed differences are small, this finding may have different consequences:

1. The findings underline the usability of the current approach that uses a rather complex but still ecologically relevant framework (modeling HSR by ASR with different feature sets) to examine the effectiveness and completeness of a given auditory feature set.

2. The physiologically based STRF feature set was not derived from humans operating on spoken language but from a different species on a different set of acoustical signals (i.e., specific vocalizations presented to zebra finches). Even though the features were largely adapted to the current task (by shifting the STRFs to cover a large range of center frequencies), it is unlikely that these features match the highly optimized features used by humans in such a task. Nevertheless, the close match to the Gabor feature performance and the advantage over standard MFCC features demonstrates the potential and relevance of such animal physiological results for understanding the human auditory system.
3. Even though the Gabor functions employed here are a very abstract representation of the "effective" STRFs employed by humans, they have the advantage of an orthonormal system which spans a maximum number of independent dimensions of the underlying data space with a fixed number of feature functions. This property is only relevant if the abstract vector space spanned by these functions bears any relevance for the perceptual task that is modeled (i.e., human speech recognition in noise). Obviously, this seems to be the case here.

Acknowledgment This work was supported by Deutsche Forschungsgemeinschaft (SFB-TRR 31).

References

Gill P, Zhang J, Woolley S, Fremouw T, Theunissen FE (2006) Sound representation methods for spectro-temporal receptive field estimation. J Comput Neurosci 21:5–20

Hirsch H, Pearce D (2000) The AURORA experimental framework for the performance evaluation of speech recognition systems under noisy conditions. In: Proceedings of ICSLP, Beijing, 2000, vol 4, pp 29–37

Jürgens T, Brand T (2009) Microscopic prediction of speech recognition for listeners with normal hearing in noise using an auditory model. J Acoust Soc Am 126:2635–2648

Kim C, Stern RM (2009) Feature extraction for robust speech recognition using a power-law nonlinearity and power-bias subtraction. In: Proceedings of Interspeech, 2009, Brighton, UK, pp 28–31

Meyer BT, Brand T, Kollmeier B (2011) Effect of speech-intrinsic variations on human and automatic recognition of spoken phonemes. J Acoust Soc Am 129:388–403

Schädler MR, Meyer BT, Kollmeier B (2012) Spectro-temporal modulation subspace-spanning filter bank features for robust ASR. J Acoust Soc Am 131:4134–4151

Sroka JJ, Braida LD (2005) Human and machine consonant recognition. Speech Commun 45:401–423

Chapter 38
Modelling Speech Intelligibility in Adverse Conditions

Søren Jørgensen and Torsten Dau

Abstract Jørgensen and Dau (J Acoust Soc Am 130:1475–1487, 2011) proposed the speech-based envelope power spectrum model (sEPSM) in an attempt to overcome the limitations of the classical speech transmission index (STI) and speech intelligibility index (SII) in conditions with nonlinearly processed speech. Instead of considering the reduction of the temporal modulation energy as the intelligibility metric, as assumed in the STI, the sEPSM applies the signal-to-noise ratio in the envelope domain (SNR_{env}). This metric was shown to be the key for predicting the intelligibility of reverberant speech as well as noisy speech processed by spectral subtraction. The key role of the SNR_{env} metric is further supported here by the ability of a short-term version of the sEPSM to predict speech masking release for different speech materials and modulated interferers. However, the sEPSM cannot account for speech subjected to phase jitter, a condition in which the spectral structure of the intelligibility of speech signal is strongly affected, while the broadband temporal envelope is kept largely intact. In contrast, the effects of this distortion can be predicted successfully by the spectro-temporal modulation index (STMI) (Elhilali et al., Speech Commun 41:331–348, 2003), which assumes an explicit analysis of the spectral "ripple" structure of the speech signal. However, since the STMI applies the same decision metric as the STI, it fails to account for spectral subtraction. The results from this study suggest that the SNR_{env} might reflect a powerful decision metric, while some explicit across-frequency analysis seems crucial in some conditions. How such across-frequency analysis is "realized" in the auditory system remains unresolved.

S. Jørgensen • T. Dau (✉)
Department of Electrical Engineering,
Centre for Applied Hearing Research, Technical University of Denmark,
Ørsteds Plads, Building 352, DK-2800, Kgs., Lyngby, Denmark
e-mail: tdau@elektro.dtu.dk

B.C.J. Moore et al. (eds.), *Basic Aspects of Hearing*,
Advances in Experimental Medicine and Biology,
DOI 10.1007/978-1-4614-1590-9_38,

1 Introduction

The key acoustical features determining speech intelligibility have traditionally been considered to be the signal-to-noise ratio (SNR), as used in the Articulation Index (AI; French and Steinberg 1947), or the modulation transfer function (MTF), as measured by the speech transmission index (STI; Steeneken and Houtgast 1980). These acoustical features have been shown to account for intelligibility remarkably well in a broad range of situations, such as speech in stationary noise, speech subjected to low- and high-pass filtering, and speech with varying amounts of reverberation. However, the AI and the STI are limited to conditions with stationary interferers, due to long-term integration of the stimuli, and fail to account for speech masking-release (MR) effects, occurring when the interferer is temporally modulated or speech-like. Moreover, these metrics fail when noisy speech is subjected to nonlinear noise reduction, such as spectral subtraction, or distorted by phase jitter, which removes the spectral structure of the speech while keeping the temporal envelope largely intact.

An extension to the STI was presented by Elhilali et al. (2003), denoted as the spectro-temporal modulation index (STMI), which includes a two-dimensional modulation filter bank, analysing the spectral modulations of the speech signal in addition to the temporal modulations. The STMI successfully accounts for phase jitter distortion; however, since the decision metric is conceptually similar to the STI, it is unclear whether it can also account for processing by spectral subtraction.

In contrast to considering the reduction in the envelope energy of speech, as performed in the STI, Jørgensen and Dau (2011) demonstrated that a metric based on the signal-to-noise ratio in the envelope domain (SNR_{env}) is both highly correlated to the intelligibility of noisy speech processed by spectral subtraction and consistent with the STI in conditions with speech in stationary noise and reverberation. The major difference with respect to the STI is the explicit consideration of the envelope noise floor, which is increased after spectral subtraction (e.g. Dubbelboer and Houtgast 2008) and proposed to be responsible for a reduced intelligibility. The SNR_{env} metric is calculated as part of the framework denoted as the speech-based envelope power spectrum model (sEPSM), inspired by the EPSM (Ewert and Dau 2000), which originally was developed to account for modulation detection and masking data. However, the SNR_{env} was also calculated from a long-term integration of the stimuli and the sEPSM must, therefore, fail in conditions with fluctuating interferers.

Here, it is suggested that a "short-term" estimation of the SNR_{env} will generalize the sEPSM framework to also account for conditions with fluctuating interferers. The hypothesis is that the SNR_{env} is increased in the dips of a fluctuating masker and that a running estimation of the SNR_{env} will account for the MR effect. The new model is evaluated with three categories of interferers and nonlinear processing: (1) speech mixed with three types of stationary noise with widely different spectral characteristic, thus evaluating the model's ability to account for differences in spectral masking; (2) speech mixed with three types of fluctuating

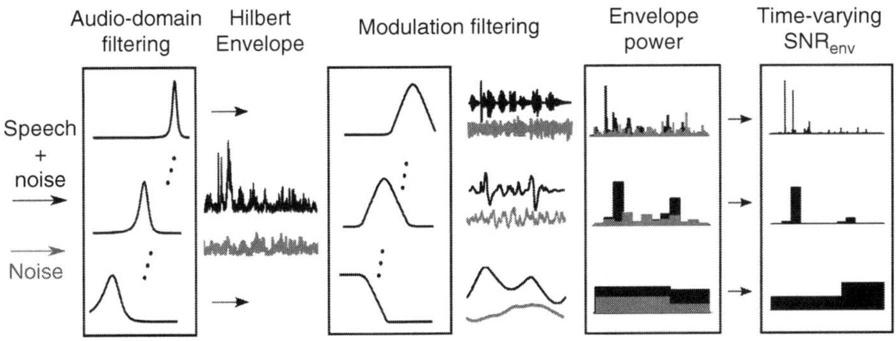

Fig. 38.1 Schematic of the short-term sEPSM. The noisy speech (*black*) and the noise alone (*grey*) are processed separately through the model. The decision metric is based on the time-varying SNR_{env}

interferers with very different temporal structure, evaluating predictions of speech MR; and (3) two conditions with nonlinearly processed noisy speech in the form of spectral subtraction and phase jitter, testing the model's ability to account for non-linear distortions.

2 Model Description

The processing structure of the modified sEPSM is illustrated in Fig. 38.1. The first stage is a bandpass filter bank consisting of 22 gammatone filters with one equivalent rectangular bandwidth (Moore and Glasberg 1983) and third-octave spacing between the centre frequencies, covering the range from 63 to 8 kHz. An absolute sensitivity threshold is included such that individual gammatone filters are included, only if the level of the stimulus at the output is above the absolute hearing threshold for normal-hearing listeners. The temporal envelope of each output is extracted via the Hilbert transform and then low-pass filtered with a cut-off frequency of 150 Hz using a first-order Butterworth filter. The resulting envelope is analysed by a modulation bandpass filter bank, which consists of eight second-order bandpass filters with octave spacing, covering the range from 2 to 256 Hz, in parallel with a third-order low-pass filter with a cut-off frequency of 1 Hz (see Jørgensen and Dau 2011).

The running temporal output of each modulation filter is divided into short segments using rectangular windows with no overlap. The duration of the windows is specific for each modulation channel and is the inverse of the centre frequency of a given modulation filter (or the cut-off frequency in the case of the 1-Hz low-pass filter). For example, the window duration in the 4-Hz modulation channel is 250 ms. For each window, the AC-coupled envelope power (variance) of the noisy speech and the noise alone are calculated separately and normalized with the corresponding long-term DC power. The SNR_{env} of a window is estimated from the envelope power as

$$\mathrm{SNR}_{\mathrm{env}} = \frac{P_{\mathrm{S+N}} - P_{\mathrm{N}}}{P_{\mathrm{N}}}, \tag{38.1}$$

where $P_{\mathrm{S+N}}$ and P_{N} denote the envelope power of the noisy speech and the noise alone after the normalization. For each modulation channel, the running $\mathrm{SNR}_{\mathrm{env}}$ values are averaged across time, thus assuming that all parts of a sentence contribute equally to intelligibility. The time-averaged $\mathrm{SNR}_{\mathrm{env}}$ values from the different modulation filters are then combined across modulation filters and across gammatone filters, using the "integration model" from Green and Swets (1988). The combined $\mathrm{SNR}_{\mathrm{env}}$ is converted to the probability of correctly recognizing the speech item using the concept of a statistically "ideal observer" (Jørgensen and Dau 2011).

3 Method

Model predictions were compared to data from the literature as well as data collected for the present study. The target speech was either Danish sentences from the DANTALE II speech material (Wagener et al. 2003), Danish sentences from the CLUE speech material (Nielsen and Dau 2009), or sentences from the TIMIT database. The data reflect either speech reception thresholds (SRTs) corresponding to the 50 % point on the psychometric function or percentage of correct scores. All subjects were normal-hearing listeners.

Three conditions of stationary interferers were considered: (1) speech-shaped noise (SSN), (2) car-cabin noise (CAR), and (3) the sound of bottles on a conveyer belt (BOTTLE). Moreover, three conditions with fluctuating interferers were considered: (1) a conversation between two people sitting in a café (CAFE), (2) SSN that was amplitude modulated by an 8-Hz sinusoid (SAM), and (3) the speech-like, but non-semantic, International Speech Test Signal (ISTS; Holube et al. 2010).

Finally, two conditions with nonlinear processing were considered: (1) speech mixed with SSN and further processed by spectral subtraction (Berouti et al. 1979) using six different values of the over-subtraction factor, ρ, and (2) clean speech distorted by phase jitter with a varying degree of the jitter constant, α (Elhilali et al. 2003).

For the predictions, the model parameters were calibrated to a close match between the predictions and the data for the unprocessed SSN condition for a given speech material. These parameters were then used for all other experimental conditions. Identical stimuli were used for the simulations as for obtaining the data, except for the conditions with phase jitter where the data were obtained using sentences from the TIMIT database (Elhilali et al. 2003), whereas the predictions were obtained using the CLUE sentences.

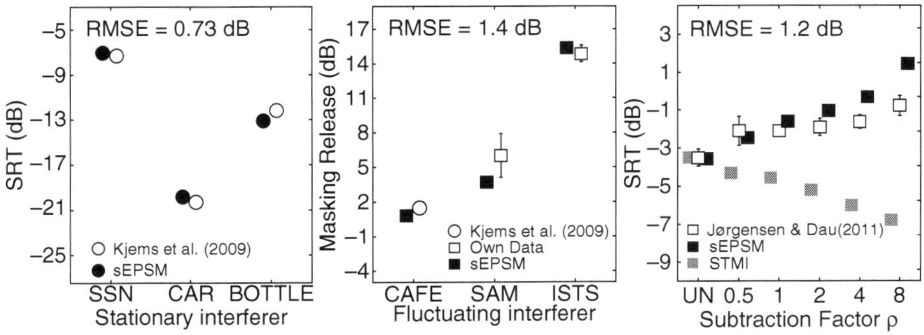

Fig. 38.2 Measured (*open symbols*) and predicted (*filled symbols*) SRTs in conditions with stationary interferers (*left panel*), masking release in conditions with fluctuating interferers (*middle panel*), and SRTs in conditions with noisy speech processed by spectral subtraction

4 Results

4.1 Conditions with Stationary and Fluctuating Interferers

The left panel of Fig. 38.2 shows SRTs obtained by Kjems et al. (2009) (open circles) and corresponding predictions obtained with the sEPSM (filled circles) in the conditions with stationary interferers. The SRTs range from −17 to −7 dB, reflecting the differences of spectral masking for the various stationary interferers. The sEPSM accounts well for the SRTs for the three stationary conditions. The root mean square error (RMSE) between the measured and simulated data amounts to 0.71 dB.

The middle panel of Fig. 38.2 shows the results for the fluctuating interferers, represented by the MR calculated as the difference between the SRT obtained in the SSN condition and a given condition with fluctuating noise. The MR is quite low for the CAFE noise compared to the other interferers, indicating that the fluctuations in this noise type are less useful for the listener. The greatest MR is found for the ISTS interferer. The sEPSM accounts for the MR effects obtained with the different interferers with an RMSE of 1.5 dB.

4.2 Conditions with Processed Noisy Speech

4.2.1 Spectral Subtraction

The right panel of Fig. 38.2 shows SRTs obtained by Jørgensen and Dau (2011; open squares) and corresponding predictions by the sEPSM (filled squares) for six conditions of ρ, where UN denotes the reference condition with no spectral subtraction. The data show an increase of the SRT with increasing ρ, demonstrating a lower

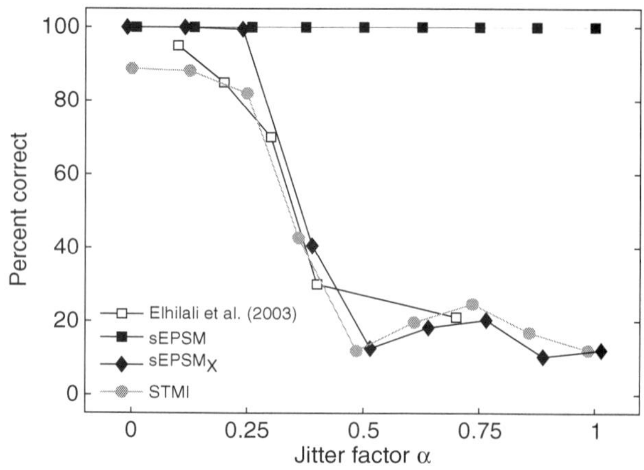

Fig. 38.3 Measured (*open squares*) and predicted (*filled symbols*) percentage of correct responses as a function of the jitter factor α

intelligibility with spectral subtraction than without the processing. The sEPSM predicts the trends in the data, although it overestimates the SRTs for $\rho=2$, 4, and 8 (RMSE = 1.2 dB). Predictions obtained with the STMI (grey squares) suggest that the intelligibility increases after spectral subtraction, in contrast to the measured data. The STMI thus fails to account for spectral subtraction, as does the STI (Jørgensen and Dau 2011).

4.2.2 Phase Jitter

Figure 38.3 shows the measured data obtained by Elhilali et al. (2003; open squares) for speech distorted by phase jitter. The intelligibility is almost unaffected for jitter factors, α, below 0.2 but sharply decreases for values above 0.2 and remains low for higher values of α. The sEPSM (filled squares) clearly fails to account for the measured data, predicting perfect intelligibility independent of α. In contrast, the STMI (grey bullets) is in good agreement with the data.

5 Model Analysis

The sEPSM was found to be insensitive to the phase jitter distortion and the underlying reason is analysed in the following. The left panel of Fig. 38.4 shows the temporal output of the 4-Hz modulation filter for a subset of the gammatone filters tuned to frequencies between 0.16 and 4 kHz for the condition without phase jitter. The middle panel shows the same representation, but for the condition with $\alpha=0.5$. The jitter has increased the temporal correlation across the audio-frequency channels. Thus, the instantaneous across-audio-frequency *variance* of the representation is decreased in this condition. The model is not sensitive to this since it includes no across-frequency analysis. The right panel of Fig. 38.4 shows the normalized

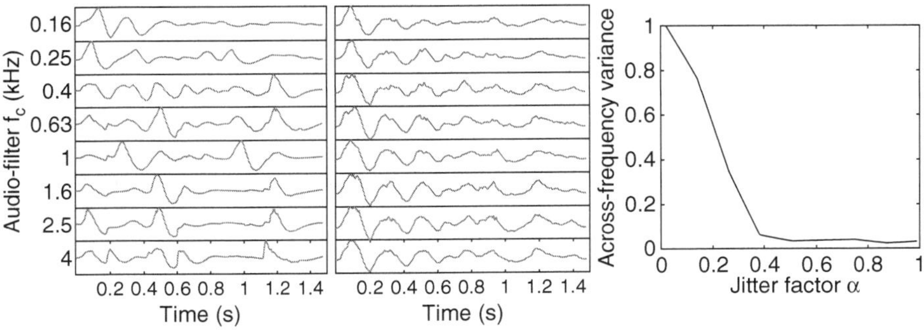

Fig. 38.4 *Left*: the internal representation at the output of the 4-Hz modulation filter for a subset of the auditory filters for noisy speech without phase jitter. *Middle*: as left, but for $\alpha=0.5$. *Right*: the normalized across (audio)-frequency variance, calculated from the internal representation in the 4-Hz modulation filter, as a function of α

across-frequency variance, averaged across time, as a function of the jitter factor. The variance decreases with increasing jitter factor, in a very similar manner as the decrease of speech intelligibility shown in Fig. 38.3.

Additional predictions were obtained with a version of the model ($\mathrm{sEPSM_X}$), where the $\mathrm{SNR_{env}}$ was weighted by the across-frequency variance of the internal representation of the stimuli in the 4-Hz modulation filter. The corresponding predictions (Fig. 38.3; filled diamonds) largely agree with the predictions based on the STMI and the measured data.

6 Discussion

It was demonstrated that the short-term estimation of $\mathrm{SNR_{env}}$ in the sEPSM framework can account for the masking effects found with different stationary interferers, as well as for the MR effect observed for fluctuating interferers. The critical element in estimating the short-term $\mathrm{SNR_{env}}$ was the modulation filter-dependent analysis windows, ranging from 4 to 1000 ms, which allows the model to evaluate the increased $\mathrm{SNR_{env}}$ in the dips of the masker.

Moreover, the model accounts for the decreased intelligibility in conditions with noisy speech processed by spectral subtraction. Jørgensen and Dau (2011) demonstrated that the $\mathrm{SNR_{env}}$ decreases as the spectral subtraction factor increases and that the decreasing $\mathrm{SNR_{env}}$ is caused by an increase of the intrinsic fluctuations in the noise part of the noisy speech, producing more modulation masking in the model. Thus, the increased noise fluctuations seem to reduce the perceptual salience of the target speech fluctuations and, thus, decrease the intelligibility for the listeners.

In contrast to the sEPSM, the STMI was shown to fail in conditions of spectral subtraction. The reason is that the STMI considers the *reduction* of the clean speech envelope power, instead of the *signal-to-noise ratio*, and therefore does not capture the increased envelope power of the noise after processing.

However, the sEPSM fails in conditions with speech distorted by phase jitter, which the STMI successfully describes. The model analysis demonstrated that an explicit across-audio-frequency analysis is required in the sEPSM framework and that a modified version of the model which evaluates the across-frequency variance of the model's internal representation accounts for the data and produces similar results as the STMI. While the STMI assumes a two-dimensional (spectral and temporal) modulation-frequency-selective process, the sEPSM is only based on a one-dimensional (temporal) modulation analysis. The latter modelling concept is consistent with other recent models of across-channel processes, such as models of comodulated masking release (Piechowiak et al. 2007) and models of computational auditory scene analysis (Elhilali et al. 2009). A more complex two-dimensional representation as the one assumed in the STMI framework may not be required.

7 Conclusion and Perspective

This study demonstrated that a modelling framework based on estimating the SNR_{env} in short time windows, following temporal modulation-frequency selectivity, can account for intelligibility in conditions with stationary and fluctuating interferers, as well as in conditions with noisy speech processed by spectral subtraction. Furthermore, with an inclusion of an across-audio-frequency mechanism, such a framework is sufficient to account for the effects of phase jitter distortion on speech intelligibility.

Including the SNR_{env} metric in more detailed models of auditory preprocessing and perception might be interesting for studying the consequences of hearing impairment on speech intelligibility.

Acknowledgements We thank Ewen MacDonald and Hedwig Gockel for helpful comments and suggestions.

References

Berouti M, Schwartz R, Makhoul J (1979) Enhancement of speech corrupted by acoustic noise. Proc IEEE Int Conf Acoust, Speech, Signal Proces (ICASSP-79), USA 4:208–211

Dubbelboer F, Houtgast T (2008) The concept of signal-to-noise ratio in the modulation domain and speech intelligibility. J Acoust Soc Am 124:3937–3946

Elhilali M, Chi T, Shamma SA (2003) A spectro-temporal modulation index (STMI) for assessment of speech intelligibility. Speech Commun 41:331–348

Elhilali M, Ma L, Micheyl C, Oxenham AJ, Shamma SA (2009) Temporal coherence in the perceptual organization and cortical representation of auditory scenes. Neuron 61:317–329

Ewert S, Dau T (2000) Characterizing frequency selectivity for envelope fluctuations. J Acoust Soc Am 108:1181–1196

French N, Steinberg J (1947) Factors governing intelligibility of speech sounds. J Acoust Soc Am 19:90–119

Green DM, Swets JA (1988) Signal detection theory and psychophysics. Peninsula Publishing, Los Altos, pp 238–239

Holube I, Fredelake S, Vlaming M, Kollmeier B (2010) Development and analysis of an International Speech Test Signal (ISTS). Int J Audiol 49:891–903

Jørgensen S, Dau T (2011) Predicting speech intelligibility based on the signal-to-noise envelope power ratio after modulation-frequency selective processing. J Acoust Soc Am 130: 1475–1487

Kjems U, Boldt JB, Pedersen MS, Lunner T, Wang D (2009) Role of mask pattern in intelligibility of ideal binary-masked noisy speech. J Acoust Soc Am 126:1415–1426

Moore BCJ, Glasberg BR (1983) Suggested formulae for calculating auditory-filter bandwidths and excitation patterns. J Acoust Soc Am 74:750–753

Nielsen JB, Dau T (2009) Development of a Danish speech intelligibility test. Int J Audiol 48:729–741

Piechowiak T, Ewert SD, Dau T (2007) Modeling comodulation masking release using an equalization-cancellation mechanism. J Acoust Soc Am 121:2111–2126

Steeneken HJM, Houtgast T (1980) A physical method for measuring speech transmission quality. J Acoust Soc Am 67:318–326

Wagener K, Josvassen JL, Ardenkjaer R (2003) Design, optimization and evaluation of a Danish sentence test in noise. Int J Audiol 42:10–17

Chapter 39
Better Temporal Neural Coding with Cochlear Implants in Awake Animals

Yoojin Chung, Kenneth E. Hancock, Sung-Il Nam, and Bertrand Delgutte

Abstract Both the performance of cochlear implant (CI) listeners and the responses of auditory neurons show limits in temporal processing at high frequencies. However, the upper limit of temporal coding of pulse-train stimuli in the inferior colliculus (IC) of anesthetized animals appears to be lower than that observed in corresponding perceptual tasks. We hypothesize that the neural rate limits have been underestimated due to the effect of anesthesia. To test this hypothesis, we developed a chronic, awake rabbit preparation for recording responses of single IC neurons to CI stimulation without the confound of anesthesia and compared these data with earlier recordings from the IC of anesthetized cats. Stimuli were periodic trains of biphasic pulses with rates varying from 20 to 1,280 pulses per second (pps). We found that the maximum pulse rates that elicited sustained firing and phase-locked responses were 2–3 times higher in the IC of awake rabbits than in anesthetized cats. Moreover,

Y. Chung, PhD (✉) • K.E. Hancock
Eaton-Peabody Laboratories,
Massachusetts Eye and Ear Infirmary, Boston, MA 02114, USA

Department of Otology and Laryngology,
Harvard Medical School, Boston, MA 02114, USA
e-mail: yoojin.chung@gmail.com

S.-I. Nam
Department of Otology and Laryngology,
Harvard Medical School, Boston, MA 02114, USA

Department of Otolaryngology, School of Medicine,
Keimyung University, Daegu, South Korea

B. Delgutte
Eaton-Peabody Laboratories,
Massachusetts Eye and Ear Infirmary, Boston, MA 02114, USA

Department of Otology and Laryngology,
Harvard Medical School, Boston, MA 02114, USA

Research Laboratory of Electronics,
Massachusetts Institute of Technology, Cambridge, MA 02139, USA

B.C.J. Moore et al. (eds.), *Basic Aspects of Hearing*,
Advances in Experimental Medicine and Biology,
DOI 10.1007/978-1-4614-1590-9_39,

about 25 % of IC neurons in awake rabbit showed sustained responses to periodic pulse trains at much higher pulse rates (>1,000 pps) than observed in anesthetized animals. Similar differences were observed in single units whose responses to pulse trains were monitored while the animal was given an injection of an ultrashort-acting anesthetic. In general, the physiological rate limits of IC neurons in awake rabbit are more consistent with the psychophysical limits in human CI subjects compared to the data from anesthetized animals.

1 Introduction

Previous studies of neural responses to cochlear implant (CI) stimulation in anesthetized animals show upper frequency limits to temporal processing. Neurons in the inferior colliculus (IC) show sustained and pulse-locked responses to periodic pulse trains only up to a few hundred pulses per second (pps). A similar pulse-rate limit is observed for neural sensitivity to interaural time differences (ITD). Although human CI listeners also show limits in the ability to discriminate the pitch or ITD of periodic pulse trains at high pulse rates, these limits are higher than those observed from IC neurons in anesthetized animals (Table 39.1).

We hypothesize that neural pulse-locking, ITD sensitivity, and sustained firing to CI stimulation at high pulse rates may be underestimated due to the effect of anesthesia. We developed an awake animal model of CIs for single-unit recording from the IC to avoid this confound. Results show higher pulse-rate limits for sustained firing and temporal pulse-locking compared to anesthetized preparations. We further demonstrate that these differences are due to the effect of anesthesia by monitoring the responses of single units in the rabbit before and after injection of an ultrashort-acting barbiturate.

2 Methods

We measured responses of IC neurons to electric pulse trains presented through 8-contact intracochlear electrode arrays (Cochlear Corp.) in both anesthetized cats and awake rabbits. Experiments on awake rabbits were performed on two female Dutch-belted rabbits that received unilateral cochlear implantations. The implanted ear was deafened during surgery with an injection of distilled water into the cochlea (Ebert et al. 2004). Although no attempt was made to deafen the unimplanted ear, auditory brainstem response thresholds were nevertheless elevated by 40–50 dB relative to normal in that ear. Rabbits were implanted with head bars and trained to sit still in the recording apparatus for 2–3 h/day. Recordings from the IC contralateral to the implanted ear were performed from 29 to 431 days postimplantation.

Table 39.1 Comparison of perceptual rate limits in human CI users and neural rate limits in the IC of anesthetized animals

Psychophysical: human CI subjects		Neural: IC neurons in anesthetized cats	
Percept lasts throughout stimulus	>2,000 pps	30–300 pps[a]	Sustained responses
Rate pitch discrimination	200–1,000 pps[b]	40–200 pps[a]	Temporal coding
ITD sensitivity	250–600 pps[c]	10–200 pps[d]	ITD sensitivity

[a]Snyder et al. (1995)
[b]Tong and Clark (1985), Townshend et al. (1987)
[c]van Hoesel (2007)
[d]Smith and Delgutte (2007), Hancock et al. (2010)

For comparison with the awake rabbit data, we reanalyzed data from anesthetized cats that were partially described in earlier reports (Hancock et al. 2010, 2012). These experiments were performed on eight cats deafened with ototoxic drugs either 1 week ($n=3$) or 6 months ($n=3$) before the experiment and which received bilateral cochlear implants at the time of experiment. For the experiments, cats were anesthetized with a combination of urethane (300 mg/kg urethane, i.p.) and either diallyl barbituric acid (75 mg/kg, i.p.) or sodium pentobarbital (37 mg/kg, i.p.). Data from both short-term (1 week) and long-term (6 months) deafened cats are combined herein because the differences in response patterns between the two groups were minimal.

In both preparations, stimuli were 300-ms trains of biphasic pulses (50 μs/phase) presented every 600 ms using a wide bipolar configuration. The current was 2 dB above the single-pulse threshold, and the stimuli were presented diotically in anesthetized cats and monaurally in awake rabbits. Pulse rate was varied in random order from 20 to 1,280 pps in half-octave steps. The single-unit recording methods were as described previously (Devore and Delgutte 2010; Hancock et al. 2010).

In one rabbit, a catheter was surgically implanted into the right jugular vein to allow the administration of an ultrashort-acting barbiturate (sodium methohexital, 5 mg/kg) while recording from single units (Kuwada et al. 1989). All procedures were approved by the animal care committees at the Massachusetts Eye and Ear Infirmary and the Massachusetts Institute of Technology.

3 Results

3.1 *Pulse-Rate Limits Are Higher in Awake Rabbits Compared to Anesthetized Cats*

Responses to electric pulse trains were measured as a function of pulse rate in 104 units in anesthetized cats and 80 units in awake rabbits. We observed clearly different response patterns in the two preparations. Results from two example units are

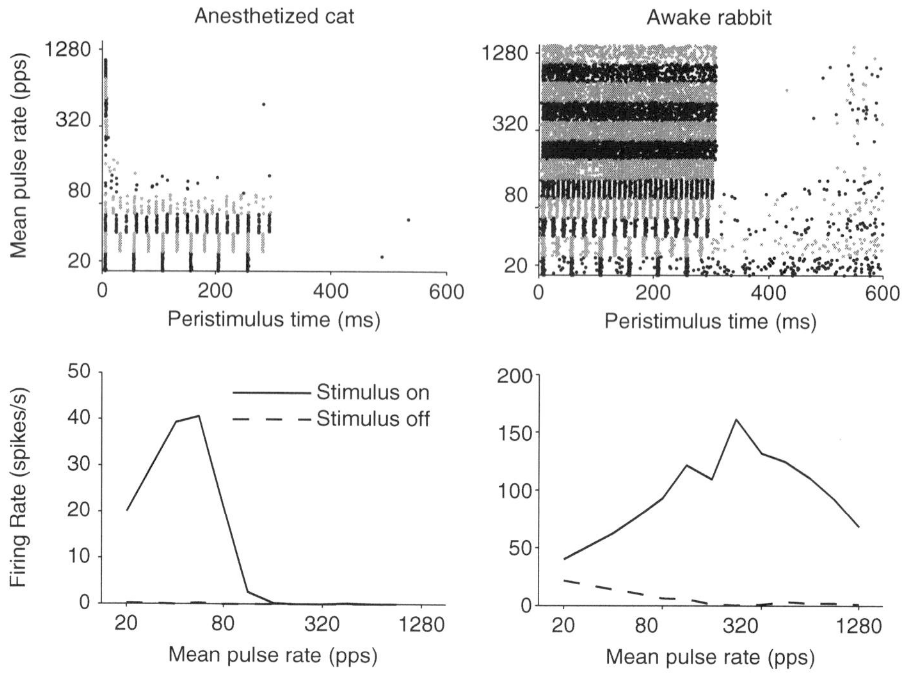

Fig. 39.1 Temporal response pattern (*top*) and average firing rates (*bottom*) as a function of pulse rate for two example neurons in anesthetized cat (*left*) and awake rabbit (*right*). *Top panels* show *dot rasters*, where each *dot* represents a spike and alternating shades of *gray* distinguish blocks of stimulus trials at different pulse rates. *Bottom panels* represent mean sustained firing rate (excluding the first ~30 ms after stimulus onset) vs. pulse rate

presented in Fig 39.1. In the neuron from anesthetized cat, there is a strong pulse-locked response at low pulse rates indicated by the periodic pattern in the dot raster. The unit fires one spike per pulse at 20 and 40 pps, but the firing rate begins to decrease at higher pulse rates until the response is limited to the onset for pulse rates >80 pps. There is no spontaneous activity either during the off period or between pulses at low pulse rates. In contrast, the neuron from an awake rabbit fires several spikes per pulse at low pulse rates (i.e., the firing rate is greater than the pulse rate) and the responses are pulse-locked up to higher pulse rates than in the neuron from anesthetized cat. Sustained responses are maintained up to the highest pulse rate tested (1,280 pps). In addition, spontaneous activity is observed both during the off period and between pulse-locked responses at low pulse rates.

Neural spike trains were cross-correlated with stimulus pulse trains to characterize the degree of temporal pulse-locking. The cross-correlograms in Fig. 39.2 are consistent with the temporal patterns shown for the same neurons in Fig. 39.1. Robust pulse-locking is evoked by low-rate pulse trains in both neurons, as shown by a prominent peak in the cross-correlogram. In the neuron from anesthetized cat, pulse-locked responses vanish above 80 pps, consistent with the limit of sustained firing. In the neuron from awake rabbit, tight pulse-locking is observed up to 224 pps. The correlogram shows multiple peaks for rates between 80 and 224 pps,

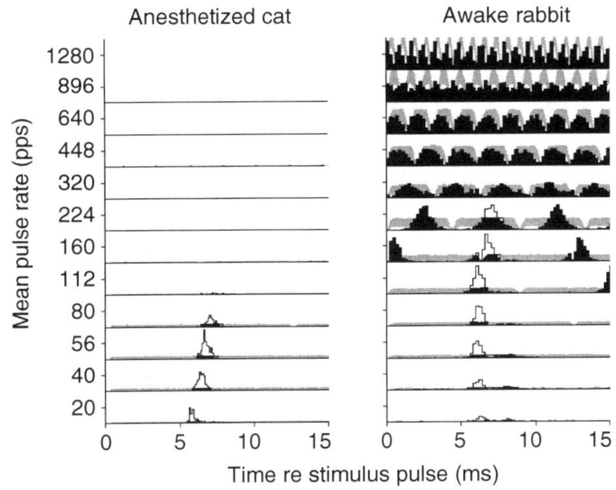

Fig. 39.2 Cross-correlograms between neural spike trains and stimulus pulse trains for the same two example neurons as in Fig. 39.1. *Gray shading* indicates 99 % confidence bounds; correlation peaks exceeding the confidence bounds are filled in *white*

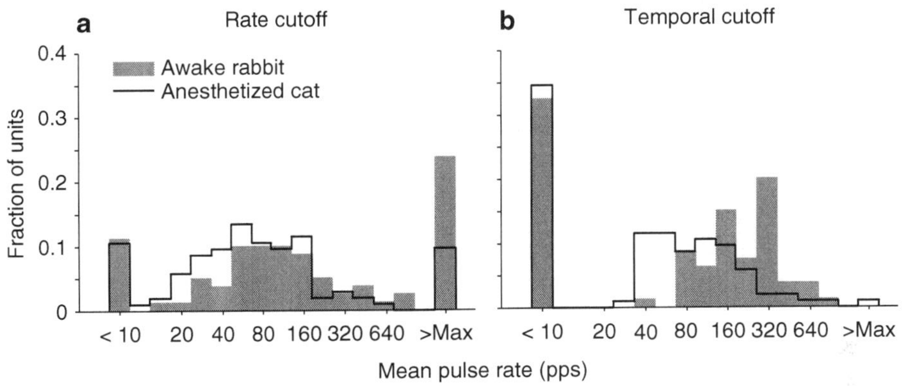

Fig. 39.3 Distributions of rate-based (**a**) and temporal cutoffs (**b**) across the IC neuron population in anesthetized cat (*gray bars*) and awake rabbit (*black lines*)

reflecting the periodicity of the stimulus. Above 224 pps, the responses become unsynchronized, as indicated by the absence of cross-correlogram peaks exceeding the 99 % confidence bound for a random spike train.

Two metrics were used to characterize the upper pulse-rate limit for sustained firing and pulse-locked responses, respectively. A neural detectability index d' was calculated for each pulse rate by comparing the firing rate during the stimulus (>30 ms after the stimulus onset) to the rate during the interstimulus silent interval (>100 ms after stimulus offset to discard rebound activity) in units of standard deviations. The cutoff for sustained firing was defined as the interpolated pulse rate where $d'=1$. This can occur either when an excitatory response becomes too low or when a suppressive response becomes too high to be statistically distinguishable from spontaneous activity. Figure 39.3a compares the distribution of cutoff pulse rates for sustained responses between anesthetized cat and awake rabbit. About 25 % of IC neurons in awake rabbit show sustained firing to the highest pulse rate tested (1,280 pps), a much higher proportion than in anesthetized cats. The median cutoff rate is about twice as large in awake rabbits (122 pps) as in anesthetized cats (65 pps). The difference is significant (Wilcoxon rank sum test, $p = 0.003$).

The temporal (pulse-locking) cutoff rate was defined as the lowest pulse rate where the area of the cross-correlogram peak above the 99 % confidence bound falls below 0.02 spikes per pulse. The distribution of temporal cutoff rates in awake rabbit is biased towards higher rates compared to the anesthetized cat distribution, and ~29 % of units in awake rabbit pulse lock at rates ≥320 pps compared to ~7 % in anesthetized cats (Fig. 39.3b). The median temporal cutoff pulse rate is about 2.5 times higher in awake rabbits (206 pps) than in anesthetized cats (87 pps) among the units that show pulse-locking at any rate. The difference is highly significant (Wilcoxon rank sum test $p < 0.001$).

3.2 Anesthesia Administration Reduces Single-Unit Pulse-Rate Limits

To ascertain whether the different effects of pulse rate on neural responses in the two preparations are due to anesthesia rather than species differences, we monitored changes in the responses of 13 single units and 3 multiunit clusters following intravenous injection of an ultrashort-acting barbiturate in one rabbit.

An example is presented in Fig. 39.4a–e for one single unit. Before injection (Fig. 39.4a), the neuron showed spontaneous activity, pulse-locked spikes up to 160 pps, and unsynchronized sustained firing at higher pulse rates, much like the example in Fig. 39.1. One minute after injection (Fig. 39.4b), pulse-locked responses were observed only up to 56 pps, whereas the response was limited to the onset at higher pulse rates. The spontaneous activity also largely disappeared. Ten minutes after injection (Fig. 39.4c), the response mostly recovered back to the preinjection pattern. The firing rate was not simply attenuated across all pulse rates following injection (Fig. 39.4d) but rather the upper rate cutoff also shifted to lower pulse rates. This change in cutoff rate is even more apparent in the normalized firing rate vs. pulse-rate curves for each time period (Fig. 39.4e).

We defined the 50 % cutoff rate as the point where the firing rate falls to 50 % of its maximum value. Figure 39.4f compares the preinjection and postinjection 50 % cutoff rates and temporal cutoff rates (defined as described earlier) across our sample. In all 16 units studied, both the 50 % cutoff rate and the temporal cutoff rate decreased after injection. In some units, pulse-locking was completely eliminated. Overall, the effects of anesthesia on pulse-rate limits of single units in rabbit IC are consistent with the differences observed between anesthetized cats and awake rabbits, suggesting anesthesia is primarily responsible for these differences.

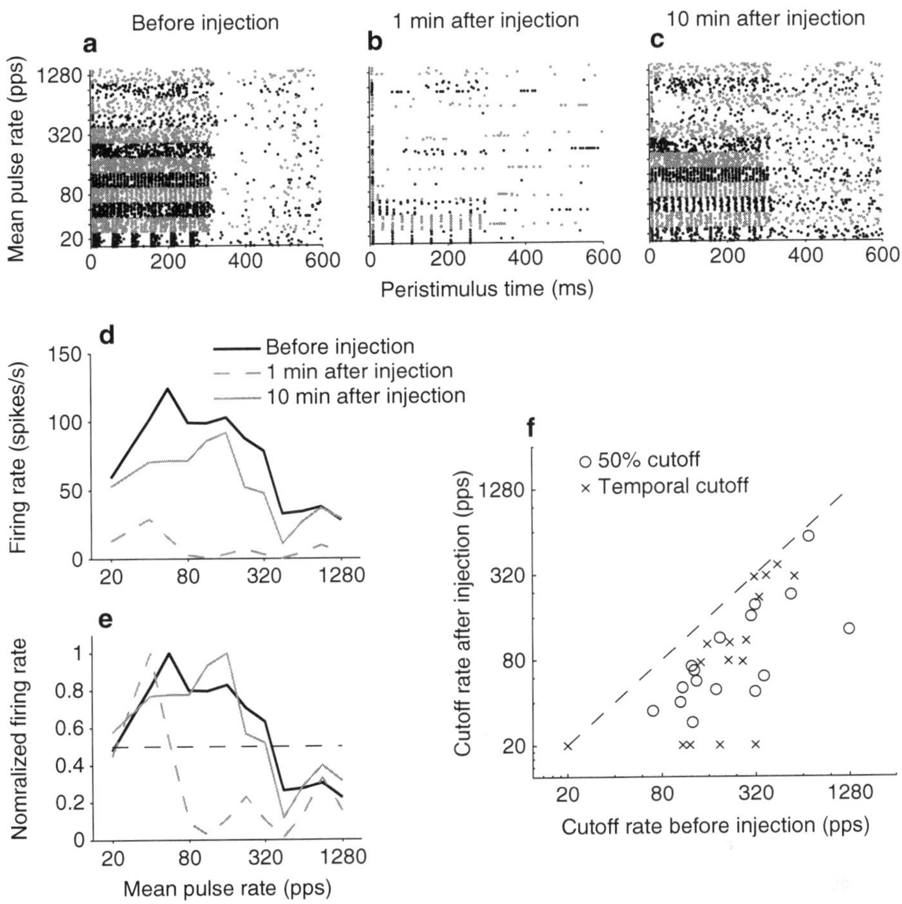

Fig. 39.4 Effect of barbiturate anesthesia in single units in rabbit. (**a**–**c**) Effect of anesthesia in an example unit from the rabbit IC. (**d**) Firing rate vs. mean pulse rate. (**e**) Normalized firing rate vs. mean pulse rate. (**f**) Comparison of rate-based cutoff and temporal cutoff before and 1–2 min after the injection

4 Summary and Discussion

Different response patterns are observed in IC units for electric pulse-train stimuli in awake rabbits compared to anesthetized cats. Cutoff pulse rates for both sustained firing and pulse-locked responses are 2–3 times higher in awake rabbits than in anesthetized cats. Effects of anesthesia in single units from the rabbit are consistent with the differences between the awake rabbit and anesthetized cat preparations, suggesting that these differences are mainly due to the effects of anesthesia.

The tendency for both spontaneous and evoked firing rates to be reduced under anesthesia has also been observed in the IC of normal-hearing rabbits (Kuwada et al. 1989) and in the auditory cortex of marmosets and guinea pigs wearing CIs (Johnson et al. 2011; Kirby and Middlebrooks 2012). On the other hand, Ter-Mikaelian et al. (2007) found minimal differences between awake and anesthetized conditions in the temporal response properties of IC neurons to amplitude-modulated tones in normal-hearing gerbils, so the effects of anesthesia we observed on pulse-locking limits may be specific to deaf animals. The reduction of firing rates by anesthesia is consistent with an enhancement of inhibition mediated by GABA receptors as suggested by Kuwada et al. (1989). However, the urethane anesthesia used in our cat preparation does not act solely through GABA receptors, but rather through a wide spectrum of neurotransmitter-gated ion channels (Hara and Harris 2002).

In earlier reports (Colburn et al. 2009; Hancock et al. 2012), we suggested that the rate limits of pulse-locked responses to electric pulse trains may be mediated by the low-voltage-activated potassium currents ($I_{K,LVA}$) present in many brainstem auditory neurons. The neurons expressing $I_{K,LVA}$ show a cumulative increase in membrane conductance after each stimulus pulse that increases the spiking threshold (Manis and Marx 1991). Barbiturates decrease presynaptic neurotransmitter release by reducing voltage-dependent calcium conductance (Werz and Macdonald 1985). The resulting reduction of excitatory drive may further limit the ability of neurons to reach threshold at high pulse rates under anesthesia.

The pulse-rate following limits of IC neurons in awake animals better agree with human performance in temporal processing tasks such as rate pitch discrimination than the cutoffs observed in anesthetized cats. Specifically, more than 25 % of the neurons in awake rabbit have temporal pulse-locking limits >300 pps, which is a typical temporal pitch limit in CI users. Whether the enhanced pulse-locking at high rates in awake animals also results in improved ITD sensitivity in bilaterally implanted animals will be of interest in future studies.

Acknowledgements Supported by NIH grants R01DC005775 and P30DC005209, Curing Kids Fund from Massachusetts Eye and Ear and a Hearing Health Foundation grant to Y. Chung. We thank Connie Miller, Melissa McKinnon, and Mike Kaplan for technical assistance.

References

Colburn HS, Chung Y, Zhou Y, Brughera A (2009) Models of brainstem responses to bilateral electrical stimulation. J Assoc Res Otolaryngol 10:91–110

Devore S, Delgutte B (2010) Effects of reverberation on the directional sensitivity of auditory neurons across the tonotopic axis: influences of interaural time and level differences. J Neurosci 30:7826–7837

Ebert CS, Fitzpatrick DC, Cullen RD, Finley CC, Bassim MK, Zdanski CJ, Coffey CS, Crocker W, Skaggs J, Marshall AF, Falk SE (2004) Responses of binaural neurons to combined auditory and electrical stimulation. Abstr Assoc Res Otolaryngol 27:485

Hancock KE, Noel V, Ryugo DK, Delgutte B (2010) Neural coding of interaural time differences with bilateral cochlear implants: effects of congenital deafness. J Neurosci 30:14068–14079

Hancock KE, Chung Y, Delgutte B (2012) Neural ITD coding with bilateral cochlear implants: effect of binaurally-coherent jitter. J Neurophysiol 108:714–728

Hara K, Harris RA (2002) The anesthetic mechanism of urethane: the effects on neurotransmitter-gated ion channels. Anesth Analg 94:313–318

Johnson L, Santina CD, Wang X (2011) Neural responses to cochlear implant stimulation in auditory cortex of awake marmoset. Abstr Assoc Res Otolaryngol 34:943

Kirby AE, Middlebrooks JC (2012) Unanesthetized auditory cortex exhibits multiple codes for gaps in cochlear implant pulse trains. J Assoc Res Otolaryngol 13:67–80

Kuwada S, Batra R, Stanford TR (1989) Monaural and binaural response properties of neurons in the inferior colliculus of the rabbit: effects of sodium pentobarbital. J Neurophysiol 61: 269–282

Manis PB, Marx SO (1991) Outward currents in isolated ventral cochlear nucleus neurons. J Neurosci 11:2865–2880

Smith ZM, Delgutte B (2007) Sensitivity to interaural time differences in the inferior colliculus with bilateral cochlear implants. J Neurosci 27:6740– 6750

Snyder R, Leake P, Rebscher S, Beitel R (1995) Temporal resolution of neurons in cat inferior colliculus to intracochlear electrical stimulation: effects of neonatal deafening and chronic stimulation. J Neurophysiol 73:449–467

Ter-Mikaelian M, Sanes DH, Semple MN (2007) Transformation of temporal properties between auditory midbrain and cortex in the awake Mongolian gerbil. J Neurosci 27:6091–6102

Tong YC, Clark GM (1985) Absolute identification of electric pulse rates and electrode positions by cochlear implant patients. J Acoust Soc Am 77:1881–1888

Townshend B, Cotter N, Compernolle D, White RL (1987) Pitch perception by cochlear implant subjects. J Acoust Soc Am 82:106–115

van Hoesel RJ (2007) Sensitivity to binaural timing in bilateral cochlear implant users. J Acoust Soc Am 121:2192–2206

Werz MA, Macdonald RL (1985) Barbiturates decrease voltage-dependent calcium conductance of mouse neurons in dissociated cell culture. Mol Pharmacol 28:269–277

Chapter 40
Relationships Between Auditory Nerve Activity and Temporal Pitch Perception in Cochlear Implant Users

Robert P. Carlyon and John M. Deeks

Abstract Cochlear implant (CI) users can derive a musical pitch from the temporal pattern of pulses delivered to one electrode. However, pitch perception deteriorates with increasing pulse rate, and most listeners cannot detect increases in pulse rate beyond about 300 pps. In addition, previous studies using irregular pulse trains suggest that pitch can be substantially influenced by neural refractory effects. We presented electric pulse trains to one CI electrode and measured rate discrimination, pitch perception, and auditory nerve (AN) activity in the same subjects and with the same stimuli. The measures of AN activity, obtained using the electrically evoked compound action potential (ECAP), replicated the well-known finding that the neural response to isochronous pulse trains at rates above about 200–300 pps is modulated, with the ECAP being larger to odd-numbered than to even-numbered pulses. This finding has been attributed to refractoriness. Behavioural results replicated the deterioration in rate discrimination at rates above 200–300 pps and the finding that pulse trains whose inter-pulse intervals (IPIs) alternate between a shorter and a longer value (e.g. 4 and 6 ms) have a pitch lower than that corresponding to the mean IPI. To link ECAP modulation to pitch, we physically modulated a 200-pps pulse train by attenuating every other pulse and measured both ECAPs and pitch as a function of modulation depth. Our results show that important aspects of temporal pitch perception cannot be explained in terms of the AN response, at least as measured by ECAPs, and suggest that pitch is influenced by refractory effects occurring central to the AN.

R.P. Carlyon (✉) • J.M. Deeks
Medical Research Council,
Cognition and Brain Sciences Unit, Cambridge CB2 7EF, UK
e-mail: bob.carlyon@mrc-cbu.cam.ac.uk

B.C.J. Moore et al. (eds.), *Basic Aspects of Hearing*,
Advances in Experimental Medicine and Biology,
DOI 10.1007/978-1-4614-1590-9_40,

1 Introduction

Cochlear implants (CIs) have allowed many previously deaf patients to understand speech well in quiet, but even the most successful recipients have poor pitch perception. This leads to a reduced enjoyment of music and impairs speech perception in noisy environments. Not surprisingly, CI companies and academic scientists have tried to improve coding of pitch in CIs, mainly by modifying the temporal pattern of electrical stimulation. Unfortunately, the neural basis of the impaired pitch perception by CI users is poorly understood, a fact that is likely to hinder attempts to remedy it. Here, we focus on two key aspects of "temporal" pitch perception by CI users and compare behavioural results with measures of auditory nerve (AN) activity obtained from ECAPs to the same stimuli and with the same subjects.

The first phenomenon concerns the pitch of pulse trains whose inter-pulse intervals (IPIs) alternate between a longer and a shorter value, such as 4 and 6 ms. CI users judge the pitch of these trains to be equal to that of an isochronous pulse train having an IPI of about 5.7 ms – longer than the average IPI of 5 ms in the "4–6" train (Carlyon et al. 2002, 2008). van Wieringen et al. (2003) proposed an explanation based on refractory effects, whereby the neural response to pulses occurring after 4-ms intervals was reduced. This would cause the neural response to be amplitude modulated, with smaller responses after 4-ms intervals interspersed between larger responses after 6-ms intervals. This modulation could cause some intervals equal to 10 ms to be conveyed to the brain, thereby reducing the pitch. To test this idea, they physically attenuated every other pulse of a 4–6 train and found some evidence that the pitch was lower when the attenuated pulses occurred after the 4-ms than after the 6-ms intervals, consistent with the physical modulation exaggerating the "neural modulation" in the former case and counteracting it in the latter. They obtained analogous findings when bandpass-filtered acoustic pulse trains were presented to normal-hearing (NH) listeners. Subsequently, Carlyon et al. (2008) showed that the compound action potential (CAP) to 4–6 pulse trains was indeed amplitude modulated both in humans and guinea pigs. However, it was not known whether the modulation in the CAP was sufficient to cause the reported changes in pitch or whether more central refractory effects might play a role.

The second phenomenon is the "upper limit" of temporal pitch. Although increasing the rate of an electric pulse train causes an increase in pitch up to about 300 pps, pitch usually asymptotes and rate discrimination deteriorates at higher rates (Shannon 1983; Townshend et al. 1987). A possible neural correlate comes from measures of the ECAP to simple isochronous pulse trains. Wilson (1997) reported that at low rates the ECAP to each pulse in the train had a roughly equal amplitude but that the ECAP to higher-rate pulse trains was typically amplitude modulated, with larger responses to odd-numbered than to even-numbered pulses. As noted above, modulation in the AN response could reduce pitch, and if the modulation were greater in the signal (higher-rate) interval than in the standard interval of a forced-choice task, this could counteract the increase in pitch due to the shorter IPI, thereby making discrimination harder.

2 Experiment 1: Effect of Modulation on Pitch Perception and on the AN Response

2.1 Rationale

Attenuating every other pulse of a pulse train reduces its pitch (McKay and Carlyon 1999; McKay et al. 1994). Here we measure the pitch change produced by attenuating every other pulse of a 200-pps pulse train by either 0.17 or 0.68 dB. We also measure the resulting modulation of the AN response, using the ECAP, thereby providing a link between AN modulation and pitch perception.

2.2 Subjects and Stimuli

Nine users of the Cochlear Corporation Freedom implant took part. The standard was a 200-pps pulse train presented in monopolar ("MP1+2") mode to an electrode near the middle of the array. The amplitude of the even-numbered pulses could be attenuated by either 0.17 or 0.68 dB. Phase duration was 25 μs with an 8-μs interphase gap. Stimuli were presented using the NIC2 interface and an L34 (behavioural experiments) or SP12 (ECAP experiments) processor.

2.3 Behavioural Experiment

In each interval of a 2-interval trial, listeners were presented with a 200-pps modulated pulse train and an unmodulated pulse train in random order. The rate of the unmodulated pulse train was selected from 100, 119, 141, 168, 200, and 238 pps at random on each trial. Stimulus duration was always 100 ms. Listeners judged which sound had the higher pitch. No feedback was provided. This was repeated for a total of between 40 and 60 times per data point for each listener. The resulting psychometric functions were fit by probit functions and the point of subjective equality (PSE) was estimated as the rate of an unmodulated pulse train that would be judged higher than the modulated pulse train on 50 % of trials. The "pitch shift" was calculated as the difference between the PSE and 200 pps in percent. The modulated pulse trains were presented at a comfortable listening level. The levels of the unmodulated pulse trains were loudness balanced to each other and to the level of the modulated pulse train to which they were compared.

2.4 ECAP Measures

ECAPs were obtained to the first 10 pulses of each (20-pulse) modulated pulse train used in the behavioural experiment. ECAP measurements were restricted to 10 pulses for technical reasons; the implications of this are discussed in Sect. 3.2. The pulse trains were presented at a rate of 4 Hz. Pulse polarity was the same throughout each pulse train but was inverted on every other presentation. For each stimulus, a total of 100 presentations for each polarity were obtained and the results for the two polarities were averaged to reduce stimulus artefacts. ECAP amplitude was defined as the difference (in μV) between the first negative (N1) and first positive (P1) peak of the waveform.

2.5 Results

ECAPs to the 2nd through 9th pulse of modulated pulse trains are shown for four example subjects by the solid lines in Fig. 40.1. The ECAPs are amplitude modulated. To summarise the depth of this modulation, we removed linear trends and then divided the mean ECAP to pulses (3,5,7,9) by that to pulses (2,4,6,8). The resulting factors were converted to a percentage and are shown for each subject and modulation depth by the second and third columns in faint type in Table 40.1. The table also shows, in bold type, the pitch shifts in percent obtained in the behavioural experiment in each condition.

3 Experiment 2: Pitch of Alternating-Interval Pulse Trains

3.1 Methods

Stimuli, procedures, and analyses were similar to those in experiment 1. In the behavioural task, psychometric functions were measured for comparisons between the "4–6" pulse train and that of isochronous pulse trains having IPIs of 2, 3, 4, 5, 6, 7, and 8 ms. The duration of all pulse trains was 100 ms, so that the 4–6 train consisted of 20 pulses. ECAPs were obtained for the first ten pulses of each stimulus used in the behavioural experiment for each subject.

3.2 Results

The results of the behavioural experiment were analysed to determine the IPI of an isochronous pulse train judged equal in pitch to the 4–6 pulse train. These values were converted to pitch shifts in percent, relative to the 5-ms mean IPI in the 4–6 stimulus. They are shown in the third column in bold type in Table 40.1. This

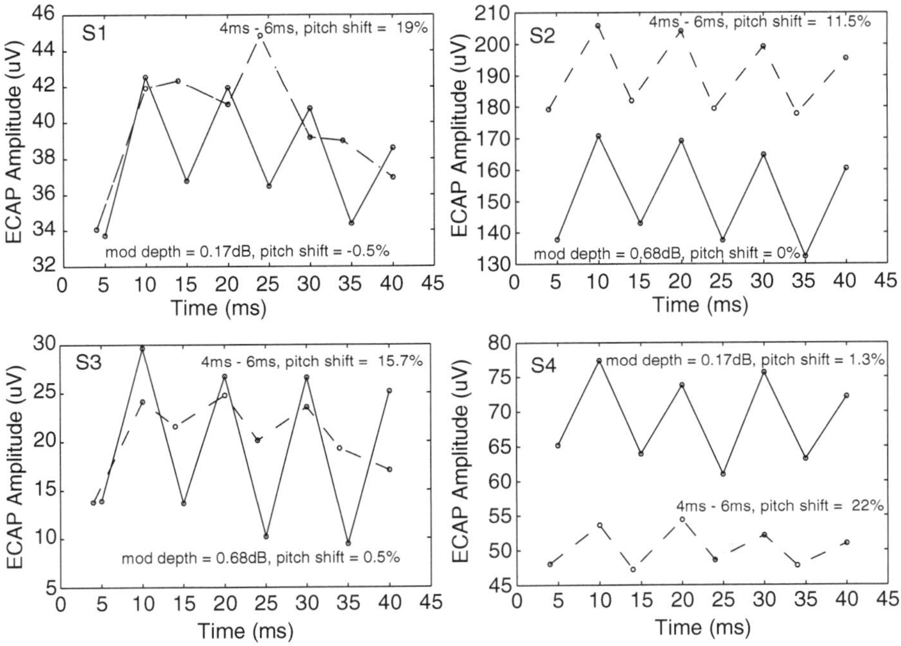

Fig. 40.1 ECAPs for 4–6 (*dashed lines*) and amplitude-modulated (*solid lines*) pulse trains Each panel shows data for one subject

Table 40.1 ECAP modulation depth in percent (*faint type*) and pitch shifts in percent (*bold type*) for the stimuli used in experiments 1 and 2. Mean data are shown only for the "4–6" stimuli (experiment 2) as not all subjects took part in experiment 1

	0.17 dB		0.68 dB		"4–6"	
Subject	Mod(%)	**Shift (%)**	Mod(%)	**Shift (%)**	Mod(%)	**Shift(%)**
S1	13.7	**−2.8**	48.9	**30.6**	−0.8	**19.1**
S2	7.4	**−2.3**	17.3	**0.0**	10.7	**11.5**
S3	32.0	**0.5**	56.3	**32.2**	16	**15.7**
S4	15.3	**1.3**	42.8	**4.6**	9.2	**22.0**
S5	3.8	**0.7**	11.2	**−0.3**	−3.8	**−7.0**
S6	8.3	**−4.7**	26.8	**30.5**	−3.8	**17.8**
S7	7.0	**0.7**	24.9	**−0.4**	5.7	**−6.4**
S8					22.9	**11.7**
S9					7.3	**18.9**
Mean					*7.6*	***12.4***

average difference of 12.4 % corresponds to an IPI of 5.7 ms, which is in good agreement with previous results (Carlyon et al. 2002, 2008).

ECAPs to the 2nd through 9th pulse of the 4–6 pulse train are shown for four subjects by the dashed lines in Fig. 40.1. The summary modulation depth values for these ECAPs are shown by the third column in faint type in Table 40.1.

Figure 40.1 and Table 40.1 show that there can be a substantial pitch shift associated with the 4–6 stimulus and that the physically modulated pulse train (from experiment 1) can produce a larger ECAP modulation but a much smaller pitch shift. It therefore seems that modulation in the AN response – at least as measured by the ECAP – is not entirely responsible for the fact that the pitch of a 4–6 pulse train is lower than the 200 Hz that would be expected from the mean IPI in that stimulus. A possible *caveat* is that behavioural responses in experiment 2 were obtained for 100-ms stimuli, whereas ECAPs were obtained only to the first 50 ms of the modulated 200-pps and unmodulated 4–6 pulse trains. However, for this to explain the differing relationship between pitch shift and ECAP modulation observed in the two experiments, the ECAP modulation depth would need to decrease over time for the modulated stimulus and for this effect to be substantially larger than for the 4–6 stimulus. No such trend was observed over the first ten pulses. In addition, additional data (not shown) from listeners S6 and S7 showed similar modulation depth for the first and second ten pulses of modulated pulse trains.

4 Experiment 3: Upper Limit of Rate Discrimination

4.1 Methods

In the behavioural experiment, eight listeners were presented, on each trial, with a standard and a signal pulse train in random order. The standard pulse train had a rate of either 100, 200, 300, 400, or 500 pps; the signal differed from the standard in that its rate was 30 % higher. The standard pulse train contained either 10 or 50 pulses and the signal pulse train contained 30 % more pulses, so that its duration was the same as that of the standards. After each trial, the subject indicated which sound had the higher pitch; if this was the signal, then the response was scored as correct. Feedback was provided after every trial. All combinations of standard duration and pulse rate were mixed within a block of 100 trials. Each subject performed at least five blocks leading to a minimum of 50 trials per condition. The standards were presented at a comfortable level. Each signal was loudness balanced to the corresponding standard.

ECAPs to the 10-pulse standard and signals were measured and analysed in the same way as in experiments 1 and 2. However, for technical reasons, the highest rate at which we could measure ECAPs was 470 pps. Therefore, this rate was used in lieu of the 520-pps signal for the 400-pps standard. No ECAPs were measured for the 500-pps standard or signal, which were therefore absent from statistical comparisons between the behavioural and ECAP measures.

4.2 Results

The results of the behavioural experiment are shown for the 10-pulse standards in Fig. 40.2a. (Results for the 50-pulse standards are not shown but were similar although slightly better overall.) They replicate previous findings obtained with

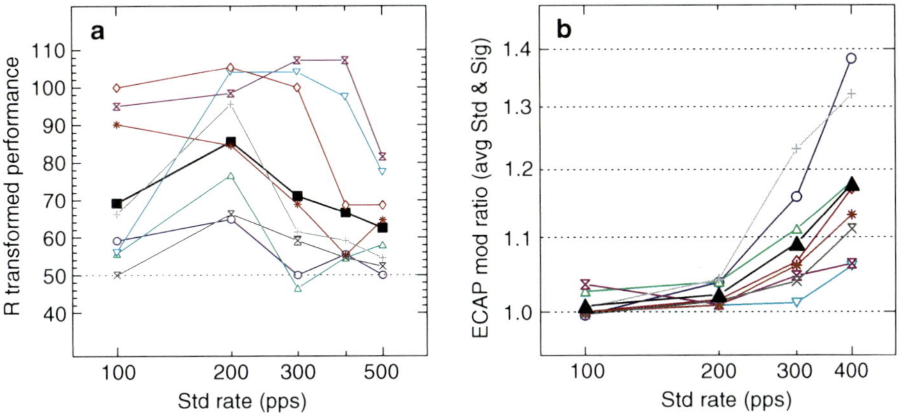

Fig. 40.2 Part **a** shows rau-transformed %-correct scores for the behavioural rate-discrimination experiment as a function of baseline rate. Part **b** shows the ECAP modulation, averaged across the standard and signal rate for each standard rate, expressed as a factor. In both plots, individual subjects' data are shown by *coloured lines* and mean data are shown by a *thick black line*

longer stimuli: on average (bold curve), performance is good for 100- and 200-pps standards and deteriorates at higher rates, and the pattern of results varies across subjects (Kong et al. 2009; Townshend et al. 1987). The ECAP modulation depth was averaged across the standard and signals for each rate (e.g. across 100 and 130 pps for the 100-pps standard) and is plotted in Fig. 40.2b. This also replicates a well-known finding that the ECAP is unmodulated at low rates and is modulated at higher rates (Wilson 1997).

Interestingly, there was a statistically significant relationship between the behavioural and ECAP measures; subjects who showed a large ECAP modulation tended to perform worse at the task. This was determined by performing a univariate ANOVA on the data for all subjects and rates, with behavioural performance as the dependent variable, rate as a fixed factor, and ECAP modulation depth as a covariate; with the effect of rate partialled out in this way, the ECAP modulation depth accounted for 18.7 % of the variance in the discrimination scores, corresponding to a correlation coefficient of $r=0.43$ ($df=26$, $p<0.05$).

Despite the statistical relationship between the ECAP and rate-discrimination measures, there is evidence that modulation in the AN response, as measured by the ECAP, cannot explain all instances where subjects performed poorly. This comes from cases where rate discrimination is poor but where the corresponding ECAP modulation is smaller than that resulting from manipulations in experiment 1 that, in turn, had only a small effect on pitch. For example, subject S2 (green curve in Fig. 40.2) scored only 46.6 % at rate discrimination with a 300-pps standard, and the ECAP modulations for this standard and signal were 8 % and 12 %, respectively. These were smaller than the 17 % ECAP modulation observed in experiment 1 by attenuating every other pulse by 0.68 dB, a manipulation that produced no change in perceived pitch. Similarly, listener S4 (brown curve) scored 55 % on the rate-discrimination task at 400 pps, with ECAP modulation depths of 9 and 14 % for the standard and signal, whereas a 0.17-dB modulation in experiment 1 produced 15 % ECAP modulation but only a 1.3 %

pitch shift. Listener S1 (dark blue curve) scored 50 % on the rate-discrimination task at 300 pps, with ECAP modulation depths of 9 and 17 % for the standard and signal, whereas a 0.17-dB modulation in experiment 1 produced a 14 % ECAP modulation but no pitch shift. A *caveat* with these comparisons is that their validity rests on the assumption that a given modulation depth will have a similar effect on pitch for the 200-pps stimuli of experiment 1 as for the higher-rate stimuli from experiment 3. However, we should also note that ECAP modulation should impair rate discrimination only to the extent that it is greater in the signal than in the standard interval. Although this was generally true, these differences were much smaller than the total ECAP modulation discussed here and shown in Table 40.1.

5 Discussion

Experiment 2 showed that modulation in the AN response, as measured by the ECAP, was insufficient to account for the fact that the pitch of a 4–6 pulse train corresponds not to 200 Hz (the reciprocal of its mean IPI) but to a value that is on average 12.6 % lower. As noted in the introduction, van Wieringen et al. (2003) showed that attenuating pulses that occur after the 4-ms intervals resulted in a lower pitch than when the pulses occurring after the 6-ms intervals were attenuated. This was consistent with the former manipulation exaggerating, and the latter counteracting, a modulation in the neural response. Taken together, these two findings suggest that modulation in the neural response exists, and can influence pitch, but at least partially arises from sites central to the auditory nerve. A *caveat* is that the ECAP is, by definition, a compound measure and that we cannot be sure that exactly the same population of neurons contributed to the ECAPs in experiments 1 and 2.

Experiment 3 showed that listeners whose ECAPs were more strongly modulated tended to perform worse than those showing less ECAP modulation. However, a comparison of the overall size of the ECAP modulations in experiments 1 and 3 suggested that the ECAP modulation was unlikely to provide a complete explanation for the poor rate-discrimination performance of some listeners at higher rates. It is possible that ECAP modulation serves as an index of an additional factor, such as AN survival, that in turn correlates with rate discrimination. The origins of poor rate discrimination at high rates may lie either in aspects of the AN response that are not reflected in modulation of the ECAP, or in processes central to the AN.

References

Carlyon RP, van Wieringen A, Long CJ, Deeks JM, Wouters J (2002) Temporal pitch mechanisms in acoustic and electric hearing. J Acoust Soc Am 112:621–633

Carlyon RP, Mahendran S, Deeks JM, Long CJ, Axon P, Baguley D, Bleeck S, Winter IM (2008) Behavioral and physiological correlates of temporal pitch perception in electric and acoustic hearing. J Acoust Soc Am 123:973–985

Kong Y-Y, Deeks JM, Axon PR, Carlyon RP (2009) Limits of temporal pitch in cochlear implants. J Acoust Soc Am 125:1649–1657

McKay CM, Carlyon RP (1999) Dual temporal pitch percepts from acoustic and electric amplitude-modulated pulse trains. J Acoust Soc Am 105:347–357

McKay CM, McDermott HJ, Clark GM (1994) Pitch percepts associated with amplitude-modulated current pulse trains in cochlear implantees. J Acoust Soc Am 96:2664–2673

Shannon RV (1983) Multichannel electrical stimulation of the auditory nerve in man. I. Basic psychophysics. Hear Res 11:157–189

Townshend B, Cotter N, van Compernolle D, White RL (1987) Pitch perception by cochlear implant subjects. J Acoust Soc Am 82:106–115

van Wieringen A, Carlyon RP, Long CJ, Wouters J (2003) Pitch of amplitude-modulated irregular-rate stimuli in electric and acoustic hearing. J Acoust Soc Am 114:1516–1528

Wilson BS (1997) The future of cochlear implants. Br J Audiol 31:205–225

Chapter 41
Robust Cortical Encoding of Slow Temporal Modulations of Speech

Nai Ding and Jonathan Z. Simon

Abstract This study investigates the neural representation of speech in complex listening environments. Subjects listened to a narrated story, masked by either another speech stream or by stationary noise. Neural recordings were made using magnetoencephalography (MEG), which can measure cortical activity synchronized to the temporal envelope of speech. When two speech streams are presented simultaneously, cortical activity is predominantly synchronized to the speech stream the listener attends to, even if the unattended, competing-speech stream is more intense (up to 8 dB). When speech is presented together with spectrally matched stationary noise, cortical activity remains precisely synchronized to the temporal envelope of speech until the noise is 9 dB more intense. Critically, the precision of the neural synchronization to speech predicts subjectively rated speech intelligibility in noise. Further analysis reveals that it is longer-latency (~100 ms) neural responses, but not shorter-latency (~50 ms) neural responses, that show selectivity to the attended speech and invariance to background noise. This indicates a processing transition, from encoding the acoustic scene to encoding the behaviorally important auditory object, in auditory cortex. In sum, it is demonstrated that neural synchronization to the speech envelope is robust to acoustic interference, whether speech or noise, and therefore provides a strong candidate for the neural basis of acoustic-background invariant speech recognition.

N. Ding
Department of Electrical and Computer Engineering,
University of Maryland, College Park, MD, USA

J.Z. Simon (✉)
Department of Electrical and Computer Engineering,
University of Maryland, College Park, MD, USA

Department of Biology, University of Maryland,
College Park, MD, USA
e-mail: jzsimon@umd.edu

B.C.J. Moore et al. (eds.), *Basic Aspects of Hearing*,
Advances in Experimental Medicine and Biology,
DOI 10.1007/978-1-4614-1590-9_41,

1 Introduction

Normal-hearing human listeners are remarkably good at understanding speech in adverse listening environments. Acoustic degradation to speech can be due to energetic or informational masking (Brungart 2001). Here, energetic masking refers to the energetic overlap between the target speech and maskers, arising dominantly in the auditory periphery. Informational masking refers to the interference caused by the perceptual similarity between the target speech and maskers, arising dominantly in the central auditory system. In this study, energetic masking of speech is studied using a stationary noise masker with a long-term spectrum matching that of speech. The stationary noise affects the audibility of speech by severely reducing its intensity contrast, i.e., the depth of the spectro-temporal modulations (see also Stone et al. 2011). Informational masking, in contrast, is exemplified by the masking caused by a competing speech signal. In this case, both speech streams are audible and intelligible, and therefore the difficulty the listeners face is to identify which speech features are from the target speech and then to selectively process them.

We recorded neural responses from normal-hearing human listeners using MEG, which can measure cortical activity precisely synchronized to the temporal envelope of speech (Ding and Simon 2012). Based on the MEG measurements, we analyzed how the neural synchronization to speech is affected by acoustic interference, speech or noise, at different intensity levels. It is demonstrated that a robust neural representation of speech is maintained in human auditory cortex, reflecting a variety of top-down and bottom-up gain control effects in human auditory cortex.

2 The Cortical Representation of Competing Speech Streams

2.1 *Experimental Procedures*

In this competing-speech experiment (Ding and Simon 2012), two stories, narrated by a male and a female speaker respectively, were mixed into a single acoustic channel with different intensity ratios. One speaker was always presented at roughly 75 dB SPL, while the other was presented at either the same level (by RMS value) or 5 or 8 dB weaker. These two speakers were referred to as the constant-intensity speaker and the varying-intensity speaker, respectively. For this set of speech mixtures, when different speakers were attended to, the target-to-masker ratio (TMR) ranges from −8 to 8 dB. Each stimulus was presented twice under each attentional condition (attend-to-male vs. attend-to-female).

For each TMR, two 1-min duration stimuli were presented, after each of which a comprehension question was asked to ensure the subjects' attention. The listeners correctly answered 71 % of the questions, and this percentage did not significantly vary with TMR ($p > 0.7$, one-way repeated-measures ANOVA). Six subjects participated in the experiment. Five of them were asked to subjectively rate speech intelligibility (as percentage) after the first listening to each stimulus. The neuromagnetic signals were recorded using a 157-channel whole-head MEG system.

2.2 Neural Reconstruction of Each Speech Stream

The temporal envelope of each speech stream was reconstructed separately, by integrating neural activity over time and MEG sensors. The reconstructed envelope $\hat{s}(t)$ is $\hat{s}(t)=\Sigma_k \Sigma_\tau r_k(t+\tau)h_k(\tau)$, where $r_k(t)$ is the recording from the kth MEG sensor and the decoder $h_k(\tau)$ is a weighting matrix for the kth MEG sensor and a time lag τ between the stimulus and response. The decoder was optimized using boosting, with ten-fold cross validation, which maximizes the accuracy of neural reconstruction, i.e., the correlation between the reconstructed envelope and the actual envelope of speech (Ding and Simon 2012). The envelope of each speaker, $s(t)$, was expressed on a linear amplitude scale. The accuracy of the neural reconstruction depends on how precisely the speech envelope is encoded in neural activity and is an index of the fidelity of neural encoding.

The chance-level reconstruction accuracy was estimated by generating pseudo-reconstructions based on unmatched stimulus-response pairs. To create a stimulus not matching the neural response, we cut the actual stimulus into eight segments, shuffled them, and concatenated the shuffled segments. A hundred pseudo-reconstructions were generated based on the shuffled stimuli, and the maximal reconstruction accuracy from the pseudo-reconstructions was used as a threshold to test the significance of normal neural reconstructions ($P<0.01$).

Beyond the correlation analysis, a more detailed relationship between $\hat{s}(t)$ and $s(t)$ is obtained by fitting $\hat{s}(t)$ as a function of $s(t)$. This function is called the amplitude-intensity function (AIF). Since the neural reconstruction is just the spatial-temporally integrated neural response, the AIF describes the relationship between the instantaneous amplitude of the neural response and the instantaneous intensity of the stimulus.

2.3 Results

The temporal envelope of each speaker in the stimulus is reconstructed separately from the cortical response. The correlation between the reconstruction and the actual envelope of speech is shown in Fig. 41.1a. The reconstruction accuracy is above chance for both speech signals ($P<0.01$ for every condition) and is significantly higher for the attended speech ($P<0.02$, 2-way repeated-measures ANOVA, factors: attention, TMR). The main effect of TMR and the interaction were both not significant. Since the same decoder is used in every TMR condition, the TMR-independent reconstruction accuracy suggests TMR-independent neural encoding of the temporal modulations of each speech stream.

The TMR-independent neural reconstruction implies neural compensation for the intensity change of the speakers. This is further investigated using the AIF, which describes the relationship between the instantaneous amplitude of the neural response and the instantaneous intensity of the stimulus envelope (Fig. 41.1b). The AIFs for the two speakers show distinct behaviors. The AIF for the varying-intensity speaker shifts leftwards as the intensity of the speaker decreases, regardless of the attentional state of the listener. A leftward shift of the AIF indicates an increase in response gain since lower intensity is needed to achieve a given response amplitude. When fitted by a line, the AIF shifts 6.0 ± 0.2 dB and 5.0 ± 1.1 dB (Mean ± SEM) for the attended and unattended speaker, respectively, as the intensity of the speaker changes by 8 dB. The AIF for the constant-intensity speaker, in contrast, is not significantly affected by the intensity change of varying-intensity speaker. Therefore, the neural representation of each speaker only adapts to the mean intensity of that speaker, rather than the mean intensity of the stimulus mixture. In other words, neural adaptation to sound intensity is auditory stream specific.

3 Cortical Representation of Speech Masked by Noise

3.1 Methods

In the speech-in-noise experiment (Ding and Simon 2013), each stimulus consisted of a 50-s duration spoken narrative. Stationary noise matching the long-term spectrum of speech was generated using a 12th-order linear predictive model and mixed into speech with one of the following six TMRs: quiet, +6, +2, −3, −6, and −9 dB. The intensity of speech was the same for all stimuli while the intensity of the noise varied. Ten subjects participated.

Each stimulus (12 in total) was presented three times. The TMR always increased or decreased every two sections (counterbalanced over subjects). The subjects were asked a comprehension question after each section. During the first presentation of each stimulus, the subjects were asked to rate the intelligibility of each stimulus. The order how the sections were presented, whether with increasing or decreasing TMR, did not affect speech intelligibility (two-way repeated-measures ANOVA, factors: TMR, Order) or the neural reconstruction of speech (the same ANOVA) and therefore was not distinguished in the analysis.

3.2 Results

To investigate how the cortical representation of speech is affected by background noise, we reconstructed the temporal envelope of the underlying clean speech, not

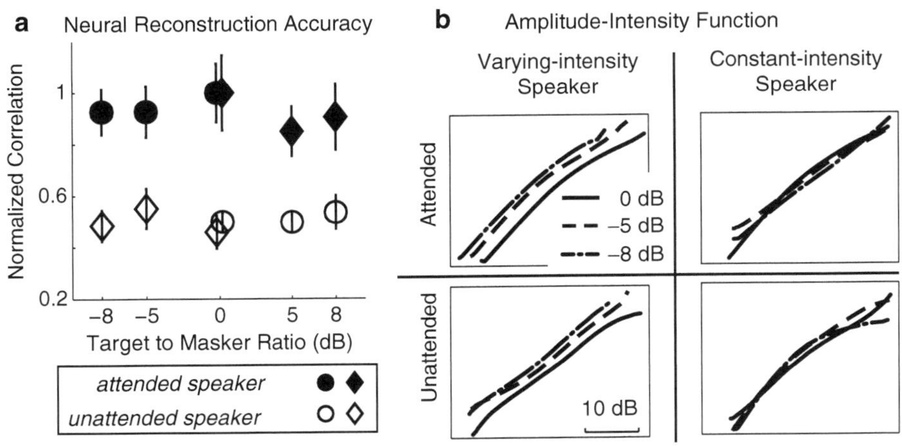

Fig. 41.1 Cortical reconstruction of each of two competing speech streams. (**a**) The correlation between neural reconstruction and the actual envelope of each speaker (*filled* and *hollow symbols* for attended and unattended speakers, respectively). The two speakers are shown by *circles* and *diamonds*, respectively. The correlation with each speaker is normalized based on the correlation at 0 dB TMR when the speaker is attended. (**b**) The AIF for each speaker, under each attentional state. The *x*-axis is the stimulus intensity and the *y*-axis is the dimensionless amplitude of the neural reconstruction on a linear scale. The level difference between the speakers is indicated by the *line* style

the actual stimulus, from the cortical response to a noisy stimulus (Fig. 41.2). The correlation between the neural reconstruction and the actual envelope of speech remains high until the TMR drops to −9 dB (Fig. 41.2). This indicates that, above −9 dB, the temporal modulations of speech are cortically encoded by phase-locked activity, regardless of the degradation caused by noise. Decoding accuracy was not affected by TMR when the −9 dB condition is excluded (2-way repeated-measures ANOVA, factors: TMR, Trial).

At the intermediately low TMR of −3 dB, the median of the rated speech intelligibility was 55 % and varied widely. At this TMR, individual subject's subjectively rated speech intelligibility is significantly correlated with neural reconstruction accuracy ($R = 0.78 \pm 0.15$, bootstrap, Fig. 41.2). No such correlation was found at high and low TMRs, because of ceiling (median > 90 %) and floor (median ≤10 %) effects in the ratings.

Stationary background noise reduces the depth of the spectro-temporal modulations, i.e., intensity contrast, of speech. Therefore, the robust neural encoding of speech suggests that the loss of stimulus contrast is compensated for by the auditory system. To demonstrate this, we estimated the AIF for each TMR condition and found the AIF to be strongly TMR dependent (Fig. 41.2), showing neural adaptation to intensity contrast. The slope of the AIF, extracted by a linear regression, increases 16 ± 2 dB (Mean ± SEM) as TMR decreases from infinity (quiet) to −6 dB.

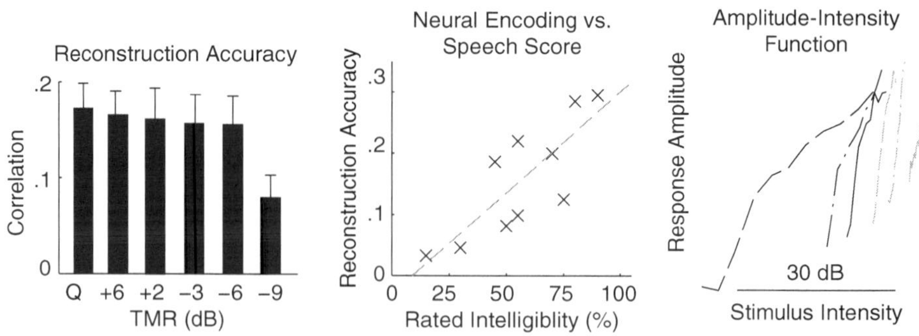

Fig. 41.2 Neural reconstruction of speech masked by stationary noise. (*Left*) The correlation between the neural reconstruction and the temporal envelope of the underlying clean speech. (*Middle*) At −3 dB TMR, individual subject's intelligibility rating is significantly correlated with the accuracy of neural reconstruction. (*Right*) The AIF for each TMR condition. The curves, from left to right, correspond to conditions with decreasing TMR

4 Time Course of Neural Encoding During Energetic and Informational Masking

4.1 Temporal Response Function

How the cortical response is generated by temporal modulations of speech can be modeled using the temporal response function (TRF). For the competing-speech experiment, the response was modeled as the sum of two TRF models, one for each speaker. The TRF was derived from a spectro-temporal response function that was estimated based on the normalized spectrogram of speech (z-score) expressed on a logarithmic scale (Ding and Simon 2012). Since the TRF is based on the normalized spectrogram, the gain of the TRF is invariant to the changes of the stimulus intensity only if neural activity is fully adapted to such changes. A TRF is estimated for each MEG sensor. Two major components are seen in the TRF, with distinct neural sources and response latencies. As an effective representation of both components, the TRF shown here is the sum of the TRFs projected to the neural sources of the two components.

4.2 Time-Dependent Gain Control

The TRFs from the competing-speech and speech-in-noise experiments are shown in Fig. 41.3. The TRF characterizes the temporal evolution of the neural response evoked by a unit power increase of the speech stream it is applied to.

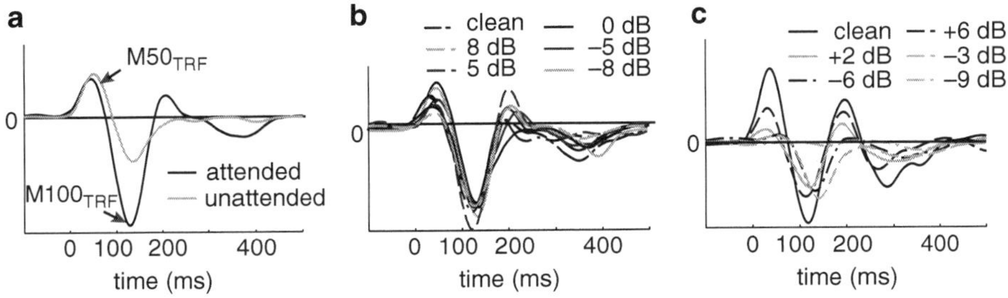

Fig. 41.3 The temporal response function (*TRF*) in the competing-speech experiment (**a**–**b**) and in speech-in-noise experiment (**c**). (**a**) The TRF for the attended and unattended speech streams. The $M100_{TRF}$ is modulated by attention, but the $M50_{TRF}$ is not. (**b**) The shape of the TRF is invariant to TMR. (**c**) In noise, the $M50_{TRF}$ weakens as the TMR decreases while the amplitude of $M100_{TRF}$ remains stable as the TMR changes from +6 to −6 dB

The earliest two peaks of the TRF have latencies near 50 and 100 ms, called the $M50_{TRF}$ and $M100_{TRF}$, respectively. In the competing-speech experiment, the amplitude of the $M100_{TRF}$ but not the $M50_{TRF}$ is significantly modulated by attention ($P < 0.03$, four-way repeated-measures ANOVA, factors: attention, hemisphere, speaker, and TMR). Furthermore, the shape of the TRF is independent of the TMR, for both the constant-intensity speaker and the varying-intensity speaker. A stimulus-invariant TRF reflects a complete neural adaptation to the mean intensity of the stimulus, since the TRF is derived from normalized speech envelope. In the speech-in-noise experiment, the $M50_{TRF}$ weakens as TMR decreases (negative correlation between $M50_{TRF}$ amplitude and TMR, $P < 0.001$, bootstrap). The $M100_{TRF}$, however, remains largely stable between +6 dB TMR and −6 dB TMR (no significant correlation between $M100_{TRF}$ amplitude and TMR).

5 Spectro-temporal Processing of Speech

5.1 Object-Based Gain Control

Neural adaptation to sound intensity occurs at multiple stages of the auditory system (Robinson and McAlpine 2009). The competing-speech experiment further suggests that the neural adaptation to sound intensity occurs separately for each auditory stream/object. In that experiment, the stable representation of the varying-intensity speaker must be maintained by neural adaptation to sound intensity, while the stable representation of the constant-intensity speaker requires no adaptation to the overall intensity of the sound mixture (which itself covaries with the intensity of the varying-intensity speaker). Therefore, the stable representation of both speakers cannot be explained by a simple mechanism of global intensity gain control, which would result in the neural representation of both speakers to be modulated in the same way based on the overall intensity of the acoustic stimulus. Instead, the results suggest object-specific intensity gain control.

5.2 Latency of Gain Control Effects During Energetic and Informational Masking

The properties of the acoustic masker, i.e., whether informational or energetic, influence the cortical processing of speech differentially. As revealed by the TRF analysis, in the competing-speech experiment, both the shorter- and longer-latency cortical responses are insensitive to the change in masker intensity. In the speech-in-noise experiment, however, only the longer-latency response $M100_{TRF}$ is resilient to the masker. This influence of masker property is straightforward to explain. During informational masking, the audibility of each stream of speech is seldom a problem and therefore each stream drives early auditory response effectively. The key question during informational masking is the selection of acoustic features belonging to the speech target, which is only reflected by the attentional modulation of long-latency responses. During energetic masking, however, background noise reduces the audibility of speech and therefore attenuates the shorter-latency response. The noise robustness of the long-latency response can only be maintained by additional active neural processing. In summary, the shorter-latency (~50 ms) response from core auditory cortex mainly reflects the audibility of a sound stream, while the longer-latency (~100 ms) response is a robust representation of the target speech stream.

5.3 Relation to Speech Intelligibility

On the one hand, the cortical synchronization to speech is more robust to acoustic interference than rated speech intelligibility. For example, at −6 dB TMR, cortical synchronization to speech is not affected by acoustic interference, whether speech or noise, but rated speech intelligibility drops to about 50 % for a speech masker and only about 10 % for a noise masker. On the other hand, the precision of cortical synchronization to speech is a good predictor of individual's intelligibility rating, when speech is masked by noise at −3 dB TMR. We explain this difference by dividing speech recognition into two consecutive processes. One is the parsing of the continuous and possibly noisy acoustic input into basic processing unit, e.g., syllables. The other is the decoding of linguistic information from each unit. We argue that the MEG-measured cortical synchronization to speech reflects the first parsing process, which is more reliable than the decoding of phonemic information from each unit (see Woodfield and Akeroyd 2010, for psychoacoustical evidence). In the presence of an intermediate amount of noise, the parsing process becomes a bottleneck for speech recognition, and therefore listeners who are better at extracting basic speech units rate speech intelligibility as higher.

6 Conclusion

We found that large-scale coherent cortical activity is precisely synchronized to the temporal modulations of speech, even in the presence of an acoustic masker twice as intense as the speech target. Two major sources of the speech-synchronized neural response are identified. One has shorter latency (~50 ms) and is from roughly core auditory cortex. The other has longer latency (~100 ms) and is from posterior association auditory cortex. The shorter-latency response is not modulated by attention and is susceptible to background noise. The longer-latency response, however, is strongly modulated by attention and is resilient against acoustic interference. In summary, the results suggest the emergence of a neural representation of the target speech stream embedded in a complex auditory scene. This auditory stream-specific representation is enhanced, from shorter-latency to longer-latency neural responses and from core to posterior auditory cortex.

Acknowledgments We thank NIH grant R01 DC 008342 for support.

References

Brungart DS (2001) Informational and energetic masking effects in the perception of two simultaneous talkers. J Acoust Soc Am 109:1101–1109

Ding N, Simon JZ (2012) The emergence of neural encoding of auditory objects while listening to competing speakers. Proc Natl Acad Sci USA 109(29):11854–11859

Ding N, Simon JZ (2013) Adaptive temporal encoding leads to a background insensitive cortical representation of speech. J Neurosci 33:5728–5735

Robinson BL, McAlpine D (2009) Gain control mechanisms in the auditory pathway. Curr Opin Neurobiol 19:402–407

Stone MA, Fullgrabe C, Mackinnon RC, Moore BCJ (2011) The importance for speech intelligibility of random fluctuations in "steady" background noise. J Acoust Soc Am 130:2874–2881

Woodfield A, Akeroyd MA (2010) The role of segmentation difficulties in speech-in-speech understanding in older and hearing-impaired adults. J Acoust Soc Am 128:EL26–EL31

Chapter 42
Wideband Monaural Envelope Correlation Perception

Joseph W. Hall III, Emily Buss, and John H. Grose

Abstract This study investigated monaural envelope correlation perception (Richards 1987) for noise bandwidths ranging from 25 to 1,600 Hz. The high-frequency side of the low band was fixed at 3,000 Hz and the low-frequency side of the high band was fixed at 3,500 Hz. When comodulated, the magnitude spectra of the pair of noise bands were either identical or reflected around the midpoint. Six listeners with normal hearing participated. Listeners showed similar performance for identical and reflected-spectrum conditions, with best performance usually occurring for bandwidths between 200 and 800 Hz. Results were considered in terms of envelope comparisons of waveforms at the outputs of multiple peripheral filters or envelope comparisons of waveforms at the outputs of central filters set to the bandwidths of the noise stimuli. Some aspects of the results were incompatible with the account based on multiple peripheral filters. However, the results of a supplementary condition involving the gating of band subregions indicated that this incompatibility could be accounted for by nonoptimal weighting of peripheral filter outputs.

1 Introduction

In a paradigm known as monaural envelope correlation perception (MECP), Richards (1987) demonstrated that listeners are able to discriminate between a stimulus consisting of two narrow bands of noise having random temporal envelopes and a stimulus in which the two bands are comodulated. That study, and most other investigations of MECP, used noise bandwidths of 100 Hz or less (e.g., Hall and

---

J.W. Hall III (✉) • E. Buss • J.H. Grose
Department of Otolaryngology/Head and Neck Surgery,
University of North Carolina School of Medicine,
170 Manning Dr., Chapel Hill, NC 27599, USA
e-mail: jwh@med.unc.edu

B.C.J. Moore et al. (eds.), *Basic Aspects of Hearing*,
Advances in Experimental Medicine and Biology,
DOI 10.1007/978-1-4614-1590-9_42,

Grose 1993; Moore and Emmerich 1990; Richards 1987, 1988a, b, 1989). A rationale for exploring these relatively narrow bandwidths is that they are narrower than the equivalent rectangular bandwidths (ERB) of auditory filters over a relatively large frequency region (Glasberg and Moore 1990; Patterson 1976). Therefore, many MECP results can be conceptualized in terms of comparisons of temporal envelopes at the outputs of frequency-separated auditory filters.

We recently explored MECP for noise bandwidths considerably wider than 100 Hz, examining a range of values up to 1,600 Hz (Buss et al. 2013). As in most MECP studies, there were two noise bands of equal spectral width, a "low" band and a "high" band. The upper edge of the low band was fixed at 2,000 Hz, and the lower edge of the high band was fixed at 2,500 Hz. In one set of conditions, the comodulated bands were generated by assigning identical amplitude/phase values to corresponding frequency bins for the low and high bands. Therefore, the low and high bands had identical temporal envelopes and spectral profiles for all bandwidths. These are referred to as "identical-spectrum" conditions. In a separate set of conditions, bands were generated that were comodulated but had *frequency-reversed* spectral profiles. These are referred to as "reflected-spectrum" conditions. The reflected-spectrum conditions were generated using a method developed by Richards (1988b) where the low and high bands have magnitude spectra that are mirror reflections of each other and phase spectra that are reversed and multiplied by negative one.

We found that MECP performance was best for noise bandwidths between 100 and 400 Hz, but percent correct remained above chance even at the 1,600-Hz bandwidth. A finding of interest was that there was no significant difference between the results of the identical- and reflected-spectrum stimuli. Two different ideas were considered to account for sensitivity to comodulation when the noise bandwidths were relatively wide. One was that the auditory system somehow combines the outputs of the peripheral auditory filters to derive wider filter bands matched in width to the noise bandwidths. The temporal envelopes at the outputs of the derived filters could then be extracted and compared. This would be consistent with the finding of comparable results for the identical- and reflected-spectrum stimuli. The other idea was that MECP could involve a matrix of comparisons between the outputs of multiple peripheral auditory filters, with good performance depending on the combination of information from the best comparisons. For the identical-spectrum noise bands, the best comparisons would presumably involve corresponding regions of the low and high bands (i.e., the lower regions of the low band with the lower regions of the high band, and the higher regions of the low band with the higher regions of the high band). For the reflected-spectrum noise bands, the best comparisons would presumably involve spectrally rotated regions of the low and high bands (i.e., the lower regions of the low band with the higher regions of the high band, and the higher regions of the low band with the lower regions of the high band). Note that a factor that would limit performance is the difference in auditory filter widths between "matched" comparison regions, due to the increase in auditory filter width with increasing stimulus frequency (e.g., Fletcher 1940; Glasberg and Moore 1990).

As noted above, MECP performance remained above chance but showed a downturn at the widest bandwidths. It is possible that this finding is related to an effect noted by Richards (1987) where MECP performance decreased at lower stimulus frequencies.

In our band-widening paradigm, the low band contained progressively lower frequencies as the noise bandwidth was increased, extending as low as 400 Hz in the widest bandwidth tested. The present study used a higher spectral region to test the possibility that no downturn in performance would occur, even at the widest bandwidths, as long as the noise bands did not include relatively low-frequency components.

2 Methods

2.1 *Listeners and Stimuli*

Six listeners with normal hearing and previous listening experience in MECP tasks were recruited. In this study, the high-frequency cutoff of the low band was fixed at 3,000 Hz, and the low-frequency cutoff of the high band was fixed at 3,500 Hz. Both identical-spectrum and reflected-spectrum noises were tested for bandwidths from 25 to 1,600 Hz. The low-frequency bands were generated in the frequency domain by assigning random values to the magnitude and phase components within the noise passband. In the comodulated identical-spectrum conditions, the high band was constructed from the same frequency-ordered draws as the low band. In the comodulated reflected-spectrum conditions, the high band was constructed from the same draws as the low band, but the Fourier components were assigned to sequential frequency bins in reverse order, and component phases were multiplied by −1. Independent random draws were used for the high-frequency noise bands in the random envelope conditions. Bands were presented for 400 ms, including 30-ms raised cosine ramps. Each band was presented at a level of 65 dB SPL. For some bandwidths of the identical-spectrum noise, some listeners reported hearing a tonal pitch corresponding to the spectral separation between matched stimulus components in the two bands (e.g., a pitch associated with 900 Hz for the 400-Hz bandwidth condition (2,600–3,000 Hz and 3,500–3,900 Hz)). A continuous, 500-Hz-wide band of masking noise centered on this difference frequency was introduced to mask this unintended cue, except for the 1,600-Hz bandwidth condition where the masking band would have overlapped with part of the low band. The masking noise had a level of 47 dB SPL.

2.2 *Procedure*

Performance was quantified as percent correct on fixed-blocks of 25 trials. At least five blocks were completed, with additional blocks obtained if there was evidence of learning. A 3AFC procedure was used, with intervals separated by 300 ms. In two intervals, the noise bands had random envelopes, and in one, chosen at random, the bands were comodulated. Listeners responded by pressing one of three buttons on a response box. Feedback was provided after each response. Conditions were completed in random order. At the beginning of each block, listeners were given the opportunity to listen to examples of random versus comodulated bands. Here, comodulated bands were always presented in a known interval (the second of three listening intervals). The listener was told to listen to as many reminder trials as desired by continuing to press buttons 1 or 3 in response to each trial and to press button 2 when ready to start an experimental block.

3 Results and Discussion

Mean results for identical- and reflected-spectrum conditions are shown in Fig. 42.1. All listeners achieved performance levels above chance in all conditions. As can be seen from the standard deviations shown in Fig. 42.1, individual differences were relatively small for the middle bandwidths. At the 200- and 400-Hz bandwidths, the total range in performance among listeners was 76–98 % correct. However, the interindividual variation was quite large at the bandwidth extremes, ranging from 41 to 82 % correct at the 25-Hz bandwidth and from 45 to 90 % correct at the 1,600-Hz bandwidth. Listeners reported that the signal interval was associated with a roughness percept for bandwidths of 100 Hz or higher. As in our previous study using lower-frequency noise bands, best performance occurred at intermediate bandwidths. Also as in our previous study, the functions for the identical- and reflected-spectrum stimuli were grossly similar. A repeated-measures ANOVA was performed with two levels of spectral profile (identical- and reflected-spectrum) and seven levels of bandwidth. The analysis was performed on rationalized arcsine (RAU) transformed percent correct scores in order to stabilize the error variance associated with proportional data (Studebaker 1985). This analysis showed a main effect of bandwidth ($F_{6,30}=28.9$, $p<0.001$), but no effect of spectral profile ($F_{1,5}=0.046$, $p=0.84$). The two-way interaction was not significant ($F_{6,30}=2.21$, $p=0.07$). The main effect of bandwidth was further evaluated with within-subjects contrasts. The quadratic contrast with bandwidth was significant ($F_{1,5}=158.0$, $p<0.001$), consistent with the trend for performance to decrease at

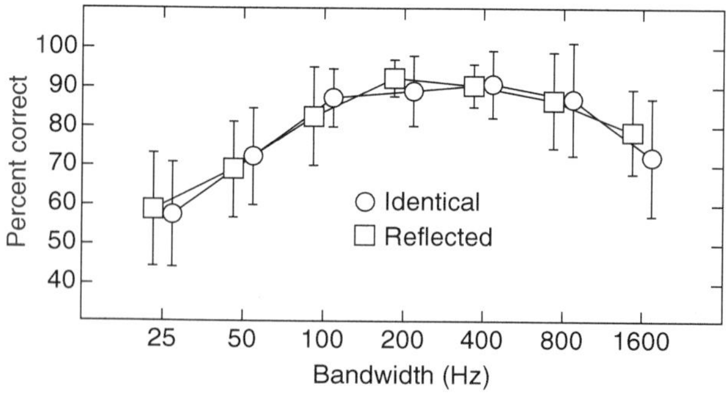

Fig. 42.1 *Circles* and *squares* show average percent correct for identical- and reflected-spectra, respectively. *Vertical lines* show plus and minus 1 standard deviation. Abscissa values are offset for clarity

bandwidth extremes (see Fig. 42.1). As a coarse test of where the performance decrease began with the increase in bandwidth, *t*-tests were done comparing data for the 200-Hz bandwidth to data for the two widest bandwidths. Comparing the 200- and 800-Hz bandwidths showed no significant decrease in performance for either the identical-spectrum noise ($t_5=0.50$; $p=0.64$) or the reflected-spectrum noise ($t_5=1.52$; $p=0.18$). However, comparing the 200- and 1,600-Hz bandwidths showed significant declines in performance for both the identical-spectrum noise ($t_5=3.83$; $p=0.012$) and the reflected-spectrum noise ($t_5=4.84$; $p=0.004$). The present results suggest that the reduction in MECP performance at the 1,600-Hz bandwidth may be a robust feature of the effect rather than one related to the inclusion of lower frequencies where MECP cues are relatively weak.

A result that could be seen as being inconsistent with an interpretation in terms of envelope comparisons between the outputs of "matched" peripheral auditory filters is the reduction in performance that occurred at the widest bandwidth for the reflected-spectrum condition. In the identical-spectrum condition, increasing the bandwidth from 800 to 1,600 Hz resulted in larger frequency separations between all corresponding frequency regions of the low and high bands. This would result in more poorly matched auditory filters for all corresponding frequencies in the 1,600-Hz case, which might account for the drop in performance between the 800- and 1,600-Hz bandwidths. In the reflected-spectrum case, band widening from 800 to 1,600 Hz introduces corresponding stimulus components into increasingly mismatched auditory filters, while maintaining stimulation of the more closely matched filters associated with the 800-Hz bandwidth condition. The added frequencies in the 1,600-Hz bandwidth condition would not necessarily hurt the performance if the listener were able to weight information optimally. One possibility is that listeners do not weight information optimally. If this interpretation is valid, better performance in the 1,600-Hz reflected-spectrum condition might occur if cues were present to encourage weighting of the better-matched filters. We attempted to examine this idea in two supplementary conditions.

3.1 Supplementary Conditions

The two supplementary conditions involved reflected-spectrum, 1,600-Hz bandwidth stimuli and used temporal gating manipulations intended to perceptually isolate particular 800-Hz portions of the lower and higher 1,600-Hz-wide bands. In one of the conditions, the aim was to perceptually isolate the *better-matched* filters associated with the low and high bands. Here, the lower half of the low band and the higher half of the high band were presented *continuously*, and the higher half of the low band and the lower half of the high band were presented only during the three listening intervals. This gating pattern was intended to promote preferential weighting of the better-matched filters. The complementary condition was also run, where the higher half of the low band and the lower half of the high band were presented continuously, and the lower half of the low band and the higher half of the high band were presented only during the listening intervals. This gating pattern was intended to promote preferential weighting of the more poorly matched filters.

An important feature of the continuous portions of the 1,600-Hz-wide bands is that they were normally random but were comodulated in the signal interval. This was accomplished by gating off the random bands while simultaneously gating on the comodulated bands. Because the total level of the bands was unchanged over time, this transition was "seamless." The random bands were also gated off in the *non-signal* listening intervals, with separate random bands simultaneously gated on, in order to insure that the difference across intervals was related to comodulation rather than to an uncontrolled feature related to gating. As in the main experiment, the listening intervals were 400 ms in duration and were separated by 300 ms. All six listeners from the main experiment completed data collection for these conditions. Assuming that the gating manipulations resulted in the intended effects, we had the following expectations for the supplementary conditions when compared to the 800-Hz and 1,600-Hz reflected-spectrum conditions of the main experiment: (1) performance for the 1,600-Hz bandwidth supplementary condition with gating intended to isolate the better-matched filters should be similar to that for the 800-Hz bandwidth condition from the main experiment; (2) performance for the 1,600-Hz bandwidth supplementary condition with gating intended to isolate the better-matched filters should be better than that for the 1,600-Hz bandwidth condition from the main experiment; and (3) performance for the 1,600-Hz bandwidth condition from the main experiment should be better than that for the 1,600-Hz bandwidth supplementary condition with gating intended to isolate the more poorly matched filters.

Figure 42.2 shows the results of these four conditions, plotted in order of expected performance. As can be seen, some aspects of the findings were consistent with expectations. A repeated-measures ANOVA was performed to compare the four reflected-spectrum conditions shown in Fig. 42.2. In the analysis, the conditions were ordered by expected performance. This analysis showed a

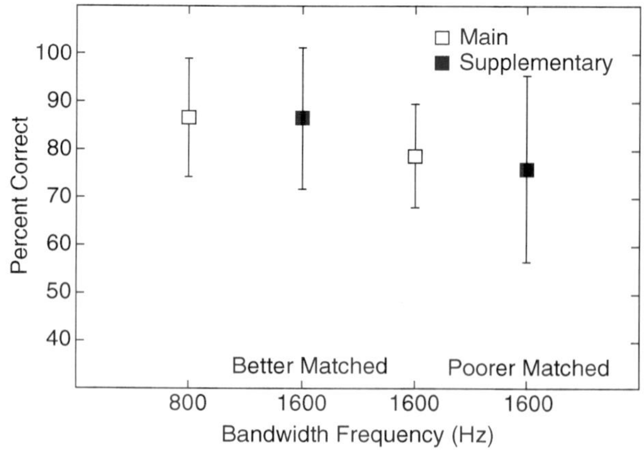

Fig. 42.2 Percent correct for the 800- and 1,600-Hz reflected-spectrum stimuli of the main experiment and for the two supplementary 1,600-Hz reflected-spectrum stimuli. Gating in the supplementary conditions was intended to isolate either the better or more poorly matched auditory filters. *Vertical lines* show plus and minus 1 standard deviation

significant effect of condition ($F_{1,5}=7.47$; $p=0.003$). Planned contrasts between adjacent conditions indicated:

(a) The 800-Hz bandwidth condition did not differ significantly from the 1,600-Hz bandwidth condition with gating intended to isolate the better-matched filters ($F_{1,5}=0.007$; $p=0.938$).
(b) The 1,600-Hz bandwidth condition with gating intended to isolate the better-matched filters was associated with better performance than the 1,600-Hz condition of the main experiment ($F_{1,5}=9.355$; $p=0.028$).
(c) The 1,600-Hz condition from the main experiment did not differ significantly from the condition with gating intended to isolate the more poorly matched filters ($F_{1,5}=0.557$; $p=0.489$).

These results could be taken as allaying concerns about interpretation in terms of comparisons between "matched" peripheral filters, suggesting that the performance decrease in the main experiment when the reflected-spectrum bandwidth increased from 800 to 1,600 Hz may be due to nonoptimal weighting, where better-matched and more poorly matched auditory filters are given similar weighting. When gating cues were provided that may discourage inclusion of more poorly matched filters, a performance benefit was found. This outcome is also compatible with interpretation in terms of a relatively wideband central filter. Here, the width of this arbitrarily wide filter could be driven by the bandwidths of the gated stimulus components (800 Hz). With an assumption of reduced sensitivity to modulation at relatively high modulation frequencies (Bernstein and Trahiotis 2002; Dau et al. 1999; Kohlrausch and Fassel 2000; Viemeister 1979), performance would be expected to be better for an 800-Hz-wide band than a 1,600-Hz-wide band.

It is not clear how to interpret the finding that the 1,600-Hz condition from the main experiment did not differ significantly from the condition with gating intended to isolate the more poorly matched filters. From the standpoint of comparisons involving peripheral filters, the expectation was that performance should have been worse for the supplementary condition where noise components associated with the poorly matched filters were present only during the listening intervals.

It is possible that further insight into the mechanisms likely to underlie MECP performance may be gleaned from examination of individual differences. In contrast to the results of most of our listeners, one listener (S2) showed very reliable, comparable performance for both supplemental conditions (84.8 % for the condition where the intent was to isolate the better-matched filters and 86.4 % for the condition where the intent was to isolate the more poorly matched filters). The results of this listener are more consistent with a wideband analysis interpretation than a peripheral filter interpretation. It is possible that multiple mechanisms may underlie performance and that there are individual differences in the balance of the mechanisms. Ongoing research is investigating the validity of the gating cue for achieving perceptual isolation of noise subregions and the nature of the stimulus cues that give rise to good performance.

Acknowledgments This research was supported by NIH NIDCD grant R01-000418.

References

Bernstein LR, Trahiotis C (2002) Enhancing sensitivity to interaural delays at high frequencies by using "transposed stimuli". J Acoust Soc Am 112:1026–1036

Buss E, Hall JW, Grose JH (2013) Monaural envelope correlation perception for bands narrower or wider than a critical band. J Acoust Soc Am 133(1):405–416

Dau T, Verhey J, Kohlrausch A (1999) Intrinsic envelope fluctuations and modulation-detection thresholds for narrow-band noise carriers. J Acoust Soc Am 106:2752–2760

Fletcher H (1940) Auditory patterns. Rev Mod Phys 12:47–65

Glasberg BR, Moore BCJ (1990) Derivation of auditory filter shapes from notched-noise data. Hear Res 47:103–138

Hall JW, Grose JH (1993) Monaural envelope correlation perception in listeners with normal hearing and cochlear impairment. J Speech Hear Res 36:1306–1314

Kohlrausch A, Fassel R (2000) The influence of carrier level and frequency on modulation and beat-detection thresholds for sinusoidal carriers. J Acoust Soc Am 108:723–734

Moore BCJ, Emmerich DS (1990) Monaural envelope correlation perception, revisited: effects of bandwidth, frequency separation, duration, and relative level of the noise bands. J Acoust Soc Am 87:2628–2633

Patterson RD (1976) Auditory filter shapes derived with noise stimuli. J Acoust Soc Am 59: 640–654

Richards VM (1987) Monaural envelope correlation perception. J Acoust Soc Am 82:1621–1630

Richards VM (1988a) Aspects of monaural synchrony detection. In: Duifhuis H, Horst J, Wit H (eds) Basic issues in hearing: proceedings of the 8th international symposium on hearing. Academic, New York, pp 317–322

Richards VM (1988b) Components of monaural envelope correlation perception. Hear Res 35:47–58

Richards VM (1989) Comparing monotic, diotic, and dichotic presentation modes in synchrony detection. J Acoust Soc Am 1(Suppl):s143

Studebaker GA (1985) A "rationalized" arcsine transform. J Speech Hear Res 28:455–462

Viemeister NF (1979) Temporal modulation transfer functions based upon modulation thresholds. J Acoust Soc Am 66:1364–1380

Chapter 43
Detection Thresholds for Amplitude Modulations of Tones in Budgerigar, Rabbit, and Human

Laurel H. Carney, Angela D. Ketterer, Kristina S. Abrams, Douglas M. Schwarz, and Fabio Idrobo

Abstract Envelope fluctuations of complex sounds carry information that is essential for many types of discrimination and for detection in noise. To study the neural representation of envelope information and mechanisms for processing of this temporal aspect of sounds, it is useful to identify an animal model that can sensitively detect amplitude modulations (AM). Low modulation frequencies, which dominate speech sounds, are of particular interest. Yet, most animal models studied previously are relatively insensitive to AM at low modulation frequencies. Rabbits have high thresholds for low-frequency modulations, especially for tone carriers. Rhesus macaques are less sensitive than humans to low-frequency modulations of wideband noise (O'Conner et al. Hear Res 277, 37–43, 2011). Rats and chinchilla also have higher thresholds than humans for amplitude modulations of noise (Kelly et al. J Comp Psychol 120, 98–105, 2006; Henderson et al. J Acoust Soc Am 75, 1177–1183, 1984). In contrast, the budgerigar has thresholds for AM detection of wideband noise similar to those of human listeners at low modulation frequencies (Dooling and Searcy. Percept Psychophys 46, 65–71, 1981). A one-interval, two-alternative operant conditioning procedure was used to estimate AM detection thresholds for 4-kHz tone carriers at low modulation frequencies (4–256 Hz). Budgerigar thresholds are comparable to those of human subjects in a comparable task. Implications of these comparative results for temporal coding of complex sounds are discussed. Comparative results for masked AM detection are also presented.

L.H. Carney, PhD (✉) • A.D. Ketterer • K.S. Abrams • D.M. Schwarz
Departments of Biomedical Engineering and Neurobiology & Anatomy, University of Rochester, 601 Elmwood Ave, Box 603, Rochester, NY, USA
e-mail: laurel.carney@Rochester.edu

F. Idrobo
Department of Psychology, Boston University, Boston, MA, USA

B.C.J. Moore et al. (eds.), *Basic Aspects of Hearing*, Advances in Experimental Medicine and Biology, DOI 10.1007/978-1-4614-1590-9_43,

1 Introduction

The importance of amplitude modulations (AM) for carrying information in complex sounds has motivated numerous psychophysical, behavioral, and physiological studies of AM detection and discrimination. Humans and birds are sensitive to sinusoidal amplitude modulation (SAM) depths as low as −25 dB (in terms of 20 log m, where m is the modulation index), whereas several other species studied are less sensitive (Fig. 43.1). In particular, a study of AM detection in rabbits has shown that they are insensitive to low-frequency modulations of tone carriers (Carney et al. 2009). Gourevitch and Eggermont (2010) found, in a physiological study of auditory cortex in cat, that low-frequency AM is most effectively coded by the timing of discharges, whereas high-frequency AM is effectively coded in terms of average discharge rates. Thus, differences across species in the ability to detect low-frequency AM may indicate differences in the ability to make use of temporal information in neural responses. A goal of this study was to identify an animal model that is able to detect low-frequency modulations of narrowband sounds.

In this study, AM detection thresholds for narrowband stimuli were estimated for the budgerigar, a vocal learner that has been used in a number of previous behavioral studies (e.g., Dooling and Searcy 1981; Dooling et al. 1989; Dent et al. 2002). Amplitude modulation detection thresholds for wideband noise stimuli (Dooling and Searcy 1981) suggested that the budgerigar's sensitivity for SAM noise is comparable to that of human. Therefore, it was hypothesized that the budgerigar would be sensitive to low-frequency amplitude-modulated tones.

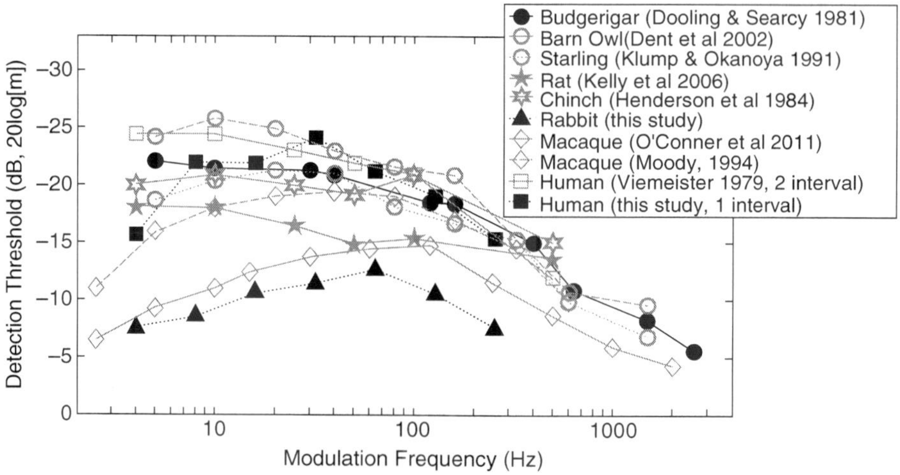

Fig. 43.1 Comparison of AM detection thresholds for sinusoidally amplitude-modulated (*SAM*) wideband noise across several species. More sensitive thresholds appear higher on the plot. Thresholds in three bird species (*circles*) are generally comparable to human thresholds (*squares*). Rabbits and macaques are less sensitive to AM for wideband SAM stimuli similar to those used in the human studies. Thresholds in a one-interval task for human, budgerigar, and rabbit are highlighted by the *solid symbols*

In addition to estimating AM detection thresholds, performance of the budgerigar in a masked modulation detection task was studied. Responses were compared to results using the same stimuli in human and rabbit. This experiment was motivated by previous studies of masked modulation in human listeners (e.g., Strickland and Viemeister 1996; Ewert and Dau 2000; Ewert et al. 2002; Nelson and Carney 2006). However, the use of reproducible narrowband noises as modulation maskers allows detailed comparisons of hit and false-alarm rates across masker waveforms as a means of identifying cues used in this detection task.

2 Methods

AM detection thresholds of budgerigar were estimated for SAM tones. Four English budgerigars were tested using a 4-kHz tone carrier modulated at frequencies ranging from 4 to 256 Hz. Stimuli were presented at 50 dB SPL.

Operant methods were used with a single-interval two-down, one-up (2D1U) adaptive tracking procedure (Levitt 1971) to estimate AM detection thresholds. Correct responses were reinforced by delivery of a single hulled millet seed. Incorrect responses were followed by a 5-s timeout with the house light extinguished. Bias was monitored throughout each session and was controlled by the delivery of two seeds for the responses on the side that was biased against for a percentage of trials that depended on the degree of bias. Sessions of approximately 250 trials were typically 10–20 min in duration and were conducted twice a day. Threshold estimates were based on the average of an even number of reversals over the last half of each track that had bias less than 0.3 and a standard deviation of modulation depth less than 3 dB.

The operant testing was done in the behavioral setup shown in Fig. 43.2. A row of three switches was mounted on the end of the enclosure. The speaker was mounted overhead. The bird started each trial by making an observing response on the center switch which initiated an acoustic stimulus. A correct reporting response for the standard stimulus was a peck on the left switch, and for the target stimulus, on the right switch. Each block of ten trials consisted of a random sequence of five modulated and five unmodulated trials, to avoid long runs of either trial type that could result in short-term bias.

3 Results

3.1 AM Detection Thresholds

At low modulation frequencies, the average AM detection thresholds for budgerigar were comparable to those of human listeners in a matched task (Fig. 43.3). At these frequencies, rabbits had the most difficulty in detecting modulation.

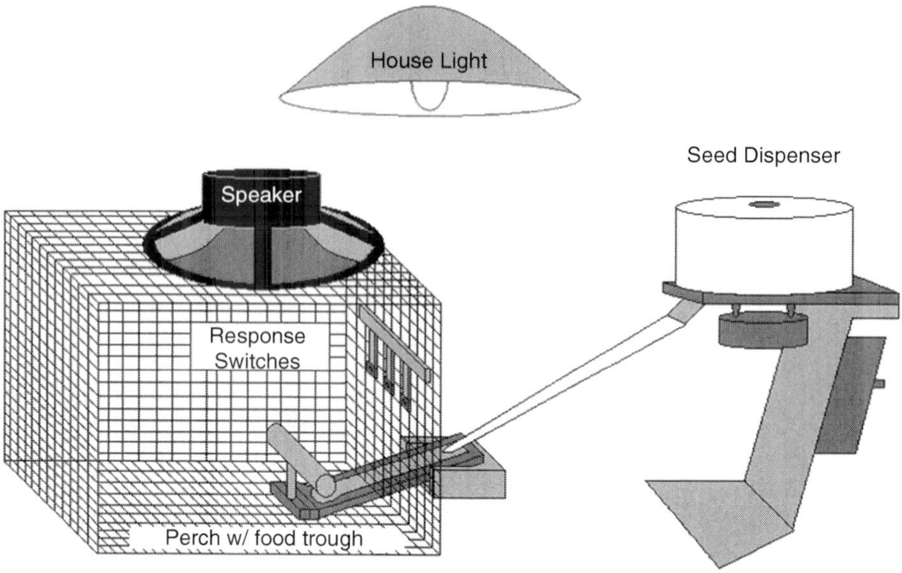

Fig. 43.2 Schematic diagram of behavioral test setup. The dimensions of the enclosure are 23 cm on each side. This apparatus was housed in a small sound-proof booth, the inner walls of which were lined with sound-absorbing foam

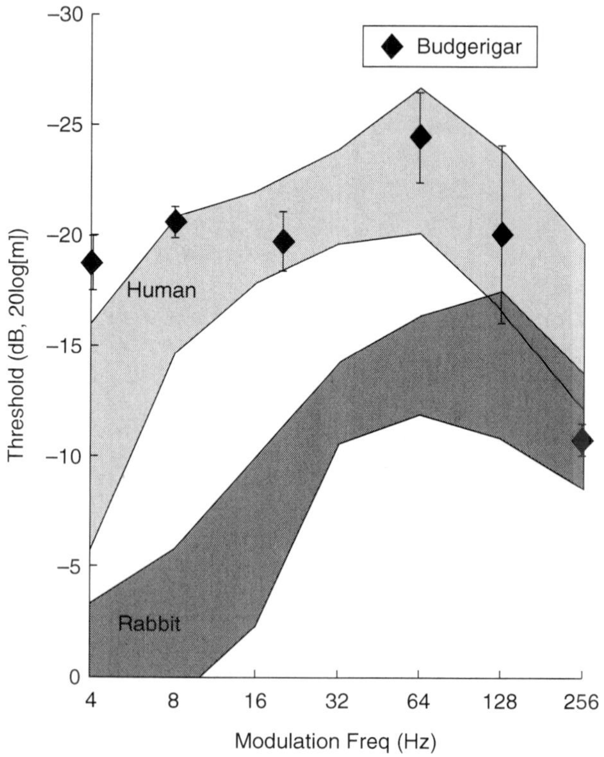

Fig. 43.3 AM detection thresholds of budgerigar ($n=4$, for 4–64 Hz; $n=2$ for 128 and 256 Hz). Budgerigar (*diamonds*) thresholds for SAM tones with 4 kHz carriers are superimposed on threshold ranges for humans ($n=3$) and Dutch-belted rabbits ($n=5$) using matched stimuli and methods. Thresholds are plotted as 20 log(m); more sensitive thresholds appear at the top of the plot. Means across birds +/− standard deviation are plotted. All stimuli were presented at 50 dB SPL

3.2 Masked Modulation Detection

Masked AM detection was studied using reproducible noises as modulation maskers. Stimuli were matched to those in a previous study of human listeners (Nelson and Carney 2006). The target modulation was 64-Hz SAM. Stimuli were generated by adding a 64-Hz sinusoid to a noise masker that was then used to modulate a 4-kHz tone carrier. The noise maskers consisted of a set of 20 reproducible ("frozen") noises, created using a 32-Hz wide Gaussian noise centered at 64 Hz. The root mean square, RMS, level of the modulation masker was 13 dB below the RMS level of a tone modulator giving 100 % modulation. This level was selected to elevate modulation detection thresholds approximately 6 dB with respect to unmasked thresholds. Stimulus level was 65 dB SPL. Masker waveforms were randomly selected for each trial from the ensemble of 20 maskers, while the AM detection threshold was estimated using a 2DIU paradigm. The combined reproducible noise and 2D1U tracking test procedures are illustrated in Fig. 43.4.

Estimated detection patterns (Fig. 43.5) consist of hit rate and false-alarm rates for each of the noise waveforms in the ensemble. Average detection patterns are shown in Fig. 43.5 for budgerigars, in comparison to rabbit and human results tested with matched stimuli and similar procedures. Masked modulation detection thresholds for the 64-Hz SAM were 0–1 dB SNR for birds, humans, and rabbits.

Patterns were consistent within each subject, as evidenced by 1st-half, 2nd-half correlations (Table 43.1) and χ^2 analyses (not shown) for both hits and false alarms.

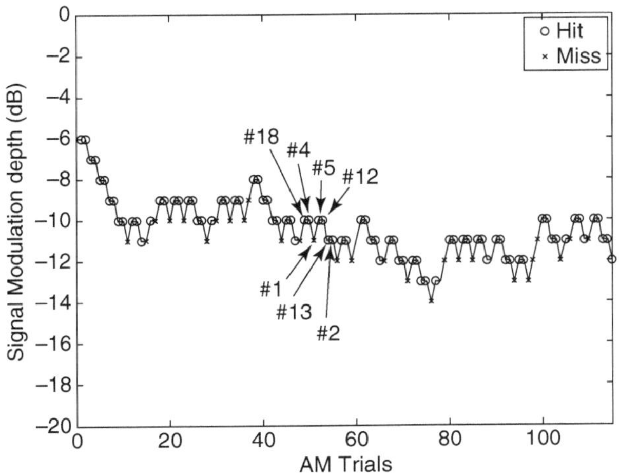

Fig. 43.4 Illustration of a 2D1U track with reproducible noise masker. Noise IDs are indicated for a few of the trials. Trials near threshold were sorted by noise ID to compute hit and false-alarm rates for each masker noise waveform in the ensemble, as follows: The distribution of modulation depths (omitting the initial 1/3 of the track) was computed for each animal. Hit rates (correct detections) and false-alarm rates were computed for each reproducible masker based on trials that had modulation depths within one standard deviation of the mean depth over the latter 2/3 of the track. False-alarm trials were assigned to the modulation depth of the nearest preceding modulated trial

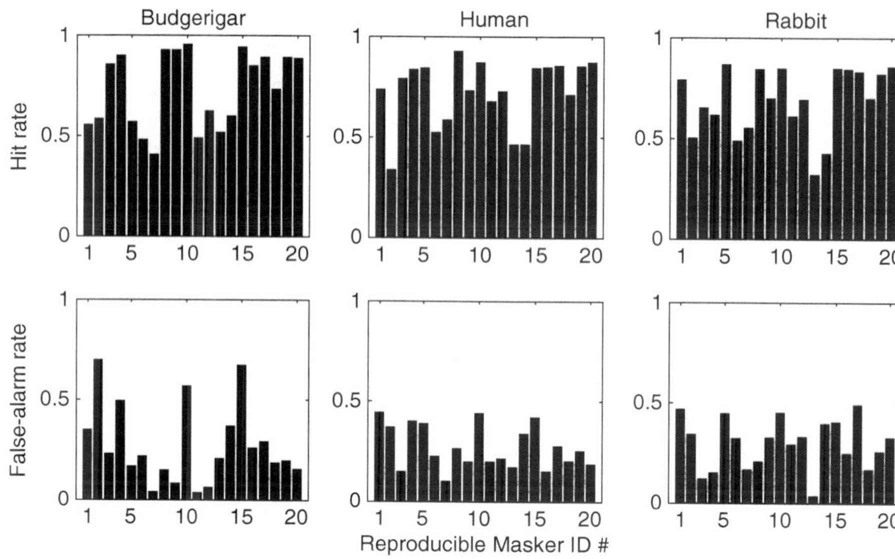

Fig. 43.5 Average reproducible noise results for three species are shown as "detection patterns," which consist of hit and false-alarm rates for each of the noise masker waveforms used

Table 43.1 1st-half, 2nd-half correlations show the consistency in individual birds of detection patterns computed from reproducible noise results (*left*). Across-subject correlations show consistency of the detection patterns across birds (*middle*). Cross-species correlations show the similarity of detection patterns across species (*right*)

1st-half, 2nd-half correlations			Across-subject correlations			Cross-species correlations		
	H	FA		H	FA		H	FA
B1	0.84	0.83	B1–B2	0.79	0.74	Bird-human	0.73	0.75
B2	0.95	0.92	B1–B3	0.67	0.51	Bird-rabbit	0.63	0.32
B3	0.97	0.95	B1–B4	0.84	0.83	Human-rabbit	0.89	0.62
B4	0.92	0.92	B2–B3	0.86	0.86			
			B2–B4	0.93	0.82			
			B3–B4	0.87	0.82			

Detection patterns were strongly correlated across birds, as is true for humans. The average patterns for birds, rabbits, and humans were used to make comparisons across species (Table 43.1).

The stimuli used for the above experiment were equalized for overall energy. Envelope energy was equalized across maskers, but energy varied across the masker + target stimuli due to interactions between the target 64-Hz modulation and the 32-Hz bandwidth masker noise, which was centered at 64 Hz. The budgerigar detection patterns for hits were significantly correlated to envelope energy ($r = 0.60$; 36 % of the variance in the patterns was explained by energy). Envelope energy did not vary across unmodulated waveforms; thus, variations in the false-alarm rates across waveforms cannot be explained by envelope energy.

4 Discussion

Amplitude modulation detection thresholds for SAM tones in the budgerigar were comparable to those of human listeners tested with a similar one-interval paradigm (Fig. 43.3). This result was especially interesting for low modulation frequencies, where some mammalian species, such as rabbits and macaques, have relatively high detection thresholds. The shape of the modulation transfer function (MTFs) for tone carriers in budgerigar is similar to that of human. The overall shape of the MTFs in bird and human are notably different from that of rabbit, especially for low modulation frequencies. The thresholds and MTFs in budgerigar suggest that this species is a good model for human AM processing, for both wideband and narrowband carriers.

In a masked modulation detection task using reproducible envelope maskers, detection performance varied significantly and consistently across masker waveforms in all three species tested (Fig. 43.5). Masked modulation thresholds for all species were tested at 64 Hz because this was a favorable modulation frequency for all three species (Fig. 43.1). Performance differences across masker waveforms were consistent within individual subjects. Variations in hit and false-alarm rates from waveform to waveform were significantly correlated across individuals within a species and across species. Masked modulation thresholds across species were within approximately 1 dB. These results suggest that similar strategies are used across these species for masked modulation detection of a 64-Hz sinusoidal amplitude modulation. Envelope energy differences in the masker-plus-target stimuli were correlated to the hit rates; however, consistent differences across maskers were also observed for false-alarm trials, which had identical envelope energy. Ongoing studies are investigating cues that can explain both hit and false-alarm rates in the masked modulation task.

Acknowledgments Kelly-Jo Koch, Paula Aronson, Asia Ingram, Tiara Jackson, Erin Keegan, Hannah Rasmussen, Erin Schnellinger, and Whitney Williams assisted with data collection and analysis. The Dent Lab at SUNY-Buffalo provided us with invaluable advice and information (Supported by NIDCD-R01-001641).

References

Carney LH, Abrams KS, Koch K-J, Zilany MSA, Idrobo F (2009). Behavioral and physiological studies of amplitude-modulation detection. Abstract, ARO, 801

Dent ML, Klump GM, Schwenzfeier C (2002) Temporal modulation transfer functions in the barn owl (*Tyto alba*). J Comp Physiol A 187:937–943

Dooling RJ, Searcy MH (1981) Amplitude modulation thresholds for the parakeet (*Melopsittacus undulatus*). J Comp Physiol 143:383–388

Dooling RJ, Okanoya K, Brown SD (1989) Speech perception by budgerigars (*Melopsittacus undulatus*): the voiced-voiceless distinction. Percept Psychophys 46:65–71
Ewert SD, Dau T (2000) Characterizing frequency selectivity for envelope fluctuations. J Acoust Soc Am 108:1181–1196
Ewert SD, Verhey JL, Dau T (2002) Spectro-temporal processing in the envelope-frequency domain. J Acoust Soc Am 112:2921–2931
Gourevitch B, Eggermont JJ (2010) Maximum decoding abilities of temporal patterns and synchronized firings: application to auditory neurons responding to click trains and amplitude modulated white noise. J Comp Neurosci 29:253–277
Henderson D, Salvi R, Pavek G, Hamernik R (1984) Amplitude modulation thresholds in chinchillas with high-frequency hearing loss. J Acoust Soc Am 75:1177–1183
Kelly JB, Cooke JE, Gilbride PC, Mitchell C, Zhang H (2006) Behavioral limits of auditory temporal resolution in the rat: amplitude modulation and duration discrimination. J Comp Psychol 120:98–105
Klump GM, Okanoya K (1991) Temporal modulation transfer functions in the European starling (*Sturnus vulgaris*). I. Psychophysical modulation detection thresholds. J Comp Physiol A 164:531–538
Levitt H (1971) Transformed up-down methods in psychophysics. J Acoust Soc Am 49:467–477
Moody DB (1994) Detection and discrimination of amplitude-modulated signal by macaque monkeys. J Acoust Soc Am 95:3499–3510
Nelson PC, Carney LH (2006) Cues for masked amplitude-modulation detection. J Acoust Soc Am 120:978–990
O'Conner KN, Johnson JS, Niwa M, Noreiga NC, Marshall EA, Sutter ML (2011) Amplitude modulation detection as a function of modulation frequency and stimulus duration: comparisons between macaques and humans. Hear Res 277:37–43
Strickland EA, Viemeister NF (1996) Cues for discrimination of envelopes. J Acoust Soc Am 99:3638–3646
Viemeister NF (1979) Temporal modulation transfer functions based upon modulation thresholds. J Acoust Soc Am 66:1364–1380

Chapter 44
Phase Discrimination Ability in Mongolian Gerbils Provides Evidence for Possible Processing Mechanism of Mistuning Detection

Astrid Klinge-Strahl, Timo Parnitzke, Rainer Beutelmann, and Georg M. Klump

Abstract Compared to humans, Mongolian gerbils (*Meriones unguiculatus*) are much more sensitive at detecting mistuning of frequency components of a harmonic complex (Klinge and Klump. J Acoust Soc Am 128:280–290, 2010). One processing mechanism suggested to result in the high sensitivity involves evaluating the phase shift that gradually develops between the mistuned and the remaining components in the same or separate auditory filters. To investigate if this processing mechanism may explain the observed sensitivity, we determined the gerbils' thresholds to detect a constant phase shift in a component of a harmonic complex that is introduced without a frequency shift.

The gerbils' detection thresholds for constant phase shifts were considerably lower for a high-frequency component (6,400 Hz) than for a low-frequency component (400 Hz) of a 200-Hz harmonic complex and increased with decreasing stimulus duration. Compared to the phase shifts calculated from the mistuning detection thresholds, the detection thresholds for constant phase shifts were similar to those for gradual phase shifts for the low-frequency harmonic but considerably lower for the high-frequency harmonic. A simulation of the processing of harmonic complexes by the gerbil's peripheral auditory filters when components are phase shifted shows waveform changes comparable to those assessed for mistuning detection Klinge and Klump (J Acoust Soc Am 128:280–290, 2010) and provides evidence that detection of the gradual phase shifts may underlie mistuning detection.

A. Klinge-Strahl • T. Parnitzke • R. Beutelmann • G.M. Klump (✉)
Animal Physiology and Behaviour Group, Department for Neuroscience, School of Medicine and Health Sciences and Cluster of Excellence Hearing for all,
Carl von Ossietzky University Oldenburg, Oldenburg 26111, Germany
e-mail: georg.klump@uni-oldenburg.de

B.C.J. Moore et al. (eds.), *Basic Aspects of Hearing*,
Advances in Experimental Medicine and Biology,
DOI 10.1007/978-1-4614-1590-9_44, © Springer Science+Business Media New York 2013

1 Introduction

Harmonicity is an important cue used by the auditory system to group components of sounds originating from a single source into an auditory object. Components that deviate from a harmonic series, for example, through a frequency shift, may either be perceived as a separate auditory object or may induce a different percept due to changes in the temporal waveform (Moore et al. 1985, 1986). It has been proposed that the processing of harmonicity cues either relies on temporal processing mechanisms like autocorrelation functions or on changes in the spectral pattern of excitation on the basilar membrane (de Cheveigné 2005). Based on our findings on the detection of a mistuned component in a harmonic complex in Mongolian gerbils (Klinge and Klump 2009, 2010), we suggested that gerbils rely on the processing of temporal patterns rather than on processing of spectral patterns of excitation. We observed that gerbils, while performing poorly in a pure-tone frequency discrimination task, were highly sensitive in detecting a frequency-shifted (mistuned) component in a sine-phase harmonic complex. Randomizing the phase of each component substantially increased mistuning detection thresholds. Simulations of the output of gerbil auditory filters when presented with mistuned harmonic complexes suggested that gradually developing phase shifts generated temporal cues in the filter outputs that could be used to detect the mistuning.

In order to test the hypothesis that gerbils rely on exploiting temporal cues related to phase shifts within and between auditory filters to detect a mistuned component, we measured behavioral thresholds for detecting a constant phase shift in a component of a harmonic complex that is introduced without a frequency shift. Detection thresholds were determined for phase shifts in a resolved low-frequency component (400 Hz) and an unresolved high-frequency component (6,400 Hz) of a 200-Hz complex for various stimulus durations. Additionally, the output of gerbil auditory filters was simulated with the harmonic and phase-shifted complexes used in the behavioral experiment as input. This allowed elucidation of possible cues for the detection of constant phase shifts of components in a harmonic complex and evaluation of whether these may be similar to those suggested for mistuning detection.

2 Material and Methods

2.1 Animals

Six normal-hearing Mongolian gerbils served as subjects.

2.2 Apparatus and Stimulus Generation

The apparatus has been described in detail by Klinge and Klump (2009). Briefly, a doughnut-shaped experimental cage mounted in a single-walled, echo-reduced, sound-attenuating booth (IAC 401-A) was used in the experiments. A loudspeaker (Canton Plus XS) presenting the stimuli was mounted about 20 cm in front of the position of the gerbil at 0° azimuth and 0° elevation.

The reference stimulus was a sine-phase harmonic complex comprised of the first 48 harmonics of 200 Hz (60 dB SPL per component). The overall level of the complex was randomly varied by ±3 dB. The ability to detect a phase shift in a component of the complex was measured for the second (400 Hz) and 32nd (6,400 Hz) harmonic. The duration of the complex stimulus was either 400 or 100 ms, including 25-ms raised cosine ramps or 50 ms, including 12.5-ms ramps. For the stimulus duration of 50 ms, only data for the 6,400-Hz component were obtained as the gerbils could not perform the task for the 400-Hz component.

2.3 Procedure

Gerbils were trained to sit and wait on an elevated platform in a Go/NoGo operant paradigm with food rewards (red feeder LED as additional reinforcer). During the session, the reference stimulus was repeatedly presented. In the test stimuli, one of the harmonics was phase shifted (test trial, Go-stimulus, phase shifts according to method of constant stimuli) or not shifted (catch trial, NoGo-stimulus). Hits and false alarms were used to calculate the sensitivity measure d′. Sessions consisted of 11 blocks of three catch trials and seven test trials presented in random order (for details see Klinge and Klump 2009).

2.4 Data Analysis

In valid sessions, the gerbils responded correctly to at least 80 % of the two test stimuli with the largest phase shift and the false-alarm rate did not exceed 20 %. A psychometric function was constructed relating d′ to the phase shift (linearly interpolating between points). The phase shift detection threshold (PSDT) was determined as the phase shift resulting in a d′ of 1.8. Data from two consecutive valid sessions in which PSDTs differed by no more than 20 % were combined to calculate the final PSDT. To eliminate training effects, the first PSDT was measured again. If the repeated PSDT differed by more than 20 % from the previous measurement, the following PSDT measurements were repeated until the threshold difference reached the similarity criterion. Only the data from the repetition were used for further data analysis.

The harmonic component (2nd or 32nd) for which PSDTs were obtained first was randomized between animals. As detection appeared to be quite difficult for the gerbils, the PSDT for the stimulus with the longest duration (400 ms) was measured first for each gerbil, followed by the 100-ms and then by the 50-ms stimuli.

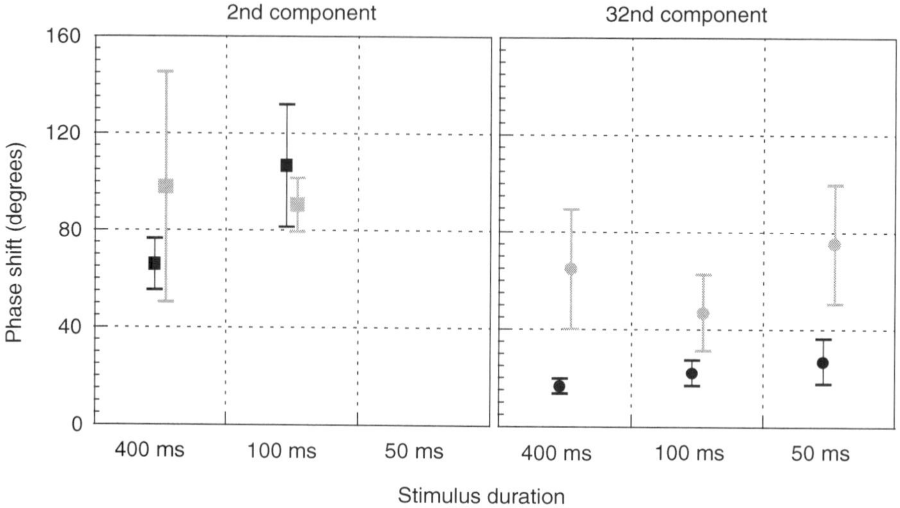

Fig. 44.1 Mean PSDTs (*black symbols*) compared to the maximum phase shift calculated from the mistuning detection thresholds (MDTs) of Klinge and Klump (2010) (*gray symbols*). The phase shift in a complex with a frequency-shifted component calculated from MDTs gradually develops in the ongoing stimulus. PSDT data could not be obtained for the 400 Hz/50 ms condition. Error bars = 2 × SEM

2.5 Control Experiment

After the main experiment, a control experiment was conducted in which we determined PSDTs for the longest stimulus duration with pink noise added as a background (spectral density 20 dB SPL at 200 Hz) to mask distortion products that could affect thresholds. The pink noise was generated by a TDT RP2 signal processor using a third-order IIR filter followed by a filter to equalize the loudspeaker frequency response and a bandpass filter between 100 Hz and 10 kHz. The RP2 output was attenuated (TDT PA5) and added to the stimulus channel before being passed to the amplifier. Only three of the six gerbils were used for this experiment.

3 Results

The results of the present experiment are shown in Fig. 44.1 (black symbols) compared to results from the mistuned harmonic experiment (gray symbols) of Klinge and Klump (2010). In the present experiment, PSDTs are generally higher for the second component (400 Hz) than for the 32nd component (6,400 Hz), and PSDTs increased with decreasing stimulus duration. A general linear mixed model (GLMM) ANOVA revealed significant main effects of harmonic component ($p<0.001$) and stimulus duration ($p<0.005$) on the PSDT and a significant two-way interaction between the two ($p<0.05$). Prior to further analysis, the dataset was divided into a group comprised of the results for the phase-shifted second component and a group comprised of the results for the phase-shifted 32nd component. A GLMM ANOVA revealed a significant main effect of duration on the threshold for both analysis groups ($p<0.05$), that is, with decreasing stimulus duration, the ability to detect a phase shift in one of the components deteriorated. However, an a posteriori pair-wise comparison within the 32nd-component group showed that the only significant difference was between the PSDTs for the 400-ms and the 50-ms stimulus durations. In the control experiment with the pink noise background, PSDTs were not different from those of the main experiment ($p=0.712$).

4 Discussion

Many experiments investigating phase effects in the human auditory system showed that humans can exploit phase cues if the frequencies of a complex stimulus are unresolved in the auditory periphery, that is, if more than one component falls within an auditory filter (e.g., Mathes and Miller 1947; Licklider 1957; Patterson 1987; Moore and Glasberg 1989). Studies in humans using two- and three-component signals with unresolved frequency components (e.g., sinusoidal amplitude modulated [SAM] tones vs. quasi-frequency modulated [QFM] tones with the carrier frequency phase shifted by 90°; Mathes and Miller 1947; Goldstein 1967; Nelson 1994) showed that phase differences result in timbre differences that may be used to distinguish between such signals based on envelope cues. Similar observations have been made for harmonic complex stimuli with a larger number of components (Licklider 1957; Plomp and Steeneken 1969; Patterson 1987; Moore and Glasberg 1989). Licklider (1957), Patterson (1987), and Moore and Glasberg (1989) demonstrated that phase shifts were better detectable by the human auditory system for higher harmonic numbers and for complexes with lower fundamental frequencies (F0s).

Here we show that the gerbil auditory system is also highly phase sensitive. Similar to results from Moore and Glasberg (1989) obtained in humans, PSDTs increased with decreasing harmonic number. However, while gerbils were still able to detect a phase shift in the second component of a 200-Hz complex, PSDTs for such a low harmonic could not be obtained in humans. Gerbils had smaller PSDTs than humans for higher harmonics of the 200-Hz complex.

Further evidence of high phase sensitivity in gerbils is provided by the observation that mistuning detection thresholds (MDTs) for gerbils were significantly higher for random-phase than for sine-phase harmonic complexes. In contrast, humans showed no significant differences in MDTs between sine-phase and random-phase harmonic complexes, thus indicating a lower importance of phase processing for mistuning detection (Klinge and Klump 2009). Similar to gerbils, several bird species showed high sensitivity for both mistuning detection and detection of phase changes (Lohr and Dooling 1998; Dooling et al. 2002; Klump and

Groß 2013, unpublished). The results suggest that gerbils (and maybe also these bird species) exploit changes in the phase relationship between components as a cue for mistuning detection whereas humans probably do not.

A possible cue that has been proposed for detecting phase differences is the presence of cubic difference tones that can produce phase-dependent changes in the internal spectral representation of sounds which, for example, could be used to distinguish between SAM and QFM tones (Nelson 1994). The results of the control experiment with a pink noise background ruled out the use of such cues by the gerbils.

The gerbil's MDTs (Klinge and Klump 2010) can be used to calculate the maximum phase shift gradually developing in ongoing mistuned stimuli and these can be compared to the PSDTs (Fig. 44.1). If we assume that gerbils exploit cues related to the phase change in the output of auditory filters to detect mistuning in a harmonic complex, then PSDTs should be similar to the phase shifts calculated from the MDTs. There were no significant differences for the second component of the harmonic complex, but for the 32nd component, PSDTs were significantly lower than the MDT phase shifts (all $p < 0.05$, student's t-test, $N_{PSDT} = 6$, $N_{MDT} = 4$ [400 ms], $N_{MDT} = 2$ [100, 50 ms]). Thus, gerbils detected a smaller phase shift when it was presented as a constant phase shift than when it gradually developed due to the mistuning. A further difference between results from the two experiments was the effect of stimulus duration on thresholds. PSDTs increased with decreasing stimulus duration, but no such trend was observed for the phase shifts calculated from the MDTs.

In order to assess possible mechanisms explaining the gerbil's ability to detect a phase shift in a component of a harmonic complex, we simulated the signal processing by gerbil auditory filters when excited by a 200-Hz complex with stimulus durations of 400, 100, and 50 ms (Fig. 44.2) without and with a phase-shifted component. Generally, the temporal waveform of the output of the various filters differs between a harmonic complex with and without a phase-shifted component. For the 32nd-component condition (right panels in Fig. 44.2), the differences in the temporal pattern evident in the fast fluctuations of the envelope (FFEs) at the filter output

Fig. 44.2 Simulation of the processing of harmonic complexes with and without phase-shifted harmonic by gerbil auditory filters (details of the construction of the filters in Klinge and Klump 2010). The *left* and *right panels* show changes in the waveform of the filter output due to phase shifting the 2nd and the 32nd component of a 200-Hz complex, respectively, for stimulus durations of 400 ms (*top*), 100 ms (*middle*), and 50 ms (*bottom*). Within each panel the left-most column shows some spectral lines (*solid lines*) of the complex, the phase-shifted harmonic (*dashed line*), and the simulated gerbil auditory filter. The second and third columns show the filter outputs over the total duration and a 10-ms segment of the stimulus if excited by a complex without a phase-shifted component. The fourth and fifth columns show the output of the same filter when either the 2nd or the 32nd harmonic in the complex is phase shifted by the respective threshold value

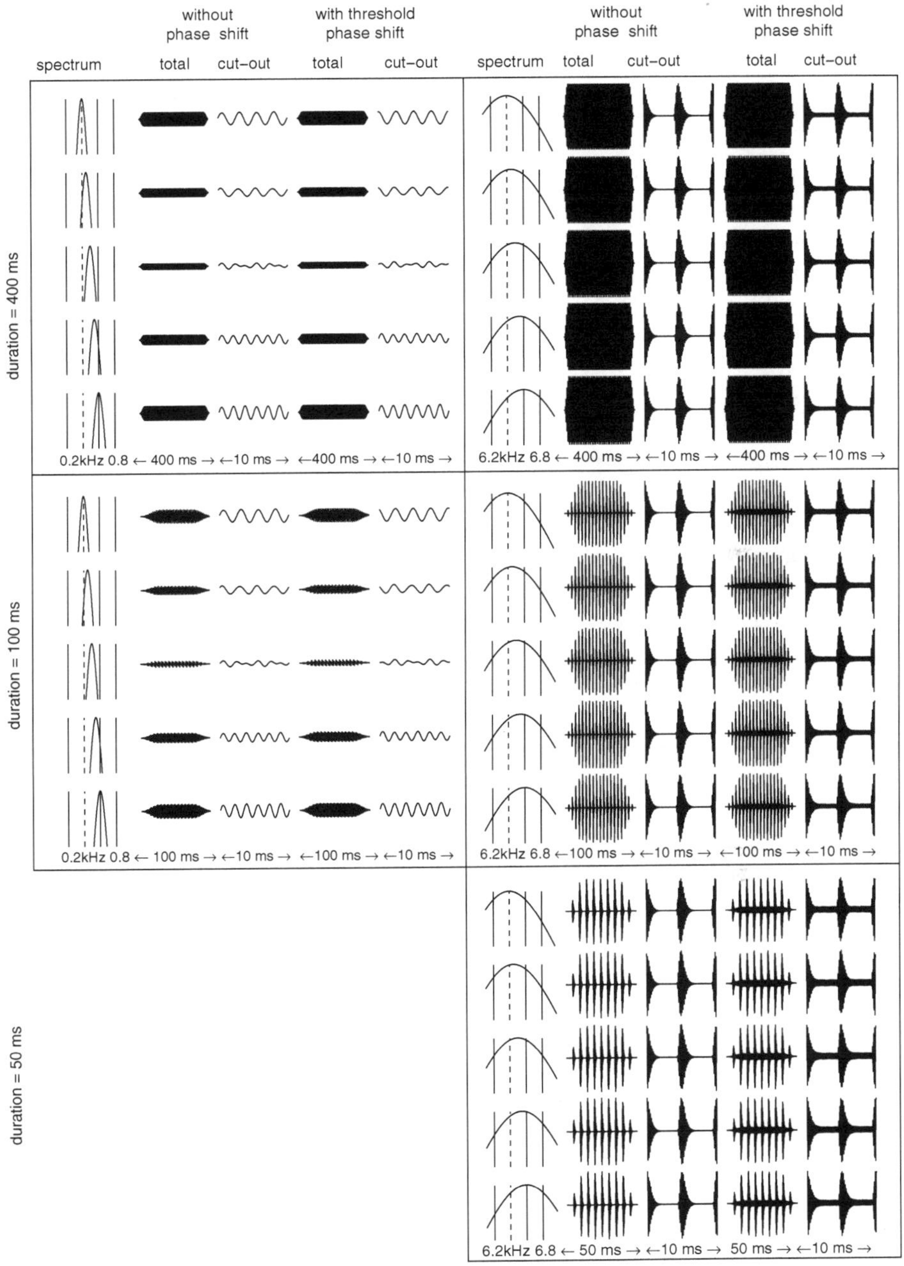
without phase shift
with threshold phase shift
spectrum
total
cut–out
duration = 400 ms
duration = 100 ms
duration = 50 ms
0.2kHz 0.8 ← 400 ms → ←10 ms → ←400 ms → ←10 ms →
6.2kHz 6.8 ← 400 ms → ←10 ms → ←400 ms → ←10 ms →
0.2kHz 0.8 ← 100 ms → ←10 ms → ←100 ms → ←10 ms →
6.2kHz 6.8 ←100 ms → ←10 ms → ←100 ms → ←10 ms →
6.2kHz 6.8 ← 50 ms → ←10 ms → 50 ms → ←10 ms →

are due to the interaction of harmonics within the filter (compare third and fifth columns). The interaction of the components starting in sine phase produces temporal waveforms having FFEs with portions of the amplitude being almost zero. Phase shifting a component results in a level increase for the low-amplitude portion. Such changes in the FFEs of the waveform may result in differences in the neural response that could be used by successive stages of the auditory system to detect the phase shift either by sequential comparisons within or simultaneous comparisons across filters. For example, temporal and rate responses of neurons from the ventral cochlear nucleus and the inferior colliculus of guinea pigs showed an asymmetry in response to temporally asymmetric FFEs (Pressnitzer et al. 2000). Moore (2002) describes the possible exploitation of the changes in the low-amplitude portion of the envelope as "listening in the dips" and proposes that this ability might be limited by the temporal resolution of the auditory system. For the second-component condition (left panels in Fig. 44.2), the changes concern phase shifts in the auditory filter containing the phase-shifted component relative to the neighboring filters. Thus, for detection the auditory system has to make simultaneous comparisons across different auditory filters. A possible neural mechanism suitable for processing such cues may rely on coincidence detectors comparing the filter outputs. Phase locking to the temporal fine structure has been shown to still be possible in this frequency range (e.g., Palmer and Russell 1986 for guinea pigs). A comparison of the output patterns of the auditory filters for the constant phase shifts of the present experiment with those for gradual phase shifts in the mistuning detection experiment (Klinge and Klump 2010) reveals similar changes in the temporal waveform of the filter output suggesting that similar processing mechanisms may account for both MDTs and PSDTs.

References

de Cheveigné A (2005) Pitch perception models. In: Plack CJ, Oxenham AJ, Fay RR, Popper AN (eds) Pitch: neural coding and perception. Springer, New York, pp 169–233

Dooling RJ, Leek MR, Gleich O, Dent ML (2002) Auditory temporal resolution in birds: discrimination of harmonic complexes. J Acoust Soc Am 112:748–758

Goldstein JL (1967) Auditory spectral filtering and monaural phase perception. J Acoust Soc Am 41:458–479

Klinge A, Klump GM (2009) Frequency difference limens of pure tones and harmonics within complex stimuli in Mongolian gerbils and humans. J Acoust Soc Am 125:304–314

Klinge A, Klump GM (2010) Mistuning detection and onset asynchrony in harmonic complexes in Mongolian gerbils. J Acoust Soc Am 128:280–290

Licklider JCR (1957) Effects of changes in the phase pattern upon the sound of a 16-harmonic tone. J Acoust Soc Am 29:780

Lohr B, Dooling RJ (1998) Detection of changes in timbre and harmonicity in complex sounds by zebra finches (*Taeniopygia guttata*) and budgerigars (*Melopsittacus undulatus*). J Comp Psychol 112:36–47

Mathes RC, Miller RL (1947) Phase effects in monaural perception. J Acoust Soc Am 19:780–797

Moore BCJ (2002) Interference effects and phase sensitivity in hearing. Philos Trans R Soc Lond A 360:833–858

Moore BCJ, Glasberg BR (1989) Difference limens for phase in normal and hearing-impaired subjects. J Acoust Soc Am 84:1351–1365

Moore BCJ, Peters RW, Glasberg BR (1985) Thresholds for the detection of inharmonicity in complex tones. J Acoust Soc Am 77:1861–1867

Moore BCJ, Glasberg BR, Peters RW (1986) Thresholds for hearing mistuned partials as separate tones in harmonic complexes. J Acoust Soc Am 80:479–483

Nelson DA (1994) Level-dependent critical bandwidth for phase discrimination. J Acoust Soc Am 95:1514–1524

Palmer AR, Russell IJ (1986) Phase-locking in the cochlear nerve of the guinea pig and its relation to the receptor potential of inner hair-cells. Hear Res 24:1–15

Patterson RD (1987) A pulse ribbon model of monaural phase perception. J Acoust Soc Am 82:1560–1586

Plomp R, Steeneken HJM (1969) Effect of phase on the timbre of complex tones. J Acoust Soc Am 46:409–421

Pressnitzer D, Winter IM, Patterson RD (2000) The responses of single units in the ventral cochlear nucleus of the guinea pig to damped and ramped sinusoids. Hear Res 149:155–166

Part VI
Auditory Cortex and Beyond

Chapter 45
Stimulus-Specific Adaptation Beyond Pure Tones

Israel Nelken, Amit Yaron, Ana Polterovich, and Itai Hershenhoren

Abstract Detecting rare and surprising events is a useful strategy for sensory systems. In the human auditory system, deviance detection is indexed by an important component of the auditory event-related potentials, the mismatch negativity (MMN). Responses of single neurons in the inferior colliculus, medial geniculate body, and auditory cortex of mammals (cats, rats, and mice) show responses that share some properties with MMN: they are evoked by rare events, are preattentive (in as much as they occur in anesthetized animals), and, at least at the level of primary auditory cortex, cannot be accounted for by simple fatigue of the incoming sensory information. Here we extend these results to deviations beyond tone frequency. Recording in rat primary auditory cortex and using oddball sequences consisting of two frozen tokens of broadband noise samples, we found differences between the responses to the same token when used as the common and when used as the deviant, showing an exquisite sensitivity to the small differences between two spectro-temporally similar sounds. Similarly, differential adaptation can be demonstrated when using two word-like stimuli that have been derived from human speech

I. Nelken (✉)
Department of Neurobiology, Silberman Institute of Life Sciences,
Hebrew University, Edmond J. Safra Campus, Givat Ram,
Jerusalem, Israel

Edmond and Lily Safra Center for Brain Sciences,
Hebrew University, Jerusalem, Israel

Interdisciplinary Center for Neural Computation,
Hebrew University, Jerusalem, Israel
e-mail: israel@cc.huji.ac.il

A. Yaron • I. Hershenhoren • A. Polterovich
Department of Neurobiology,
The Silberman Institute of Life Sciences,
The Interdisciplinary Center for Neural Computation,
Hebrew University, Jerusalem, Israel

B.C.J. Moore et al. (eds.), *Basic Aspects of Hearing*,
Advances in Experimental Medicine and Biology,
DOI 10.1007/978-1-4614-1590-9_45,

but adapted to the rat auditory system. Thus, differential adaptation to common and rare sounds is present also with sounds whose complexity mirrors that of natural environments.

1 Introduction

Stimulus-specific adaptation (SSA) is the reduction in the responses to a common sound which does not generalize, or generalizes only partially, to a second, rare sound that is presented in the same sequence. SSA has been recently demonstrated in a number of mammalian and nonmammalian species, including cats (Ulanovsky et al. 2003), rats (Taaseh et al. 2011), mice (Anderson et al. 2009), gerbils (Bauerle et al. 2011), and barn owls (Reches and Gutfreund 2008). SSA at the single-neuron level has been initially described in auditory cortex but has been since demonstrated also in rat inferior colliculus (Malmierca et al. 2009; Zhao et al. 2011), in rat medial geniculate body (Antunes et al. 2010; Bauerle et al. 2011), and in mouse MGB (Anderson et al. 2009).

The mechanisms underlying the generation of SSA have been under intense study recently. While SSA is present at least from the IC and up, it is at best weak in the lemniscal divisions of the IC (the central nucleus, ICc) and the MGB (the ventral division, vMGB). In contrast, in the non-lemniscal divisions of the IC and the MGB, SSA is extremely strong, with some units barely responding to the common sound beyond its first few presentations (Antunes et al. 2010; Bauerle et al. 2011; Perez-Gonzalez et al. 2005). Since on the one hand cooling cortex does not modify SSA in the thalamus (Antunes et al. 2010) but on the other hand SSA in rat auditory cortex tends to increase from the thalamo-recipient layers and out into both the infra- and supra-granular layers (Szymanski et al. 2009), current results suggest that SSA is generated at least twice, once in the non-lemniscal IC and a second time in primary auditory cortex, with a possible third locus in non-lemniscal MGB.

The computations underlying SSA are of great interest as well. Attempts to model SSA to frequency deviants rely on a tonotopic, narrowly tuned input that feeds into a layer of more widely tuned neurons (Mill et al. 2011; Taaseh et al. 2011). This model, in which adaptation occurs within narrow-frequency channels, can be fitted to some of the current results. Furthermore, neurons in non-lemniscal IC and MGB tend to have relatively wide, sometimes badly defined frequency response areas, suggesting the presence of integration across frequency.

In primary auditory cortex (A1), the ability of such a model to account for the full pattern of the published results is less clear. Neurons in A1 tend to have wider tuning width than their MGB input (Miller et al. 2001), suggesting again the presence of integration across frequency that could underlie SSA. However, Taaseh et al. (2011) demonstrated that the responses to a rare tone in the presence of a common one is almost twice as large as that predicted by the model of adaptation in narrow-frequency channels. Thus, there seems to be additional mechanisms, beyond adaptation in narrow-frequency channels, which enhance the responses to deviant tones in rat A1.

Part of the initial interest in SSA was due to the possibility that it is the neural correlate of mismatch negativity (MMN), an important component of the human event-related potentials (ERPs) which is sensitive to deviance. While a substantial amount of data supports the view that SSA is not MMN (Farley et al. 2010), the fact that SSA and MMN both depend largely in the same way on many stimulus parameters suggests that SSA is upstream of MMN – a hypothesis supported by the recent discovery of deviance sensitivity in midlatency potentials (Grimm and Escera 2012).

The main contribution of this chapter is a demonstration that SSA is more than just "frequency-specific adaptation," in the sense that SSA can be elicited by stimuli with complex spectral structure, and is present even in the absence of substantial spectral differences between the common and the rare sounds. These findings demonstrate that the mechanisms underlying SSA may be operative in real-world conditions, with complex sounds that include both spectral and temporal variations.

2 Methods

For detailed methods, see Taaseh et al. (2011). We used six adult female Sabra rats weighing 220–250 g. The joint ethics committee (IACUC) of the Hebrew University and Hadassah Medical Center approved the study protocol for animal welfare (protocols NS-06-10041-3, NS-08-11349-3). The Hebrew University is an AAALAC International accredited institute.

Electrophysiological recordings were conducted from the left auditory cortex. A craniotomy was performed over the estimated location of the left auditory cortex. We recorded extracellularly from the auditory cortex using 4–6 glass-coated tungsten electrodes (Alpha-Omega Ltd., Nazareth-Illit, Israel) with tip separations of ~600 μm. Electrical signals were amplified and filtered between 3 Hz and 8 kHz to obtain both LFP and action potentials. All experiments were conducted in a sound-proof chamber (IAC, Winchester, UK). Sounds were stored as computer files, transduced to voltage signals by a sound card (HDSP9632, RME, Germany), attenuated (PA5, TDT), and played through a sealed speaker (EC1, TDT) into the right ear canal of the rat.

3 Results

The two white noise tokens used here were synthesized in MATLAB using the function randn. Ten tokens were generated, and the two tokens with the largest squared difference between their power spectra were selected for use in the oddball

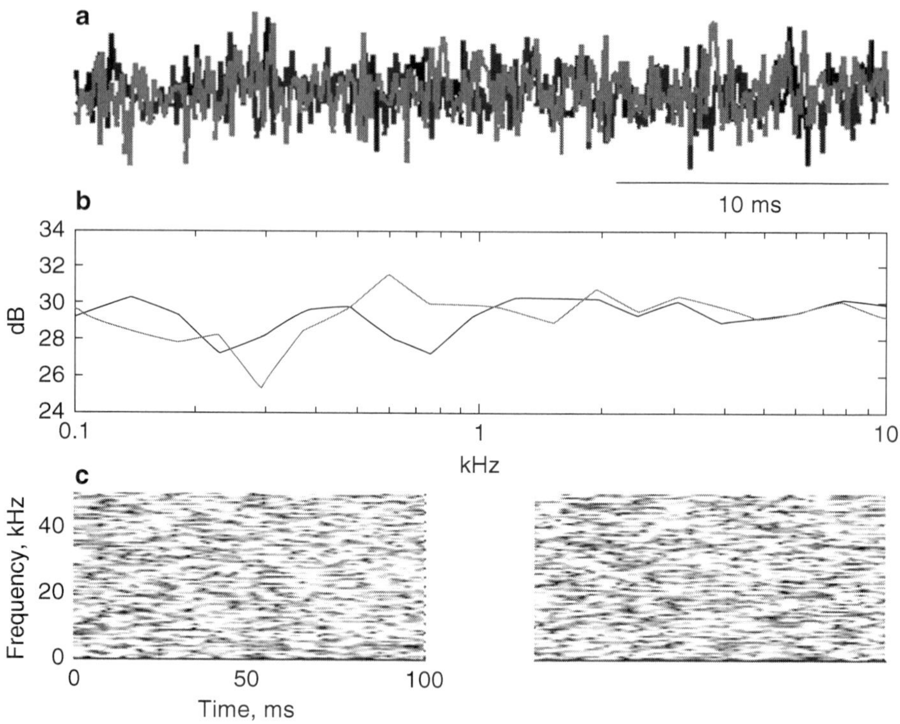

Fig. 45.1 The two noise tokens used in the SSA experiments. (**a**) Waveforms of the first 30 ms of each of the two tokens. (**b**) Power spectra of the two tokens, calculated in 1/3 octave bands. (**c**) Spectrograms of the two tokens

sequences. The two tokens are presented in Fig. 45.1. While their temporal waveforms (Fig. 45.1a), power spectra (Fig. 45.1b), and spectrograms (Fig. 45.1c) are certainly not identical, they are nevertheless very similar to each other.

The two word-like stimuli are presented in Fig. 45.2. They were synthesized using the default settings in the text-to-speech synthesizer, Festival (available in Linux, Fedora 14). The two words have been modified with the help of the vocoder STRAIGHT (Kawahara et al. 2008) using additional processing routines written by us. The frequency content of the two sounds was shifted to above 1 kHz, and the pitch contour was set to a constant 350 Hz. These modifications resulted in sounds that had some features of speech, notably strong spectro-temporal modulations in the speech range. We specifically equalized the peak energy (see Fig. 45.2a) and the overall power spectra (Fig. 45.2b) of the two stimuli. Their spectro-temporal modulations are, however, different (Fig. 45.2c).

The two white noise tokens were used as the common and rare stimuli in oddball sequences. Two sequences were used, with each token playing both roles. The responses of three multiunit clusters to the two tokens are displayed in Fig. 45.3a–c. Figure 45.3a shows the responses of a multiunit cluster that responded essentially

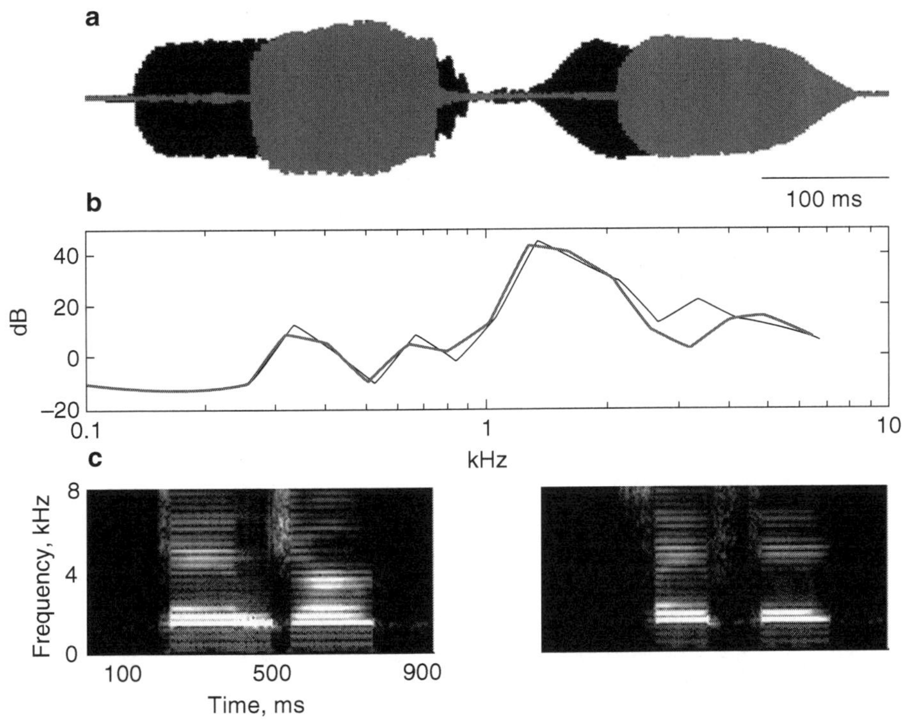

Fig. 45.2 The two word-like stimuli used in the SSA experiments. (**a**) Waveforms of the two stimuli (*black*: "danger," *gray*: "safety"). (**b**) Power spectra of the two stimuli, calculated in 1/3 octave bands (same color convention). (**c**) Spectrograms of the two stimuli (*left*: "danger," *right*: "safety")

identically to the two tokens when common and when rare. Figure 45.3b shows the responses of a multiunit cluster that showed substantial reduction in the response to a token when common, relative to the response to the same token when rare. This is a somewhat extreme example – in most multiunit clusters, the effect was more subtle, as illustrated in Fig. 45.3c. The differences between the responses to a token when common and when rare are significant but small. Note in addition that the enhanced responses to a token when common and when rare had to a large extent the same time course. On the other hand, the responses to the two different tokens could follow quite different time courses (and have quite different overall magnitude).

The word-like stimuli had a longer duration and were presented at a substantially lower rate (ISI of 1,200 ms) than the white noise tokens. Nevertheless, they elicited differential responses when used in oddball sequences, with each of the stimuli tested both as standard and as deviant. Examples of local field potentials recorded in response to the word-like stimuli are shown in Fig. 45.4. The major component of the responses to both words occurred at about the time of the major amplitude transient into the first vowel. The peak-to-peak amplitudes of the responses were

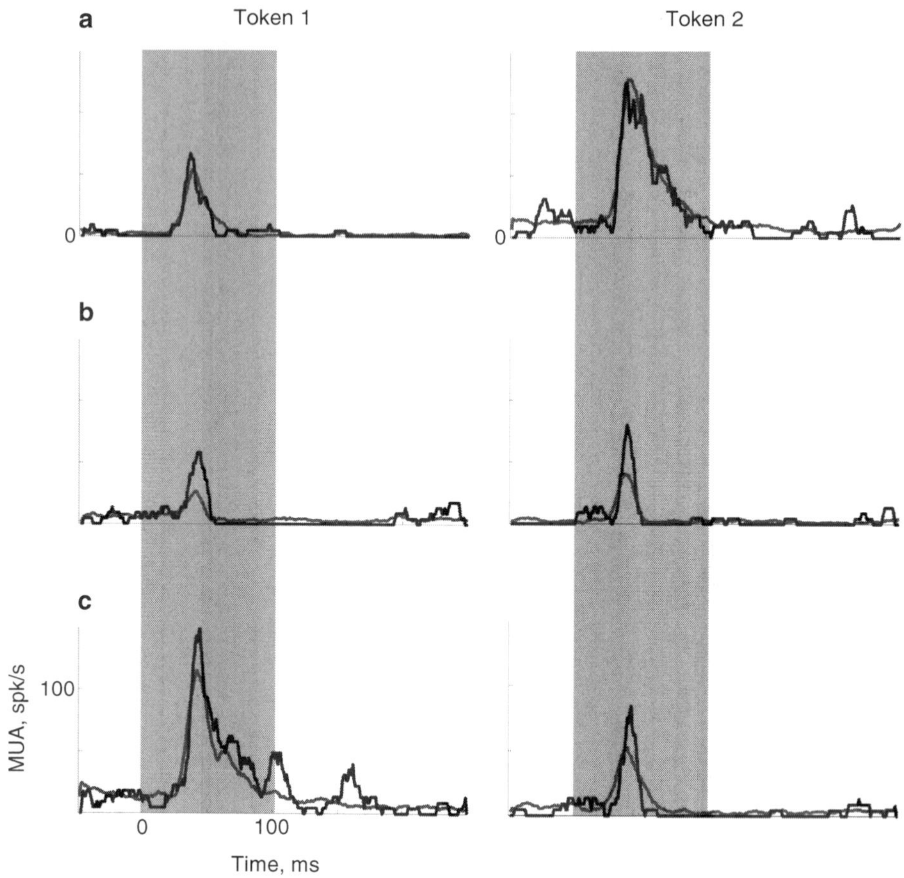

Fig. 45.3 MUA elicited by the two noise tokens when used as common (*gray lines*) and rare (*black lines*) sounds in oddball sequences. (**a–c**) responses at three recording sites

larger when stimuli were rare than when common, although both the response itself and the contrast between common and rare responses were larger for the word-like stimulus derived from "safety" than for the stimulus derived from "danger." The response to "safety" (top) may have included another response component at offset, although its significance is difficult to establish.

4 Discussion and Conclusions

The single-neuron deviance detection system in A1 has been studied mostly with pure tones and frequency deviants. The differential responses to common and rare sounds demonstrated here suggest that deviance detection in A1 is sensitive to

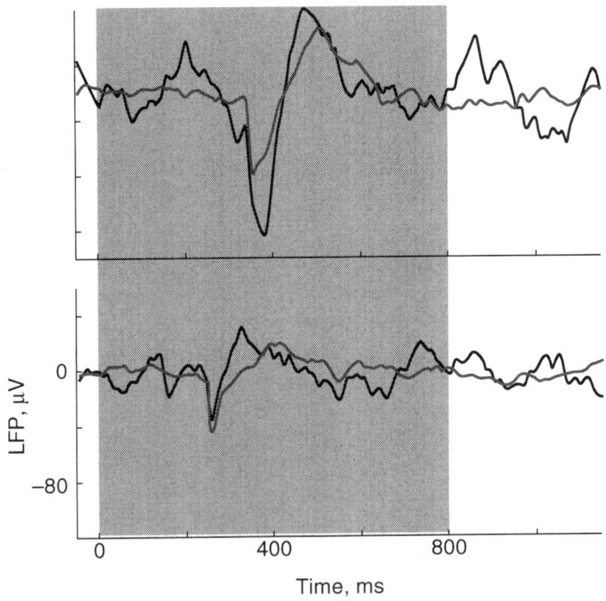

Fig. 45.4 LFP elicited by the two word-like stimuli (*top*: "safety," *bottom*: "danger") when used as standards (*gray lines*) and deviants (*black lines*)

spectro-temporal features as subtle as the differences between two white noise tokens and is capable of distinguishing two word-like stimuli with very similar average spectra (but differing in their spectro-temporal modulations).

While the sensitivity of SSA to subtle spectro-temporal differences is remarkable, the mechanisms that underlie SSA to these complex stimuli may be to a large extent the same mechanisms that underlie SSA to pure tones. For example, while very similar, the two white noise tokens nevertheless differ in their spectral content – some frequencies have average levels that are greater in one than in the other. It could be that these differences in spectro-temporal structure were sufficient to evoke SSA.

To test this hypothesis, we need to know more about the spectro-temporal features of the noise that elicited the responses. Such information can be extracted using an appropriate variant of reverse correlation. Unfortunately, the overall duration of the noise tokens was only 100 ms, and we used only two of them. Thus, we do not have enough information to extract the relevant stimulus features from the responses.

The word-like stimuli illustrated another feature of SSA– the fact that adaptation seems to affect individual temporal components of the responses. This finding extends the observation that SSA occurs close to, or at, response onset. When using word-like stimuli, with relatively long durations and complex spectro-temporal structures, the dominant response components may occur away from stimulus onset. SSA for the word-like stimuli occurred at about the same time as responses to physical events within the sound. This is different from the behavior of MMN, which is locked to the onset of deviance rather than to the time at which a response component occurred.

The findings presented here suggest that SSA in A1 is more than just "frequency-specific adaptation" and that SSA contributes to the detection of small changes in complex, ethologically valid stimuli. At the same time, these findings emphasize the differences between SSA and MMN. Thus, this chapter supports a hierarchical view of deviance detection in which SSA in A1 determines some, but not all, MMN properties (Grimm and Escera 2012).

Acknowledgments This work was supported by grants to I.N. from the Israeli Science Foundation (ISF), the German-Israeli Foundation (GIF), the US-Israel Binational Foundation (BSF), the Israeli Ministry of Health to under the framework of ERA-Net NEURON, and the Gatsby Charitable Foundation.

References

Anderson LA, Christianson GB, Linden JF (2009) Stimulus-specific adaptation occurs in the auditory thalamus. J Neurosci 29:7359–7363

Antunes FM, Nelken I, Covey E, Malmierca MS (2010) Stimulus-specific adaptation in the auditory thalamus of the anesthetized rat. PLoS One 5:e14071

Bauerle P, von der Behrens W, Kossl M, Gaese BH (2011) Stimulus-specific adaptation in the gerbil primary auditory thalamus is the result of a fast frequency-specific habituation and is regulated by the corticofugal system. J Neurosci 31:9708–9722

Farley BJ, Quirk MC, Doherty JJ, Christian EP (2010) Stimulus-specific adaptation in auditory cortex is an NMDA-independent process distinct from the sensory novelty encoded by the mismatch negativity. J Neurosci 30:16475–16484

Grimm S, Escera C (2012) Auditory deviance detection revisited: evidence for a hierarchical novelty system. Int J Psychophysiol 85:88–92

Kawahara H, Morise M, Takahashi T, Nisimura R, Irino T, Banno H (2008) Tandem-STRAIGHT: a temporally stable power spectral representation for periodic signals and applications to interference-free spectrum, F0, and aperiodicity estimation. ICASSP 2008:3933–3936

Malmierca MS, Cristaudo S, Perez-Gonzalez D, Covey E (2009) Stimulus-specific adaptation in the inferior colliculus of the anesthetized rat. J Neurosci 29:5483–5493

Mill R, Coath M, Wennekers T, Denham SL (2011) A neurocomputational model of stimulus-specific adaptation to oddball and Markov sequences. PLoS Comput Biol 7:e1002117

Miller LM, Escabi MA, Schreiner CE (2001) Feature selectivity and interneuronal cooperation in the thalamocortical system. J Neurosci 21:8136–8144

Perez-Gonzalez D, Malmierca MS, Covey E (2005) Novelty detector neurons in the mammalian auditory midbrain. Eur J Neurosci 22:2879–2885

Reches A, Gutfreund Y (2008) Stimulus-specific adaptations in the gaze control system of the barn owl. J Neurosci 28:1523–1533

Szymanski FD, Garcia-Lazaro JA, Schnupp JW (2009) Current source density profiles of stimulus-specific adaptation in rat auditory cortex. J Neurophysiol 102:1483–1490

Taaseh N, Yaron A, Nelken I (2011) Stimulus-specific adaptation and deviance detection in the rat auditory cortex. PLoS One 6:e23369

Ulanovsky N, Las L, Nelken I (2003) Processing of low-probability sounds by cortical neurons. Nat Neurosci 6:391–398

Zhao L, Liu Y, Shen L, Feng L, Hong B (2011) Stimulus-specific adaptation and its dynamics in the inferior colliculus of rat. Neuroscience 181:163–174

Chapter 46
Mapping Tonotopy in Human Auditory Cortex

Pim van Dijk and Dave R.M. Langers

Abstract Tonotopy is arguably the most prominent organizational principle in the auditory pathway. Nevertheless, the layout of tonotopic maps in humans is still debated. We present neuroimaging data that robustly identify multiple tonotopic maps in the bilateral auditory cortex. In contrast with some earlier publications, tonotopic gradients were not found to be collinearly aligned along Heschl's gyrus; instead, two tonotopic maps ran diagonally across the anterior and posterior banks of Heschl's gyrus, set at a pronounced angle. On the basis of the direction of the tonotopic gradient, distinct subdivisions of the auditory cortex could be clearly demarcated that suggest homologies with the tonotopic organization in other primates. Finally, we applied our method to tinnitus patients to show that – contradictory to some pathophysiological models – tinnitus does not necessarily involve large-scale tonotopic reorganization. Overall, we expect that tonotopic mapping techniques will significantly enhance our ability to study the hierarchical functional organization of distinct auditory processing centers in the healthy and diseased human brain.

1 Introduction

Tonotopy is a key organizational feature that pervades all levels of the mammalian central auditory system. Electrophysiological recordings in numerous species have shown the existence of multiple tonotopically organized cortical fields with

P. van Dijk, PhD • D.R.M. Langers, PhD (✉)
Department of Otorhinolaryngology/Head and Neck Surgery,
University of Groningen, University Medical Center Groningen,
Groningen, 9700 RB, The Netherlands

Graduate School of Medical Sciences
Research School of Behavioural and Cognitive Neurosciences,
University of Groningen, Groningen, 9700 RB, The Netherlands
e-mail: audio-fmri@langers.nl

B.C.J. Moore et al. (eds.), *Basic Aspects of Hearing*,
Advances in Experimental Medicine and Biology,
DOI 10.1007/978-1-4614-1590-9_46,

abutting "high-to-low-to-high" gradients of characteristic frequency (Kaas and Hackett 2000; Bendor and Wang 2008). Because various functional specializations (like speech) and morphological differentiations (like Heschl's gyrus, HG) complicate the translation of animal results to humans, various researchers have started to map the tonotopic organization noninvasively in healthy human subjects as well.

Although the coarse spatial resolution and the limited number of reconstructable dipoles in magneto- and electroencephalography (MEG/EEG) limit the ability to map cortical frequency representations in detail, these techniques initially revealed effective low-to-high frequency gradients extending in the lateral-to-medial direction along HG (Romani et al. 1982; Verkindt et al. 1995). Positron emission tomography (PET) and functional magnetic resonance imaging (fMRI) studies offer more precise spatial resolution and the ability to sample large numbers of cortical sites simultaneously, and subsequently confirmed that the higher frequencies were represented more (postero)medially than the lower frequencies (Lauter et al. 1985; Wessinger et al. 1997). Later, the existence of multiple gradients in each hemisphere was studied, and evidence was reported for two oppositely oriented tonotopic gradients along HG (Formisano et al. 2003). At the same time, this view was challenged by findings involving as many as six complexly organized gradients per hemisphere (Talavage et al. 2004).

Around 2010, most studies agreed on the existence of at least one dominant lateral-to-medial low-to-high frequency progression in primary auditory cortex on medial HG, but the layout of additional tonotopic gradients was the subject of some debate (for a meta-analysis, see Woods and Alain 2009). This issue is of considerable importance because the identification of tonotopic progressions might enable the delineation of distinct cortical subfields in individuals and thus help establish the hierarchical organization and unravel the various types of processing in human auditory cortex, both in health and disease.

In recent years, an increasing amount of converging evidence has appeared suggesting that tonotopic progressions in humans run across rather than along HG (Humphries et al. 2010; Striem-Amit et al. 2011; Da Costa et al. 2011). In the current study, we investigated this organization further, both in normal-hearing subjects and tinnitus patients in which tonotopic representations have been proposed to be disturbed.

2 Methods

Twenty healthy subjects (21–60 y, mean 33 y) and 20 tinnitus patients (26–60 y, mean 46 y) participated in this fMRI study on the basis of written informed consent. The two groups did not significantly differ with regard to average hearing thresholds, which were normal up to 8 kHz according to pure-tone audiometry.

The fMRI paradigm was identical to the one described in a previous report (Langers and van Dijk 2012). In short, subjects performed an engaging visual/emotional task while at the same time tone sequences were presented at octave frequencies from 250 Hz to 8 kHz at a level of approximately 40 dB HL. Task-irrelevant, unattended, low-level stimuli were employed to avoid excessive spread of sound-evoked activation. Per subject, 120 high-resolution fMRI images were obtained using sparse clustered-volume acquisitions to avoid interference from acoustic scanner noise.

Standard fMRI preprocessing steps were performed, including motion correction, normalization to a standard stereotaxic space, and moderate smoothing using a 5-mm kernel. At the subject level, a regression model was fitted that included responses to the sound stimuli and activation related to task performance, as well as various nuisance variables (residual head motion effects, scanner baseline, and drift). The individual responses to the six different sound frequencies were fed into a group-level statistical model to assess sound-evoked responses.

A region of interest (ROI) was defined comprising 60 cm^3 of significantly responsive brain tissue, constituting two coherent clusters of approximately equal size in the bilateral superior temporal lobes. The corresponding activation data were collected into an aggregate $6 \times 300{,}680$ matrix (6 frequencies; 7,517 voxels from 40 subjects each) that was subsequently decomposed into principal components in order to extract the most prevalent response characteristics. The two strongest components were studied further, each yielding a single 6-element frequency response profile and two corresponding 7,517-element spatial response maps (one average for each subject group).

The significance of differences between the maps of the two groups was assessed using permutation testing, i.e., by repeatedly reassigning subjects to two equal-sized random subgroups to obtain the null distributions of outcome measures.

3 Results

Extensive significant activation was found covering the superior surface of the bilateral temporal lobes. In addition, focal subcortical activation was found in the inferior colliculi and medial geniculate bodies. The activity in the bilateral auditory cortices was used to define a ROI, and the contained responses were analyzed using principal component analysis. The frequency profiles and spatial maps of the first two components are shown in Fig. 46.1.

The first principal component summarizes the most typical frequency profile and the corresponding magnitudes in all voxels. This component revealed a uniformly positive response to all frequencies, with some roll-off towards high frequencies that indicates a smaller overall responsiveness to high frequencies compared to low frequencies. The corresponding map showed positive coefficients for all voxels, peaking in the general vicinity of the diagonally running HG and declining towards the clusters' edges. These panels show similar behavior to what would be obtained by straightforward averaging (over all voxels to obtain a mean response profile or over all frequencies to obtain a mean response map).

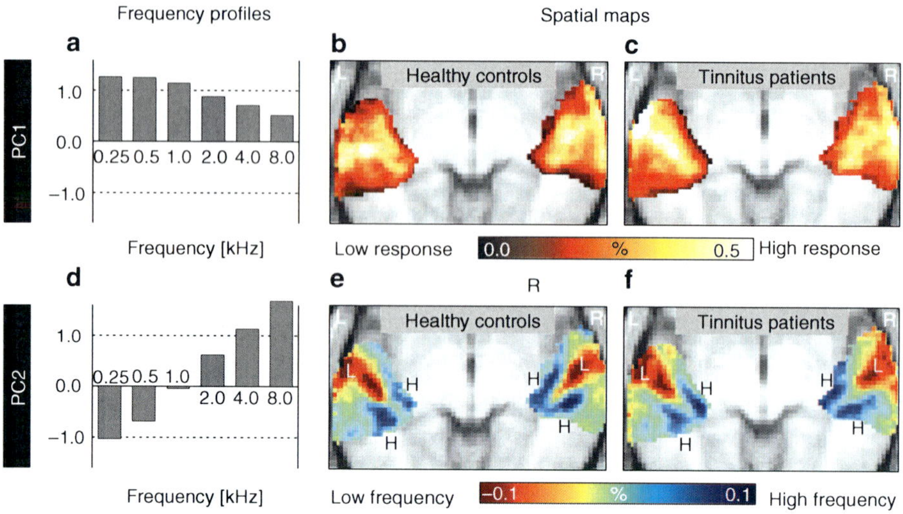

Fig. 46.1 Two principal components were extracted, each consisting of a single frequency profile (*left*) and corresponding spatial maps for both subject groups (*right*). The first component (PC1, *top*) summarized the overall sound-evoked activation, with positive responses to all frequencies (falling off towards the highest frequencies) at all locations in auditory cortex (falling off towards the edges of the activation clusters). Interestingly, the second component (PC2, *bottom*) summarized preferential activation to low- or high-frequency stimuli, with a monotonic frequency profile and negative/positive coefficients in lateral/medial cortical locations, indicating low- (*L*) and high- (*H*) frequency preferences, respectively

More interestingly, the second component summarizes the residual after the first component is removed from the activation data. In other words, it shows how voxels most typically tend to deviate from the average behavior that is captured by the first component. This component's frequency profile shows a monotonic increase, ranging from negative values for low frequencies to positive values for high frequencies. The corresponding map shows negative coefficients on the crest of lateral HG and positive coefficients on the rostral and caudal banks of medial HG. This indicates that the most medially located voxels display a disproportionately large response to high frequencies and a disproportionately weak response to low frequencies and can therefore be classified as "high-frequency voxels"; conversely, the lateral crest of HG contains "low-frequency voxels." Therefore, this component's map can be interpreted to represent gradual variations in frequency tuning as a function of cortical position and thus reflects the tonotopic organization in human auditory cortex.

Although we currently report group averages only due to space constraints, we observed that the reported behavior could be traced back to similar tonotopic mappings in individuals quite consistently (Langers et al. 2012).

In Fig. 46.2, the direction of the tonotopic gradient is shown by means of a color code. For the purpose of this plot, frequency tuning was not quantified by the second component's map; instead, the ratio of a voxel's coefficient in the second and first

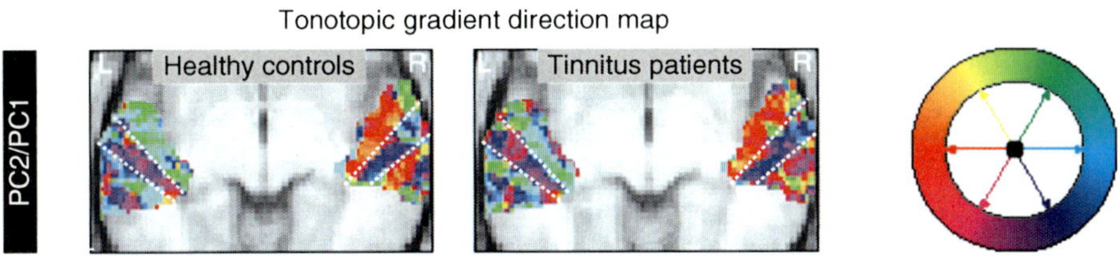

Fig. 46.2 The direction of the tonotopic progression's gradient in the axial x,y-plane was color-coded, revealing the existence of multiple strips of cortex that run diagonally parallel to HG. Within each such strip, the tonotopic gradient direction is reasonably homogenous, but it abruptly changes when crossing from one strip to the next (*dotted white lines*)

component was used, which is more representative of the shape of the response profile, regardless of the overall level of responsiveness to sound.

The gradient can be seen to be oriented in a more or less homogeneous direction inside a number of "strips" that run more or less parallel to HG. At the same time, gradient direction abruptly switches from one strip to the next. For example, on the rostral bank of HG, the low-to-high tonotopic gradient is oriented in a medial direction, whereas on its caudal bank, it points caudally. On the planum temporale, further gradient reversals are visible.

We checked for differences in the tinnitus group by analyzing the histograms of voxel values that could be extracted from the first and second component's maps, as well as their ratios, and comparing them to bootstrapped null distributions. No significant differences were found to exist compared to healthy controls. These analyses showed, in particular, that auditory cortex as a whole did not display elevated responsiveness to sound in tinnitus patients nor did it show an overrepresentation of high frequencies. The details of these comparisons are reported in Langers et al. (2012).

4 Discussion

In the present study, robust tonotopic maps were demonstrated in response to minimally salient tone stimuli. The tonotopic gradient was found to remain similarly oriented within parallel strips of cortex but abruptly changed direction from one strip to the next. Extrapolating results from primates, in which primary cortical fields were found to show separate tonotopic progressions with gradient directions that were set under a pronounced angle, we surmise that the locations of these reversals correspond with borders between neighboring cortical subfields of human auditory cortex. Our results suggest the presence of at least two distinct subfields on the rostral and caudal banks of HG, rHG, and cHG (Fig. 46.3). We hypothesize that these contain the human homologues of the core fields R and AI in primates.

Our results are in good agreement with the large number of studies that positioned high-frequency areas on medial HG and low-frequency areas on lateral HG, if the limited spatial resolution of these studies is taken into account: because the

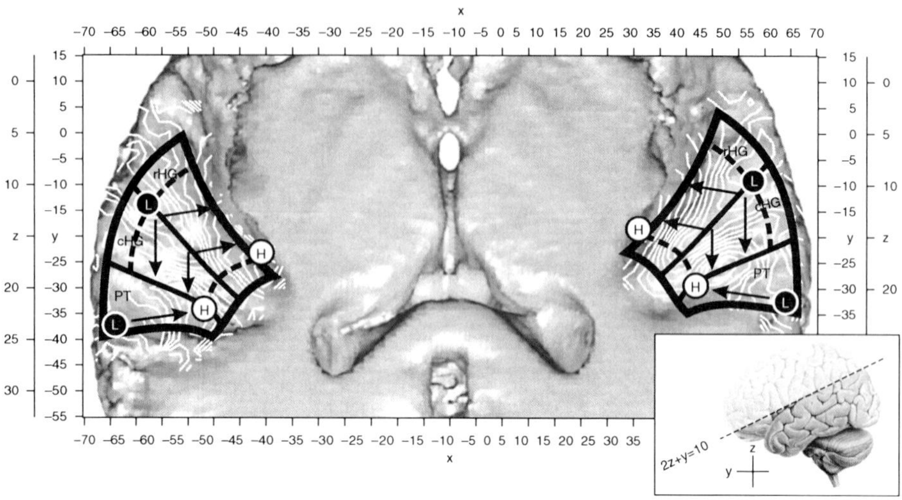

Fig. 46.3 In the auditory cortices of both hemispheres, viewed here from the top, multiple subfields could be distinguished on the basis of sharp reversals in the local direction of the tonotopic gradient. One such field, located on the rostral bank of HG (*rHG*), and featuring a gradient in a medial direction, presumably corresponds with field R in primates. Another such field, located on the caudal bank of HG (*cHG*), with a gradient in a caudal direction, then corresponds with field A1. Further gradients were found in the planum temporale (*PT*). Low- and high-frequency endpoints of the tonotopic progressions are indicated by *circles* (labeled *L* and *H*)

two medial high-frequency endpoints are positioned relatively closely together, the gradients in rHG and cHG may have been indistinguishable and apparently merged into a single gradient that well aligns with the axis of HG. Given the present detail, however, our results suggest that these supposed primary auditory fields actually line up across HG instead of along HG, while at the same time the gradients within a field are diagonally oriented and set at a pronounced angle. This general arrangement is in close correspondence with the results of various recent studies (Humphries et al. 2010; Striem-Amit et al. 2011; Da Costa et al. 2011) and well fits the organization in primates (Kaas and Hackett 2000; Bendor and Wang 2008). Moreover, the proposed fields reasonably agree with at least a subset of gradients that were proposed by Talavage et al. (2004). At first sight, our results are more difficult to reconcile with a prevalent view in which two abutting gradients span a pair of fields that extend along HG, as demonstrated by Formisano et al. (2003), for instance. This interpretation was based on the frequency profile as obtained along a curve that ran diagonally along HG. When consulting their original maps, evidence for a second high-frequency endpoint on the medial aspect of rHG is visible, however.

As a potential clinical application, we looked for differences between the tonotopic organization in tinnitus patients as compared to normally hearing controls. Various models regarding tinnitus pathophysiology state that tinnitus arises as a result of aberrant plastic changes, either in the form of abnormal homeostatic gain or in the form of tonotopic reorganization (Bartels et al. 2007). This is thought to subsequently lead to abnormal levels of spontaneous neural activity or synchronicity, which may be perceived as a phantom sound. Applying our method to matched groups of subjects, we found no significant differences in tonotopic organization. This may be attributed to inadequate temporal or spatial sensitivity in fMRI or to the particular characteristics of our patients who had normal hearing thresholds. Nevertheless, our findings indicate that macroscopic tonotopic reorganization is not required for tinnitus, at least in patients with near normal absolute thresholds.

In summary, we found robust tonotopic gradients in bilateral auditory cortex. On the basis of the direction of the gradient, two distinct fields could be identified on HG, one on its caudal bank (likely homologous to AI) and one on its rostral bank (likely homologous to R). Although more detailed, this organization retrospectively agrees with the majority of publications in the literature. Supported by various other recent publications, we commend that the common view that these primary cortical fields extend along HG is abandoned in favor of a view in which they fold across HG.

References

Bartels H, Staal MJ, Albers FWJ (2007) Tinnitus and neural plasticity of the brain. Otol Neurotol 28:178–184

Bendor D, Wang X (2008) Neural response properties of primary, rostral, and rostrotemporal core fields in the auditory cortex of marmoset monkeys. J Neurophysiol 100:888–906

Da Costa S, van der Zwaag W, Marques JP, Frackowiak RSJ, Clarke S, Saenz M (2011) Human primary auditory cortex follows the shape of Heschl's gyrus. J Neurosci 31:14067–14075

Formisano E, Kim DS, Di Salle F, van de Moortele PF, Ugurbil K, Goebel R (2003) Mirror-symmetric tonotopic maps in human primary auditory cortex. Neuron 40:859–869

Humphries C, Liebenthal E, Binder JR (2010) Tonotopic organization of human auditory cortex. Neuroimage 50:1202–1211

Kaas JH, Hackett TA (2000) Subdivisions of auditory cortex and processing streams in primates. Proc Natl Acad Sci USA 97:11793–11799

Langers DRM, van Dijk P (2012) Mapping the tonotopic organization in human auditory cortex with minimally salient acoustic stimulation. Cereb Cortex 22:2024–2038

Langers DRM, de Kleine E, van Dijk P (2012) Tinnitus does not require macroscopic tonotopic map reorganization. Front Syst Neurosci 6:2

Lauter JL, Herscovitch P, Formby C, Raichle ME (1985) Tonotopic organization in human auditory cortex revealed by positron emission tomography. Hear Res 20:199–205

Romani GL, Williamson SJ, Kaufman L (1982) Tonotopic organization of the human auditory cortex. Science 216:1339–1340

Striem-Amit E, Hertz U, Amedi A (2011) Extensive cochleotopic mapping of human auditory cortical fields obtained with phase-encoding FMRI. PLoS One 6:e17832

Talavage TM, Sereno MI, Melcher JR, Ledden PJ, Rosen BR, Dale AM (2004) Tonotopic organization in human auditory cortex revealed by progressions of frequency sensitivity. J Neurophysiol 91:1282–1296

Verkindt C, Bertrand O, Perrin F, Echallier JF, Pernier J (1995) Tonotopic organization of the human auditory cortex: N100 topography and multiple dipole model analysis. Electroencephalogr Clin Neurophysiol 96:143–156

Wessinger CM, Buonocore MH, Kussmaul CL, Mangun GR (1997) Tonotopy in human auditory cortex examined with functional magnetic resonance imaging. Hum Brain Mapp 5:18–25

Woods DL, Alain C (2009) Functional imaging of human auditory cortex. Curr Opin Otolaryngol Head Neck Surg 17:407–411

Chapter 47
Cortical Activity Associated with the Perception of Temporal Asymmetry in Ramped and Damped Noises

André Rupp, André Spachmann, Anna Dettlaff, and Roy D. Patterson

Abstract Human listeners are very sensitive to the asymmetry of time-reversed pairs of ramped and damped sounds. When the carrier is noise, the hiss component of the perception is stronger in ramped sounds and the drumming component is stronger in damped sounds (Akeroyd and Patterson 1995). In the current study, a paired comparison technique was used to establish the relative "hissiness" of these noises, and the ratings were correlated with (a) components of the auditory evoked field (AEF) produced by these noises and (b) the magnitude of a hissiness feature derived from a model of the internal auditory images produced by these noises (Irino and Patterson 1998). An earlier AEF report indicated that the peak magnitude of the transient N100m response mirrors the perceived salience of the tonal perception (Rupp et al. 2005). The AEFs of 14 subjects were recorded in response to damped/ramped noises with half-lives between 1 and 64 ms and repetition rates between 12.5 and 100 ms. Spatio-temporal source analysis was used to fit the P50m, the P200m, and the sustained field (SF). These noise stimuli did not produce a reliable N100m. The hissiness feature from the auditory model was extracted from a time-averaged sequence of summary auditory images as in Patterson and Irino (1998). The results show that the perceptual measure of hissiness is highly correlated with the hissiness feature from the summary auditory image, and both are highly correlated with the magnitude of the transient P200m. There is a significant

A. Rupp (✉) • A. Spachmann • A. Dettlaff
Section of Biomagnetism, Department of Neurology, University of Heidelberg, Im Neuenheimer Feld 400, Heidelberg 69120, Germany
e-mail: andre.rupp@uni-heidelberg.de; aspachma@students.mail.uni-mannheim.de

R.D. Patterson
Department of Physiology, Development and Neuroscience, Centre for the Neural Basis of Hearing, University of Cambridge, Downing Site, Cambridge, Cambridgeshire CB2 3EG, UK
e-mail: rdp1@cam.ac.uk

B.C.J. Moore et al. (eds.), *Basic Aspects of Hearing*, Advances in Experimental Medicine and Biology, DOI 10.1007/978-1-4614-1590-9_47,

but weaker correlation with the SF and a nonsignificant correlation with the P50m. The results suggest that regularity in the carrier effects branching at an early stage of auditory processing with tonal and noisy sounds following separate spatio-temporal routes through the system.

1 Introduction

The fine structure of environmental sounds like speech and musical instruments is typically asymmetric in time. Temporal asymmetry affects the timbre of a sound (Patterson 1994a, b), its pitch (Hartmann 1978), and its loudness (Stecker and Hafter 2000). The sensitivity of human listeners to temporal asymmetry has been demonstrated by comparing "ramped" and "damped" sounds – sinusoids or noises modulated by periodically rising (ramped) or falling (damped) exponential functions (Patterson 1994a, b; Akeroyd and Patterson 1995; Patterson and Irino 1998). The ramped version is perceived to have a stronger tonal (sinusoid) or hiss (noise) component and a weaker drumming component than the damped version for a wide range of half-lives and modulation rates. The perceptual distinction was explained with summary statistics derived from stabilised auditory images (SAIs) (Patterson and Allerhand 1995) of the stimuli. Subsequently, there have been physiological demonstrations of asymmetry in the neural responses to sounds with temporal asymmetry, for example, in the cochlear nucleus (Pressnitzer et al. 2000), the inferior colliculus (Neuert et al. 2001), and the single units of the primary auditory cortex (Lu et al. 2001). This chapter describes how magnetoencephalography was employed to measure the aggregate cortical response (AEF) to temporal asymmetry in noise bursts and how the results compare with a new measure of the corresponding perceptually asymmetry and a measure of perceptual asymmetry based on the activity in SAIs of the sounds.

2 Methods

Bursts of ramped and damped noise were created with 12 combinations of modulation period (MP) and half-life (HL); the combinations were MP = 12.5 ms, HL = 1, 2, 4 ms; MP = 25 ms, HL = 1, 4, 16 ms; MP = 50 ms, HL = 1, 8, 32 ms; and MP = 100 ms, HL = 1, 16, 64 ms. These HL values produced a broad range of discrimination performance for the listeners in Akeroyd and Patterson (1995, Fig. 2, p. 2467). The bursts were all 700 ms in duration. The burst with MP = 25 ms, HL = 4 ms was presented at a level of 65 dB SPL. The levels of the remaining bursts were adjusted to produce a similar loudness by increasing their level 3 dB for each period doubling and decreasing their level by 1.5 dB for each half-life doubling as in Akeroyd and Patterson (1995). The stimuli were presented diotically via ER-3 earphones (Etymotic Research Inc.) equipped with 90-cm plastic tubes and foam earpieces. Fourteen normal-hearing listeners (26.6 y, ±5.85 y) participated in the study.

The Bradley-Terry-Luce (BTL) technique was used to establish the relative hissiness of the noises using the procedure suggested in David (1988). The listeners were presented with all combinations of the noise bursts in pairs and asked to choose the noise with the stronger hiss on each trial. A repeated-measures ANOVA with dependent factors "ramped-damped", "period", and "half-life" was used to assess the hissiness ratings.

Auditory evoked fields were recorded using a Neuromag-122 (Elekta Neuromag Oy) whole-head system, with a sampling rate of 1,000 Hz and a passband from DC to 330 Hz. Sounds were presented with an ISI chosen randomly from the range 700 to 800 ms. Each stimulus condition was presented about 190 times and the stimulus was generated online with an unique noise source for each scan. During the recordings subjects watched a silent movie and they were asked not to pay attention to the sounds. Spatio-temporal analysis (Scherg 1990) was performed using the BESA®5.2 package (BESA Software GmbH). Separate source models with one equivalent dipole in each hemisphere were used to analyse the P50m, the P200m, and the sustained field (SF). Source waveforms were derived to compute grand-average waveforms for each of the 24 stimulus conditions separately. The specific representations of the P50m and the P200m were analysed using zero-phase 1st-order bandpass filters with passbands from 1 to 80 Hz and 1 to 30 Hz, respectively. The SF was analysed using a first-order low-pass filter of 30 Hz. The magnitudes of these AEF components were used for further correlation analyses. Individual AEF magnitudes could not be derived for each subject reliably, so mean values were derived from grand-average source waveforms. Bootstrap-based standard errors of the mean were calculated to assess the scatter of the AEFs.

Simulated hiss correlates were computed using the physiological route of AIM (Patterson et al. 1995) which simulates the peripheral activation along the basilar membrane with a non-linear transmission line filterbank (Giguère and Woodland 1994). We used 100 channels to cover the range from 100 to 6,000 Hz. Within each channel, neural transduction was simulated using the Meddis hair cell (1988). Finally, strobed temporal integration was used to construct stabilised auditory images (SAIs) (Patterson et al. 1995). The temporal profile of the SAI is referred to as a summary SAI. For each stimulus condition, 50 independent summary SAIs were averaged and the level of the summary SAI activity between −15 and −10 ms was averaged to produce a "SAI hiss correlate" value.

3 Results

Figure 47.1 shows the hissiness ratings averaged across subjects: The effects of asymmetry (ramped-damped), half-life, and modulation were significant ($F_{(1,13)}=111.65$, $P<0.0001$; $F_{(2,26)}=106.43$, $P<0.0001$; and $F_{(3,18)}=25.42$, $P<0.0001$, respectively). All of the interactions were significant (ramped-damped*period: $F_{(3,39)}=18.46$, $P<0.0001$; ramped-damped*half-life: $F_{(2,26)}=29.26$, $P<0.0001$;

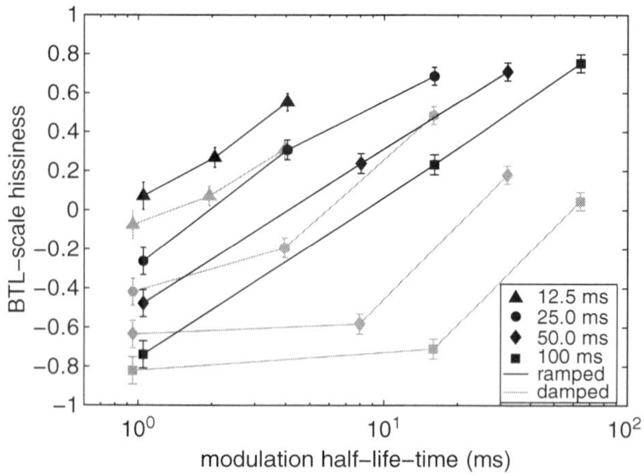

Fig. 47.1 Average hissiness ratings as a function of half-life for ramped (*black lines*) and damped (*grey lines*) noises averaged over subjects; the parameter is modulation period. The error bars show standard errors of the mean

period*half-life: $F_{(6,78)}=25.12$, $P<0.0001$; ramped-damped*period*half-life: $F_{(6,78)}=12.53$, $P<0.0001$).

The gradiometer waves exhibited strong P50m and P200m components as well as a substantial sustained field. Stable spatio-temporal models were readily fitted to these AEFs in both hemispheres for all 14 subjects. The N100m could not be derived reliably from the waves so it was omitted from further analysis. Figure 47.2 shows six grand-average source waveforms for conditions with a wide range of hissiness ratings (left column) and the corresponding average summary SAIs (right column). Figure 47.3 shows the bandpass filtered grand-average waveforms to demonstrate the correspondence of hissiness values with P200 onset and offset responses.

Figure 47.4 shows scatter plots of hissiness against three AEF measures. There are significant correlations between hissiness and P200m magnitude ($r=.85$, $P<0.0001$) (Fig. 47.4b) and between hissiness and SF magnitude ($r=.56$, $P<0.01$). The correlation between hissiness and the P50m is not significant ($r=.33$, ns). Figure 47.5 shows scatter plots and regression lines for the SAI hiss correlate against hissiness rating and P200m magnitude, the strongest of the correlations in Fig. 47.4. Both of the plots show strong correlations.

4 Discussion

The hissiness ratings obtained with the BTL technique are compatible with the ramped/damped discriminations described in Akeroyd and Patterson (1995) and Irino and Patterson (1996). The hissiness rating do have the advantage, however, of differentiating smaller differences than ramped-damped paired comparisons, and it is this that makes it possible to examine the correlations between the hissiness perception and neuromagnetic responses on the one hand and simulated hissiness values derived

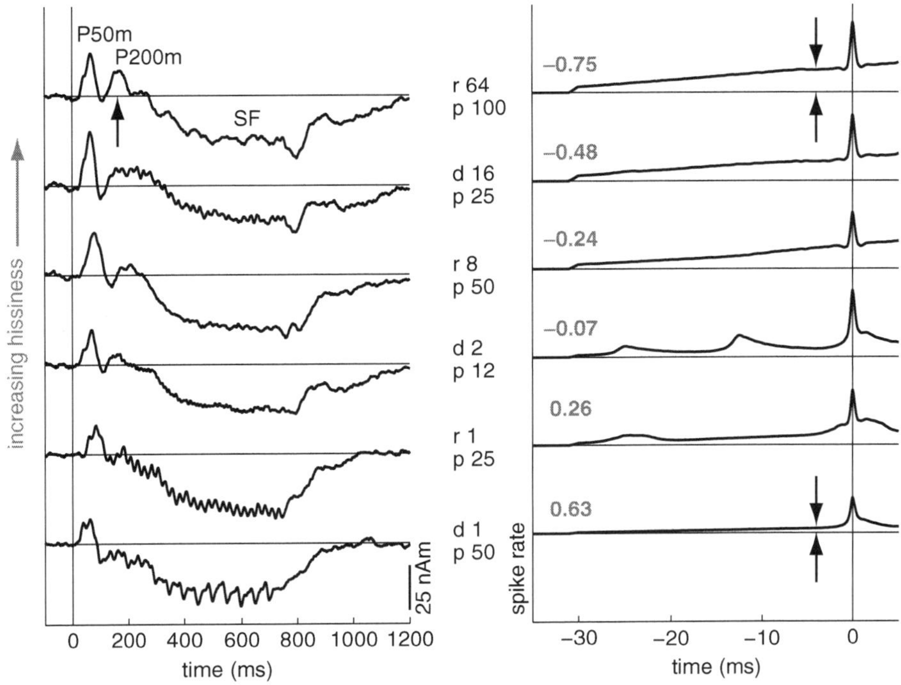

Fig. 47.2 Correspondence of AEF magnitude and summary SAI hiss correlates. *Left panel*: Grand-average source waveforms (pooled over both hemispheres) of the sustained field model for a subset of conditions ordered from high to low hissiness. The P200m magnitude increases with higher hissiness values. The right panel depicts the summary SAI for the corresponding stimulus conditions. As indicated by the *arrows*, SAI activity increases for higher hissiness values

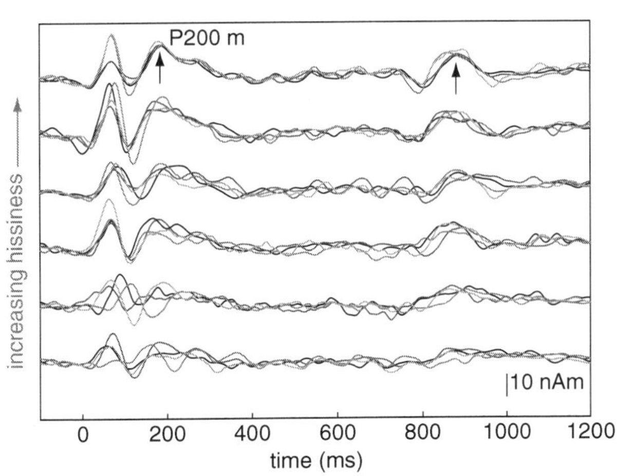

Fig. 47.3 Grand-average P200m source waveforms (1–30 Hz) ordered from low at the *bottom* to high at the *top* (pooled over hemispheres). Large hissiness values are associated with larger P200m amplitudes. Note that there are also P200m offset responses whose magnitude corresponds with that of the P200m onsets ($r = .82$, $P < 0.0001$)

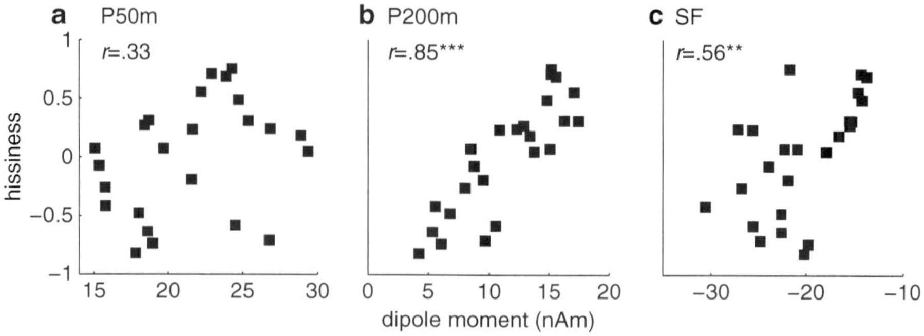

Fig. 47.4 Scatter plots to illustrate the relationship between hissiness value and the P50m (**a**), the P200m (**b**), and the sustained field (**c**)

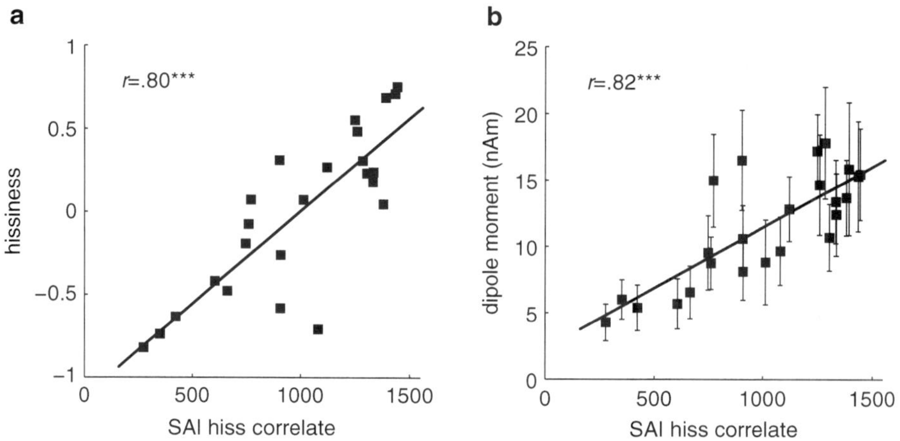

Fig. 47.5 Scatter plots to illustrate the relationship between SAI hiss correlate and hissiness (**a**) and the dipole moment of the P200m (**b**). The error bars show standard errors of the P200m mean values

from SAI on the other hand. The strong pairwise correlations between perceived hissiness, the P200m magnitude, and the SAI hiss correlate indicate that this neuromagnetic component reflects temporal integration. The SF magnitude is also correlated with perceived hissiness, but the correlation is not as strong as with the P200m.

In summary, neuromagnetic responses to damped and ramped noise bursts contain a P200m component that could be used as an objective correlate of the hissiness in the perception of the sounds. The P200m might serve as non-invasive measure of the physiological activity described in models of pitch and timbre processing in humans and so assist in the evaluation of such models.

Acknowledgement Author AR was supported by the German Academic Exchange Service.

References

Akeroyd MA, Patterson RD (1995) Discrimination of wideband noises modulated by a temporally asymmetric function. J Acoust Soc Am 98:2466–2474

David HA (1988) The method of paired comparisons. Oxford University Press, New York

Giguère C, Woodland PC (1994) A computational model of the auditory periphery for speech and hearing research. I. Ascending path. J Acoust Soc Am 95: 331–342

Hartmann WM (1978) The effect of amplitude envelope on the pitch of sine wave tones. J Acoust Soc Am 63:1105–1113

Irino T, Patterson RD (1996) Temporal asymmetry in the auditory system. J Acoust Soc Am 99: 2316–2331

Lu T, Liang L, Wang W (2001) Neural representations of temporally asymmetric stimuli in the auditory cortex of awake primates. J Neurophysiol 85:2364–2380

Neuert V, Pressnitzer D, Patterson RD, Winter IM (2001) The responses of single units in the inferior colliculus of the guinea pig to damped and ramped sinusoids. Hear Res 159:36–52

Patterson RD (1994a) The sound of a sinusoid: spectral models. J Acoust Soc Am 96:1409–1418

Patterson RD (1994b) The sound of a sinusoid: time-interval models. J Acoust Soc Am 96: 1419–1428

Patterson RD, Allerhand M (1995) Time-domain modeling of peripheral auditory processing: a modular architecture and a software platform. J Acoust Soc Am 98:1890–1894

Patterson RD, Irino T (1998) Modeling temporal asymmetry in the auditory system. J Acoust Soc Am 104:2967–2979

Pressnitzer DP, Winter IM, Patterson RD (2000) The responses of single units in the ventral cochlear nucleus of the guinea pig to damped and ramped sinusoids. Hear Res 149:155–166

Rupp A, Uppenkamp S, Bailes J, Gutschalk A, Patterson RD (2005) Time constants in temporal pitch extraction: a comparison of psychophysical and neuromagnetic data. In: Pressnitzer D, de Cheveigné A, McAdams S, Collet L (eds) Auditory signal processing: physiology, psychoacoustics and models. Springer, New York, pp 145–153

Scherg M (1990) Fundamentals of dipole source potential analysis. In: Grandori F, Hoke M, Romani GL (eds) Auditory evoked magnetic fields and potentials. Advances in audiology, vol 5. Karger, Basel, pp 40–69

Stecker GC, Hafter ER (2000) An effect of temporal asymmetry on loudness. J Acoust Soc Am 107:3358–3368

Chapter 48
Cortical Representation of the Combination of Monaural and Binaural Unmasking

Stefan Uppenkamp, Christian H. Uhlig, and Jesko L. Verhey

Abstract The audibility of a target tone is improved by introducing either amplitude modulations that are coherent across different frequency channels of the masker (comodulation masking release, CMR) or interaural phase differences that are different for target and masker (binaural masking-level difference, BMLD). Although the two effects are likely to be based on different processing strategies, they both result in improved figure-background decomposition for a target-in-noise situation. In this study, we analyzed the combination of CMR and BMLD for a target tone in a masker with six 48-Hz-wide noise bands, distributed over a wide frequency range from 216 Hz to 2.78 kHz. Psychoacoustical detection thresholds for the tones in noise were determined for two masker conditions (comodulated or unmodulated bands) and two interaural phase differences of the target tone (0 or 180°). The mean results indicate that the effects of unmasking add independently. The lowest thresholds are found for the dichotic signal embedded in a modulated

S. Uppenkamp (✉)
Medizinische Physik, Carl von Ossietzky University,
Carl-von-Ossietzky-Str. 9-11, 26111 Oldenburg, Germany

Forschungszentrum Neurosensorik,
Carl von Ossietzky University, 26111 Oldenburg, Germany
e-mail: stefan.uppenkamp@uni-oldenburg.de

C.H. Uhlig
Medizinische Physik, Carl von Ossietzky University,
Carl-von-Ossietzky-Str. 9-11, 26111 Oldenburg, Germany

Neurologische Klinik, Universität Heidelberg,
Im Neuenheimer Feld 400, 69120 Heidelberg, Germany

J.L. Verhey
Department of Experimental Audiology, Otto-von-Guericke University of Magdeburg,
Leipziger Str. 44, 39120 Magdeburg, Germany

Forschungszentrum Neurosensorik, Carl von Ossietzky University,
26111 Oldenburg, Germany

B.C.J. Moore et al. (eds.), *Basic Aspects of Hearing*,
Advances in Experimental Medicine and Biology,
DOI 10.1007/978-1-4614-1590-9_48,

masker with an overall threshold difference of about 16 dB compared to the unmodulated condition with no binaural cues. Based on the psychoacoustic results, a set of 12 signal-masker configurations was selected individually to explore the representation of the audibility of the test tone in brain activation maps by means of auditory functional MR imaging. The comparison of the results for the combination of CMR and BMLD with the results for the separate effects indicates a large overlap of the activated brain regions, where a largely extended area is activated, covering primary auditory cortex and adjacent regions. The result is in agreement with previous fMRI studies on auditory masking, identifying specific regions in the auditory cortex representing a change of the audibility of a target tone in a noise masker, irrespective of the overall sound pressure level of the stimulus.

1 Introduction

Comodulation masking release (CMR) denotes the observation of lower masked thresholds for a target embedded in a noise masker, when the masker has modulations that are coherent across frequency channels. The binaural masking-level difference (BMLD) denotes the reduction of the masked threshold for a target in noise, when target and noise differ in their interaural phase relations. Both effects of unmasking are contributing to the formation of auditory objects. Recent psychoacoustic results indicate that the effects add independently, suggesting a serial processing order for the neural correlates of CMR and BMLD (Epp and Verhey 2009).

In a previous study, functional magnetic resonance imaging (fMRI) had been utilized to explore a cortical correlate of psychoacoustical masking (Ernst et al. 2008). The results suggested a partial separation of regions sensitive to changes of the signal-to-noise ratio (S/N) and therefore the audibility of a target in a masker signal, and regions specifically sensitive to changes of the overall level of an acoustic stimulus. Regions sensitive to S/N changes were shown to be comparatively small, mainly around the lateral edge of Heschl's gyrus (HG) and largely overlapping with pitch-sensitive regions identified in several previous studies (Patterson et al. 2002; Krumbholz et al. 2003; Penagos et al. 2004; Bendor and Wang 2005; Schönwiesner and Zatorre 2008), while the effect of overall level is widely distributed over the auditory cortex, with much involvement of more primary regions of auditory cortex and most of the temporal planes (Ernst et al. 2008). A similar result was obtained in an fMRI study using a typical CMR paradigm, with the expected respective shift of all thresholds towards a smaller S/N due to the effect of unmasking in the comodulated condition (Ernst et al. 2010).

The questions for the current study were:

1. How is the combination of CMR und BMLD represented in the activation maps obtainable in a functional MRI experiment employing the BOLD effect?
2. Is there a universal representation of the audibility of a target tone in noise, irrespective of the method used to manipulate this audibility, either monaural in the case of CMR or binaural in the case of BMLD?

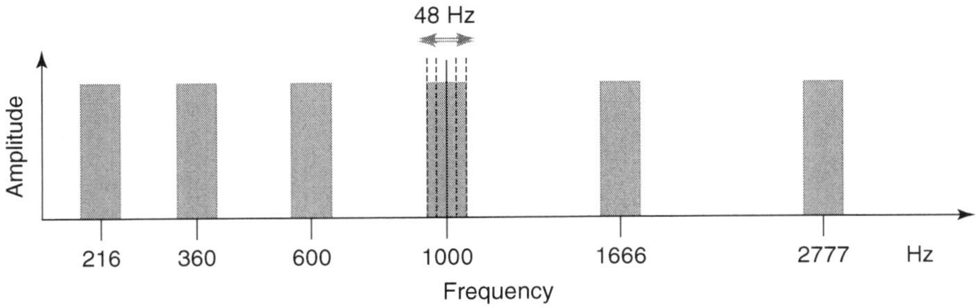

Fig. 48.1 Sketch of stimulus configuration in the frequency domain

2 Methods

The masker was generated in the frequency domain as the sum of six bands of Gaussian noise with center frequencies 216 Hz, 360 Hz, 600 Hz, 1 kHz, 1.67 kHz, and 2.78 kHz and a bandwidth of 48 Hz each. Masker duration was 500 ms. For the uncorrelated (UN) condition, the intrinsic modulations in each masker band caused by the narrow bandwidth were independent from each other, resulting in a comparatively flat envelope for the overall time signal. Since the spacing between the narrow frequency bands is large, the sound of the overall signal resembles that of a stationary inharmonic complex tone. For the comodulated (CM) condition, all noise bands had the same envelope, i.e., the resulting masker had a distinct irregular modulation of the overall time signal. The masker signals were always presented diotically (N_0 configuration). The target signal was a sinusoidal tone of 250-ms duration. The signal frequency was chosen from trial to trial with an average frequency of 1 kHz and a frequency jitter of ±24 Hz, to avoid adaption effects for the target. A sketch of the spectrum of the stimulus is shown in Fig. 48.1. The target signals were presented either diotically (S_0) or with an interaural phase difference of 180° (S_π).

Twenty-one listeners (21–47 years, mean 25 years, 12 females) participated in the study. None of the listeners had a history of neurological illness, head injury, or hearing impairment in the explored frequency range. Written informed consent was obtained from all participants. The study was approved by the local ethics committee of the University of Oldenburg.

3 Detection Thresholds

All psychoacoustic measurements were carried out in a double-walled sound booth. Masked thresholds for the target in the presence of the UN and the CM masking noise were determined using an adaptive three-alternative, forced-choice procedure (1 up 2 down algorithm for the target level, minimum step size 1 dB) for all four combinations of target-masker configuration: S_0N_0-UN, $S_\pi N_0$-UN, S_0N_0-CM, and $S_\pi N_0$-CM. The

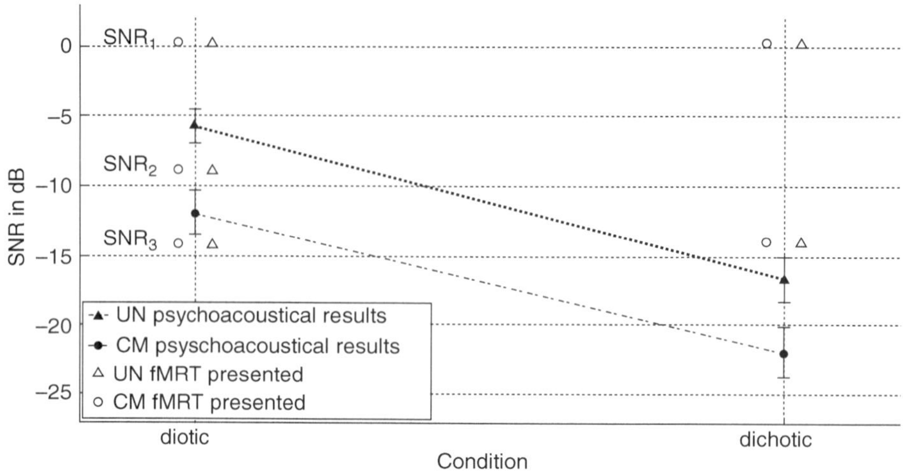

Fig. 48.2 Results from detection threshold measurements. The *filled symbols* represent the mean detection thresholds for all four signal-masker combinations, i.e., unmodulated (*triangles*) and comodulated (*filled circles*) masker for a diotic configuration (S_0N_0) and for a target tone with an interaural phase difference of 180° ($S_\pi N_0$); *error bars* show standard deviation across all 21 participants. The *open symbols* indicate those signal-to-noise conditions (SNR_1, SNR_2, SNR_3) which are chosen for following fMRI experiment

digital signals were played via D/A converters RME ADI-8DS, a headphone amplifier TDT-HB7, and headphones Sennheiser HD 650 at a fixed overall level of 57 dB SPL for the masker. The mean results for all 21 participants for the four different target-masker conditions are shown in Fig. 48.2 (filled symbols connected by dashed lines). In summary, the size of the CMR effect for this signal-masker configuration is about 6 dB, irrespective of the interaural phase for the target signal, while the BMLD effect is about 10 dB, irrespective of the masker being uncorrelated or comodulated. Although the CMR and BMLD were somewhat different from listener to listener (the standard deviation across listeners is indicated by error bars), the overall observation was largely consistent across listeners. This apparent lack of interaction between CMR and BMLD is consistent with the previous interpretation of largely independent processing of monaural and binaural cues involved in the release from masking (Epp and Verhey 2009).

4 Functional MRI Experiment

4.1 Stimulus Selection, Procedure, and Data Analysis

The psychoacoustic results were used to choose three signal-to-noise ratios for each listener to explore the effect of SNR on an fMRI correlate of target audibility. The first S/N (SNR_1) was selected so that the target signal was audible for both noise conditions and both interaural phase conditions. This S/N was set to 6 dB above the masked threshold for the S_0N_0-UN configuration, i.e., the most effective masker condition. The second S/N (SNR_2) was selected so that the target signal in the S_0N_0 configuration was not audible for the uncorrelated noise but audible for the comodulated noise ("halfway" between S_0N_0-UN and S_0N_0-CM thresholds). This S/N was picked to search for a correlate of the CMR effect alone, irrespective of binaural cues. The third S/N (SNR_3) was selected so that the target was inaudible in both S_0N_0 configurations but audible for both $S_\pi N_0$ configurations ("halfway" between S_0N_0-CM and $S_\pi N_0$-UN thresholds), to identify a correlate of the BMLD effect, largely independent of the effect of the modulation of the masker signal.

Thirteen different sound conditions in total were used during the fMRI recording for each of the 21 listeners. These included the conditions S_0N_0-UN at SNR_1, SNR_2, and SNR_3; S_0N_0-CM at SNR_1, SNR_2, and SNR_3; $S_\pi N_0$-UN at SNR_1 and SNR_3; $S_\pi N_0$-CM at SNR_1 and SNR_3; N_0-UN and N_0-CM alone with no target signal; and finally a condition with no signal presentation as a baseline control. The selected signal-to-noise ratios are shown as additional symbols in Fig. 48.2, to illustrate the relationship of the fMRI design relative to the results from the preceding detection experiments. As mentioned above, signal-to-noise ratios were chosen individually based on the results for the masked thresholds determined before. A subgroup of 16 listeners showed largely homogeneous results in their detection thresholds. The mean selected S/N for these listeners were $SNR_1=0.1$ (1.1) dB, $SNR_2=-9.2$ (1.1) dB, and $SNR_3=-14.7$ (1.8) dB. Five more listeners showed bigger deviations from this general trend, mainly with respect to the size of the CMR effect, which was very small for some of the participants. They were still included the fMRI study, with correspondingly different signal-to-noise ratios for the experimental conditions (not listed in detail here).

Each stimulus presentation was a sequence of twelve 400-ms noise bursts together with the respective target at 1 kHz and jittered by ±24 Hz from burst to burst. After each presentation of a stimulus block (i.e., during the respective scan acquisition), the participants indicated by button presses whether they heard the slightly varying tone in the masking noise. This task was introduced to maintain the participants' attention. Stimuli were played via MR-compatible insert earphones (Sensimetrics S14) using a fixed masker level of 60 dB SPL. Throughout the full fMRI experiment, each condition was repeated 36 times, giving a total of 468 brain images. The order of conditions was fully randomized, and the complete session was divided into three runs with 156 scans each. Functional MRI data were acquired using a Siemens Sonata 1.5 T MRI system. For the functional data, 21 axial EPI slices (in-plane resolution 3 × 3 mm; thickness 5 mm, echo time TE = 63 ms to maximize BOLD contrast) were acquired covering most of the cortex, including the whole of the temporal lobes and frontal regions. Sparse imaging with clustered volume acquisition was used (Hall et al. 1999). The total volume acquisition time TA was 2.7 s. On each trial, there was a 5-s stimulus interval followed by the 2.7-s scanning interval, making a total repetition time of TR = 7.7 s. A T_1-weighted high-resolution anatomical image (176 sagittal slices, TR = 2.11 s, TE = 4.38 ms) was also collected for each subject.

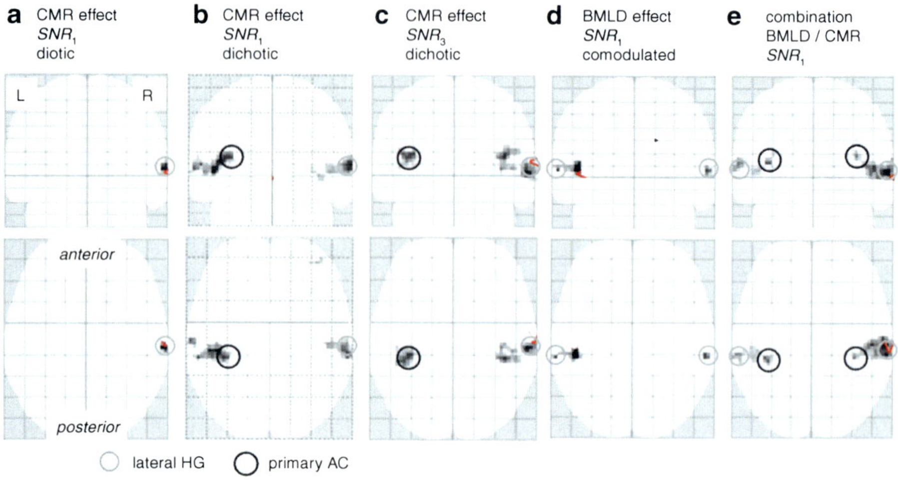

Fig. 48.3 Examples of fMRI activation maps for the specific effects of unmasking for a CMR configuration (**a–c**), a BMLD configuration (**d**), and the combined effect (**e**)

All anatomical and functional data were analyzed using statistical parametric mapping (SPM5, http://www.fil.ion.ucl.ac.uk/spm). The preprocessing of the functional brain images included realignment of subject motion, normalization to a standard EPI template, and smoothing with a Gaussian filter of 5-mm full width at half maximum for all directions. To begin with, a fixed-effects model for the whole group of 21 participants was used to derive a region of interest for further random-effects analysis. This region of interest was based on the general effect of sound presentation (i.e., contrast between all sound conditions combined vs. silence) and covered essentially all of the expected auditory regions in cortex (most of the temporal lobes) as well as inferior colliculus in the brainstem. For all participants, a general linear model describing the time series of the fMRI signal was then estimated on an individual basis. The respective contrast images for the comparisons of conditions characterizing the CMR effect, the BMLD effect, or the combination were fed into a second level analysis searching for consistent activation across the whole group of listeners.

4.2 Results from fMRI Experiment

Figure 48.3 shows, for the coronal and axial orientations in "glass brain" view, examples for those areas in auditory cortex that exhibit consistent activation across all participants for the specific effects of introducing comodulation for the masker (CMR effect, Fig. 48.3a–c), the specific effect of introducing an interaural phase difference of 180° for the target signal (BMLD effect, Fig. 48.3d), and for the combination of both effects (Fig. 48.3e). In Fig. 48.3a–c, contrast images were calculated for the signal-plus-masker configuration with CM vs. UN masker, at those S/N for which the target was audible (i.e., SNR_1 for the diotic and dichotic conditions and SNR_3 for the dichotic condition only). The change from unmodulated to comodulated masker increases the audibility of the target in all cases, due to the effect of unmasking. This perceptive change in audibility is accompanied by activation in lateral HG in at least one hemisphere and for the dichotic conditions (Fig. 48.3b, c) also in primary auditory regions. This result is in line with previous reports of the involvement of lateral HG in processing the audibility of a tonal target, as well as the involvement of primary auditory cortex in the processing of amplitude modulations.

The question is whether a similar result can also be demonstrated when the effect of unmasking, i.e., the increase of the audibility of the tonal target, is based on a binaural cue due to interaural phase differences of the target only, rather than on the monaural cue of a change in the envelope of the masker. One example for the effect of the binaural cue is shown in Fig. 48.3d, where the masker was kept constant, and only the interaural phase difference for the tonal target was changed. As expected, lateral HG is now activated in both hemispheres, indicating the effect of a change of audibility of the tone, while the involvement of primary auditory regions is now largely missing, as any specific effect of amplitude modulations is not included in this contrast. The comparison of the results in Fig. 48.3a–c with the results in Fig. 48.3d indicates that those regions in auditory cortex representing the audibility of a tonal target, i.e., lateral HG, are consistently activated, irrespective of the particular way to manipulate the audibility.

Finally, Fig. 48.3e shows the effect of combining monaural and binaural unmasking cues, a "mixed" contrast between the conditions $S_\pi N_0$-CM and S_0N_0-UN at SNR_1 (0.1 dB on average), representing the maximum difference of 16 dB in masked threshold that was achieved during this study. Significant BOLD activation in lateral HG and in primary auditory cortex can now be observed in a similar way in both hemispheres. This essentially reflects the superposition of the CMR-specific effect and the BMLD-specific effect, in line with the psychoacoustic results reported in Sect. 3.

5 Discussion

Psychoacoustic detection thresholds for a tone-in-noise masking experiment showed no interaction between the effects of CMR and BMLD. Both effects contribute independently to unmasking, thereby increasing the audibility of the target. This finding provides additional evidence for the hypothesis that the effects are based on independent processors that probably come in serial order (Epp and Verhey 2009).

Based on the psychoacoustic results, a set of 12 different signal-masker conditions was selected on an individual basis, to identify a cortical correlate of unmasking and the combination of CMR and BMLD by means of fMRI. Overall, the observed effects in the BOLD response are comparatively weak, since many of the stimulus conditions employed in our design have been close to or even below threshold for the detection of the target in the noise masker, and therefore, the perceptual difference between conditions was comparatively small. Still, two main areas involved in the effects of interest have been identified consistently across the whole group of 21 participants in a second-level analysis of the respective contrast images.

Activation in lateral Heschl's gyrus appears to be largely determined by the audibility of the tone in the background noise, irrespective of whether this audibility is varied by the effect of BMLD or the effect of CMR. It is therefore interpreted as a general perceptual correlate of the tone rather than a correlate of the particular processing underlying the effect of unmasking. In contrast, fMRI activation in primary auditory regions appears to be more related to the difference between comodulated and unmodulated masker signals. This activation is not detected in the analysis of the isolated BMLD effect. It is therefore more likely representing the specific processing of amplitude modulated stimuli, rather than the audibility of the target tone itself. This interpretation would be in line with previous reports on AM-specific activation in fMRI activation maps (Giraud et al. 2000; Ernst et al. 2010). Simultaneous consideration of the CMR and BMLD effects in the data analysis results in an fMRI activation map that is largely determined by the independent combination of both effects separately, in line with the idea of serial processing for the neural correlates of CMR and BMLD.

Acknowledgment This study was supported by a grant from the Deutsche Forschungsgemeinschaft (UP 10/2-2).

References

Bendor D, Wang X (2005) The neuronal representation of pitch in primate auditory cortex. Nature 436:1161–1165

Epp B, Verhey JL (2009) Superposition of masking releases. J Comput Neurosci 26:393–407

Ernst SMA, Verhey JL, Uppenkamp S (2008) Spatial dissociation of changes of level and signal-to-noise ratio in auditory cortex for tones in noise. Neuroimage 43:321–328

Ernst SMA, Uppenkamp S, Verhey JL (2010) Cortical representation of release from auditory masking. Neuroimage 49:835–842

Giraud AL, Lorenzi C, Ashburner J, Wable J, Johnsrude I, Frackowiak R, Kleinschmidt A (2000) Representation of the temporal envelope of sounds in the human brain. J Neurophysiol 84: 1588–1598

Hall DA, Haggard MP, Akeroyd MA, Palmer AR, Summerfield AQ, Elliott MR, Gurney EM, Bowtell RW (1999) "Sparse" temporal sampling in auditory fMRI. Hum Brain Mapp 7:213–223

Krumbholz K, Patterson RD, Seither-Preisler A, Lammertmann C, Lütkenhöner B (2003) Neuromagnetic evidence for a pitch processing center in Heschl's gyrus. Cereb Cortex 13: 765–772

Patterson RD, Uppenkamp S, Johnsrude IS, Griffiths TD (2002) The processing of temporal pitch and melody information in auditory cortex. Neuron 36:767–776

Penagos H, Melcher JR, Oxenham AJ (2004) A neural representation of pitch salience in nonprimary human auditory cortex revealed with functional magnetic resonance imaging. J Neurosci 24:6810–6815

Schönwiesner M, Zatorre RJ (2008) Depth electrode recordings show double dissociation between pitch processing in lateral Heschl's gyrus and sound onset processing in medial Heschl's gyrus. Exp Brain Res 187:97–105

Chapter 49
Processing of Short Auditory Stimuli: The Rapid Audio Sequential Presentation Paradigm (RASP)

Clara Suied, Trevor R. Agus, Simon J. Thorpe, and Daniel Pressnitzer

Abstract Human listeners seem to be remarkably able to recognise acoustic sound sources based on timbre cues. Here we describe a psychophysical paradigm to estimate the time it takes to recognise a set of complex sounds differing only in timbre cues: both in terms of the minimum duration of the sounds and the inferred neural processing time. Listeners had to respond to the human voice while ignoring a set of distractors. All sounds were recorded from natural sources over the same pitch range and equalised to the same duration and power. In a first experiment, stimuli were gated in time with a raised-cosine window of variable duration and random onset time. A voice/non-voice (yes/no) task was used. Performance, as measured by d', remained above chance for the shortest sounds tested (2 ms); d's above 1 were observed for durations longer than or equal to 8 ms. Then, we constructed sequences of short sounds presented in rapid succession. Listeners were asked to report the presence of a single voice token that could occur at a random position within the sequence. This method is analogous to the "rapid sequential visual presentation"

C. Suied (✉)
Département d'études cognitives, Equipe Audition,
Ecole Normale Supérieure, Paris, France

Laboratoire de Psychologie de la Perception (UMR CNRS 8158),
Université Paris Descartes, Paris, France

Département Action et Cognition en Situation Opérationnelle,
Institut de Recherche Biomédicale des Armées, Brétigny-sur-Orge, France
e-mail: clara.suied@irba.fr

T.R. Agus • D. Pressnitzer
Département d'études cognitives, Equipe Audition,
Ecole Normale Supérieure, Paris, France

Laboratoire de Psychologie de la Perception (UMR CNRS 8158),
Université Paris Descartes, Paris, France

S.J. Thorpe
Centre de Recherche Cerveau et Cognition,
Université Toulouse 3, CNRS. CHU Purpan, Pavillon Baudot, Toulouse, France

B.C.J. Moore et al. (eds.), *Basic Aspects of Hearing*,
Advances in Experimental Medicine and Biology,
DOI 10.1007/978-1-4614-1590-9_49,

paradigm (RSVP), which has been used to evaluate neural processing time for images. For 500-ms sequences made of 32-ms and 16-ms sounds, d' remained above chance for presentation rates of up to 30 sounds per second. There was no effect of the pitch relation between successive sounds: identical for all sounds in the sequence or random for each sound. This implies that the task was not determined by streaming or forward masking, as both phenomena would predict better performance for the random pitch condition. Overall, the recognition of familiar sound categories such as the voice seems to be surprisingly fast, both in terms of the acoustic duration required and of the underlying neural time constants.

1 Introduction

In this chapter, we describe a new timbre recognition task, using familiar sound sources such as the human voice, which focuses on the time course of recognition. The aim is to develop performance measures that are directly applicable to future physiological investigations of the neural bases of timbre processing.

In a first experiment, we extracted short snippets from natural recordings and measured the ability of listeners to recognise them, as a function of the duration of the snippets (Gray 1942; Robinson and Patterson 1995). This provides an estimate of the minimal duration of the acoustic features required for recognition; here, it is also used as a baseline for the second series of experiments.

In this second series, we chose sound durations that were short but recognisable when presented in isolation, and we constructed sequences of those short sounds in rapid succession. The method is similar to the classic visual task termed rapid sequential visual presentation (RSVP, e.g. Subramaniam et al. 2000). In RSVP, among other things, it is found that short stimulus-onset asynchrony (SOA) causes a drop in performance for the identification of a target image embedded in the sequence. Physiological data have also been collected with RSVP; the SOAs for which performance falls to chance have been shown to reflect neural time constants in the recognition process (Keysers et al. 2001).

2 Experiment 1: Timbre Recognition at Short Durations

2.1 Methods

Sixteen participants took part in the study (19–38 years old). They all reported normal hearing.

Similar to Agus et al. (2012), samples of recorded sound sources were extracted from the RWC database (Goto et al. 2003). They consisted of sung voices and musical instruments, all produced at 12 different pitches (A3 to G#4). The target set was the human voice (/a/ and /i/ vowels, sung by a male tenor singer). The distractor set consisted of seven different instruments (bassoon, clarinet, oboe, piano, saxophone, trumpet, and trombone). The oboe's range does not extend to a note at A3, so one was created by resampling its A#3 down a semitone.

Stimuli were then gated in time by multiplication with a raised-cosine window. The duration of the window could be 2, 4, 8, 16, 32, 64, or 128 ms. The starting point of the window was chosen randomly between 0 and 100 ms in the original 256-ms sample. The precise segment of the sound that was presented to the listener was thus different on each trial. No attempt was made to synchronise the window with the period of the stimulus. Finally, stimulus intensities were normalised by their root-mean-square values and divided by the square root of their durations (Robinson and Patterson 1995).

Stimuli were presented through an RME Fireface digital-to-analogue converter at a 16-bit resolution and a 44.1-kHz sample rate. They were presented diotically through Sennheiser HD 250 Linear II headphones. Presentation level was calibrated at 70 dB(A) for the 128-ms stimuli. Listeners were tested individually in a double-walled IAC sound booth.

In each trial, participants heard a single sound, which could be either extracted from a voice or a musical instrument (/a/ or /i/ for the target voices, or any instrument for the distractors). They had to indicate whether the sound they just heard was from the voice category (yes/no). Visual feedback was provided after each response. Voices were presented on 50 % of the trials. The seven gate durations were presented in random order. For each gate condition, 50 repetitions were collected per participant. In a given trial, the sound source and its pitch were chosen randomly. The only cues that participants could use to perform the task were timbre cues, pitch being random and loudness being approximately the same for all stimuli.

Performance was evaluated with the d' statistics of signal detection theory (MacMillan and Creelman 2005), with signal trials being all those containing a vocal sound and non-signal trials being all those with a distractor. Two participants were excluded from subsequent analyses as they performed poorly at the longer durations tested. All analyses are presented on the remaining 14 participants.

2.2 Results

Results of Experiment 1 are displayed in Fig. 49.1a. As expected, performance is good for long durations and decreases as duration gets shorter. A repeated-measures ANOVA with duration as the within-subjects variable confirmed a significant main effect of duration [$F(6,78)=166.76$, $p<0.001$].

Voices could be recognised significantly better than chance right from the shortest duration tested, i.e. a raised-cosine window of 2-ms length in total [$t(13)=4.3$, $p<0.001$].

2.3 Interim Discussion

The present results confirm and extend previous findings (Gray 1942; Robinson and Patterson 1995): here, recognition of very short sounds still occurs (1) for a relatively

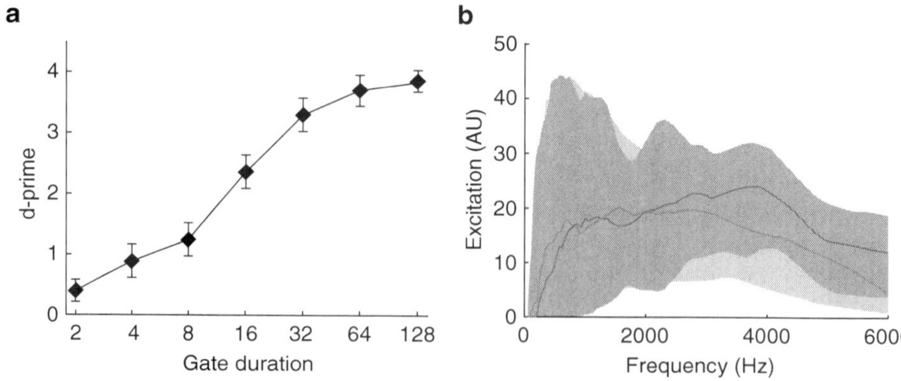

Fig. 49.1 (**a**) Results for Experiment 1. Voice recognition performance is plotted as mean d' over the group of listeners ($N=14$) as a function of the gate duration. Given the number of trials, the maximum possible d' is 4.1. *Error bars* represent the standard error about the mean. (**b**) Median excitation patterns and interquartile range (voice: *dark gray*; instruments: *light gray*) for the 8-ms gate duration (see text for details)

diverse set of natural recordings (2) for durations smaller than the pitch period (3) with gate onsets chosen randomly within the sounds. It should be noted that the minimum duration observed should, in all likelihood, depend on the stimulus set investigated. In particular, we selected a voice-detection task that seems to be particularly easy for human listeners, at least for longer sounds (Agus et al. 2012). In any case, the good performance obtained for relatively short sounds ($d'>1$ for 8 ms and above) enables the design of the rapid sequential paradigm of Experiments 2 and 3.

We calculated excitation patterns for the target and distractor categories (gammatone filterbank followed by half-wave rectification, low-pass filtering, and logarithmic compression; Patterson et al. 1995). Results are illustrated in Fig. 49.1b, for the 8-ms gate duration; results with other gate durations are not shown but they were similar. While it is clear that there must be some spectral cues that listeners used to perform the task, at least for the shorter durations, it does not appear that these cues are trivially simple given the large overlap between interquartiles (see also Agus et al. 2012, for further acoustic analyses of the sound set).

3 Experiments 2 and 3: Rapid Audio Sequential Presentation (RASP)

3.1 Methods

The 14 listeners of Experiment 1 were divided in two groups: 8 listeners took part in Experiment 2 and 6 listeners took part in Experiment 3.

Gated sounds were generated as in Experiment 1. These sounds were presented in sequences that differed in terms of their SOAs and number of sounds.

In Experiment 2, all sequences had a fixed duration of 500 ms. In half of the blocks, they were composed of 32-ms sounds; in the other half of the blocks, 16-ms sounds were used. For the 32-ms sequences, the presentation rate (sounds per second, or equivalently 1/SOA) varied between 5.3 and 30 Hz on a logarithmic scale. For the 16-ms sequences, the rate varied between 5.3 and 60 Hz. Nontarget trials were composed of musical instruments only. For target trials, one sound of the sequence was a voice sound, at a random position but not first or last. Sound sources in the sequences were chosen randomly from the distractor and target sets, as appropriate, with pitch also chosen randomly for each sound.

In Experiment 3, two experimental factors were tested: the number of sounds and the pitch relation between sounds. Sequences could have either a fixed duration (500 ms, as in Experiment 2) or a fixed number of sounds (7 sounds). The pitch for each sound in the sequence could either be drawn randomly for each sound (as in Experiment 2) or pitch was held constant for all sounds throughout a sequence (the pitch value was drawn randomly for each sequence). For this experiment, only the 32-ms sounds were used.

The same apparatus as in Experiment 1 was used.

Participants had to indicate whether each sequence included a voice sound. Target sequences containing a voice sound were presented 50 % of the time. Visual feedback was provided after each response. Prior to the experiments, participants performed a short training session, including easier sequences composed of 64-ms sounds.

In Experiment 2, the two types of sequences (16 and 32 ms) were tested in separate blocks, always in the same order, with the 32-ms sequences first. The 16-ms sequences were included in the design as it appeared that listeners performed the task above chance for the shortest possible SOA (30 Hz) with the 32-ms sounds (see Sect. 3.2.1). Within a block, the presentation rates were randomised. For each condition, 60 repetitions were collected per participant.

In Experiment 3, the two types of sequences were tested in separate blocks: sequences with a fixed duration and sequences with a fixed number of sounds. Within each block, we compared the two pitch conditions, fixed or random. The 6 (rate) × 2 (pitch) conditions were presented in a random order. For each condition, 60 repetitions were collected per participant.

Performance was again evaluated with d'. The assumption was made that listeners performed their judgment on a "voice" signal dimension, without any correction for the number of non-voice sounds presented (which could vary from trial to trial because of the presentation rate). For Experiment 2, two participants had an unusual pattern of results: their sequence performance was poor, for all presentation rates, even though they had good performance for the recognition of sounds presented in isolation (Experiment 1). These participants were excluded from further analysis.

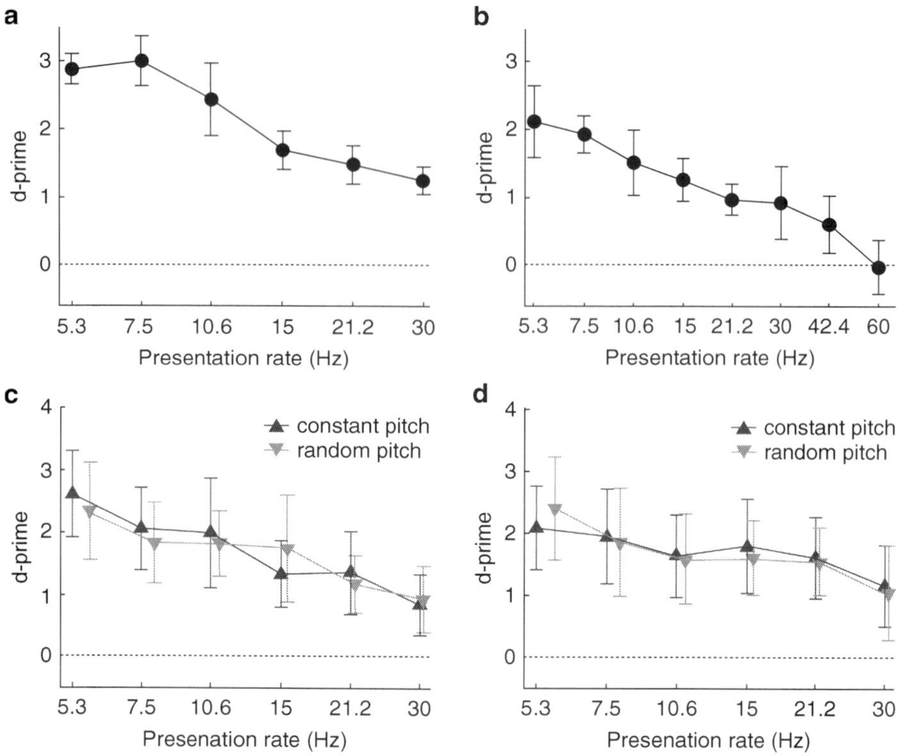

Fig. 49.2 (**a**) Results for Experiment 2. Mean *d′* ($N=6$) and standard error about the mean for the sequences composed of 32-ms sounds. (**b**) Same as **a**, for the sequences composed of 16-ms sounds. (**c**) Results for Experiment 3. Sequences with fixed duration (500 ms, as in **a**). (**d**) Same as (**c**), for the sequences with a fixed number of sounds (7)

3.2 Results

3.2.1 Experiment 2

Results are displayed in Fig. 49.2a, b. As expected, as the rate of the sequence increased, performance decreased. For the 32-ms and 16-ms sound duration sequences separately, data were analysed with a repeated-measures ANOVA with presentation rate as the within-subjects variable. For both type of sequences, the ANOVA revealed a significant main effect of rate [32 ms: $F(5,25)=21.35$, $p<0.001$; 16 ms: $F(7,35)=14.97$, $p<0.001$].

Sequences could be processed significantly better than chance at the fastest possible rate for the 32-ms sequences, i.e. 30 Hz [$t(5)=11.9$, $p<0.001$]. For the 16-ms sequences, the rate of 30 Hz was above chance as well [$t(5)=3.2$, $p<0.05$], 42.4 Hz was close to significance [$t(5)=2.5$, $p=0.06$], and 60 Hz was at chance [$t(5)=-0.6$, $p=0.6$].

3.2.2 Experiment 3

Results are displayed in Fig. 49.2c, d. A repeated-measures ANOVA was run for each sequence type (fixed duration or fixed number of sounds) with presentation rate and pitch condition as within-subjects variables. For both types of sequences, the ANOVA revealed a significant main effect of rate [fixed duration: $F(5,25)=26.21$, $p<0.001$; fixed number of sounds: $F(5,25)=12.18$, $p<0.001$]. Pitch condition did not have a significant effect on performance, neither as a main effect [fixed duration: $F(1, 5)=0.1$, $p=0.8$; fixed number of sounds: $F(1, 5)=0.3$, $p=0.6$] nor as an interaction with rate [fixed duration: $F(5,25)=1.6$, $p=0.2$; fixed number of sounds: $F(5,25)=0.5$, $p=0.8$].

Finally, to compare the two types of blocks, we performed a repeated-measures ANOVA including all the data and using the sequence type as a within-subjects variable. This ANOVA showed that there was no significant effect of the sequence type [$F(1,5)=0.04$, $p=0.9$]: performance was the same for a given SOA irrespective of the number of distractor sounds included in the sequence. A small but significant interaction was observed between sequence type and the rate [$F(5,25)=3.7$, $p<0.05$].

3.3 *Interim Discussion*

The results of Experiment 3 rule out two important mechanisms potentially influencing performance in a RASP task. First, streaming may play a role, as performance should improve when the target sound can be streamed out of the sequence. Second, total or partial forward masking may limit the detectability of each sound in the sequence or distort its audible spectrum, thus reducing performance. Both explanations would predict better performance if pitch is varied from sound to sound within the sequence; streaming should become easier, and forward masking should be alleviated as harmonics of consecutive sounds are separated in frequency. However, this is not what was observed; performance was effectively the same for constant-pitch and variable-pitch sequences. Thus, it seems that performance was mostly limited by factors related to timbre recognition.

4 Discussion and Conclusions

The present study investigates various time constants involved in timbre recognition. Together with the results of Agus et al. (2012), we observe that with vocal target sounds and musical instruments sounds, timbre recognition (1) is fast (first reliable response time at 255 ms, Agus et al. (2012)), (2) remains reliable for short-duration sounds (2 ms), and (3) is maintained for sequences with high presentation rates (30 Hz).

The present experiment combined a rapid sequential presentation paradigm with a task of timbre recognition. There have been previous investigations of RSVP-like paradigms in the auditory modality but using pure tones or verbal material (e.g. Woods and Alain 1993; Duncan et al. 1997). It is likely that the maximum rate will depend to some extent on the stimulus set, as it is the case in vision (Subramaniam et al. 2000; Keysers et al. 2001). It is still noticeable that the high rate achieved by listeners here is at the boundary of the existence region for pitch (Pressnitzer et al. 2001). Also, it is faster than the rates thought to convey semantic information in speech (Drullman et al. 1994).

The behavioural time constants of timbre recognition could, in the future, be tested with neurophysiological techniques. In the visual modality, for instance, the RSVP paradigm has been used in single-unit recordings to show that selectivity to complex stimuli such as faces is already expressed in the first volley of spikes (Keysers et al. 2001). Based on our findings, it is possible that similar results would be observed in some auditory areas, such as the voice-selective ones suggested by brain imaging (Belin et al. 2000).

Finally, a tentative comparison can be made with another visual task: that of visual search and the associated effect of pop-out. Perhaps because of our choice of target (voice), performance was immune to the number of distractors in the RASP sequence. Auditory pop-out has been investigated by Cusack and Carlyon (2003) using basic features such as frequency modulation or duration. Here, we may have observed a pop-out caused by complex features, such as those learnt by listeners to recognise voices among crowded auditory environments.

Acknowledgements This work was supported by the Fondation Pierre Gilles de Gennes pour la Recherche.

References

Agus TA, Suied C, Thorpe SJ, Pressnitzer D (2012) Fast recognition of musical sounds based on timbre. J Acoust Soc Am 131:4124–4133

Belin P, Zatorre RJ, Lafaille P, Ahad P, Pike B (2000) Voice-selective areas in human auditory cortex. Nature 403:309–312

Cusack R, Carlyon RP (2003) Perceptual asymmetries in audition. J Exp Psychol Hum Percept Perform 29:713–725

Drullman R, Festen JM, Plomp R (1994) Effect of temporal envelope smearing on speech reception. J Acoust Soc Am 95:1053–1064

Duncan J, Martens S, Ward R (1997) Restricted attentional capacity within but not between sensory modalities. Nature 387:808–810

Goto M, Hashiguchi H, Nishimura T, Oka R (2003) RWC music database: music genre database and musical instrument sound database. In: 4th international conference on Music Information Retrieval, Baltimore, 2003

Gray GW (1942) Phonemic microtomy: the minimum duration of perceptible speech sounds. Speech Monogr 9:75–90

Keysers C, Xiao DK, Foldiak P, Perrett DI (2001) The speed of sight. J Cogn Neurosci 13:90–101
Macmillan NA, Creelman CD (2005) Detection theory: a user's guide, 2nd edn. Lawrence Erlbaum Associates, Mahwah
Patterson RD, Allerhand MH, Giguere C (1995) Time-domain modeling of peripheral auditory processing: a modular architecture and a software platform. J Acoust Soc Am 98:1890–1894
Pressnitzer D, Patterson RD, Krumbholz K (2001) The lower limit of melodic pitch. J Acoust Soc Am 109:2074–2084
Robinson K, Patterson RD (1995) The stimulus duration required to identify vowels, their octave, and their pitch chroma. J Acoust Soc Am 98:1858–1865
Subramaniam S, Biederman I, Madigan S (2000) Accurate identification but no priming and chance recognition memory for pictures in RSVP sequences. Vis Cogn 7:511–535
Woods DL, Alain C (1993) Feature processing during high-rate auditory selective attention. Percept Psychophys 53:391–402

Chapter 50
Integration of Auditory and Tactile Inputs in Musical Meter Perception

Juan Huang, Darik Gamble, Kristine Sarnlertsophon, Xiaoqin Wang, and Steven Hsiao

Abstract Musicians often say that they not only hear but also "feel" music. To explore the contribution of tactile information to "feeling" music, we investigated the degree that auditory and tactile inputs are integrated in humans performing a musical meter-recognition task. Subjects discriminated between two types of sequences, "duple" (march-like rhythms) and "triple" (waltz-like rhythms), presented in three conditions: (1) unimodal inputs (auditory or tactile alone); (2) various combinations of bimodal inputs, where sequences were distributed between the auditory and tactile channels such that a single channel did not produce coherent meter percepts; and (3) bimodal inputs where the two channels contained congruent or incongruent meter cues. We first show that meter is perceived similarly well (70–85 %) when tactile or auditory cues are presented alone. We next show in the bimodal experiments that auditory and tactile cues are integrated to produce coherent meter percepts. Performance is high (70–90 %) when all of the metrically important notes are assigned to one channel and is reduced to 60 % when half of these notes are assigned to one channel. When the

The authors Xiaoqin Wang and Steven Hsiao contributed equally to this study.

J. Huang (✉)
The Solomon H. Snyder Department of Neuroscience,
Zanvyl Krieger Mind/Brain Institute, The Johns Hopkins University,
Baltimore, MD 21205, USA

Laboratory of Auditory Neurophysiology, Department of Biomedical Engineering,
The Johns Hopkins University, Baltimore, MD 21205, USA
e-mail: jhuang7@jhu.edu

D. Gamble • X. Wang
Laboratory of Auditory Neurophysiology, Department of Biomedical Engineering,
The Johns Hopkins University, Baltimore, MD 21205, USA

K. Sarnlertsophon • S. Hsiao
The Solomon H. Snyder Department of Neuroscience,
Zanvyl Krieger Mind/Brain Institute, The Johns Hopkins University,
Baltimore, MD 21205, USA

B.C.J. Moore et al. (eds.), *Basic Aspects of Hearing*,
Advances in Experimental Medicine and Biology,
DOI 10.1007/978-1-4614-1590-9_50,

important notes are presented simultaneously to both channels, congruent cues enhance meter recognition (90 %). Performance dropped dramatically when subjects were presented with incongruent auditory cues (10 %), as opposed to incongruent tactile cues (60 %), demonstrating that auditory input dominates meter perception. These observations support the notion that meter perception is a cross-modal percept with tactile inputs underlying the perception of "feeling" music.

1 Introduction

When listening or playing music, we not only hear music with our ears but also have the experience of "feeling" music from loudspeakers, strings, keyboards, or percussion drums. The neural basis of what it means to "feel" music is not understood. In this study we explored whether inputs from the somatosensory system contribute to the experience of "feeling" music. In addition to pitch and timbre, music is distinguished by the delicate temporal processing of the sequence of notes that gives rise to rhythm, tempo, and meter, which is the focus of the present study. Meter is defined as the abstract temporal structure that corresponds to periodic regularities of music (Palmer and Krumhansl 1990). It is based on the perception of regular beats that are equally spaced in time (Cooper and Meyer 1960). The meter is perceived as "duple" (or march-like) when a musical measure (the primary cycle) is subdivided into two or four beats and "triple" (or waltz-like) when subdivided into three beats (Randel 1986). Whether a piece of music is perceived as a duple or a triple often depends on the emphasis (increased duration, change in frequency, or amplitude) placed on the first beat (downbeat) of a measure in music (Keller and Repp 2005) and is strongly influenced by the probability of when the accent cues occur at key metrically positioned notes within a measure, the primary cycle, in music (Palmer and Krumhansl 1990; Hannon and Johnson 2005; Povel and Essens 1985).

While auditory cues are clearly important for signaling meter, the meter perception can be influenced by inputs from other sensory or motor modalities. For example, people tend to tap, dance, or drum to the strong beats of a musical rhythm, demonstrating the close relationship between movement and rhythm (Drake et al. 2000; Brochard et al. 2008). Brochard et al. showed that meter could be perceived purely in the tactile modality. Trainor and her colleagues (Trainor et al. 2009; Phillips-Silver and Trainor 2005, 2007) have shown that body movements or even electrical stimulation of the vestibular system during the metrically important notes can be used to disambiguate whether a tone sequence was duple or triple. These results suggest that music metrical pattern perception may be a multimodal process that integrates vestibular, somatosensory, and auditory inputs. However, whether musical meter perception is a cross-modal grouping process in which disassembled meter information presented from different sensory modalities is grouped to form an integrated meter perception remains unknown.

2 Methods and Materials

2.1 *Participants*

Twelve healthy musically trained participants (9 females; mean age = 19.3 ± 1.6 years; years of playing musical instruments = 7.75 ± 3.5) took part in the experiments. They were first tested for meter perception using the Montreal Battery of Evaluation of Amusia (MBEA) (Peretz and Hyde 2003) with all subjects performing above 94 % correct. The testing procedures were performed in compliance with the policies and procedures of the IRB of the Johns Hopkins University.

2.2 *Experiment Setup*

Auditory stimuli were delivered to the left ear of participants from circumaural sealed headphones (HDA 200, Sennheiser, Old Lyme, CT) via a Crown D-75A amplifier (Crown Audio and IOC Inc., Elkhart, IN). Tactile stimuli were delivered along the axis perpendicular to the left index finger of participants by a circular contact (8 mm diameter) connected to a Chubbuck motor (Chubbuck 1966). The participant placed his or her hand through an entry hole (lined with foam) and rested their fingers on a support platform mounted directly below the Chubbuck and contact probe.

2.3 *Stimulus*

Stimuli were sequences consisting of 24 temporal units that were 500 ms each in duration. Fifteen of the temporal units were note units and nine were silent. Each note consisted of a 350 ms sinusoidal tone (220 Hz (A3)) or a vibration (220 Hz) followed by a 150 ms silent period. The onset and offset of each of the tones and vibrations were ramped on and off within a 35 ms time window. Silent units consisted of 500 ms of silence. Each temporal unit simulated a beat in a musical note sequence, which were equally spaced points in time. Sequences were retained and were classified into "duple tending" (duple) and "triple tending" (triple) in a manner adapted from previous studies (Hannon and Johnson 2005; Povel and Essens 1985). Stimuli were generated digitally and converted to analog (PCI-6229, National Instruments, Austin, TX; sampling rate = 44.1 kHz) and delivered to either the headphone or Chubbuck tactile stimulator.

2.4 Procedure

Four meter-recognition experiments were conducted. In experiment 1, we examined subjects' meter perception of unimodal auditory or tactile presentation of a set of complete sequences. In experiments 2 and 3, we tested to what degree to the inputs from the auditory and tactile channels are integrated. In these two experiments, a whole sequence was disassembled and assigned partly to the auditory channel and partly to the tactile channel. In experiment 4, we tested how subjects could deal with consistent or conflicting meter cues between auditory and tactile channels (see figure). These four experiments were tested in mixed conditions to avoid adaption. Test conditions were organized into unimodal and bimodal test blocks. Participants were asked to judge whether the units were presented in groups of "three" or "four" by responding on a custom-written computer interface to a "triple" or a "duple" push button.

Examples of trials with triple sequences are shown in Fig. 50.1. Exp. 1: unimodal trial, a whole triple sequence was assigned to either auditory or tactile modality. Exp. 2 and 3: bimodal trials, red and blue bars together composed a whole triple sequence. Exp. 4 bimodal trials, red channel contained a whole triple sequence, and blue channel contained additional triple (congruent) or duple (incongruent) M notes. Red and blue channels were assigned to either auditory and tactile modalities or tactile and auditory modalities, respectively. The testing procedure was described in detail in a recently published paper (Huang et al. 2012).

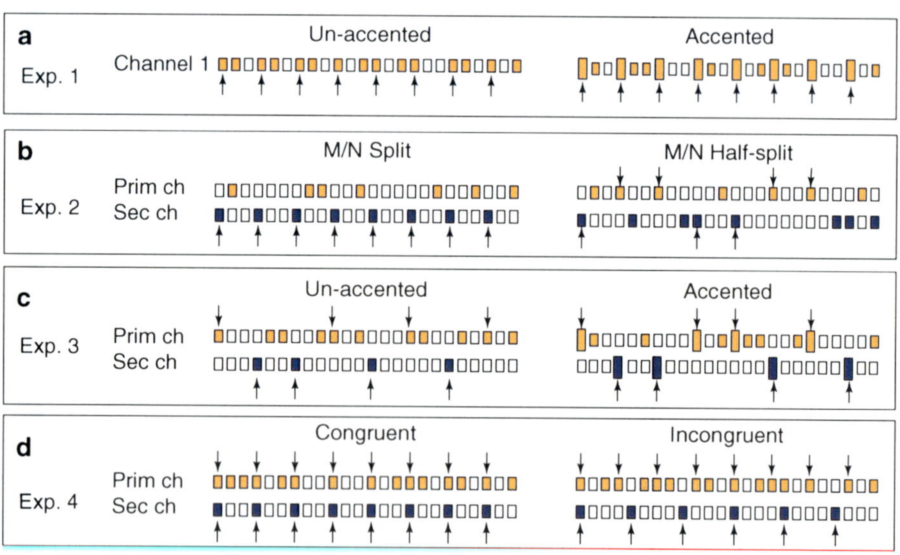

Fig. 50.1 Example stimuli from each experiment. Only triple sequences are shown here. Each empty bar represents a 500 ms silence unit. Each colored bar represents a note unit, a pure tone/sinusoidal vibration at the pitch level of A4 (220 Hz) and duration at 350 ms followed by a 150 ms silence

3 Results

3.1 *Experiment 1: Meter Perception via Unimodal Stimulation*

The mean correct response for unaccented triple sequences through auditory and tactile stimuli alone was 82 and 75 %, respectively. Adding amplitude cues increased performance to 90 and 84 % for auditory and tactile stimuli, respectively. Results for duple perception showed the same trend. The comparison of d' values for the two test conditions shows that accented cues may be weighted more heavily in auditory than in tactile modality.

3.2 *Experiment 2: Bimodal Integration in Meter Perception*

In the M/N split condition, there was little, if any, metric information within a single channel with performance at chance level (50 %) for triple sequences and performance slightly above chance (60 %) for duple sequences. When the sequences were presented bimodally, subjects were able to perceive meter clearly for both triple meter and duple meter. For triple sequences, the bimodal enhancement was 19 % when the red channel was auditory and 39 % when the red channel was tactile. For duple sequences, bimodal enhancement was 7 % for auditory and 22 % for tactile conditions. The results from the M/N split condition show that meter is integrated across touch and audition and that M notes play a larger role when presented through the auditory system.

The second condition (M/N half split) used a less structured distribution of notes with the M and N notes split evenly between the two channels. Subjects did not perceive triple meter in the unimodal conditions, with performance at 43 and 45 % for the auditory and tactile modalities, respectively, confirming that meter information was not present within a single channel. In the bimodal condition, correct response to triple sequences increased to 59 and 63 %, respectively. The comparison of d' between unimodal and bimodal conditions showed that adding the remaining notes from the other channel produced reliable recognition of meter pattern. The results provide evidence of cross-modal grouping of disassembled meter cues from information presented in auditory and tactile modalities to form an integrated meter perception.

3.3 *Experiment 3: Asymmetry Between Auditory and Tactile Stimulation in Meter Perception*

In experiment 3, the red channel contained all of the N notes and half of the M notes with the blue channel containing the other half of the M notes. This asymmetric distribution allowed us to observe modality-dependent characteristics of meter perception. Unaccented unimodal control conditions produced chance-level performance for triple perception and slightly above chance-level performance for duple perception. The presence of the other half of the M notes in the blue channel from auditory inputs significantly increased meter perception performance for the unimodal tactile condition by 20 % for triple perception and 23 % for duple perception. However, the presence of the other half of the M notes from the tactile modality did not significantly change subjects' performance. These results indicate that the contribution of metric cues in meter perception is modality dependent and that the auditory modality obviously plays a bigger role than the tactile modality. Accented metric cues significantly enhanced the discriminability of meter when the red channel was the auditory input, and auditory input significantly enhanced the discriminability of meter when the red channel was from the tactile input. The results show that the roles of auditory and tactile stimulation in meter perception are asymmetric with auditory input dominating.

3.4 Experiment 4: Interference Between Auditory and Tactile Channels in Meter Perception

In the congruent condition, while the meter cues were the same for both channels, performance was high, with correct responses to triple sequences at 83 and 96 % for auditory and tactile red channels, respectively, and 68 and 93 % for duple sequences. In the incongruent condition, performance for triple sequences dropped to 70 and 11 % for auditory and tactile red channels, respectively, and 54 and 2 % for duple sequences.

4 Discussion

Music is generally considered to be an auditory experience. However, it is often accompanied by other sensory stimuli (e.g., proprioceptive, vestibular stimuli) and motor actions. One of the most prominent sensory modalities that contribute to music perception is the tactile modality. As we know, mechanical vibrations to the skin allow musicians to "feel" vibrations in their instruments when playing music. This "feeling" of music can also been seen in listeners tapping their hands and feet to the rhythm.

In the present study, we tested young adults with some musical training in a meter discrimination task. Subjects can perceive the implied meter patterns from auditory or tactile sequences with ambiguous rhythms (unaccented condition) at an average accuracy rate of about 82 % (auditory) and 75 %. This performance is slightly better than what Hannon and Johnson found in their auditory studies, which could be explained by our subjects having musical training and being older than those tested in the Hannon et al. studies (Hannon and Johnson 2005). We showed that unimodal tactile meter perception behaves like auditory meter perception and that performance increases significantly when accent cues are added to key metrical notes. These results demonstrate that meter can be perceived through passive touch and that tactile and auditory meter perception share similar characteristics.

In the next set of experiments, we tested the degree to which auditory and tactile inputs are integrated in processing meter. If, for example, the sensory systems process information independently, then presenting the inputs bimodally should not affect meter perception. We found that meter information is extracted from combined auditory and tactile input. Performance rose from chance with unimodal input when there were no meter cues to 70–90 % with bimodal input. It should be noted that subjects performed all of the experiments without feedback, demonstrating that auditory-tactile integration for meter perception is an automatic process. Previous studies have shown that sensory grouping could not occur across sensory modalities (for review, see Spence and Chen 2012). Results from the present study provide evidence of cross-modal grouping of disassembled meter information from auditory and tactile modalities to form an integrated meter perception.

We further explored whether auditory or tactile input was dominant in meter perception by altering the balance between the auditory and tactile input of metric cues. We found that the presence of metrically important notes from the auditory modality has a significant larger influence on meter perception than those from the tactile modality, indicating that auditory input plays a dominant role in music perception. We also found that accent cues enhanced meter perception more strongly when presented from the auditory modality. This dominance could be due to the level of stimulation that we used in the current study. It is not clear if the dominance of audition over touch would persist if a larger area of the body and gut which is full of Pacinian corpuscles were activated by tactile input. Thus, intensity cues cannot be ruled out as playing a role in the dominance of audition over touch in our study. Other possibilities need to be also considered, as previous studies on audiovisual interactions have shown auditory dominance for the processing of rhythmic temporal stimuli (Guttman et al. 2005), and auditory inputs seem to be the least susceptible to influences from other sensory inputs in the perception of temporal events (Bresciani et al. 2008).

In the last set of experiments, we tested whether meter is processed along separate or common pathways. We found that while congruent stimulation enhanced meter perception, incongruent metrical cues inhibited meter perception. Again, we found that the integration between audition and touch was asymmetrical with auditory cues being weighted more strongly than tactile cues. We believe that the conflict is resolved using mechanisms similar to the mechanisms of selective attention.

The neural mechanisms of meter perception are not well understood. There are many similarities shared by auditory and tactile systems that might contribute to metrical cue integration. Physically, auditory and tactile stimuli in these experiments are mechanical vibrations. In this study we used 220 Hz vibratory stimuli, which is the optimal range for activating the Pacinian afferents. The Pacinian afferent system has been proposed as being critical for processing tactile temporal input and plays an important role for encoding vibratory inputs necessary for tool use

(Hsiao and Gomez-Ramirez 2011; Johnson and Hsiao 1992). Tactile inputs also activate the low-frequency rapidly adapting afferents (RA) which are important for coding flutter (Talbot et al. 1968). There is evidence that the processing of low frequencies may be similar in audition and touch (Bendor and Wang 2006). Based on our results, we suggest that the brain treats the stimulus sequences from the two channels as one stream rather than as two independent streams. How the tactile inputs interact with auditory inputs is not understood. One possibility is that the integration is simply due to energy summation, but this does not explain the asymmetrical effects observed between the auditory and tactile inputs that we found in experiments 3 and 4. A more likely explanation is that meter is processed along a common central neural pathway that receives inputs from both systems that is modulated by attention.

The interaction between hearing and touch in signal detection, frequency discrimination, and sensory illusion is well documented. Several studies report that there are central connections linking the auditory and tactile systems. Candidate areas where the integration could take place are the cerebellum (Ivry 1996), premotor cortices, auditory cortex (Bengtson et al. 2009), as well as the superior prefrontal cortex (Fraisse 1982; Grahn and Brett 2007). Studies have suggested that auditory cortex is involved in tactile temporal processing and auditory rhythm perception activates dorsal prefrontal cortex, cerebellum, and basal ganglia (Zatorre et al. 2007). Our findings demonstrate that musical meter perception involves the integration of somatosensory and auditory input. We hypothesize that the brain areas in which neurons respond to both auditory and tactile stimulation, such as MT, may be involved in auditory-tactile integration processing. Multisensory-driven neural activity might be the neural basis underlying cross-modal grouping for perception of music meter.

Acknowledgments This work was supported by a JHU Brain Science Institute grant (X.W. and S.H.), NIH grants DC03180 (X. W.), NS34086 (S.H.), and NSF grant of China #30670697 (J.H.)

References

Bendor D, Wang X (2006) Cortical representations of pitch in monkeys and humans. Curr Opin Neurobiol 16:391–399

Bengtson SL, Ullén F, Ehrsson HH, Hashimoto T, Kito T, Naito E, Forssberg H, Sadato N (2009) Listening to rhythms activates motor and premotor cortices. Cortex 45:62–71

Bresciani JP, Dammeier F, Ernst MO (2008) Tri-modal integration of visual, tactile and auditory signals for the perception of sequences of events. Brain Res Bull 75:753–760

Brochard R, Touzalin P, Després O, Dufour A (2008) Evidence of beat perception via purely tactile stimulation. Brain Res 1223:59–64

Chubbuck JG (1966) Small motion biological stimulator. Johns Hopkins Applied Physics Laboratory Technical Digest May–June 18–23

Cooper G, Meyer LB (1960) The rhythmic structure of music. The University of Chicago, Chicago

Drake C, Penel A, Bigand E (2000) Tapping in time with mechanically and expressively performed music. Music Percept 18:1–24

Fraisse P (1982) Rhythm and tempo. In: Deutsch D (ed) The psychology of music. Academic, New York, pp 149–180

Grahn JA, Brett M (2007) Rhythm and beat perception in motor areas of the brain. J Cogn Neurosci 19:893–906

Guttman SE, Gilroy LA, Blake R (2005) Hearing what the eyes see: auditory encoding of visual temporal sequences. Psychol Sci 16:228–235

Hannon EE, Johnson SP (2005) Infants use meter to categorize rhythms and melodies: implications for musical structure learning. Cogn Psychol 50:354–377

Hsiao SS, Gomez-Ramirez M (2011) Touch. In: Gottfried JA (ed) The neurobiology of sensation and reward. CRC Press, Lausanne, pp 141–160

Huang J, Gamble D, samlertsophon K, Wang XQ, Hsiao S (2012) Feeling music: integration of auditory and tactile inputs in musical meter perception. PloS one, Vol. 7, No. 10, doi:10.1371/journal.pone.0048496

Ivry RB (1996) The representation of temporal information in perception and motor control. Curr Opin Neurobiol 6:851–857

Johnson KO, Hsiao SS (1992) Neural mechanisms of tactual form and texture perception. Annu Rev Neurosci 15:227–250

Keller PE, Repp BH (2005) Staying offbeat: sensorimotor syncopation with structured and unstructured auditory sequences. Psychol Res 69(4):292–309

Palmer C, Krumhansl CL (1990) Mental representations for musical meter. J Exp Psychol Hum Percept Perform 16:728–741

Peretz I, Hyde K (2003) Varieties of musical disorders: the Montreal battery of evaluation of amusia. Ann N Y Acad Sci 999:58–75

Phillips-Silver J, Trainor LJ (2005) Feeling the beat: movement influences infant rhythm perception. Science 308:1430

Phillips-Silver J, Trainor LJ (2007) Hearing what the body feels: auditory encoding of rhythmic movement. Cognition 105:533–546

Povel DJ, Essens P (1985) Perception of temporal patterns. Music Percept 2:411–440

Randel DM (1986) The New Harvard dictionary of music. Belknap, Cambridge

Spence C, Chen YC (2012) Intramodal and Cross-modal perceptual grouping. In: Stein BE (ed) The new handbook of multisensory processing. The MIT Press, Cambridge

Talbot WH, Darian-Smith I, Kornhuber HH, Mountcastle VB (1968) The sense of flutter-vibration. J Neurophysiol 31:301–334

Trainor LJ, Gao X, Lei JJ, Lehtovaara K, Harris LR (2009) The primal role of the vestibular system in determining musical rhythm. Cortex 45:35–43

Zatorre RJ, Chen JL, Penhune VB (2007) When the brain plays music: auditory-motor interactions in music perception and production. Nat Rev Neurosci 8:547–558

Chapter 51
A Dynamic System for the Analysis of Acoustic Features and Valence of Aversive Sounds in the Human Brain

Sukhbinder Kumar, Katharina von Kriegstein, Karl J. Friston, and Timothy D. Griffiths

Abstract Certain sounds, for example, the squeal of chalk on a blackboard, are perceived as highly unpleasant. Functional magnetic resonance imaging (fMRI) in humans shows responses in the amygdala and auditory cortex to aversive sounds. Dynamic causal modelling (DCM) of the interaction between auditory cortex and the amygdala revealed that evoked responses to aversive sounds are relayed to the amygdala via the auditory cortex. There is a complex interaction between the auditory cortex and amygdala involving effective connectivity in both directions. While acoustic features modulate forward connections from auditory cortex to the amygdala, the valence modulates effective connectivity from the amygdala to the auditory cortex. The results support interaction between the auditory cortex and amygdala where stimuli are first processed to a higher (object) level in the auditory cortex before assignment of valence in the amygdala.

S. Kumar, PhD (✉) • T.D. Griffiths
Auditory Group, Institute of Neuroscience, Medical School,
Newcastle University, Framlington Place,
Newcastle upon Tyne NE2 4HH, UK

Neural mechanisms of human communication, Wellcome Trust Centre for Neuroimaging,
12, Queen Square, London WC1N 3BG, UK
e-mail: sukhbinder.kumar@ncl.ac.uk

K. von Kriegstein
Neural mechanisms of human communication,
Max Planck Institute for Human Cognitive and Brain Sciences,
Stephanstrasse 1A, Leipzig 04103, Germany

K.J. Friston, Ph.D
Functional Imaging Lab, Wellcome Trust Centre for Neuroimaging,
University College London (UCL),
12, Queen Square, London WC1N 3BG, UK

B.C.J. Moore et al. (eds.), *Basic Aspects of Hearing*,
Advances in Experimental Medicine and Biology,
DOI 10.1007/978-1-4614-1590-9_51,

1 Introduction

Certain sounds such as squeal of chalk on a blackboard are perceived as highly unpleasant. In humans, previous work implicates both the auditory cortex (Fecteau et al. 2007) and amygdala (Zald and Pardo 2002) in the analysis of unpleasant sounds. Animal work suggests that the aversive stimulus reaches directly (from the auditory thalamus, LeDoux et al. 1984, 1990b) to the amygdala bypassing the auditory cortex. An argument for indirect path via auditory association cortex has also been made (Phelps and LeDoux 2005), but the empirical evidence for either of the pathway in humans is lacking. By using event-related functional magnetic resonance imaging (fMRI) and analysis of connectivity between the auditory cortex and the amygdala using dynamic causal modelling (DCM), we answer the following key questions:

1. How does information about unpleasant stimuli reach the amygdala?
2. How does the coupling between the auditory cortex and the amygdala changes as a function of acoustic features and perceived unpleasantness of sounds? Specifically:
   (a) Which pathway, from the auditory cortex to the amygdala or vice versa, is modulated by the acoustic features?
   (b) Which pathway, from the auditory cortex to the amygdala or vice versa, is modulated by the valence?

2 Methods

2.1 Stimuli and Modelling of Acoustic Bases for Unpleasantness

A set of 74 sounds (each about 2 s duration), which included highly unpleasant sounds (scraping sounds, e.g. knife on a bottle and animal cries) and neutral sounds (such as bubbling water), were used. In a previous study (Kumar et al. 2008), we showed how the acoustic features (spectral frequency, F, and temporal modulation frequency, f) extracted using a model of the auditory system (Shamma 2003) can predict perceived unpleasantness of these sounds.

2.2 MRI Data Collection and Analysis

MRI data (13 subjects) were acquired continuously (TR, 2.73 s) on Siemens 3 T Allegra scanner. In the scanner, using a button press, subjects rated each sound on a scale from 1 (neutral) to 5 (highly unpleasant). Data analysis was carried out using SPM8 (http://www.fil.ion.ucl.ac.uk/spm/software/spm8/).

Images were first preprocessed (realigned, normalized to stereotactic space and smoothed by an isotropic Gaussian kernel of 8 mm full width at half maximum).

After preprocessing, general linear model (GLM) analysis was performed for each subject. The design matrix for this analysis consisted of five regressors: (1) Stimulus onsets convolved with a haemodynamic response function and four parametric regressors: stimulus onsets modulated by (2) spectral frequency (*F*), (3) temporal modulation frequency (*f*), (4) the interaction between the spectral frequency and temporal modulation frequency ($F \times f$) and (5) ratings of perceived unpleasantness (valence). To determine the response that is uniquely explained by the acoustic features and valence, the regressors were orthogonalized such that variance explained by valence was orthogonal to that explained by the acoustic features. Statistical parameter maps, implementing whole-brain random-effects analysis, at second level were created by using t-tests on contrast parameter estimates from the GLM of each subject.

2.3 Dynamic Causal Modelling

Dynamic causal modelling (DCM) (Friston et al. 2003) is a technique to identify causal interactions between two or more brain areas. A key feature of DCM is that it employs a generative or forward model of how the observed data (fMRI signal in the present case) were generated. The generative model used in DCM for fMRI consists of two parts: The first part (neural model) models the dynamics of neural activity and the second part converts the neural activity into fMRI signal.

The dynamics of the neural model is described by the following bilinear differential equation:

$$\frac{dz}{dt} = Az + \sum_{j=1}^{j=m} u_j B^j z + Cu \tag{51.1}$$

where z is an n-dimensional state vector (with one state variable per region), t is continuous time and u_j is the jth experimental input. The neural model consists of three sets of parameters: (1) A ($n \times n$) represents the influence one region has over other in the absence of external stimulation, (2) $B^{(j)}$ ($n \times n$) models the modulatory effects that an external stimulus has on connection strengths, and (3) C ($n \times 1$) represents direct influence of a stimulus on a given region. The conventional GLM analysis assumes that a stimulus has only a direct influence on a region. DCM in that sense can be regarded as a generic technique, with GLM being a particular instantiation of DCM in which the coupling parameters (the parameter sets A and B in the above equation) are taken to be zero.

Integration of the above differential equation gives neural activity. However, in fMRI the neural activity is measured indirectly by measuring associated changes in the oxygenation of blood flow (blood oxygenation level-dependent (BOLD) signal). In order for the forward model to predict the measured BOLD signal, the neural activity obtained from (1) needs to be converted to BOLD signal. This is achieved by the second part of the forward model, which convolves the neural activity with a haemodynamic model of the neurovascular coupling.

The parameters of both parts of the forward model are estimated using variational Bayes (Friston et al. 2003) to give the posterior density over parameters and the model evidence. The model evidence can be further used to select the best model from a set of models.

2.4 Volumes of Interest (VOIs) for DCM Analysis

The DCM analysis was carried out on four volumes of interest (VOIs): right amygdala (38, −6, 24), left amygdala (−22, −2, 12), right auditory cortex (48, −14, −12) and left auditory cortex (−50, 6, −6). These four areas showed correlation with either acoustic features or valence of unpleasant sounds at group level in the GLM analysis. The VOIs for each subject were chosen based on subject-specific maxima that were (1) closest to and (2) fall within the same anatomical region as the group-level maxima. The activity of all voxels within a VOI was summarized to a single time series by computing principal component analysis (PCA) of time series from all voxels lying within 4 mm of subject-specific maxima. Bayesian model comparison at the group level (random effects) was performed using methods suggested by (Stephan et al. 2009) implemented in SPM8.

3 Results

3.1 Relation Between Acoustic Features and Perceived Unpleasantness

Figure 51.1 shows representation of 6 of the 74 sounds in spectral frequency (F, y-axis) and temporal modulation frequency (f, x-axis) space. The mean unpleasantness rating for these sounds is also shown (in the top-right corner of each plot). The figure shows that sounds with high unpleasantness have high spectral frequencies and low temporal modulation frequencies. In the imaging analyses below, we sought to determine how the brain activity varies as function of both the acoustic features (F and f) and the perceived unpleasantness.

3.2 General Linear Model (GLM) Analysis

Figure 51.2 (red colour) shows activity in the amygdala which correlates with interaction between spectral frequency, F, and temporal modulation frequency, f.

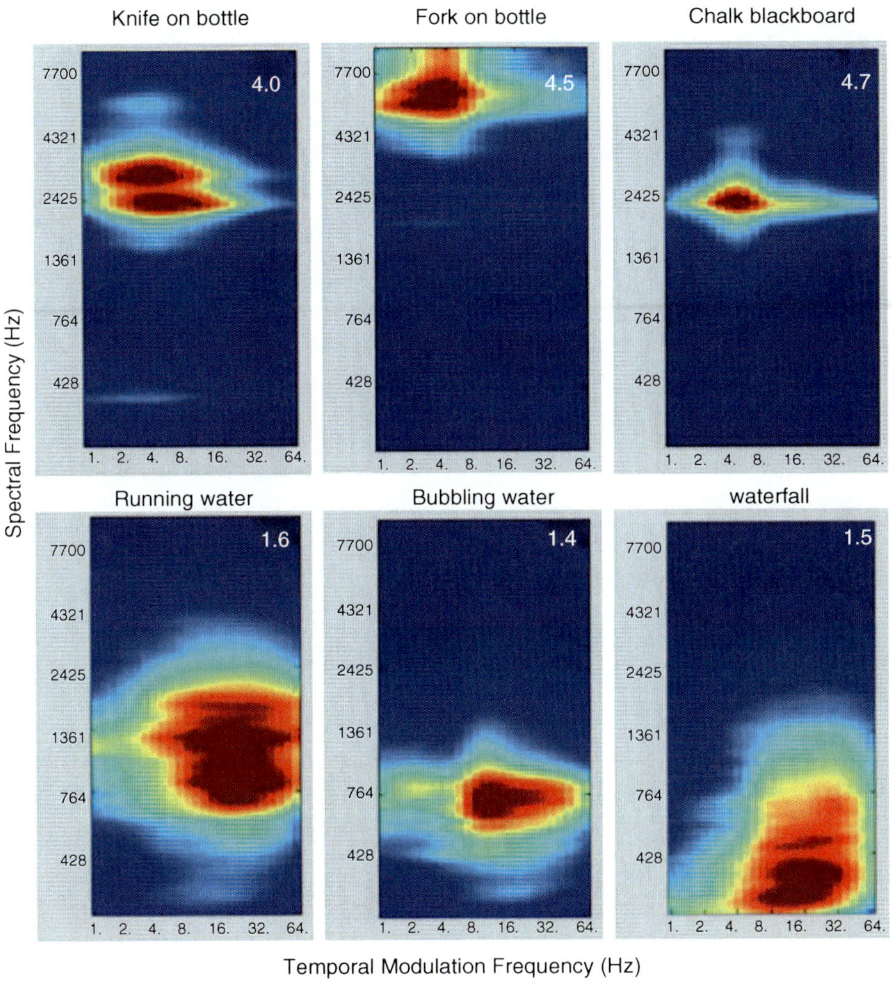

Fig. 51.1 Representation of sounds with high (*first row*) and low (*bottom row*) unpleasantness rating in spectral frequency-temporal modulation frequency space. The mean rating for each sound is shown in the *top-right corner* of each figure. Rating scale (1–5; 1 (neutral), 5 (high unpleasantness))

The effect is observed bilaterally [−22 −2 −12; $t(12)=4.98$; 24 −8 −18; $t(12)=4.13$; $p<0.001$ (uncorrected)]. Analysis of the distribution of activity in different nuclei of the amygdala (Eickhoff et al. 2005) showed that activity in the left amygdala is shared between the superficial (57 %) and basolateral (29 %) nuclei of the amygdala. The cluster in the right amygdala is mostly in the basolateral (79 %) amygdala, but a part (21 %) also lies in the superficial nucleus.

Figure 51.2 (blue colour) shows activity in the right amygdala [38, −6, −24; $t(12)=3.96$; $p<0.001$ (uncorrected)] that correlates positively with the rating of unpleasantness. This cluster of activity is located mostly (88 %) in the basolateral

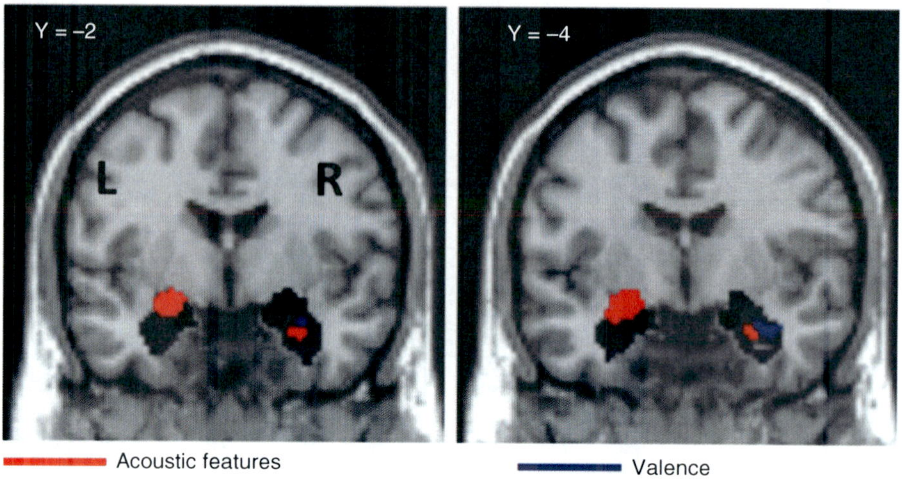

Fig. 51.2 Activity in the amygdala that correlates with acoustic features (interaction between spectral frequency and temporal modulation frequency) and rating of unpleasantness. Activity is thresholded at $p<0.005$ (uncorrected) for display purpose

nucleus of the amygdala. No activity was observed in the left amygdala even at liberal thresholds ($p=0.01$ uncorrected).

In the auditory cortex, decrease in the temporal modulation frequency elicited responses bilaterally in the anterior part of STG/upper bank of STS. Activity in the right STG [48, −14, −12; $t(12)=5.17$; $p<0.001$] and left STG [−50, 6, −6; $t(12)=7.02$; $p<0.001$] correlated with the interaction between spectral frequency and temporal modulation and valence, respectively.

3.3 *Connectivity Analysis Using Dynamic Causal Modelling*

3.3.1 How Does Stimulus Information Reach the Amygdala?

In order to answer this question, we created a model space consisting of three models (Fig. 51.3a). In these models, the amygdala either receives direct input from the ascending auditory pathway (model M1, direct) or via the auditory cortex (model M2, via auditory cortex) or both subcortical and cortical inputs (model M3, both). In the model M1, the stimulus bypasses the auditory cortex and first reaches the amygdala which then drives the auditory cortex. In the model M2, the stimulus is processed in the auditory cortex before it reaches the amygdala. The last model (M3)

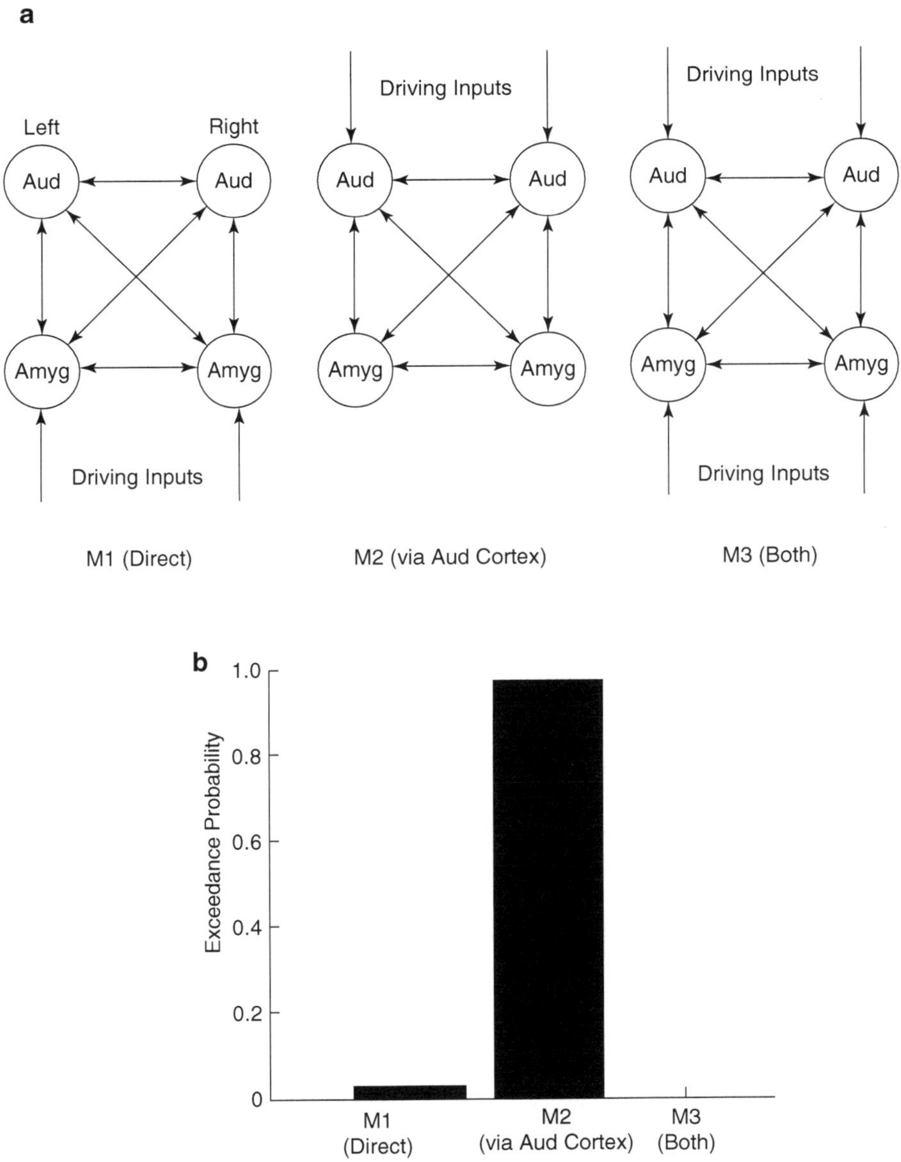

Fig. 51.3 (**a**) Structure of models to determine if the stimulus bypasses the auditory cortex to reach the amygdala. (**b**) Evidence for the models shown in (a). The model M2, in which the stimulus is processed in the auditory cortex before it reaches the amygdala, is the winning model (exceedance probability = 0.97)

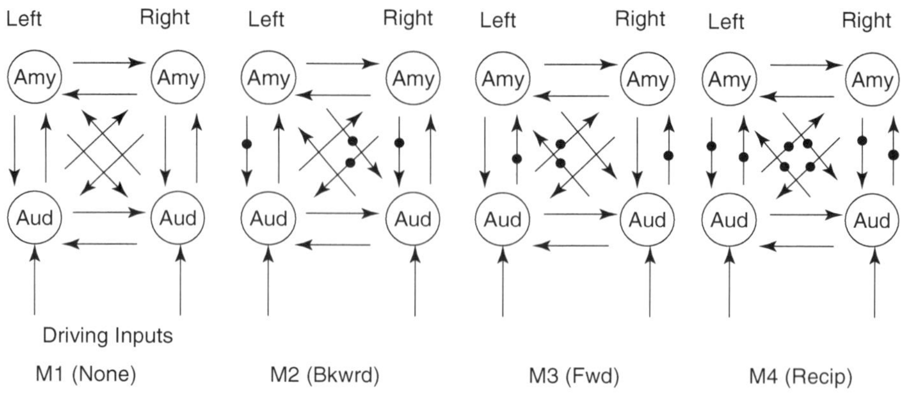

Fig. 51.4 Model space for analysis of the modulatory effects of acoustic structure and valence. The *dot* indicates the connection/s modulated by acoustic features or valence

specifies the possibility of auditory cortex and the amygdala driven independently by the inputs. The only difference in all three models is the location of driving inputs. These models were estimated for 13 subjects and compared using Bayesian model comparison at group level (random effects). The model exceedance probabilities (probability that the model has a higher posterior probability than all of the models in model space) of the three models are shown in Fig. 51.3b. These results show that the model in which stimulus is first processed in the auditory cortex before it reaches the amygdala (model M2) is the best model (exceedance probability = 0.97).

3.3.2 Which Pathway, from the Auditory Cortex to the Amygdala or Vice Versa, Is Modulated by the Acoustic Features?

To test if the acoustic features modulate backward (amygdala to auditory cortex) or forward (auditory cortex to amygdala) connections, a model space consisting of four models (Fig. 51.4) is created. In model M2 (backward), only backward connections are modulated (marked by dots in the figure), whereas in model M3 (forward), only forward connections are modulated. The model M4 (reciprocal) specifies modulation of connections in both direction. Finally, model M1(none) is a control model in which acoustic features have no modulatory effect in either direction. The driving inputs (as per the evidence from previous section) are at the auditory cortex. The evidence (exceedance probabilities) of the models is shown in Fig. 51.5a. The plot shows that the model M3, in which acoustic features modulate forward connections, has the highest exceedance probability (0.64).

3.3.3 Which Pathway, from the Auditory Cortex to the Amygdala or Vice Versa, Is Modulated by Valence?

The structure of models in this comparison is the same as in Fig. 51.4, but here the modulation is by valence rather than acoustic features. The results of Bayesian

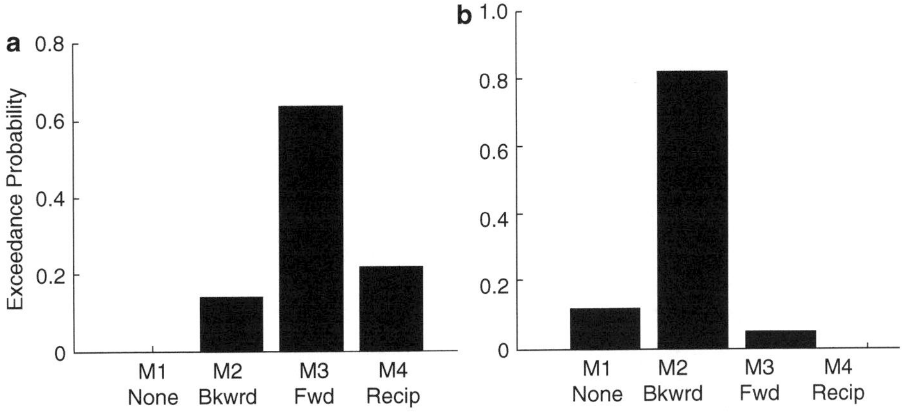

Fig. 51.5 (**a**) Evidence for the models shown in Fig. 51.4, where the modulatory input corresponds to the acoustic features (interaction between spectral and temporal modulation frequencies). The model M3, in which the forward connections from the auditory cortex to the amygdala is the best model (exceedance probability = 0.64). (**b**) Model exceedance probability for the models shown in Fig. 51.4 where the modulatory input is the rating of unpleasantness. The model M2, in which the backward connections from the amygdala to the auditory cortex are modulated, is the best model (exceedance probability = 0.83). *Bkwrd* backward, *Fwd* forward, *Recip* reciprocal

model comparison are shown in Fig. 51.5b. The model M2, in which the backward connections from the amygdala to the auditory cortex are modulated by perceived unpleasantness, is the winning model (exceedance probability = 0.83).

4 Discussion

4.1 How Does the Stimulus Reach the Amygdala?

Anatomical and functional studies have shown that the basolateral complex of the amygdala acts as a sensory interface of the amygdala and receives inputs from both the auditory thalamus (LeDoux et al. 1984, 1990a) and from association areas of the auditory cortex (Aggleton et al. 1980).These studies show that aversive stimuli can reach the amygdala via the auditory thalamus or cortex. Evidence for which pathway is used to relay the aversive stimulus to the amygdala in humans is lacking. Using dynamical causal modelling, we show that the information about aversive stimuli is relayed to the amygdala via the auditory cortex. This is consistent with the idea the stimuli are first processed to a high level in the auditory cortex before it is passed to the amygdala for affective evaluation (Rolls 2007).

4.1.1 How Does the Amygdala Interact with the Auditory Cortex?

Using DCM, we tested how the coupling between the amygdala and auditory cortex is modulated as a function of perceived unpleasantness and acoustic features.

We created a set of four models in which valence or acoustic features could either modulate forward, backward or both connections. Our results show dissociation between the modulatory effect of valence and acoustic features. While valence modulates the backward connections from the amygdala to the auditory cortex, the acoustic features modulate the forward connections from the auditory cortex to the amygdala.

5 Conclusions

The overall model of processing of aversive sounds from the above analysis can be summarized as follows: The stimulus is first processed to a high level in the auditory cortex (STG) which is then passed to amygdala for affective evaluation (valence assignment). The amygdala, in turn, modulates representation of the stimulus in the auditory cortex in accordance with the valence of stimulus.

Acknowledgement The authors acknowledge the funding from Wellcome Trust for this research.

References

Aggleton JP, Burton MJ, Passingham RE (1980) Cortical and subcortical afferents to the amygdala of the rhesus monkey (Macaca mulatta). Brain Res 190:347–368

Eickhoff SB, Stephan KE, Mohlberg H, Grefkes C, Fink GR, Amunts K, Zilles K (2005) A new SPM toolbox for combining probabilistic cytoarchitectonic maps and functional imaging data. Neuroimage 25:1325–1335

Fecteau S, Belin P, Joanette Y, Armony JL (2007) Amygdala responses to nonlinguistic emotional vocalizations. Neuroimage 36:480–487

Friston KJ, Harrison L, Penny W (2003) Dynamic causal modelling. Neuroimage 19:1273–1302

Kumar S, Forster HM, Bailey P, Griffiths TD (2008) Mapping unpleasantness of sounds to their auditory representation. J Acoust Soc Am 124:3810–3817

LeDoux JE, Cicchetti P, Xagoraris A, Romanski LM (1990a) The lateral amygdaloid nucleus: sensory interface of the amygdala in fear conditioning. J Neurosci 10:1062–1069

LeDoux JE, Farb C, Ruggiero DA (1990b) Topographic organization of neurons in the acoustic thalamus that project to the amygdala. J Neurosci 10:1043–1054

LeDoux JE, Sakaguchi A, Reis DJ (1984) Subcortical efferent projections of the medial geniculate nucleus mediate emotional responses conditioned to acoustic stimuli. J Neurosci 4:683–698

Phelps EA, LeDoux JE (2005) Contributions of the amygdala to emotion processing: from animal models to human behavior. Neuron 48:175–187

Rolls ET (2007) Emotion explained. Oxford University Press, Oxford

Shamma S (2003) Encoding sound timbre in the auditory system. IETE J Res 49(2):145–156

Stephan KE, Penny W, Daunizeau J, Moran R, Friston KJ (2009) Bayesian model selection for group studies. Neuroimage 46:1004–1017

Zald DH, Pardo JV (2002) The neural correlates of aversive auditory stimulation. Neuroimage 16:746–753

Part VII
Auditory Scene Analysis

Chapter 52
Can Comodulation Masking Release Occur When Frequency Changes Could Promote Perceptual Segregation of the On-Frequency and Flanking Bands?

Jesko L. Verhey, Bastian Epp, Arkadiusz Stasiak, and Ian M. Winter

Abstract A common characteristic of natural sounds is that the level fluctuations in different frequency regions are coherent. The ability of the auditory system to use this comodulation is shown when a sinusoidal signal is masked by a masker centred at the signal frequency (on-frequency masker, OFM) and one or more off-frequency components, commonly referred to as flanking bands (FBs). In general, the threshold of the signal masked by comodulated masker components is lower than when masked by masker components with uncorrelated envelopes or in the presence of the OFM only. This effect is commonly referred to as comodulation masking release (CMR). The present study investigates if CMR is also observed for a sinusoidal signal embedded in the OFM when the centre frequencies of the FBs are swept over time with a sweep rate of one octave per second. Both a common change of different frequencies and comodulation could serve as cues to indicate which of the stimulus components originate from one source. If the common fate of frequency components is the stronger binding cue, the sweeping FBs and the OFM with a fixed centre frequency should no longer form one auditory object and the CMR should be abolished. However, psychoacoustical results with normal-hearing listeners show that a CMR is also observed with sweeping components. The results are consistent with

J.L. Verhey (✉)
Department of Experimental Audiology,
Otto-von-Guericke University of Magdeburg,
Leipziger Str. 44, Magdeburg 39120, Germany

Forschungszentrum Neurosensorik,
Carl von Ossietzky University, 26111 Oldenburg, Germany
e-mail: jesko.verhey@med.ovgu.de

B. Epp
Department of Electrical Engineering, DTU,
Kgs. Lyngby, DK-2800, Denmark

A. Stasiak • I.M. Winter
Department of Physiology, Development and Neuroscience,
Downing Street, Cambridge CB2 3EG, UK

B.C.J. Moore et al. (eds.), *Basic Aspects of Hearing*,
Advances in Experimental Medicine and Biology,
DOI 10.1007/978-1-4614-1590-9_52,

the hypothesis of wideband inhibition as the underlying physiological mechanism, as the CMR should only depend on the spectral position of the flanking bands relative to the inhibitory areas (as seen in physiological recordings using stationary flanking bands). Preliminary physiological results in the cochlear nucleus of the Guinea pig show that a correlate of CMR can also be found at this level of the auditory pathway with sweeping flanking bands.

1 Introduction

An important task of the auditory system in complex acoustical scenes is to separate the sounds from different sound sources. In order to separate the sounds, the auditory system makes use of different signal properties, such as a common change in frequency over time (common fate, Chalikia and Bregman 1989; however, see Carlyon 1994) or correlated intensity fluctuations across frequency (Nelken et al. 1999).

A psychoacoustical phenomenon indicating sensitivity of the auditory system to the temporal coherence of intensity fluctuations across frequency is comodulation masking release (CMR, Hall et al. 1984; Verhey et al. 2003, for review). CMR is an enhanced ability to detect a sinusoid in the presence of a masker if the masker has the same temporal intensity fluctuations in different frequency regions, i.e. is comodulated. A common type of CMR experiment is the flanking band experiment, where the signal is masked by a narrowband masker centred at the signal frequency and, in the comodulated (CM) condition, by one or more additional off-frequency masker, commonly referred to as the flanking bands (FBs), with the same envelope as the on-frequency masker. Two comparison conditions have been proposed to quantify CMR in this type of experiment: (1) the uncorrelated (UN) masking condition with the same masker spectrum as in the comodulated (CM) condition but with uncorrelated level fluctuation of the masker bands and (2) the "reference" (RF) condition, where only the on-frequency masker (OFM) is present. Since CMR occurs for widely spaced masker bands, it is commonly assumed that at least part of the CMR is due to a comparison of the outputs of different auditory channels tuned to the respective masker bands. Dip listening, equalisation cancellation and cross-correlation have been proposed as possible across-channel mechanisms (Verhey et al. 2003, for a review).

Physiological findings at the level of the cochlear nucleus indicate that a broadly tuned channel, sensitive to comodulation, provides an inhibitory input to a narrowly tuned channel centred at the signal frequency (Pressnitzer et al. 2001; Neuert et al. 2004). Wideband inhibition reduces the response to the masker in the comodulated condition, i.e. it may be interpreted as a physiological realisation of a cancellation process. On the other hand, wideband inhibition enhances the signal representation for comodulated masker bands in the masker envelope minima. Thus, the wideband-inhibitor hypothesis also may be interpreted as a physiological correlate of the psychoacoustical dip-listening hypothesis. The wideband inhibition hypothesis for CMR implies that CMR should still be observed when the

FBs change frequency over time, provided that they still fall within the response range of the wideband inhibitor. In addition, since wideband inhibition is a process occurring at an early stage of the auditory pathway, it should not be abolished when other cues (such as a common change in instantaneous frequency of the components) indicate a different binding of the different stimulus components. The present study tests this hypothesis by measuring CMR in humans (using psychoacoustics) and the physiological correlate of CMR at the level of the cochlear nucleus in the Guinea pig (using single-cell recording) with FBs that are frequency swept over time.

2 Materials and Methods

2.1 Psychoacoustics

The masker duration was 500 ms including 50-ms raised-cosine ramps at masker on- and offset. The masker consisted of one or five sinusoidally amplitude-modulated (SAM) components. The modulation rate was 12 Hz and the modulation depth 100 %. Each component had a level of 60-dB SPL. The carrier of the OFM was a 700-Hz tone. The FBs were either also SAM tones (base condition) or SAM sweeps (sweep conditions). Two FBs were positioned below (lower flanking bands) and two above the signal frequency (higher flanking bands). The centre frequencies of the FBs were 300, 400, 1,000 and 1,100 Hz for the base stimuli. For the sweep stimuli, the carriers of the two FBs below the OFM were falling sweeps, while those of the two upper FBs were rising sweeps. Their instantaneous frequencies at the beginning of the stimulus were the same as for the base stimuli. The absolute sweep rate of the FBs was either one or two octaves/second. The FBs were either absent (RF condition) or had the same modulation as the OFM (CM condition), or the modulator was 180° out of phase with the modulator of the OFM. This latter condition will be referred to as the codeviant (CD) condition.

The target signal was a 700-Hz pure tone that was temporally centred in the masker and was added in quadrature phase to the OFM. The total duration of the target signal was 250 ms and the rise–fall time was 50 ms ($\cos^2$ gate). The time signals and a schematic plot of the spectra are shown in Fig. 52.1.

Seven normal-hearing listeners participated. A three-alternative, forced-choice procedure with adaptive signal-level adjustment was used to determine the masked threshold of the signal. The intervals in a trial were separated by gaps of 500 ms. Listeners had to indicate which of the intervals contained the target signal. Visual feedback was provided after each response. The level of the target was adjusted according to a two-down, one-up rule to estimate the 70.7 % point of the psychometric function. The initial step size was 4 dB. After every second reversal the step size was halved until a step size of 1 dB was reached. The run was then continued for another six reversals. The mean level at the last

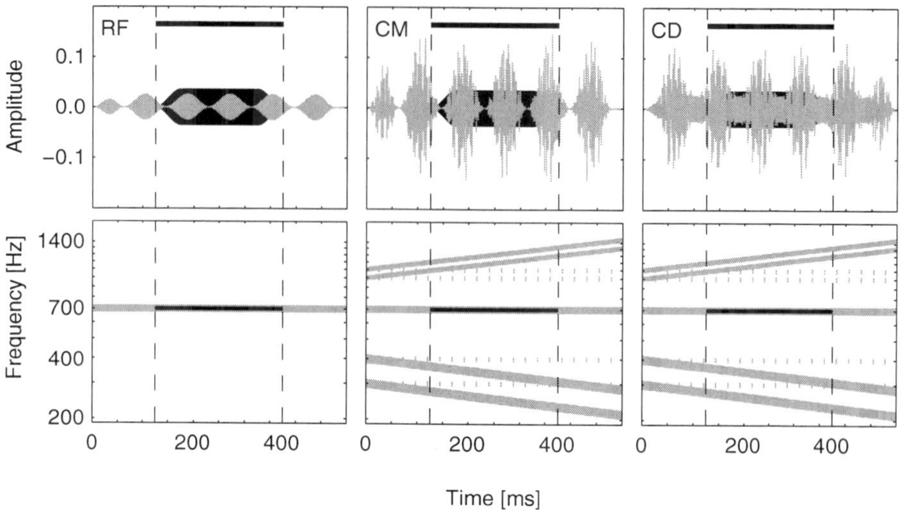

Fig. 52.1 Examples of the waveforms (*top row*) and schematic plots of the spectrograms (*bottom row*) of the stimuli for the conditions RF (*left column*), CM (*middle column*) and CD (*right column*). The masker is shown in *grey* and the signal in *black*. The signal level relative to that of the OFM was 6 dB. The spectrograms for the CM and the CD conditions are shown with sweeping FBs. In addition, the *dotted lines* indicate the spectrogram of the FBs with fixed centre frequencies (base stimuli). The temporal position of the signal is indicated by the *vertical dashed lines* and, in the top panels, with a *horizontal filled bar*

six reversals was used as an estimate of the threshold. The final individual threshold estimate was taken as the mean over four threshold estimates.

2.2 Physiology

Experiments were performed on anaesthetised, pigmented Guinea pigs weighing between 345 and 550 g (for details, see Neuert et al. (2004)). The experiments were performed under the terms and conditions of the project licence issued by the United Kingdom Home Office to the last author (IMW). Single units in the dorsal cochlear nucleus (DCN) were recorded extracellularly with glass-coated tungsten microelectrodes. In contrast to the psychoacoustical experiment, the centre frequency of the OFM was normally set equal to the BF of the unit. The number and spacing of the FBs were identical in the CM and the CD conditions and were chosen according to the response map of the unit. The carriers of the FBs were rising (upper FBs) or falling (lower FBs) sweeps with a sweep rate of one octave per second. If possible, the FBs were placed directly into the inhibitory sidebands of the response area of the unit. The frequency separation between the OFM and the nearest FB was at least twice as large as the spacing between adjacent FBs. An example of the positioning of the OFM and flanking bands is shown in Fig. 52.2.

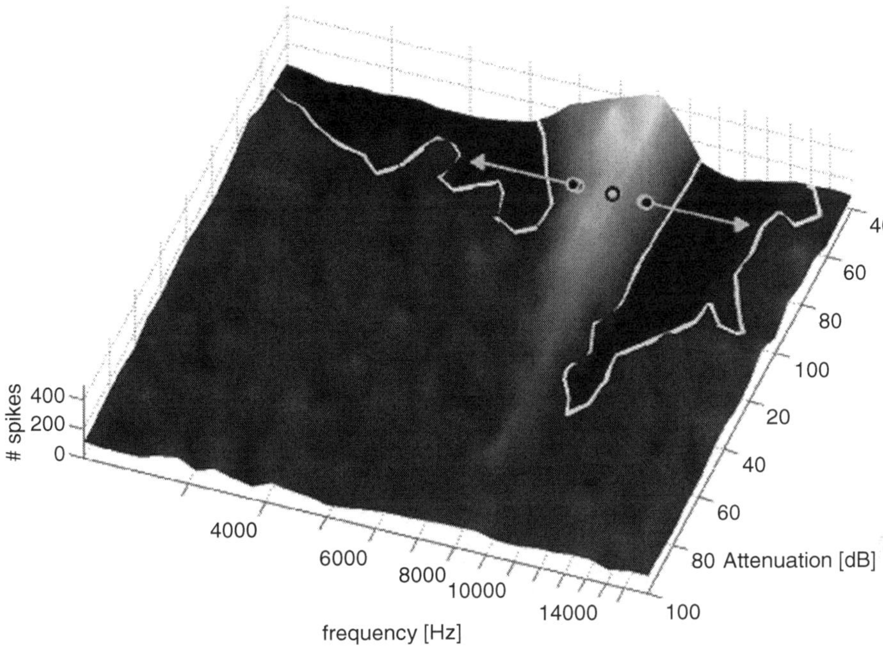

Fig. 52.2 Response map of a unit showing inhibition (unit 1514007). Since the unit had very low spontaneous activity (<2 spikes/s), the response map was measured in the presence of a low-level excitatory tone burst, positioned at 10 dB above BF threshold. The *solid white line* demarcates the boundaries of the sideband inhibition of the unit. The *black circle filled with grey* indicates the position of the OFM. The *grey circles filled with black* indicate the frequencies of the FBs at the beginning of the stimulus. The *arrows* indicate the frequency ranges covered by the sweeping FBs

3 Results

Figure 52.3 shows the results of the psychoacoustical experiment for the base stimuli and the sweep stimuli. Thresholds for the CD conditions were highest and were essentially the same for the base and sweep conditions (difference = 1 dB). The threshold for the RF condition was 9 dB lower than for the CD conditions. The lowest threshold of 34 dB was for the CM condition in the base condition. The CMR was 27 dB using the CD condition as a reference. For sweeping FBs the CMR was reduced to 21 dB.

Figure 52.4 shows the response of one unit in the DCN for conditions RF (left), CM (middle) and CD (right). The addition of the FBs resulted in a strong reduction of the response to the masker, of 35 % for the CM condition and 28 % for the CD condition (compared to the RF condition). Adding a signal at a level of 0 dB relative to the level of the OFM reduced the response by 10 % for the CM condition, i.e. more than for the other two conditions. Thus, the unit showed a physiological correlate of CMR.

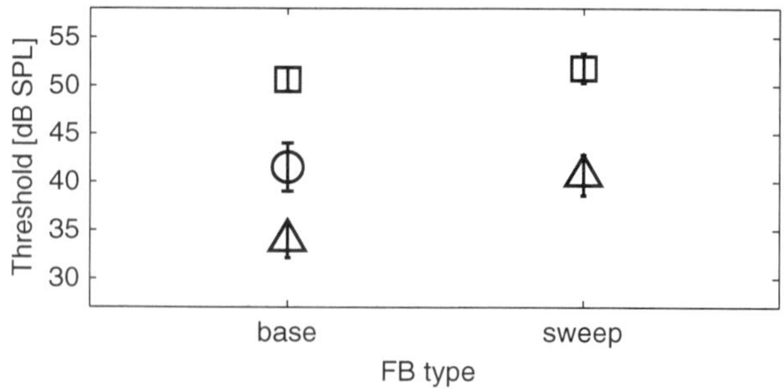

Fig. 52.3 Average thresholds are lower in the CM condition for both base and sweep stimuli. Data from seven normal-hearing listeners. The CD (*squares*), CM (*triangles*) and RF (*circle*) conditions. *Error bars* indicate ± one standard error

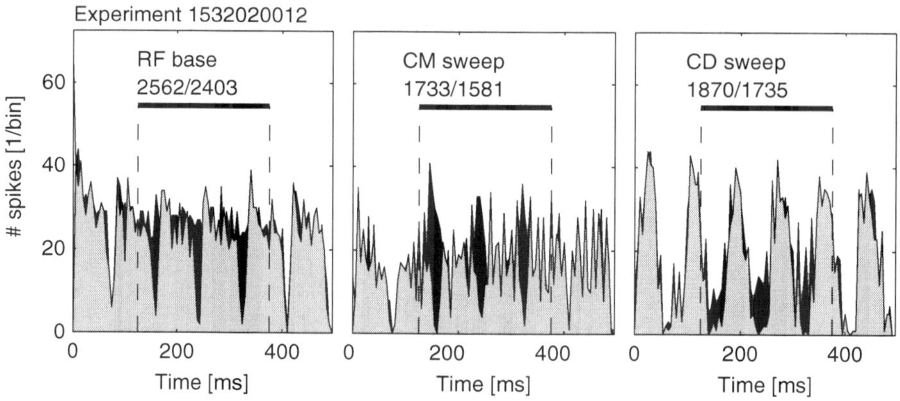

Fig. 52.4 Response of a DCN unit (1532020) showing the largest responses to the signal in the CM sweep condition. Signal and OFM were set at 6.4 kHz. At the beginning of the stimulus, the sweeping FBs (2 upper, 2 lower) were 100 Hz apart and the spacing between the OFM and the nearest FB was 800 Hz. Both the OFM and FBs were set at a level that was 43 dB above the pure tone threshold. The response of the unit to the masker alone (no-signal condition) is shown in *grey*, while the change in response due to the addition of the tone at a level equal to the level of the OFM is shown in *black*. The temporal position of the signal is indicated by the *vertical dashed lines* and *horizontal filled bar*. The total number of spikes for the signal plus masker relative to that for the masker only is shown below the name of the condition

4 Discussion

The psychoacoustical results indicate that CMR occurs, but is reduced, when the carrier frequencies of the FBs are frequency sweeps. For sweeping components, the filter characteristics of auditory channels with a bandwidth as observed in auditory nerve fibres would change the envelope of the FBs at the outputs of these filters, i.e. the output of the filters tuned to the frequencies of the FBs would no longer show the same level fluctuations as the output of the filter tuned to the OFM. Thus, the data indicate that a broadly tuned mechanism extracts the comodulation of the FBs. The physiological data indicate that this preservation of CMR can also be found in the response of DCN units.

The psychoacoustical data on their own cannot rule out the possibility that some of the processes underlying CMR are located at a higher level of the auditory pathway. Some recent psychoacoustical studies combining other object binding cues such as onset synchrony with comodulation have led to the claim that CMR cannot be based solely on a low-level process, since in some conditions CMR can be abolished when other object binding cues indicate that the OFM and the FBs belong to different auditory objects (e.g. Dau et al., 2009; Grose et al., 2009). Processing at the level of the cochlear nucleus may also contribute to these other object binding cues used in these studies (see Roberts et al. 2007, for common onset). Further studies are necessary to investigate to what extent the detrimental effects of these other object binding cues can be found in the neural responses at the level of the cochlear nucleus.

5 Summary and Conclusions

A masking release due to comodulation (CMR) was observed for sweeping FBs and an OFM with a fixed centre frequency. Thus, comodulation can lead to improved signal detection even when change in frequency might be expected to lead to perceptual segregation of the OFB and FBs. This result is consistent with the hypothesis of wideband inhibition as the physiological mechanism underlying CMR.

Acknowledgement This work was supported by the Deutsche Forschungsgemeinschaft (DFG, SFB trr 31).

References

Carlyon R (1994) Further evidence against an across-frequency mechanism specific to the detection of frequency modulation (FM) incoherence between resolved frequency components. J Acoust Soc Am 95:949–961

Chalikia MH, Bregman AS (1989) The perceptual segregation of simultaneous auditory signals: pulse train segregation and vowel segregation. Percept Psychophys 46:487–496

Dau T, Ewert S, Oxenham A (2009) Auditory stream formation affects comodulation masking release retroactively. J Acoust Soc Am 125:2182–2188

Grose J, Hall JW, Buss E (2009) Within- and across-channel factors in the multiband comodulation masking release paradigm. J Acoust Soc Am 125:282–293

Hall JW, Haggard MP, Fernandes MA (1984) Detection in noise by spectro-temporal pattern analysis. J Acoust Soc Am 76:50–56

Nelken I, Rotman Y, Bar Yosef O (1999) Responses of auditory-cortex neurons to structural features of natural sounds. Nature 397:154–157

Neuert V, Verhey JL, Winter IM (2004) High responses of dorsal cochlear nucleus neurons to signals in the presence of modulated maskers. J Neurosci 23:5789–5797

Pressnitzer D, Meddis R, Delahaye R, Winter IM (2001) Physiological correlates of comodulation masking release in the mammalian ventral cochlear nucleus. J Neurosci 21:6377–6386

Roberts B, Holmes SD, Bleeck S, Winter IM (2007) Wideband inhibition modulates the effect of onset asynchrony as a grouping cue. In: Kollmeier B, Klump G, Hohmann V, Langenmann U, Mauermann M, Uppenkamp S, Verhey J (eds) Hearing-from sensory processing to perception. Springer, Berlin/Heidelberg, pp 333–341

Verhey JL, Pressnitzer D, Winter IM (2003) The psychophysics and physiology of comodulation masking release. Exp Brain Res 153:405–417

Chapter 53
Illusory Auditory Continuity Despite Neural Evidence to the Contrary

Lars Riecke, Christophe Micheyl, and Andrew J. Oxenham

Abstract Many previous studies have shown that a tone that is momentarily interrupted can be perceived as continuous if the interruption is completely masked by noise. It has been suggested this "continuity illusion" occurs only when peripheral neural responses contain no evidence that the signal was interrupted. In this study, we used a combination of psychophysical measures and computational simulations of peripheral auditory responses to examine whether the continuity illusion can be experienced under conditions where peripheral neural responses contain evidence that the signal did not continue through the masker. Our results provide an example of a salient continuity illusion despite evidence of an interruption in the peripheral representation, indicating that the illusion may depend more on global features of the interrupting sound, such as its long-term specific loudness, than on its fine-grained temporal structure.

1 Introduction

The auditory continuity illusion refers to the illusory perception of a physically interrupted sound as "continuing through" another sound that occludes the interruption (Miller and Licklider 1950). The results of numerous studies during the last 40 years have led to the widespread view that an important prerequisite for the occurrence of the illusion is that peripheral auditory responses to the occluding

L. Riecke (✉)
Department of Cognitive Neuroscience,
Faculty of Psychology and Neuroscience, Maastricht University,
Maastricht 6200 MD, The Netherlands
e-mail: l.riecke@maastrichtuniversity.nl

C. Micheyl • A.J. Oxenham
Department of Psychology, University of Minnesota,
Minneapolis, MN 55455, USA

B.C.J. Moore et al. (eds.), *Basic Aspects of Hearing*,
Advances in Experimental Medicine and Biology,
DOI 10.1007/978-1-4614-1590-9_53, © Springer Science+Business Media New York 2013

sound contain no evidence that the signal was absent (e.g., Duifhuis 1980; Kluender and Jenison 1992; Plack and Oxenham 2000; Riecke et al. 2008; Warren et al. 1972, 1988). In particular, the results of these studies suggest that the illusion does not occur if the peripheral excitation produced by the occluder does not completely mask the excitation that would have been produced by the signal, had the latter physically continued through the occluder (Bregman 1990, p. 352; Warren et al. 1972). We refer to this as the "peripheral masking" theory of the continuity illusion (for reviews, see Bregman 1990; Petkov and Sutter 2010; Warren 1999). The interpretation of the previous findings has focused on time-averaged measures, such as the long-term power spectrum of the target and masker, or their excitation patterns, whereas relatively little attention has been paid to the fine-grained temporal characteristics of these sounds, i.e., their temporal fine structure and fast fluctuations in their temporal envelope. In this study, we investigated the importance of fine-grained temporal information in the generation of the continuity illusion.

2 Method

2.1 *Stimuli, Tasks, and Procedure*

To test the influence of differences in fine-grained temporal information in the generation of the continuity illusion, we used harmonic complex tones with positive and negative phase curvatures (Schroeder 1970) as occluders. These so-called Schroeder-phase complexes (abbreviated M+ and M–) have identical long-term power spectra but different phase spectra, set specifically to evoke different temporal response patterns in the auditory periphery, taking into account the curvature of the phase response of cochlear filters (Carlyon and Datta 1997a, b; Kohlrausch and Sander 1995; Lentz and Leek 2001; Oxenham and Dau 2001a; Smith et al. 1986). This is illustrated in Fig. 53.1a, c, which shows simulated temporal response patterns of an auditory nerve (AN) fiber to M+ and M– complexes consisting of the first 33 harmonics of a 100-Hz fundamental frequency.

These simulations were produced using a model of peripheral processing, including cochlear filters with phase responses estimated based on psychophysical data in humans (Oxenham and Dau 2001b) and cochlear compression (for details, see Riecke et al. 2012). It can be seen that the M+ complex evokes a markedly "peakier" temporal response pattern than the M– complex—an effect due to the positive phase curvature of M+ compensating for the negative phase curvature of cochlear filters—whereas the negative phase curvature of M– adds to the negative phase curvature of the cochlear filters, resulting in an accentuated phase-dispersion effect (Recio and Rhode 2000; Summers et al. 2003). This effect has been used to explain why the masked threshold of a pure tone embedded in an M– complex is markedly higher than its masked threshold when embedded in an M+ complex having the same overall physical level (Kohlrausch and Sander 1995; Lentz and Leek 2001; Oxenham and Dau 2001a; Smith et al. 1986; see Fig. 53.1b, d).

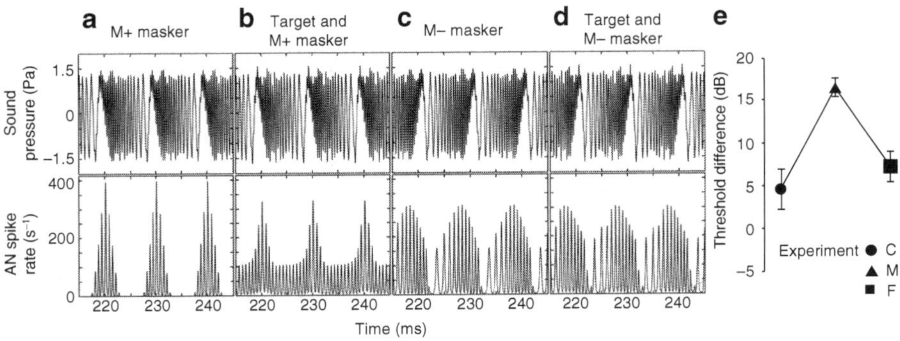

Fig. 53.1 Simulated peripheral activities and psychophysical results. (**a–d**) The *upper row* shows sound waveforms for a portion of the stimuli denoted above. As shown by the *lower row*, the M+ complex evoked "peakier" simulated peripheral responses than the M– complex (compare panels **a** and **c**). Thus, the target should be considerably easier to detect in the M+ complex than in the M– complex (compare panels **b** and **d**). (**e**) Psychophysical thresholds (mean ± s.e.m. across 11 listeners) obtained with the M– complex were markedly higher than those obtained with the M+ complex, in each experiment. Importantly, this difference was substantially smaller in experiment *C* than in experiment *S* and rather similar in experiments *C* and *F*. Thus, thresholds for the continuity illusion agreed better with the pattern of forward-masked thresholds than with that of simultaneous-masked thresholds (modified from Riecke et al. (2012))

Based on these simulation results, we predicted that if the continuity illusion depends on a fine-grained analysis of peripheral activity, then the level of the target below which the continuity illusion occurs (the "continuity-illusion threshold") should be considerably lower for the M+ masker than for the M– masker, in line with the differences in simultaneous-masked threshold. Conversely, if the continuity illusion depends on a relatively coarse representation of the masker (e.g., a representation of its overall level or specific loudness), the illusion thresholds measured with the M+ and M– maskers should be reasonably similar, i.e., their difference should be considerably smaller than the differences in masked threshold.

This prediction was tested in three experiments. In the "continuity-illusion" experiment (C), a 1,500-Hz target tone (T) and a masker (M+ or M–), each lasting 200 ms, were played in an alternating sequence (TMTMTMT), and the listener's task was to indicate whether the target was heard as continuing through the masker or not. The level of the target was varied adaptively to determine the continuity-illusion threshold (defined as the target level corresponding to the 50 % point on the psychometric function). The level of the masker was fixed at 50-dB SL per harmonic (relative to the listener's hearing threshold at the frequency of the target).

In the "simultaneous-masking" experiment (S), the 200-ms target was presented simultaneously with, and temporally centered in, the 600-ms masker in one of two consecutive observation intervals (selected at random with equal probability), while the other observation interval contained only the masker. The task of the listener was to indicate which interval contained the target. The goal of this experiment was to provide a direct measure of the amount of simultaneous masking produced by the masker.

Lastly, the "forward-masking" experiment (F) was similar to experiment S, except that a 10-ms target was presented 5 ms after a 200-ms masker. The aim of this experiment was to provide an estimate of the strength of the long-term internal excitation produced by the masker (Carlyon and Datta 1997b). Thresholds in experiments S and F were measured with an adaptive two-alternative forced-choice procedure and a two-down one-up tracking rule estimating the 70.7 % correct point on the psychometric function (Levitt 1971). The level of the signal was varied, while the level of the masker was kept constant, as in experiment C.

In all adaptive-tracking procedures, the step size was initially set to 6 dB and reduced to 3 dB after the second reversal in the direction of change of the target level (from increasing to decreasing or vice versa) and to 0.5 dB after the fourth reversal. Each measurement began with the target level set sufficiently high (62-dB SL) so that the listener did not hear the target continue through the masker in experiment C and that the listener could easily detect the target in experiments S and F. The procedure was terminated after the tenth reversal. Threshold was computed as the arithmetic mean of the target levels at the last six reversals. For each listener and each masker condition, nine, six, and six thresholds were measured in experiments C, S, and F, respectively, in fully randomized order. Only in experiments S and F, visual feedback was provided to the listener after each trial.

Target and maskers were ramped on and off with 20-ms linear ramps, except for experiment F, in which the 10-ms target was ramped with 5-ms ramps (no steady state). In experiment C, the amplitude midpoints of the ramps of consecutive elements (T and M) were made to coincide so that the audibility of the gaps in the TMTMTMT sequence was reduced. The two observation intervals in experiments S and F were separated by a 500-ms gap.

Stimuli were generated digitally and presented monaurally via a soundcard and headphones (MDR-V900HD, Sony) in a sound-attenuating chamber. Stimulus presentation and response collection were controlled using the AFC software package (developed by Stephan Ewert at University of Oldenburg).

2.2 Participants

Eleven paid volunteers (five females, ages 20–35 years) participated in the study after providing written informed consent. Except for one participant who had a slightly elevated threshold (35-dB HL) at 4,000 Hz, all participants had normal hearing (pure-tone hearing thresholds less than 20-dB HL at octave frequencies between 125 Hz and 4 kHz, including the frequency of the target) and no history of hearing disorders.

3 Results

As shown by Fig. 53.1e, the M+ complex yielded lower thresholds than the M– complex, both overall ($F_{1, 10}=44.05$, $p<0.001$) and for each experiment taken separately (C: $t_{10}=1.94$, $p=0.049$; S: $t_{10}=15.07$, $p<0.001$; F: $t_{10}=4.05$, $p=0.0018$). Importantly, this threshold difference was significantly smaller in experiment C than in experiment S ($t_{10}=-7.34$, $p<0.001$), but did not differ significantly between experiments C and F ($t_{10}=-0.97$, $p=0.36$).

4 Discussion

Using maskers with identical long-term power spectra but different temporal peripheral representations, we found that differences in continuity-illusion thresholds are (1) considerably smaller than differences in simultaneously masked threshold and (2) relatively similar to differences in forward-masked thresholds.

Our first finding shows that although the real tone was much easier to detect in the M+ complex than in the M– complex, this difference existed much less for the illusory tone. Together with our simulation results, this indicates that the illusion can occur under conditions where peripheral neural responses contain evidence that the target was interrupted. Our second finding shows that the occluder had similar effects on the continuity illusion and forward masking. Based on the notion that forward masking depends primarily on the long-term internal excitation evoked by the masker (Carlyon and Datta 1997b; Wojtczak and Oxenham 2009), this suggests that the illusion depends on a neural representation of the average excitation evoked by the occluder—possibly related to the specific loudness of the occluder, i.e., its loudness within the critical band centered on the frequency of the target sound (Mauermann and Hohmann 2007).

Recent studies have demonstrated the occurrence of the continuity illusion under conditions where the target was not interrupted by any sound (Remijn et al. 2007, 2008) or the occluder was briefly interrupted (Haywood et al. 2011). Two other studies investigating dynamic aspects of the illusion found that a modulated target was perceived as continuing through an interrupting noise, although listeners were unable to track the phase of the illusory modulation through the interruption (Carlyon et al. 2004; Lyzenga et al. 2005). Together with these previous findings, our results provide accumulating evidence against the peripheral masking theory (Warren et al. 1972) and for the notion that the fine-grained temporal structure of the target and occluder are not preserved during the illusion (but see also Plack and Watkinson 2010).

The proposed role of the occluder's long-term features (e.g., its specific loudness) for the continuity illusion implies that the illusion arises at or beyond neural processing stages that integrate the fine-grained temporal structure of the peripheral sound representation. This outcome suggests a relatively central origin, in line with claims from previous studies (Elfner and Homick 1967; Hellstrom and Young 1989; Petkov et al. 2007; Schreiner 1980).

Acknowledgments We thank Gesine Malschofsky for helping with the data acquisition. L.R. is supported by the Netherlands Organization for Scientific Research (Veni grant 451.11.014). A.J.O. and C.M. are supported by NIH grants R01 DC07657 and R01 DC05216. Some of the data presented in this chapter have been published elsewhere (Riecke et al. 2012).

References

Bregman AS (1990) Auditory scene analysis: the perceptual organization of sound. MIT Press, Cambridge

Carlyon RP, Datta AJ (1997a) Masking period patterns of Schroeder-phase complexes: effects of level, number of components, and phase of flanking components. J Acoust Soc Am 101:3648–3657

Carlyon RP, Datta AJ (1997b) Excitation produced by Schroeder-phase complexes: evidence for fast-acting compression in the auditory system. J Acoust Soc Am 101:3636–3647

Carlyon RP, Micheyl C, Deeks JM, Moore BC (2004) Auditory processing of real and illusory changes in frequency modulation (FM) phase. J Acoust Soc Am 116:3629–3639

Duifhuis H (1980) Level effects in psychophysical two-tone suppression. J Acoust Soc Am 67:914–927

Elfner L, Homick JL (1967) Continuity effects with alternately sounding tones under dichotic presentation. Percept Psychophys 2:34–36

Haywood NR, Julie Chang IC, Ciocca V (2011) Perceived tonal continuity through two noise bursts separated by silence. J Acoust Soc Am 130:1503

Hellstrom LI, Young ED (1989) Physiological responses to the pulsation threshold paradigm. II: representations of high-pass noise in average rate measures of auditory-nerve fiber discharge. J Acoust Soc Am 85:243–253

Kluender KR, Jenison RL (1992) Effects of glide slope, noise intensity, and noise duration on the extrapolation of FM glides through noise. Percept Psychophys 51:231–238

Kohlrausch A, Sander A (1995) Phase effects in masking related to dispersion in the inner ear. II. Masking period patterns of short targets. J Acoust Soc Am 97:1817–1829

Lentz JJ, Leek MR (2001) Psychophysical estimates of cochlear phase response: masking by harmonic complexes. J Assoc Res Otolaryngol 2:408–422

Levitt H (1971) Transformed up-down methods in psychoacoustics. J Acoust Soc Am 49(Suppl 2):467+

Lyzenga J, Carlyon RP, Moore BC (2005) Dynamic aspects of the continuity illusion: perception of level and of the depth, rate, and phase of modulation. Hear Res 210:30–41

Mauermann M, Hohmann V (2007) Differences in loudness of positive and negative Schroeder-phase tone complexes as a function of the fundamental frequency. J Acoust Soc Am 121: 1028–1039

Miller GA, Licklider JCR (1950) The intelligibility of interrupted speech. J Acoust Soc Am 22: 167–173

Oxenham AJ, Dau T (2001a) Towards a measure of auditory-filter phase response. J Acoust Soc Am 110:3169–3178

Oxenham AJ, Dau T (2001b) Reconciling frequency selectivity and phase effects in masking. J Acoust Soc Am 110:1525–1538

Petkov CI, Sutter ML (2010) Evolutionary conservation and neuronal mechanisms of auditory perceptual restoration. Hear Res 271:54–65

Petkov CI, O'Connor KN, Sutter ML (2007) Encoding of illusory continuity in primary auditory cortex. Neuron 54:153–165

Plack CJ, Oxenham AJ (2000) Basilar-membrane nonlinearity estimated by pulsation threshold. J Acoust Soc Am 107:501–507

Plack CJ, Watkinson RK (2010) Perceived continuity and pitch shifts for complex tones with unresolved harmonics. J Acoust Soc Am 128:1922–1929

Recio A, Rhode WS (2000) Basilar membrane responses to broadband stimuli. J Acoust Soc Am 108:2281–2298

Remijn GB, Nakajima Y, Tanaka S (2007) Perceptual completion of a sound with a short silent gap. Perception 36:898–917

Remijn GB, Perez E, Nakajima Y, Ito H (2008) Frequency modulation facilitates (modal) auditory restoration of a gap. Hear Res 243:113–120

Riecke L, Van Opstal AJ, Formisano E (2008) The auditory continuity illusion: a parametric investigation and filter model. Percept Psychophys 70:1–12

Riecke L, Micheyl C, Oxenham AJ (2012) Global not local masker features govern the auditory continuity illusion. J Neurosci 32:4660–4664

Schreiner C (1980) Encoding of alternating acoustical signals in the medial geniculate body of guinea pigs. Hear Res 3:265–278

Schroeder MR (1970) Synthesis of low peak-factor signals and binary sequences with low autocorrelations. IEEE Trans Inf Theory 16:85–89

Smith BK, Sieben UK, Kohlrausch A, Schroeder MR (1986) Phase effects in masking related to dispersion in the inner ear. J Acoust Soc Am 80:1631–1637

Summers V, de Boer E, Nuttall AL (2003) Basilar-membrane responses to multicomponent (Schroeder-phase) signals: understanding intensity effects. J Acoust Soc Am 114:294–306

Warren RM (1999) Auditory perception: a new analysis and synthesis. Cambridge University Press, Cambridge

Warren RM, Obusek CJ, Ackroff JM (1972) Auditory induction: perceptual synthesis of absent sounds. Science 176:1149–1151

Warren RM, Wrightson JM, Puretz J (1988) Illusory continuity of tonal and infratonal periodic sounds. J Acoust Soc Am 84:1338–1342

Wojtczak M, Oxenham AJ (2009) On- and off-frequency forward masking by Schroeder-phase complexes. J Assoc Res Otolaryngol 10:595–607

Chapter 54
High-Acuity Spatial Stream Segregation

John C. Middlebrooks

Abstract In a complex auditory scene, location in space is one of several acoustic features that permit listeners to segregate competing sequences of sounds into discrete perceptual streams. Nevertheless, the spatial acuity of stream segregation is unknown. Moreover, it is not clear whether this is really a spatial effect or whether it reflects a binaural process that only indirectly involves space. We employed "rhythmic masking release" as an objective measure of spatial stream segregation. That task revealed spatial acuity nearly as fine as listeners' discriminations of static locations (i.e., their minimum audible angles). Tests using low-pass, high-pass, and varying-level conditions in the horizontal dimension demonstrated that binaural difference cues provide finer acuity than does any monaural cue and that low-frequency interaural delay cues give finer acuity than do high-frequency interaural level differences. Surprisingly, stream segregation in the vertical dimension, where binaural difference cues are negligible, could be nearly as acute as that in the horizontal dimension. The results show a common spatial underpinning to performance. Nevertheless, a dissociation across conditions between localization acuity and masking-release thresholds suggests that spatial stream segregation is accomplished by brain systems discrete from those responsible for sound-localization judgments.

1 Introduction

In his classic paper, Cherry (1953) listed "the voices come from different directions" as a major factor in solving the "cocktail party problem" of hearing out a sound of interest in a complex auditory scene. "Stream segregation" is the term that

J.C. Middlebrooks
Departments of Otolaryngology, Neurobiology & Behavior,
Cognitive Sciences, and Biomedical Engineering,
University of California at Irvine, Room 116, Medical Sciences E,
Irvine, CA 92697-5310, USA
e-mail: j.midd@uci.edu

B.C.J. Moore et al. (eds.), *Basic Aspects of Hearing*,
Advances in Experimental Medicine and Biology,
DOI 10.1007/978-1-4614-1590-9_54,

describes the phenomenon in which a listener can assign multiple interleaved sequences of sounds to perceptually discrete streams, which typically would correspond to distinct sources. Fundamental frequency and spectral and temporal envelope are among the acoustic features that facilitate stream segregation (e.g., Moore and Gockel 2002). Tests of the importance of sound-source location, however, have given somewhat mixed results. "Obligatory streaming" tests require a listener to fuse multiple sounds into a single perceptual stream; that is, spatial stream segregation *hurts* performance. A surprisingly weak effect of location is seen in tests of obligatory streaming, which generally demonstrate spatial stream segregation only when sounds activate discrete peripheral channels, as when they are presented to different ears or in different acoustic hemifields (Phillips 2007; Stainsby et al. 2011). In contrast, successful stream segregation *enhances* task performance in tests of "voluntary streaming." It is voluntary streaming that helps a listener solve the cocktail party problem and voluntary streaming that was the topic of the present study.

Our psychophysical task was based on "rhythmic masking release" (RMR; Turgeon et al. 2002; Sach and Bailey 2004). Listeners were asked to discriminate between one of two rhythms formed by "target" sequences of noise bursts from one location in the presence of an interleaved "masker" sequence from a second location. Performance with broadband sounds showed rather impressive spatial acuity. Various conditions of passband, varying sound level, and horizontal-versus-vertical dimension probed the acoustic cues underlying performance. Some degree of spatial stream segregation was observed in essentially every condition. Nevertheless, a dissociation across conditions between RMR acuity and acuity for pairs of static sources (minimum audible angles; MAAs) suggests that spatial stream segregation and sound-source localization rely on different brain structures. The experiments presented here are covered in greater detail in Middlebrooks and Onsan (2012).

2 General Methods

The listener sat in a double-walled sound booth that was lined with absorbent foam. Small calibrated loudspeakers were positioned at 2.5 or 5° intervals of azimuth in the interaural horizontal plane and 5 or 10° intervals of elevation in the frontal midline. Stimuli were sequences of independent (i.e., nonfrozen) bursts of Gaussian noise filtered by broadband (400–16,000 Hz), low-band (400–1,600 Hz), or high-band (4,000–16,000 Hz) filters. Listeners sat facing the 0° loudspeaker in the 0° azimuth and in the elevation conditions or sat with the 0° loudspeaker 40° to his or her right side in the 40° conditions.

Normal-hearing listeners performed a spatial version of RMR. The task was to discriminate between two rhythms. Rhythm 1 was XX00XX00, and Rhythm 2 was XX00X0X0, where each X represents a brief sound burst and each 0 indicates a silent interval of equal duration. The bursts or silent intervals were presented at a base rate of 10/s, 100 ms from onset to onset. The target was one rhythm repeated four times without interruption, a total of 16 sound bursts. The masker consisted of the complementary pattern. That is, Rhythm 1 (XX00XX00) was masked by 00XX00XX, and Rhythm 2 (XX00X0X0) was masked by 00XX0X0X. The aggregate of target and masker (i.e., when both were presented from the same location) was a uniform 10/s pattern of 32 sound bursts. On each trial, one of the two rhythms was presented and the listener pressed one of two response keys. Trial-by-trial feedback was provided. The MAA procedure was similar to the RMR procedure. The listener heard a single noise burst at the target location (300 ms onset to onset) followed by a burst at a different location, or in the reverse order, and was asked to report whether the second sound was to the left or right of the first.

For the azimuth conditions, the target was fixed at 0° or 40° relative to the listener's midsagittal plane, and the maskers varied in random order to the left and right of the target. For the elevation conditions, the target was fixed at 0°, and the maskers varied above and below the target. For each listener and condition, a discrimination index (d') was computed based on the proportion of hits and false alarms across 51–60 trials (depending on the condition). Values of d' were plotted against masker location, and the interpolated locations at which the plot crossed the criterion of $d'=1$ to the left and right of the target (or above and below) were taken as the RMR thresholds.

3 Horizontal Stream Segregation

3.1 Rationale and Methods

The RMR task required listeners to evaluate the relative timing of sound bursts within a sequence of sounds from one source, which required the listener to segregate the masker and target streams. For that reason, we interpret the RMR thresholds as measures of the spatial acuity of stream segregation.

Listeners derive the locations of sound sources from spatial cues that result from the interaction of incident sound with the head and external ears (reviewed by Middlebrooks and Green 1991). We tested conditions in which the individual sound bursts, 20 ms in duration, were as follows: (1) broadband, in which all spatial cues were available; (2) low band, in which the dominant spatial cue would be interaural time delays (ITDs) in temporal fine structure; and (3) high band, in which the major cues would result from shadowing of sound by the head or from ITDs in sound envelopes. Results in broadband, low-band, and high-band conditions were obtained from seven listeners.

3.2 Results

Horizontal stream segregation showed remarkably fine acuity. Figure 54.1a shows the performance of one listener in the broadband condition when the target was at 0°. Performance was at a chance level when the masker was colocated with the target,

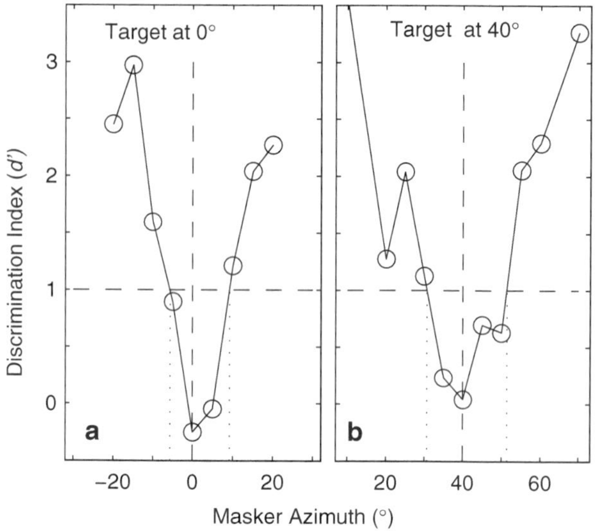

Fig. 54.1 Rhythm discrimination by one listener. The *dashed line* at $d' = 1$ indicates the criterion for threshold rhythm discrimination (From Middlebrooks and Onsan 2012)

but performance improved rapidly with increasing target and masker separation. Criterion performance was achieved for masker locations 6° to the left or 9° to the right of the target. Across the seven listeners, thresholds did not differ significantly between left and right maskers ($p > .05$; paired Wilcoxon signed rank test), so left- and right-sided thresholds were combined. The median threshold in the 0° broadband condition was 8.1°.

Acuity was somewhat coarser when the target was located 40° to the listener's right side (Fig. 54.1). Thresholds for this listener were 9° and 11° for maskers to the right or left of the target. The median across listeners in the broadband 40° target condition was 11.2°, which was significantly larger than in the 0° target condition ($p < .0005$). The 40° target condition is particularly interesting because the target and maskers all were located within the same sound hemifield. Several recent reports have emphasized that localization acuity is finest for targets around the frontal midline and have argued that localization might rely on the relative activity between left- and right-tuned neural populations (e.g., McAlpine and Grothe 2003; Stecker et al. 2005; Phillips 2007) . The present results demonstrate that high-acuity spatial stream segregation can be accomplished even when both target and masker activate primarily right-tuned neurons.

Performance across the broadband, low-band, and high-band conditions for the seven listeners is shown in Fig. 54.2. In both 0° and 40° target conditions, performance varied significantly across passband conditions ($p < .0005$ and $p < .02$ for 0° and 40° targets, respectively; Kruskal-Wallis test). That passband dependence, however, almost entirely reflected the relatively poor performance in the high-band condition. There was no significant difference between broadband and low-pass conditions ($p > .05$) but significantly broader thresholds for pair-wise comparison of high-band versus broadband conditions ($p < .005$ and $p < .05$ for 0° and 40° targets, with Bonferroni correction). We interpret these results to indicate that, in broadband conditions, listeners relied on ITDs in low-frequency temporal fine structure to segregate target from masker sounds.

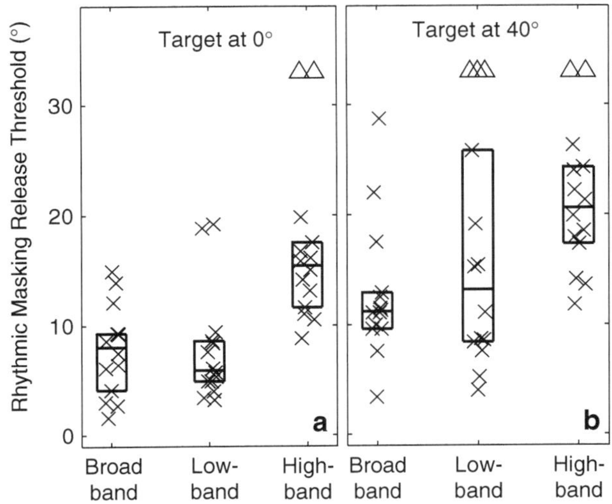

Fig. 54.2 Rhythmic masking release (RMR) thresholds across 7 listeners. Each *symbol* indicates a leftward or rightward threshold for one listener. *Triangles* indicate cases in which d' was <1 at target/masker separation of 30°. *Boxes* indicate 25th, 50th, and 75th percentiles (From Middlebrooks and Onsan 2012)

Despite the worse performance in high-band conditions, substantial spatial stream segregation was evident in those conditions. Middlebrooks and Onsan (2012) evaluated several potential spatial cues that might have accounted for the performance in the high-band condition and concluded that listeners appeared to rely on interaural level difference (ILDs), possibly with some synergy from ITDs in high-frequency sound envelopes.

In each of the three passband conditions, we measured the acuity with which listeners could distinguish the locations of pairs of single sounds presented at 300-ms onset-to-onset intervals (their MAAs). Results are summarized in Fig. 54.3. In every condition, MAAs were significantly smaller than RMR thresholds ($p<.0001$ for all conditions). Despite that significant overall difference, we note that there are several instances in the broadband and low-pass conditions in which RMR thresholds were quite close to MAAs, suggesting that some of the listeners experienced stream segregation at the limit of their localization acuity. The distributions of MAAs did not differ significantly across passbands for the 0° target condition ($p>.05$); for the 40° target condition, MAAs were slightly larger in the high-band than in the broadband condition ($p<.05$).

4 Stream Segregation in the Absence of Interaural Difference Cues

4.1 Rationale and Methods

Study of RMR in the horizontal dimension demonstrates the importance of interaural difference cues. Those results, however, do not tell us whether stream segregation relies on some sort of central-auditory-system spatial representation or whether

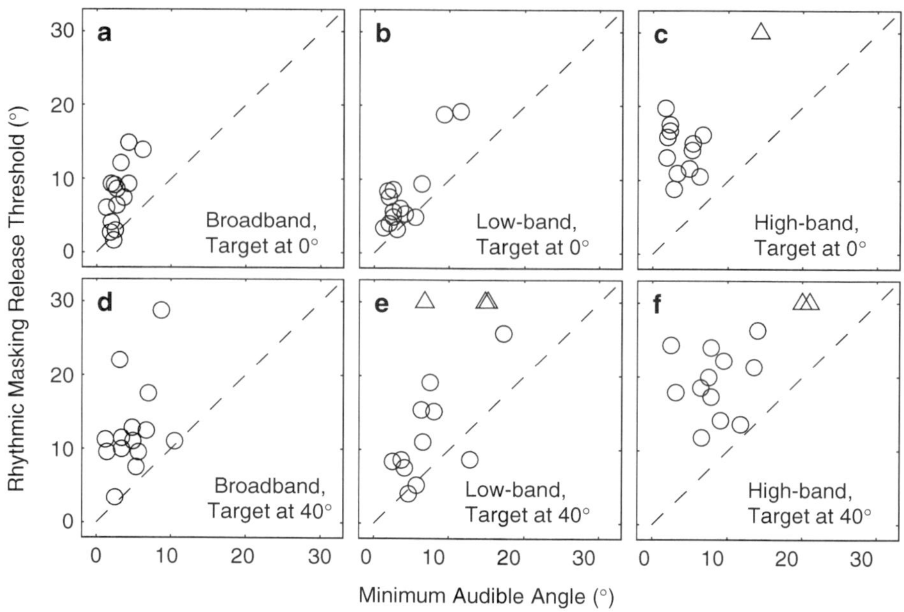

Fig. 54.3 Comparison of RMR and minimum audible angle (MAA) for 7 listeners. *Triangles* indicate cases in which d' was <1 at the greatest target/masker separation tested (From Middlebrooks and Onsan 2012). Stimulus bands and target locations differ among panels, as indicated

segregation arises directly from sensitivity to interaural differences. Interaural cues to the elevations of sounds in the vertical midline are negligible. Instead, listeners perform vertical localization on the basis of spectral shape cues provided by the interaction of incident sound with the direction-dependent filter properties of the head and external ears (reviewed by Middlebrooks and Green 1991). Measures of RMR for targets and maskers in the vertical midline provided a test of spatial stream segregation in the absence of interaural difference cues.

The tests of RMR in the vertical midline were identical to those in the horizontal plane except for the sound-source locations and the sound-burst durations. In pilot experiments, we noticed a dramatic dependence of performance on burst durations. For that reason, we measured vertical RMR thresholds for 10-, 20-, and 40-ms durations. In all conditions, the target source was at 0° elevation. Masker source locations ranged from −10 to +30° in the 10-ms condition and from −30° to +50° in the 20- and 40-ms conditions. Vertical RMR thresholds were computed only for upward (positive) masker elevations in the 10-ms condition and for both upward and downward elevations in the 20- and 40-ms conditions; the plots show two symbols (i.e., upward and downward thresholds) for each listener in the 20- and 40-ms conditions. We also measured MAAs in the vertical midline. All sounds were broadband. Ten listeners completed tests of the 10-ms condition, nine completed the 20- and 40-ms condition, and six of those listeners completed all three conditions.

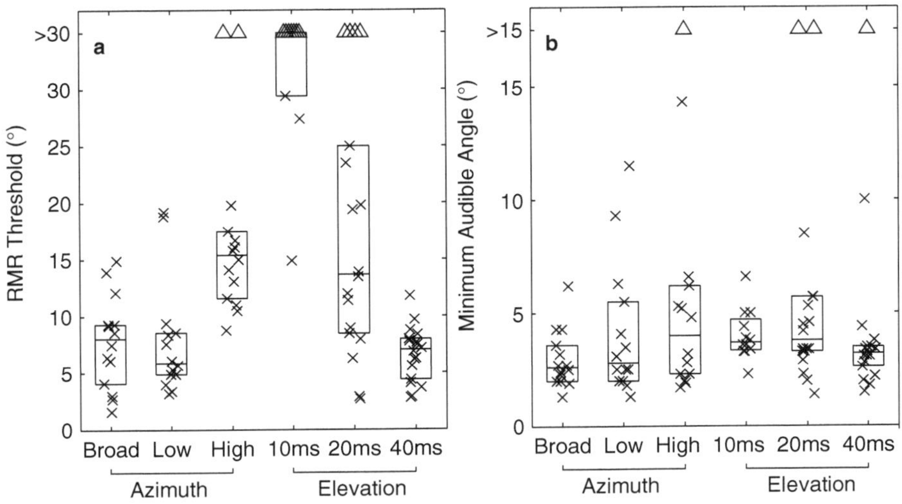

Fig. 54.4 RMR thresholds (panel **a**) and MAAs (panel **b**) for azimuth (with target at 0°) and elevation (From Middlebrooks and Onsan 2012)

4.2 *Results and Discussion*

Performance depended markedly on the duration of the sound bursts constituting the target and masker sequences. The RMR thresholds in elevation are summarized in Fig. 54.4a along with the azimuth thresholds for 0° targets and three pass bands, which are replotted for comparison. When the bursts were 10 ms in duration, only three of the ten listeners reached criterion discrimination performance within a target/masker separation of 30°. Performance was better in the 20-ms condition, but the between-listener variance was wide, ranging from inability to do the task at a 30° separation down to <5°. Performance was uniformly good in the 40-ms conditions, with thresholds tightly clustered around a median of 7.1°.

In contrast to the RMR thresholds, MAAs were relatively insensitive to sound-burst duration. Vertical MAAs narrowed significantly with increasing duration ($p < .005$, Kruskal-Wallis), but that was due to the significant difference between the 10- and 40-ms conditions ($p < .005$); comparisons of 10- versus 20-ms and 20- versus 40-ms conditions showed no significant difference ($p > .05$).

Previous studies of localization in the vertical midline have shown impaired performance for short-duration sounds, particularly at high sound levels (Hartmann and Rakerd 1993; Hofman and van Opstal 1998; Macpherson and Middlebrooks 2000). That phenomenon might account in part for the impaired vertical stream segregation in the 10- and 20-ms stream segregation conditions, although that alone does not account for why RMR thresholds varied with duration when MAA did not. One might speculate that there is a form of "sluggishness" in vertical spatial hearing. Localization performance based on spectral cues is challenging even for static sources. It might be that listeners simply cannot discriminate the spectra of brief sounds at a rate of 100 ms onset to onset, whereas they could do the discrimination in the 300-ms onset-to-onset conditions of the MAA task.

5 Summary and Conclusions

We employed a measure of "voluntary" spatial stream segregation that quantified some of the abilities needed for hearing out a conversation in a complex auditory scene. High-acuity stream segregation in the horizontal dimension appeared to rely on interaural difference cues, with low-frequency ITDs providing significantly finer acuity than did high-frequency ILDs. Nevertheless, spatial stream segregation in the vertical dimension demonstrated high acuity in conditions in which interaural cues were negligible; vertical RMR presumably relies on spectral shape cues. One interpretation of the results might be that the various spatial cues are somehow combined to form a central representation of auditory space. That "common" spatial representation might be accessed both for sound localization per se and for spatial stream segregation. Contrary to that hypothesis, however, is the observation that RMR thresholds varied markedly across conditions of pass band and burst duration, whereas little variation in MAAs was observed across those conditions. That leads to an alternative hypothesis, which we favor, in which spatial stream segregation is accomplished by one or more brain pathways that are distinct from the pathway(s) for localization. The wide variation in RMR thresholds across passband and burst-duration condition might indicate that the stream segregation pathways do not utilize the representations of various spatial cues as effectively and as uniformly as does the putative localization pathway. We are exploring these, and other, hypotheses with further psychophysical studies as well as with animal-physiological studies involving single-neuron recordings from the auditory cortex.

Acknowledgment The author's work was supported by NIH grant RO1 DC000420.

References

Cherry CE (1953) Some experiments on the recognition of speech, with one and two ears. J Acoust Soc Am 25:975–979

Hartmann WM, Rakerd B (1993) Auditory spectral discrimination and the localization of clicks in the sagittal plane. J Acoust Soc Am 94:2083–2093

Hofman PM, van Opstal JA (1998) Spectro-temporal factors in two-dimensional human sound localization. J Acoust Soc Am 103:2634–2648

Macpherson EA, Middlebrooks JC (2000) Localization of brief sounds: effects of level and background noise. J Acoust Soc Am 108:1834–1849

McAlpine D, Grothe B (2003) Sound localization and delay lines – do mammals fit the model? Trends Neurosci 26:347–350

Middlebrooks JC, Green DM (1991) Sound localization by human listeners. Annu Rev Psychol 42:135–159

Middlebrooks JC, Onsan ZA (2012) Stream segregation with high spatial acuity. J Acoust Soc Am 132(6):3896–3911

Moore BCJ, Gockel H (2002) Factors influencing sequential stream segregation. Acta Acustica 88: 320–332

Phillips DP (2007) A perceptual architecture for sound localization in man. Hear Res 238:124–132

Sach AJ, Bailey PJ (2004) Some characteristics of auditory spatial attention revealed using rhythmic masking release. Percept Psychophys 66:1379–1387

Stainsby TH, Fullgrabe C, Flanagan HJ, Waldman SK, Moore BCJ (2011) Sequential streaming due to manipulation of interaural time differences. J Acoust Soc Am 130:904–914

Stecker GC, Harrington IA, Middlebrooks JC (2005) Location coding by opponent neural populations in the auditory cortex. PLoS Biol 3(e78):520–528

Turgeon M, Bregman AS, Ahad PA (2002) Rhythmic masking release: contribution of cues for perceptual organization to the cross-spectral fusion of concurrent narrow-band noises. J Acoust Soc Am 111:1819–1831

Chapter 55
How Early Aging and Environment Interact in Everyday Listening: From Brainstem to Behavior Through Modeling

Barbara Shinn-Cunningham, Dorea R. Ruggles, and Hari Bharadwaj

Abstract We recently showed that listeners with normal hearing thresholds vary in their ability to direct spatial attention and that ability is related to the fidelity of temporal coding in the brainstem. Here, we recruited additional middle-aged listeners and extended our analysis of the brainstem response, measured using the frequency-following response (FFR). We found that even though age does not predict overall selective attention ability, middle-aged listeners are more susceptible to the detrimental effects of reverberant energy than young adults. We separated the overall FFR into orthogonal envelope and carrier components and used an existing model to predict which auditory channels drive each component. We find that responses in mid- to high-frequency auditory channels dominate envelope FFR, while lower-frequency channels dominate the carrier FFR. Importantly, we find that which component of the FFR predicts selective attention performance changes with age. We suggest that early aging degrades peripheral temporal coding in mid-to-high frequencies, interfering with the coding of envelope interaural time differences. We argue that, compared to young adults, middle-aged listeners, who do not have strong temporal envelope coding, have more trouble following a conversation in a reverberant room because they are forced to rely on fragile carrier ITDs that are susceptible to the degrading effects of reverberation.

B. Shinn-Cunningham (✉) • D.R. Ruggles • H. Bharadwaj, M.S.
Department of Biomedical Engineering, Boston University Center for Computational Neuroscience and Neural Technology,
677 Beacon St., Boston, MA 02215, USA
e-mail: shinn@bu.edu

B.C.J. Moore et al. (eds.), *Basic Aspects of Hearing*,
Advances in Experimental Medicine and Biology,
DOI 10.1007/978-1-4614-1590-9_55,

1 Introduction

The cacophony of voices, noises, and other sounds that bombards our ears in many social settings makes it challenging to focus selective auditory attention. Various acoustic cues allow us to group sound components into perceptual objects to which we can direct attention (Darwin 1997; Shinn-Cunningham 2008; Shamma and Micheyl 2010; Shamma et al. 2011). In most common settings, reflected sound energy intensifies the problem of separating sound sources and selecting the source of interest by blurring the sound features that support source segregation and selection.

Many listeners report difficulties in everyday situations demanding selective attention, especially as they age (Leigh-Paffenroth and Elangovan 2011; Noble et al. 2012). We wondered if these problems are most evident when reverberant energy challenges the auditory system. We designed a task in which listeners had to focus spatial attention on a center, target speech stream in a mixture of three otherwise identical streams of spoken digits, and then varied the level of reverberation (Ruggles et al. 2011; Ruggles and Shinn-Cunningham 2011). By design, listeners are likely to rely on interaural timing differences (ITDs) to perform this task (Ruggles et al. 2011). Since reverberant energy causes interaural decorrelation, we found, as expected, that selective attention performance got worse with reverberation. We also found that individual ability on our task was correlated both with perceptual sensitivity to frequency modulation (FM) and overall strength of the frequency-following response (FFR; see also Strelcyk and Dau 2009). However, we had too few middle-aged listeners to explore age effects.

Here, we recruited additional middle-aged listeners so that we could look for aging effects. We extended our analysis of the FFR by separating the response into the portion phase locking to the stimulus envelope (FFR_{ENV}) and that phase locking to the stimulus carrier (FFR_{CAR}; similar to approaches described in Aiken and Picton 2008; Gockel et al. 2011). We used existing brainstem response models (Dau 2003; Harte et al. 2010) to investigate which acoustic frequencies contribute to FFR_{ENV} and FFR_{CAR}.

2 Methods

2.1 Subjects

A total of 22 listeners ranging in age from 20.9 to 54.7 years participated in the experiments. All listeners had average audiometric hearing thresholds of 20-dB HL or better for frequencies from 250 to 8,000 Hz and left-right ear asymmetry of 15 dB or less at all frequencies. Of the 22 listeners, 17 were participants in earlier studies; the newly recruited five all were over 40 years of age. All gave informed consent and were paid for their participation.

2.2 FFR Measurement

FFRs were measured in response to a /dah/syllable presented in positive polarity for 2,000 trials and in inverted polarity for 2,000 trials (Ruggles et al. 2011). Trials containing eyeblinks or other artifacts were removed, leaving at least 1,800 clean trials for each subject, condition, and stimulus polarity. The time series from each trial was windowed with a first-order Slepian taper (Thomson 1982) and the Fourier transform was computed. We generated distributions of phase-locking values (PLV) for different conditions using a bootstrapping procedure to produce 200 independent PLVs, each computed from a draw (with replacement) of 800 trials (Ruggles et al. 2011). We broke the PLV into orthogonal envelope and carrier components (FFR_{ENV} and FFR_{CAR}) at every frequency from 30 to 3,000 Hz. FFR_{ENV} was calculated with equal draws from responses to each polarity, treating positive- and negative-polarity trials identically. FFR_{CAR} was determined with equal draws from responses to each polarity but inverting the phase of negative-polarity trials (see also Aiken and Picton 2008; Gockel et al. 2011). For each harmonic of 100 Hz, we computed the proportion of the total FFR in FFR_{ENV} and in FFR_{CAR}.

2.3 FFR Modeling

We used an existing model of brainstem responses (Dau 2003; Harte et al. 2010) to analyze the sources of the different components of the FFR. We presented the model with our /dah/ syllable, then calculated the FFR by summing model outputs across peripheral channels with CFs spanning the range from 100 up to 10,000 Hz. At each harmonic (multiple of 100 Hz), we then computed the proportion of the total FFR phase locked to the envelope and the proportion phase locked to the carrier (FFR_{ENV} and FFR_{CAR}). We then considered the output of each peripheral channel to explore which acoustic frequencies contributed to which components of the FFR. Finally, we analyzed the relative strength of the contribution of each peripheral channel to FFR_{ENV} at the fundamental frequency (100 Hz).

2.4 Spatial Attention Task

Subjects were asked to report a sequence of four digits appearing to come from in front while ignoring two competing digit streams, spoken by the same talker, from +15° to −15° azimuth (Ruggles and Shinn-Cunningham 2011). Spatial cues were simulated using a rectangular-room model with three different wall characteristics (Ruggles and Shinn-Cunningham 2011). Prior to statistical analyses, percent correct scores were transformed using a rationalized arcsine unit (RAU; Studebaker 1985). In the task, listeners report one of the three presented words nearly 95 % of the time; errors arise because of failures of selective attention, rather than memory limitations (Ruggles and Shinn-Cunningham 2011). Therefore, percent scores in the range 0.33–1.0 were linearly transformed to 0–1.0 (scores $<$ 0.33 set to 0) prior to applying the transform.

2.5 FM Detection Task

Listeners indicated which of three 750-Hz tones (interstimulus interval 750 ms) contained 2-Hz frequency modulation (Strelcyk and Dau 2009). A two-down, one-up adaptive procedure (step size 1 Hz) estimated the 70.7 % correct FM threshold. Individual thresholds were computed by averaging the last 12 reversals per run, then averaging across six runs.

3 Results

3.1 Generators of FFR_{ENV} and FFR_{CAR}

Figure 55.1 compares measurements and model predictions of the relative strengths of FFR_{ENV} and FFR_{CAR} at harmonics of a periodic input (F0 = 100 Hz). The lowest frequencies of the FFR are dominated by FFR_{ENV} and the higher harmonics are dominated by FFR_{CAR}. Both FFR components approach the noise floor in the empirical measurements by 800–900 Hz, which may help explain why the percentages of FFR_{ENV} and FFR_{CAR} in the total FFR both asymptote to 0.5 as frequency increases

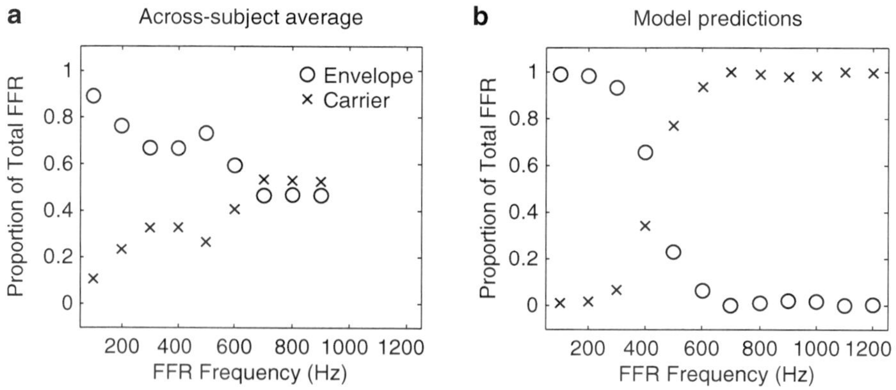

Fig. 55.1 Proportion of total FFR contained in FFR_{ENV} and in FFR_{CAR} at each harmonic of 100 Hz from (**a**) experimental measures and (**b**) model predictions

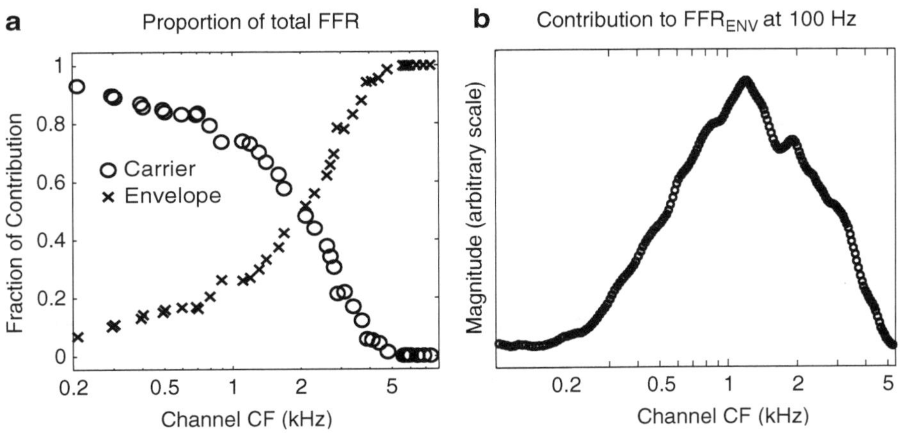

Fig. 55.2 (**a**) Relative strength of FFR_{ENV} and FFR_{CAR} generated by each peripheral channel as a function of characteristic frequency (*CF*). (**b**) Relative contribution of each CF channel to strength of FFR_{ENV} at stimulus F0 of 100 Hz

and why the measured FFR_{ENV} does not drop as completely or as rapidly as the modeled FFR_{ENV} as frequency increases.

Modeling results also suggest that different acoustic frequencies contribute to FFR_{ENV} and FFR_{CAR}. In the model, peripheral channels with the lowest characteristic frequencies (CFs) tend to contribute to FFR_{CAR} and peripheral channels with the highest CFs contribute to FFR_{ENV}, with a crossover point of about 2 kHz (Fig. 55.2a). The model also predicts that the channels that contribute the most to the 100-Hz FFR_{ENV} for our /dah/ syllable have CFs in the mid-to-high frequency range, around 1 kHz (Fig. 55.2a).

3.2 *Effects of Reverberation and Age on Selective Attention*

Selective attention performance decreases as reverberant energy increases, reaching chance levels for all but five listeners in the highest reverberation level (Fig. 55.3; chance performance is one-third; modeling performance as a binomial distribution of 600 independent trials, we computed the 95 % confidence interval around this level).

We quantified the fidelity of envelope temporal structure encoding for each listener as the FFR_{ENV} at 100 Hz. To quantify coding of the temporal fine structure in the input stimulus, we took the average of FFR_{CAR} for four harmonics (600–900 Hz, henceforth denoted $FFR_{CAR\text{-}AV}$). Importantly, these two statistics are not significantly correlated ($r=0.03$, $p=0.905$, $N=22$), supporting the modeling prediction that each component reflects different aspects of temporal coding precision driven by different tonotopic portions of the auditory pathway.

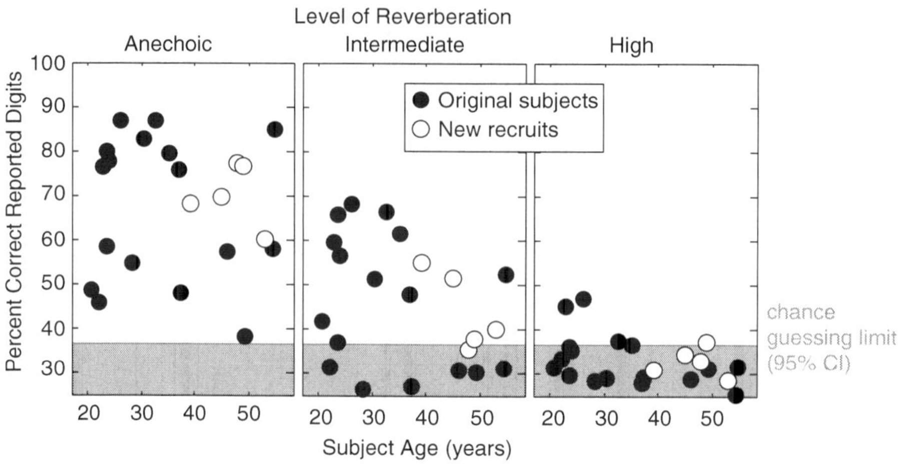

Fig. 55.3 Percentage of target digits correctly reported as a function of individual listener age for the three room conditions. *Open symbols* show subjects not in Ruggles et al. (2011)

We performed a multi-way, repeated-measures ANOVA on the selective attention results with factors of reverberation, age, $FFR_{ENV-100}$, and FFR_{CAR-AV} (treating reverberation as categorical and all other factors as continuous). Although there is no statistically significant effect of age on selective attention performance (Fig. 55.1a; $F(1, 16)=1.42$, $p=0.251$), there is a significant interaction between age and reverberation ($F(1, 16)=5.88$, $p=0.025$) and a significant main effect of reverberation ($F(1, 16)=155.17$, $p=7.01\times10^{-11}$). Although age does not predict how well an individual performs overall, the toll that reverberation takes increases with age.

3.3 Relationship Between FFR Components and Performance

Consistent with previous results showing that the total FFR strength at 100 Hz (a measure dominated by envelope phase locking; see Fig. 55.1) predicted selective attention ability (Ruggles et al. 2011), we find a significant main effect of $FFR_{ENV-100}$ on performance ($F(1, 16)=5.03$, $p=0.040$). Importantly, however, there is a significant interaction between age and $FFR_{ENV-100}$ ($F(1, 16)=4.64$, $p=0.048$). There is also a significant interaction between age and $FFR_{CAR-AVE}$ ($F(1, 16)=4.64$, $p=0.047$), with no main effect of $FFR_{CAR-AVE}$ ($F(1, 16)=0.216$, $p=0.649$). The regression coefficients of the ANOVA analysis reveal that the younger a listener is, the better $FFR_{ENV-100}$ is in predicting selective attention, whereas $FFR_{CAR-AVE}$ is a better predictor the older the listener. These results suggest that $FFR_{ENV-100}$ and $FFR_{CAR-AVE}$ reflect different perceptual cues that each aid in selective auditory attention but that are weighted differently as listeners age.

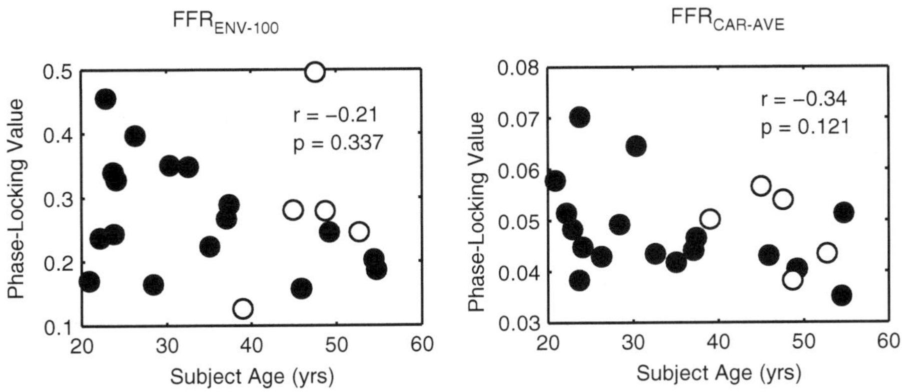

Fig. 55.4 (**a**) $FFR_{ENV\text{-}100}$ as a function of age. (**b**) $FFR_{CAR\text{-}AVE}$ as a function of age. *Open symbols* show subjects not in Ruggles et al. (2011)

3.4 *Individual Differences in FFR*

Figure 55.4 plots $FFR_{ENV\text{-}100}$ and $FFR_{CAR\text{-}AVE}$ as a function of age. While both components tend to decrease as age increases, age is not significantly correlated with either $FFR_{ENV\text{-}100}$ or with $FFR_{CAR\text{-}AVE}$. Notably, a good percentage of the younger adult listeners have strong FFRs (particularly for $FFR_{ENV\text{-}100}$), whereas nearly all the older listeners have weak FFRs. Thus, most of the variance in the FFRs is from the younger listeners and cannot be explained by age alone.

3.5 *Relationship Between FM Detection Threshold and Performance*

We previously found that FM detection threshold, a measure thought to reflect coding fidelity of temporal fine structure (Moore and Sek 1996), was also related to attention performance (Ruggles et al. 2011). This relationship remains significant with our additional subjects, as shown in Fig. 55.5.

4 Discussion

Some previous studies have found that aging reduces FFR strength (Clinard et al. 2010); however, not all studies have found group age effects (Vander Werff and Burns 2011). Moreover, even studies that find age-related group differences have not consistently found corresponding age-related differences in auditory perceptual abilities (Clinard et al. 2010). The current study helps explain these discrepant

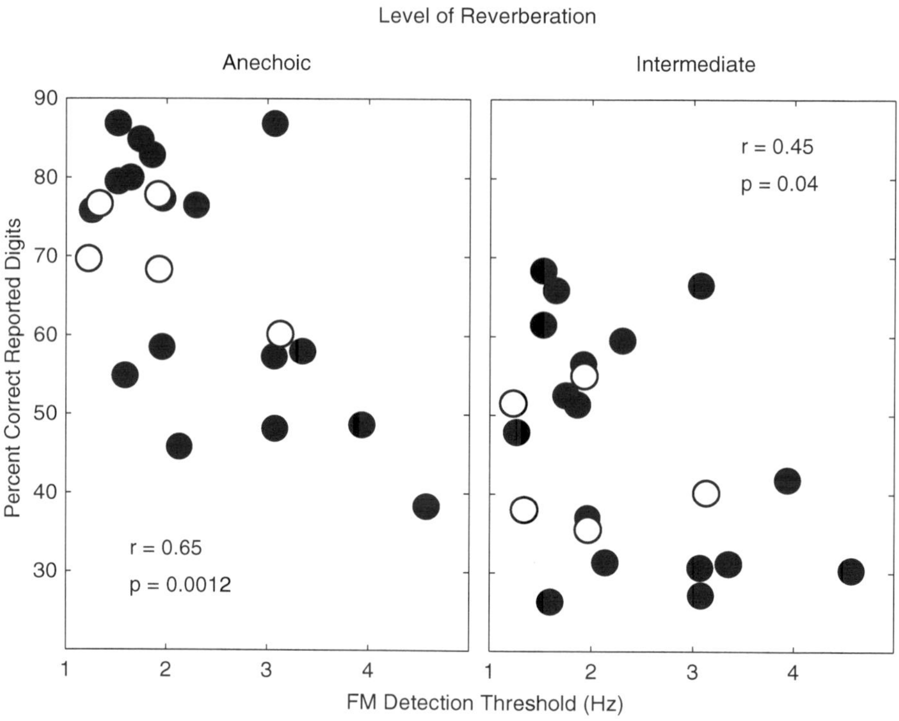

Fig. 55.5 Percentage of target digits correctly reported as a function of FM threshold for the two levels of reverberation where performance is above chance. *Open symbols* show subjects not in Ruggles et al. (2011)

findings, in that there is a large variation in brainstem responses even among young adults. By looking at individual subjects and considering different components of the FFR, we find reliable interactions between aging, perceptual ability, and specific components of the FFR.

Our results suggest that the FFR envelope component at the fundamental frequency of the stimulus tends to become weak as listeners reach middle age, possibly because the neural response to suprathreshold sound at acoustic frequencies in the mid-to-high frequency range (e.g., around 1,000 Hz) is reduced in overall strength. Physiological results show that noise exposure can reduce the magnitude of neural responses that are suprathreshold, even when thresholds are "normal" (Kujawa and Liberman 2009). These changes may come about because low-spontaneous-rate nerve fibers are particularly vulnerable to damage (Schmiedt et al. 1996).

In our task, performance is primarily limited by the ability to successfully direct spatial auditory attention, which may help explain why performance depends on the fidelity of envelope temporal coding. Envelope ITD cues in high-frequency sounds are known to carry spatial information; however, a number of classic laboratory experiments establish that for wideband, anechoic sounds, low-frequency carrier

ITDs perceptually dominate over high-frequency spatial cues (Wightman and Kistler 1992; Macpherson and Middlebrooks 2002). The current results suggest that in reverberant settings, high-frequency ITD cues, encoded in signal envelopes, may be more important for spatial perception of wideband sounds than past laboratory studies suggest.

In anechoic conditions, temporal fine structure cues and temporal envelope cues both provide reliable information for directing selective spatial auditory attention. However, in reverberant settings, interaural decorrelation of temporal fine structure is more severe than interaural decorrelation of envelope structure; thus, high-frequency envelope ITD cues may be crucial to spatial perception in everyday settings. This possibility points to the importance of providing high-frequency amplification in assistive listening devices, which have typically focused on audibility of frequencies below 8 kHz.

Our results hint that middle-aged listeners, who have generally weak encoding of mid- to high-frequency temporal cues, rely on temporal fine structure cues to direct selective spatial auditory attention. This reliance on carrier ITD cues, which are relatively fragile in ordinary listening environments, may explain why middle-aged listeners report difficulty when trying to converse in everyday social settings. In contrast, younger listeners appear to give great perceptual weight to envelope ITD cues when directing selective attention, providing them with a more reliable cue for selective spatial auditory attention.

Acknowledgments This work was sponsored by the National Institutes of Health (NIDCD R01 DC009477 to BGSC and NIDCD F31DC011463 to DR) and the National Security Science and Engineering Faculty Fellowship to BGSC.

References

Aiken SJ, Picton TW (2008) Envelope and spectral frequency-following responses to vowel sounds. Hear Res 245:35–47

Clinard CG, Tremblay KL, Krishnan AR (2010) Aging alters the perception and physiological representation of frequency: evidence from human frequency-following response recordings. Hear Res 264:48–55

Darwin CJ (1997) Auditory grouping. Trends Cogn Sci 1:327–333

Dau T (2003) The importance of cochlear processing for the formation of auditory brainstem and frequency following responses. J Acoust Soc Am 113:936–950

Gockel HE, Carlyon RP, Mehta A, Plack CJ (2011) The frequency following response (FFR) may reflect pitch-bearing information but is not a direct representation of pitch. J Assoc Res Otolaryngol 12:767–782

Harte JM, Ronne F, Dau T (2010) Modeling human auditory evoked brainstem responses based on nonlinear cochlear processing. In: Proceedings of the 20th international congress on acoustics, Sydney, 2010

Kujawa SG, Liberman MC (2009) Adding insult to injury: cochlear nerve degeneration after “temporary” noise-induced hearing loss. J Neurosci 29:14077–14085

Leigh-Paffenroth ED, Elangovan S (2011) Temporal processing in low-frequency channels: effects of age and hearing loss in middle-aged listeners. J Am Acad Audiol 22:393–404

Macpherson EA, Middlebrooks JC (2002) Listener weighting of cues for lateral angle: the duplex theory of sound localization revisited. J Acoust Soc Am 111:2219–2236
Moore BC, Sek A (1996) Detection of frequency modulation at low modulation rates: evidence for a mechanism based on phase locking. J Acoust Soc Am 100:2320–2331
Noble W, Naylor G, Bhullar N, Akeroyd MA (2012) Self-assessed hearing abilities in middle- and older-age adults: a stratified sampling approach. Int J Audiol 51:174–180
Ruggles D, Shinn-Cunningham B (2011) Spatial selective auditory attention in the presence of reverberant energy: individual differences in normal-hearing listeners. J Assoc Res Otolaryngol 12:395–405
Ruggles D, Bharadwaj H, Shinn-Cunningham BG (2011) Normal hearing is not enough to guarantee robust encoding of suprathreshold features important in everyday communication. Proc Natl Acad Sci 108:15516–15521
Schmiedt RA, Mills JH, Boettcher FA (1996) Age-related loss of activity of auditory-nerve fibers. J Neurophysiol 76:2799–2803
Shamma SA, Micheyl C (2010) Behind the scenes of auditory perception. Curr Opin Neurobiol 20: 361–366
Shamma SA, Elhilali M, Micheyl C (2011) Temporal coherence and attention in auditory scene analysis. Trends Neurosci 34:114–123
Shinn-Cunningham BG (2008) Object-based auditory and visual attention. Trends Cogn Sci 12: 182–186
Strelcyk O, Dau T (2009) Relations between frequency selectivity, temporal fine-structure processing, and speech reception in impaired hearing. J Acoust Soc Am 125:3328–3345
Studebaker GA (1985) A "rationalized" arcsine transform. J Speech Hear Res 28:455–462
Thomson DJ (1982) Spectrum estimation and harmonic-analysis. Proc IEEE 70:1055–1096
Vander Werff KR, Burns KS (2011) Brain stem responses to speech in younger and older adults. Ear Hear 32:168–180
Wightman FL, Kistler DJ (1992) The dominant role of low-frequency interaural time differences in sound localization. J Acoust Soc Am 91:1648–1661

Chapter 56
Energetic and Informational Masking in a Simulated Restaurant Environment

John F. Culling

Abstract Participants were seated at a central table for two in a virtual restaurant, simulated over headphones. They listened to the person across the table. Speech reception thresholds (SRTs) were measured as a function of the number of interfering sources distributed across other tables in the room; those sources were either speech-shaped noises or competing speech. The restaurant either was enclosed by acoustically reflective surfaces or was anechoic. Variations in SRT for speech-shaped noises were accurately predicted ($r = 0.96$) by a model of spatial release from masking based on the additive combination of better-ear listening and binaural unmasking. However, SRTs for interfering voices followed a different pattern. A single interfering voice gave a lower SRT than a single speech-shaped noise source (by 6.3 dB in anechoic conditions and 1.2 dB in reverberant conditions). This difference can be attributed to the effects of dip listening and to the exploitation of differences between voices in fundamental frequency (F0). SRTs for two interfering voices were markedly higher than for a single voice, particularly when the interfering voice was the same as the target voice. Multiple speech interferers produced more masking than multiple noise interferers. This effect can be attributed to informational masking (IM). These results indicate that current models require some elaboration before they will produce accurate predictions of intelligibility in noisy social environments.

1 Introduction

Models of the intelligibility of speech in continuous background noise have proved very effective. They have accurately predicted spatial release from masking (SRM) by background noise in different listening situations, including noise

J.F. Culling, B.Sc., D.Phil.
School of Psychology, Cardiff University, Tower Building, Park Place, Cardiff CF10 3AT, UK
e-mail: cullingj@cf.ac.uk

B.C.J. Moore et al. (eds.), *Basic Aspects of Hearing*, Advances in Experimental Medicine and Biology,
DOI 10.1007/978-1-4614-1590-9_56, © Springer Science+Business Media New York 2013

from up to three interfering sources in various spatial configurations and in various forms of reverberation (Beutelmann and Brand 2006; Beutelmann et al. 2010; Lavandier and Culling 2010; Jelfs et al. 2011; Lavandier et al. 2012). However, a number of other factors are involved when the interfering noise is composed of one or more competing voices (Hawley et al. 2004). These factors include exploitation of dips in masker energy, exploitation of differences in F0, and informational masking (IM). Moreover, interactions may exist between these factors and the listening situation (number and spatial distribution of interferers and room reverberation).

The amplitude modulation of interfering speech allows listeners to "dip listen", catching glimpses of the target voice during brief drops in masking energy. A large release from masking can be obtained using noise modulated by a square-wave function (e.g. de Laat and Plomp 1983), but more modest effects are attributable to the modulation of a speech interferer. To quantify this benefit, "speech-modulated" noise is often used (Festen and Plomp 1990; Hawley at al. 2004), in which the broadband envelope of speech is extracted and used to modulate a noise source. The models of Rhebergen (2005) and Beutelmann et al. (2010) have achieved some success at predicting dip listening and its interaction with SRM.

Differences in F0 enable listeners to perceptually separate speech from interfering speech (Brokx and Nooteboom 1992). This effect appears to be determined by the periodicity of the interfering sound (de Cheveigné et al. 1995; Deroche and Culling 2011a, b) and by the magnitude of F0 difference (Brokx and Nooteboom 1992; Culling and Darwin 1993). Reverberation can negatively influence the effect (Culling et al. 1994, 2003; Deroche and Culling 2011b).

The linguistic content of interfering speech can impair listeners' comprehension of target speech. This form of IM probably operates on several levels. Using IEEE sentences (Rothauser et al. 1969), Hawley et al. (2004) found a similar enhancement of SRM for both speech and reversed speech, compared to noise and speech-modulated noise. They attributed this enhancement to spatial release from IM. Since this enhanced SRM also occurred for reversed speech, it seemed to be independent of intelligible linguistic content but may have been caused by forms of low-level phonetic interference. On the other hand, Brungart (2001) found that intruding words from the interfering voice were a frequent source of error using the coordinate response measure, especially when the voice of the target speaker was also used as the interferer. He concluded that IM was influenced by the similarity/distinctiveness of target and masker.

The present experiment was designed to evaluate the practical significance of these factors, which may be needed for models of SRM to accurately predict intelligibility in a room with competing voices. Listeners were therefore placed in a relatively realistic simulated listening situation. They were seated at the centre table in a virtual restaurant and SRTs were measured as a function of the number of other sound sources in the room. These sound sources were either speech-shaped noises, whose effects should be well predicted by existing models, or competing voices.

2 Method

A virtual room was created using an adapted version of the image model (Allen and Berkley 1979). The model calculated each ray path between a given source location and the listener's head. Each ray's angle of incidence at the head was used to select an appropriate head-related impulse response (HRIR) from the KEMAR database collected by Gardner and Martin (1995). These HRIRs were scaled and delayed according to the rays' path lengths and the absorbencies of the surfaces with which the rays had interacted and were added together to form a binaural room impulse response (BRIR). For the reverberant room, the absorption coefficients of the internal surfaces were 0.05 for the walls, 0.07 for the floor, and 0.9 for the ceiling. For the anechoic room, they were all 1.0. BRIRs were calculated between locations in a room (6.4 m×6.4 m×2.5 m) that would correspond to the locations of diners in a restaurant with nine tables for two. The BRIRs had a measured reverberation time (Schroeder 1965) of 350 ms. Figure 56.1 shows the restaurant plan. These BRIRs were used to create virtual simulations of background restaurant noise with varying levels of occupancy.

Target speech sentences were taken from the IEEE sentences (Rothauser et al. 1969) recorded at M.I.T. (voice CW). A silent lead-in of 0.5 s was added to each sound file. These were convolved with the BRIR for one individual at the centre table talking to the other individual at that table. Forty-eight different samples of interfering speech were made from the same voice (CW) or a different one (DA); for each sample two sentences were trimmed to remove leading and trailing silence, concatenated together, and then gated to create 4.5 s of continuous speech. Interfering noise was generated by filtering white noise with a 512-point FIR filter designed to match the long-term excitation pattern (Moore and Glasberg 1983) of the target

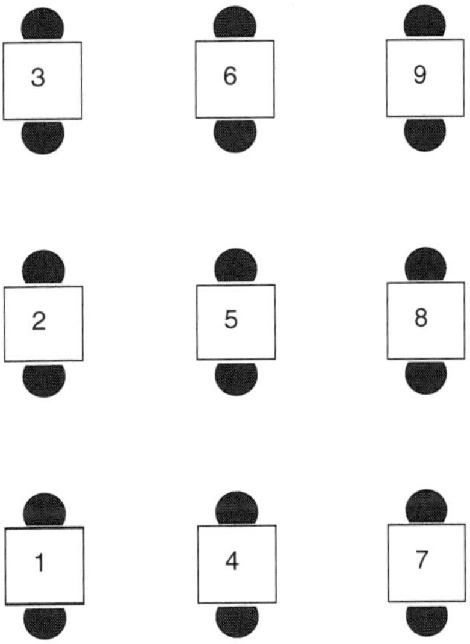

Fig. 56.1 Seating layout of the virtual restaurant

voice. Complexes of multiple interferers were created by (1) randomly selecting one of the two seats at each of the other tables in the room and (2) convolving an interfering sound with the BRIRs between these seats and the listener's at the centre table, scaling and adding together 1, 2, 4, or 8 of these virtual interferers to make an interfering complex. Interferers were scaled by 0, 3, 6, or 9 dB to compensate for the number of interferers. The average gain of the room was also factored out. Interfering complexes were either composed entirely of speech-shaped noises or entirely from speech samples. There were 48 different 1-voice interferers, 24 different 2-voice complexes, 12 different 4-voice complexes, and 6 different 8-voice complexes. Interfering sources were located either (a) on table 1; (b) on tables 1 and 9; (c) on tables 1, 3, 7, and 9; or (d) on tables 1, 2, 3, 4, 6, 7, 8, and 9.

Speech reception thresholds (SRTs) were measured using an adaptive method similar to that used by Hawley et al. (2004) in which listeners increase the level of the first sentence until they judge it intelligible. The method differed in two important respects from that used in previous studies. First, stimulus uncertainty was maintained by selecting the interfering complexes at random from the set available for that condition in the preparation of each stimulus. Second, the adaptive algorithm automatically rejected completely incorrect transcripts as responses to the first target sentence. This procedure prevented participants starting the adaptive track at an inappropriately low target-to-masker ratio after inadvertently transcribing one of the interfering sentences. Sixteen participants each attended a single 90-min session. Following two practice measurements, the 16 conditions were presented in a pseudorandom order, which was rotated for each successive participant. Target speech materials remained in the same order. Participants were advised that the target speech would be heard located in the middle, would begin about half a second after the interfering speech, and whether it was the same or a different voice from the interferers.

In one experiment, interfering sources either were the same male voice as the target or were speech-shaped noises. There were 16 conditions (2 interferer types × 2 levels of reverberation (anechoic/reverberant) × 4 numbers of interfering sources). In a second, all interfering sources were a different male voice from the target voice, and there were eight conditions (2 levels of reverberation, anechoic/reverberant × 4 numbers of interfering sources).

3 Results

Figure 56.2 shows the results from the first experiment. Overall, SRTs increased with the number of interfering voices ($F(3, 45) = 433$, $p < 0.0001$), were higher in reverberation than in an anechoic space ($F(1, 15) = 276$, $p < 0.0001$), and were higher for speech interferers than for noise interferers ($F(1,15) = 129$, $p < 0.0001$). However, some strong interactions were present.

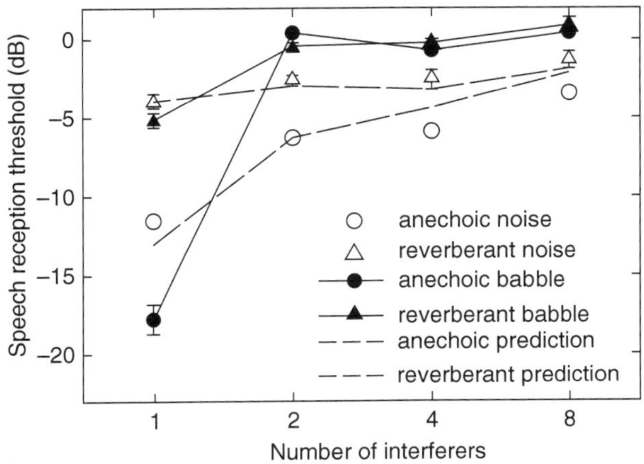

Fig. 56.2 Speech reception threshold as a function of number of interferers for different masker and room types. Error bars indicate one standard error

The effect of the number of interfering sources was much stronger for speech interferers than for noise interferers ($F(3, 45) = 66$, $p < 0.0001$); for speech interferers in anechoic conditions, SRT increased by 18 dB when a second interferer was introduced, even though the 3 dB increase in level from adding a second source had been compensated at the point of stimulus preparation. For noise, this increase was only 5 dB. Further increases in the number of interferers had little effect. In reverberant conditions, the effect was significantly smaller ($F(3, 45) = 44$, $p < 0.0001$). Overall, reverberation had marginally less effect on SRTs with speech interferers than with noise interferers ($F(1, 15) = 16.5$, $p < 0.05$). The effect of the number of interfering sources was also stronger in anechoic than in reverberant conditions ($F(3, 45) = 130$, $p < 0.0001$), because SRTs for a single interferer were substantially elevated by reverberation, while SRTs for multiple interferers were less elevated.

The effects of reverberation were well predicted by the Jelfs et al. (2011) model for noise interferers (dashed lines in Fig. 56.1). In contrast, the model overestimated SRTs for a single speech interferer and underestimated SRTs for multiple speech interferers.

Figure 56.3 shows the results when the target voice was a different male voice from the interferers. For comparison, the results using the same voice as the target are reproduced from Fig. 56.2 as grey lines. There is very good agreement between the two experiments for 1 and 8 interferers, but for 2 and 4 interferers, there seems to be an advantage for listening to different voices. A between-subjects analysis of variance comparing the second experiment with the first tested the effect of having a different voice for target and interferer. SRTs were lower overall with different voices ($F(1, 22) = 76$, $p < 0.02$). This effect was significant only in the two-interferer case ($F(1, 88) = 36$, $p < 0.0001$) and was larger in anechoic conditions, producing interactions between interferer voice and number of interferers ($F(3, 22) = 39$, $p < 0.0001$) and a three-way interaction including reverberation ($F(3, 66) = 27$, $p < 0.0001$).

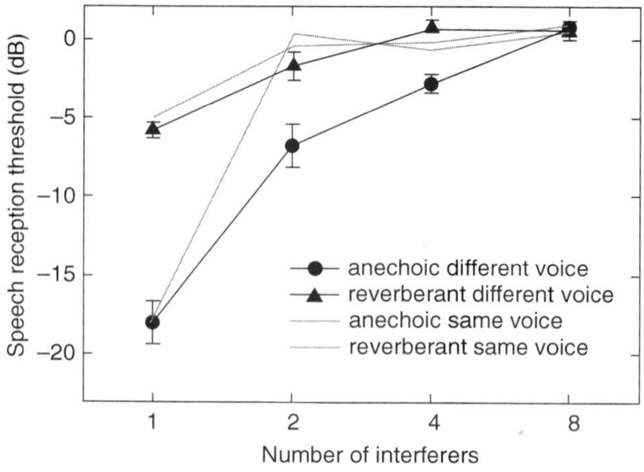

Fig. 56.3 Speech reception thresholds as a function of number of interferers for different rooms and for same-voice interferers (*grey*, reproduced from Fig. 56.2) and different voice interferers (*black*). Error bars indicate one standard error

4 Discussion

The most striking feature of the results is the dramatic increase in SRT with two speech interferers rather than one. In anechoic conditions, this increase was 18 dB. A number of factors probably contribute to this increase, including a loss of SRM, an inability to exploit F0 differences between multiple concurrent sources, reduced opportunity for dip listening, and increased IM.

When only one interferer is present, SRM is strong because the interferer comes from one side, while the target speech is in front. A second interferer from the opposite hemifield reduces SRM. The benefit of better-ear listening to SRM can be largely abolished in the latter case (Hawley et al. 2004; Culling et al. 2004). Binaural unmasking will be less affected; the Jelfs et al. model predicts that better-ear listening will be reduced by 5 dB, but binaural unmasking by only 0.4 dB. The combined influence of these two effects can be seen in the results for noise interferers, which are well predicted by the model.

F0 differences can produce large improvements in SRT, but this effect is highly dependent on the periodicity of the interfering sound (Deroche and Culling 2011b). Periodicity is reduced when (intonated) speech is heard in a reverberant space or if multiple periodic interferers are present. Thus, a second interfering voice in anechoic conditions produces a large increase in SRT, but if reverberation is already present, the increase should be less pronounced, because the interferer is already somewhat aperiodic. These expectations seem consistent with the pattern of data observed.

Modulation of interfering sources allows listeners to dip listen. When two interferers are present, the modulation of the composite interfering sound is reduced, attenuating the dip-listening effect. When this effect is simulated with speech-modulated noise, the increase in SRT is about 3 dB (Hawley et al. 2004). Again, these effects are reduced in reverberation, this time because reverberation fills in the dips in masker level.

IM is evident in the higher thresholds observed for speech when there are two or more interferers. With one interferer, IM may be absent at threshold, because the target speech is distinguished by its much lower level. Interestingly, there is no effect of reverberation with multiple speech interferers. There may be counteracting effects here; reverberation interferes with dip listening and exploitation of F0 differences, but these effects are countered by a reduction in IM. Such a reduction could occur for two reasons. First, the target speech may be more distinct from the interferer, because it is closer and less reverberant. Second, the reverberation makes the interferers less intelligible, which reduces the potential for linguistic interference. Interferers made from a different person's voice are more distinct from the target voice, and while this seems to reduce IM for a two-voice interferer, the second experiment indicates that IM for a larger number of voices may be unaffected by vocal distinctiveness.

SRM in steady noise is predictable. Rhebergen and Versfeld (2005) achieved some success in modelling speech intelligibility in modulated noise and Beutelmann et al. (2010) did so in the context of its interaction with SRM and reverberation. However, the present data indicate that modelling real listening situations with multiple, spatially separated speech interferers and reverberation requires more sophisticated models that can predict the effects of F0 differences and IM, as well as their interactions with SRM, reverberation, and dip listening.

References

Allen JB, Berkley DA (1979) Image method for efficiently simulating small-room acoustics. J Acoust Soc Am 65:943–950

Beutelmann R, Brand T (2006) Prediction of speech intelligibility in spatial noise and reverberation for normal-hearing and hearing-impaired listeners. J Acoust Soc Am 120:331–342

Beutelmann R, Brand T, Kollmeier B (2010) Revision, extension and evaluation of a binaural speech intelligibility model. J Acoust Soc Am 127:2479–2497

Brokx JPL, Nooteboom SG (1992) Intonation and the perceptual separation of simultaneous voices. J Phon 10:23–36

Brungart DS (2001) Informational and energetic masking effects in the perception of two simultaneous talkers. J Acoust Soc Am 109:1101–1109

Culling JF, Darwin CJ (1993) Perceptual separation of concurrent vowels: within and across formant grouping by F0. J Acoust Soc Am 93:3454–3467

Culling JF, Hodder KI, Toh CY (2003) Effects of reverberation on perceptual segregation of competing voices. J Acoust Soc Am 114:2871–2876

Culling JF, Hawley ML, Litovsky RY (2004) The role of head-induced interaural time and level differences in the speech reception threshold for multiple interfering sound sources. J Acoust Soc Am 116:1057–1065

Culling JF, Summerfield Q, Marshall DH (1994) Effects of simulated reverberation on binaural cues and fundamental frequency differences for separating concurrent vowels. Speech Comm. 14:71–96

de Cheveigné A, McAdams S, Laroche J, Rosenberg M (1995) Identification of concurrent harmonic and inharmonic vowels: a test of the theory of harmonic cancellation and enhancement. J Acoust Soc Am 97:3736–3748

de Laat JAPM, Plomp R (1983) The reception threshold of interrupted speech. In: Kinke R, Hartman R (eds) Hearing: physiological bases and psychophysics. Springer, Berlin, pp 359–363

Deroche M, Culling JF (2011a) Narrow noise band detection in a complex masker: masking level difference due to harmonicity. Hear Res 282:225–235

Deroche M, Culling JF (2011b) Voice segregation by difference in fundamental frequency: evidence for harmonic cancellation. J Acoust Soc Am 130:2855–2865

Festen JM, and Plomp R (1990) Effects of fluctuating noise and interfering speech on the speech-reception threshold for impaired and normal hearing. J Acoust Soc Am 88:1725–1736

Gardner WG, Martin KD (1995) HRTF measurements of a KEMAR. J Acoust Soc Am 97:3907–3908

Hawley ML, Litovsky RY, and Culling JF (2004) The benefit of binaural hearing in a cocktail party: Effect of location and type of masker. J Acoust Soc Am 115:833–843

Jelfs S, Lavandier M, Culling JF (2011) Revision and validation of a binaural model for speech intelligibility in noise. Hear Res 275:96–104

Lavandier M, Culling JF (2010) Prediction of binaural speech intelligibility against noise in rooms. J Acoust Soc Am 127:387–399

Lavandier M, Jelfs S, Culling JF, Watkins AJ, Raimond AP, Makin SJ (2012) Binaural prediction of speech intelligibility in reverberant rooms with multiple noise sources. J Acoust Soc Am 131:218–231

Moore BCJ, Glasberg BR (1983) Suggested formulae for calculating auditory-filter bandwidths and excitation patterns. J Acoust Soc Am 74:750–753

Rhebergen KS, Versfeld NJ (2005) A Speech Intelligibility Index-based approach to predict the speech reception threshold for sentences in fluctuating noise for normal-hearing listeners. J Acoust Soc Am 117:2181–2192

Rothauser EH, Chapman WD, Guttman N, Nordby KS, Silbiger HR, Urbanek GE, Weinstock M (1969) I.E.E.E. recommended practice for speech quality measurements. IEEE Trans Aud Electroacoust 17:227–246

Schroeder MR (1965) New method of measuring reverberation time. J Acoust Soc Am 37: 409–412

Chapter 57
A Computational Approach to the Dynamic Aspects of Primitive Auditory Scene Analysis

Makio Kashino, Eisuke Adachi, and Haruto Hirose

Abstract Recent psychophysical and physiological studies demonstrated that auditory scene analysis (ASA) is inherently a dynamic process, suggesting that the system conducting ASA constantly changes itself, incorporating the dynamics of sound sources in the acoustic scene, to realize efficient and robust information processing. Here, we propose computational models of ASA based on two computational principles of ASA, namely, separation in a feature space and temporal regularity. We explicitly introduced learning processes, so that the system could autonomously develop its selectivity to features or bases for analyses according to the observed acoustic data. Simulation results demonstrated that the models were able to predict some essential features of behavioral properties of ASA, such as the buildup of streaming, multistable perception, and the segregation of repeated patterns embedded in distracting sounds.

1 Introduction

Human listeners have a remarkable capability of organizing complex acoustic signals into coherent perceptual streams that usually correspond to sound sources. This process is called auditory scene analysis (ASA). Several lines of behavioral

M. Kashino (✉)
Human Information Laboratory, NTT Communication Science Laboratories,
NTT Corporation, 3-1 Morinosato-Wakamiya, Atsugi, Kanagawa 243-0198, Japan

Department of Information Processing, Tokyo Institute of Technology,
4259 Nagatsuta, Midori-ku, Yokohama, Kanagawa 226-8502, Japan
e-mail: kashino.makio@lab.ntt.co.jp

E. Adachi • H. Hirose
NTT Communication Science Laboratories, NTT Corporation,
3-1 Morinosato-Wakamiya, Atsugi, Kanagawa 243-0198, Japan

B.C.J. Moore et al. (eds.), *Basic Aspects of Hearing*,
Advances in Experimental Medicine and Biology,
DOI 10.1007/978-1-4614-1590-9_57,

data have shown that ASA is inherently a dynamic process. The formation of streams is not instantaneous; rather, streaming builds up over time and can be reset by sudden changes in the acoustic signal or the listener's movement (Kondo et al. 2012). Prolonged listening to a repeated sound sequence produces multistable perception, i.e., spontaneous switching among different perceptual forms (Hupé and Pressnitzer 2012; Kashino and Kondo 2012; Winkler et al. 2012). Preceding context affects streaming (Snyder et al. 2008). The listeners can detect repetitions of a sound pattern (Agus et al. 2010; McDermott et al. 2011) and exploit it to form or hear out streams (Andreou et al. 2011). Physiological data have also revealed the dynamic aspects of ASA. Above all, various forms of short-term neural plasticity have been found in the auditory cortex, such as habituation (Micheyl et al. 2005), stimulus-specific adaptation (Ulanovsky et al. 2004), and task-dependent changes of spectrotemporal receptive fields (Fritz et al. 2005).

Such behavioral and neural dynamics in ASA are, in our view, not trivial epiphenomena but have functional significance. The system conducting ASA constantly changes itself, incorporating the dynamics of sound sources in the acoustic scene, to realize efficient and robust information processing. This idea is consistent with the concept of predictive coding (Winkler et al. 2012). Different kinds of predictions may be generated at different levels of auditory processing. It is necessary to understand in more detail the nature of predictions actually used in ASA and computational principles underlying them.

Here, we consider two basic principles for ASA and examine whether computational models based on the principles can predict the dynamic aspects of ASA. The principles are population separation and temporal coherence (Shamma et al. 2011). The former means that the perceptual organization of sounds into streams is determined by the spatial overlap between responsible neural populations on a feature axis, such as tonotopy, periodicity (pitch), spectral shape (timbre), spatial location, and so on. The latter means that the formation of streams depends on the temporal coherence of responses of neural populations selective to various sound attributes (e.g., frequency, pitch, timbre, and location). In a strict sense, temporal coherence refers to the synchronization of different neural activities, but it could be extended to the consistent co-occurrence of activities that are not necessarily synchronous, so that the concept could be applicable to the detection of repeated regular patterns. It should be noted that there is an important difference; the formation of streams based on temporal synchronization can be instantaneous but that based on spectrotemporal regularity takes some time (repetitions) to build up. In both cases, however, the essence is to compute global correlation, which is complementary to local channeling.

In our models, we explicitly introduce adaptive or learning processes to those principles. The system autonomously develops its selectivity to features or bases for analyses according to the observed acoustic data. At this moment, the models proposed here are not linked tightly to physiological implementation. Also, we aim at predicting the essential aspects of behavioral data qualitatively, rather than simulating the quantitative characteristics of behavioral data precisely.

2 Population Separation in a Feature Space: Bayesian Inference with a Gaussian Mixture Model

We modeled streaming based on the population separation in a feature space using a Gaussian Mixture Model (GMM). We assumed that the neural activities produced by the acoustic signal coming from a single sound source have a Gaussian distribution on a feature axis (e.g., tonotopy) in the auditory system. When there are multiple sources, the distribution of the neural activities is a weighted sum of multiple Gaussian distributions each corresponding to a source:

$$P(x_i \mid \pi, \theta_1, \ldots, \theta K) = \sum_{k=1}^{K} \pi_k P(x_i \mid \theta_k) \tag{57.1}$$

where x_i, $i=1, \ldots, N$, are observed neural activities; K is the number of sources; θ_k, $k=1,\ldots, K$, are parameters (mean and variance) of component Gaussian distributions; and π_k, $k=1, \ldots, K$, are the mixture weights. The computational goal of ASA in this framework is to estimate the parameters of the GMM, θ and π, which best match the distribution of neural activities corresponding to each source. In the case of streaming in particular, the goal is simply to judge the number of sources from the observed neural activities. For example, in the classic ABA tone sequence, two hypotheses may be generated: both A and B coming from a single source (one stream) or A and B coming from different sources (two streams).

The system judges which hypothesis is more plausible based on the observations. The GMM parameters, θ and π, are estimated using the Bayesian inference. Posterior probability ("belief") about perceptual state (Z, π, θ) after the data X are observed is computed by

$$P(Z, \pi, \theta \mid X) = \frac{P(X \mid Z, \theta) P(Z \mid \pi) P(\theta) P(\pi)}{P(X)} \tag{57.2}$$

where Z is a set of hypotheses, $P(\theta)P(\pi)$ is the prior belief about sources (i.e., the Gaussian mixture parameters and weights), and $P(X \mid Z, \theta)P(Z \mid \pi)$ is the likelihood. Using this Bayesian updating, the GMM parameters are renewed each time a new data point X is observed.

We applied this model to the ABA sequence with various combinations of frequency separation (Δf) and tone SOA (ΔT) (Fig. 57.1). Note that the actual value of Δf is not critical here, and Δf could be replaced with any feature that is represented on a continuum. The memory span of the system was set to 200 (arbitrary unit, common to ΔT; if ΔT is 10, data for 20 tones (=200/10) are stored in the system). Of special interest here is the effect of W, one of the parameters of $P(\theta)$. In theory, the smaller the value of W, the stronger is the tendency for the system to model the observed data with a small number of Gaussian distributions with large variances.

This prediction was confirmed by the simulation results; with $W=0.1$ (Fig. 57.1a), the integration-segregation boundary was similar to the behavioral coherence boundary in van Noorden (1975), and with $W=10$, it was similar to the behavioral fission boundary. Thus, the parameter W may be considered as corresponding to the response set or volitional control of the listener. We introduced random fluctuations in W ($W=10w$, where w is derived by applying a low-pass filter to a Gaussian noise with a mean of 0 and a standard deviation of 1). The fluctuation of W is not totally unrealistic, because neural activities are stochastic and sensory observation is not always clean. With an appropriate parameter setting, the fluctuated W qualitatively simulated the "ambiguous zone" in van Noorden (1975) (Fig. 57.1d).

In this model, the statistical power increases with the number of observed data, leading to the "segregation" judgment. Figure 57.2 shows the probability of "segregation" judgment as a function of the number of observed tones (equivalent to time). Without the fluctuation of W, the change of Δf resulted in simple horizontal shifts of the same curve (Fig. 57.2, left), which is inconsistent with the behaviorally observed buildup of streaming. With the fluctuated W, on the other hand, the simulated curves resembled the behavioral ones, in terms of the dependence of the curve shape on Δf (Fig. 57.2, right).

The fluctuation of W also results in multistable perception. The same combination of Δf and ΔT can yield different perceptual states due to the transient boundary

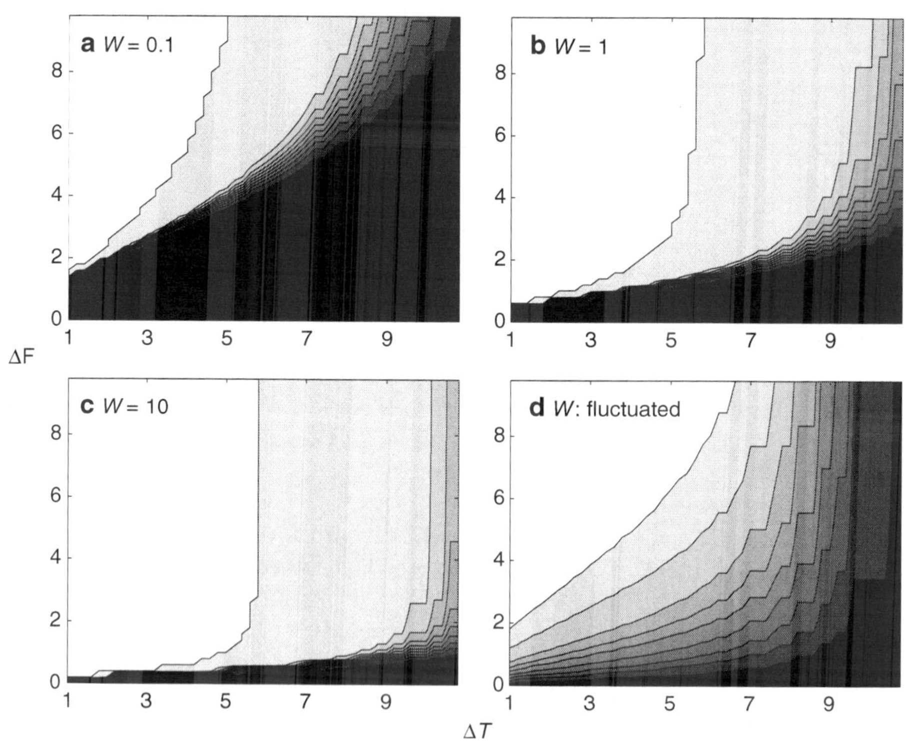

Fig. 57.1 Probability of "segregation" judgment (coded as brightness) as a function of tone SOA (ΔT) and frequency separation (ΔF), calculated with three values (0.1 (**a**), 1 (**b**), and 10 (**c**)) of W and fluctuated W (**d**). The units of ΔF and ΔT are arbitrary

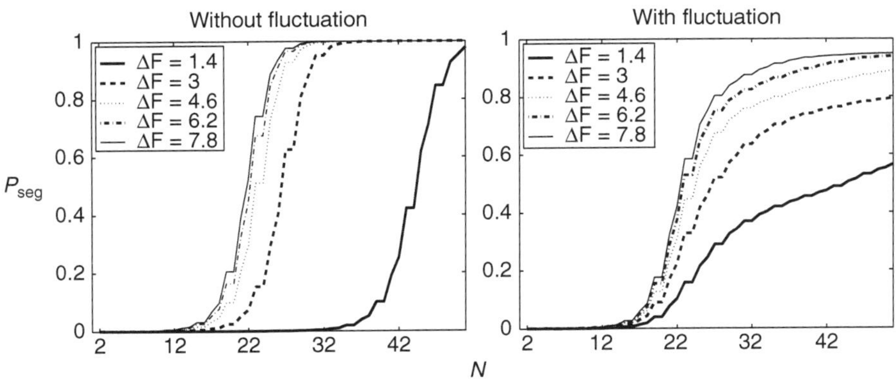

Fig. 57.2 Simulated buildup of streaming with (*left*) and without (*right*) the fluctuation of W. Probability of "segregation" judgment is plotted as a function of the number of observed tones N (equivalent to time). The unit of ΔF is arbitrary

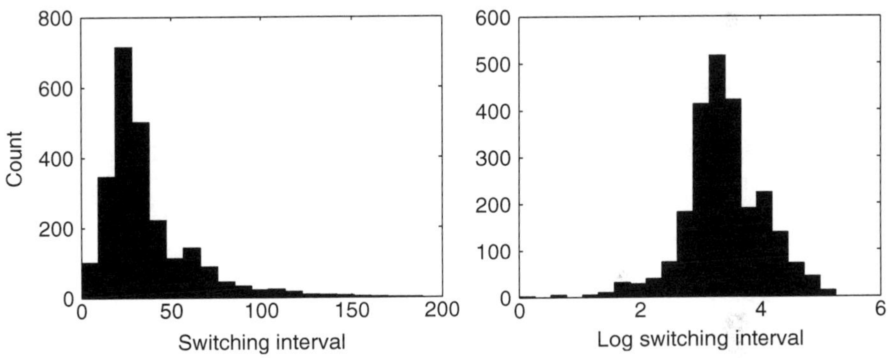

Fig. 57.3 Distribution of switching intervals simulated with fluctuated W. The data in the left panel are replotted on a logarithmic horizontal axis (*right*)

shift caused by the fluctuation of W. Figure 57.3 shows the histogram of switching intervals simulated by the model with the fluctuated W. The log-normal distribution was the best fit among several probability distributions including the gamma and normal distributions. This is consistent with behavioral data (Kashino et al. 2007).

3 Temporal Coherence and Regularity: Nonnegative Matrix Factorization

We modeled streaming based on temporal coherence and repeated co-occurrence using nonnegative matrix factorization (NMF) (Lee and Seung 1999). NMF is a way to obtain "sparse" representations. Sparse representations are representations that account for most or all information of a signal with a linear combination of a small number of elementary signals (parts). Sparse representations of an acoustic signal by NMF are obtained by factorizing the observed spectrogram Y into two nonnegative matrices, H (parts) and U (time-varying weights of the parts), so that $Y \simeq HU$. The spectrotemporal structure of components that co-occur repeatedly tends to be represented in H. The auditory scene can be represented as a time-varying combination of the parts.

In our model, the acoustic signal was analyzed using the cochlear model of Lyon (Slaney 1988), and the salient onsets of frequency components were detected from the spectrotemporal pattern. Then a temporal window of 250 ms was applied starting at each onset to generate a set of partial spectra, from which spectrotemporal patterns (parts) co-occurring repeatedly were extracted by NMF. With a shorter (e.g., 30 ms) temporal window, the model would have extracted only synchronized frequency components as parts (corresponding to temporal coherence in a strict sense). Finally, irregular components were filtered out by taking the minima of the parts.

We applied this model to the embedded repetition stimuli (McDermott et al. 2011). In input signals, target sounds were repeated, each time mixed with a different "distractor" sound (Fig. 57.4a). Both target and distractor sounds were generated by the same process so that they had spectrotemporal structures that resembled those in natural sounds. Behavioral data showed that human listeners were able to segregate and identify the targets if the sounds occurred more than once across different mixtures, even when the same target sounds were impossible to segregate in single mixtures (McDermott et al. 2011). The listeners' performance saturated at five repetitions. Figure 57.4b shows a set of parts extracted by the NMF model. Here the target sounds were repeated five times. The number of parts was predetermined as six. Figure 57.4c shows the time-varying weights of the six parts. Figure 57.4d shows two examples of the parts after the filtering of irregular components. Note the similarity between the filtered parts and the target.

4 Discussion

Separation in a feature space based on statistical learning is in a sense equivalent to a bank of band-pass filters of which the tuning changes according to the statistical structures of recent inputs. This contrasts with the auditory filters, whose tuning is thought to be largely predetermined irrespective of input statistics. The short-term neural plasticity found in the auditory cortex may be understood from the viewpoint of statistical learning.

The auditory filters can also be considered as a means for obtaining sparse representations, largely irrespective of recent inputs. The auditory filters are effective in obtaining sparse representations for speech and harmonic complex sounds, but not for signals having a continuous spectrum. On the other hand, NMF acquires sparse representations autonomously depending on the structures of observed data and is effective for signals having a dense spectrum such as the embedded repetition stimuli (McDermott et al. 2011). This may provide another way of thinking about the short-term neural plasticity found in the auditory cortex.

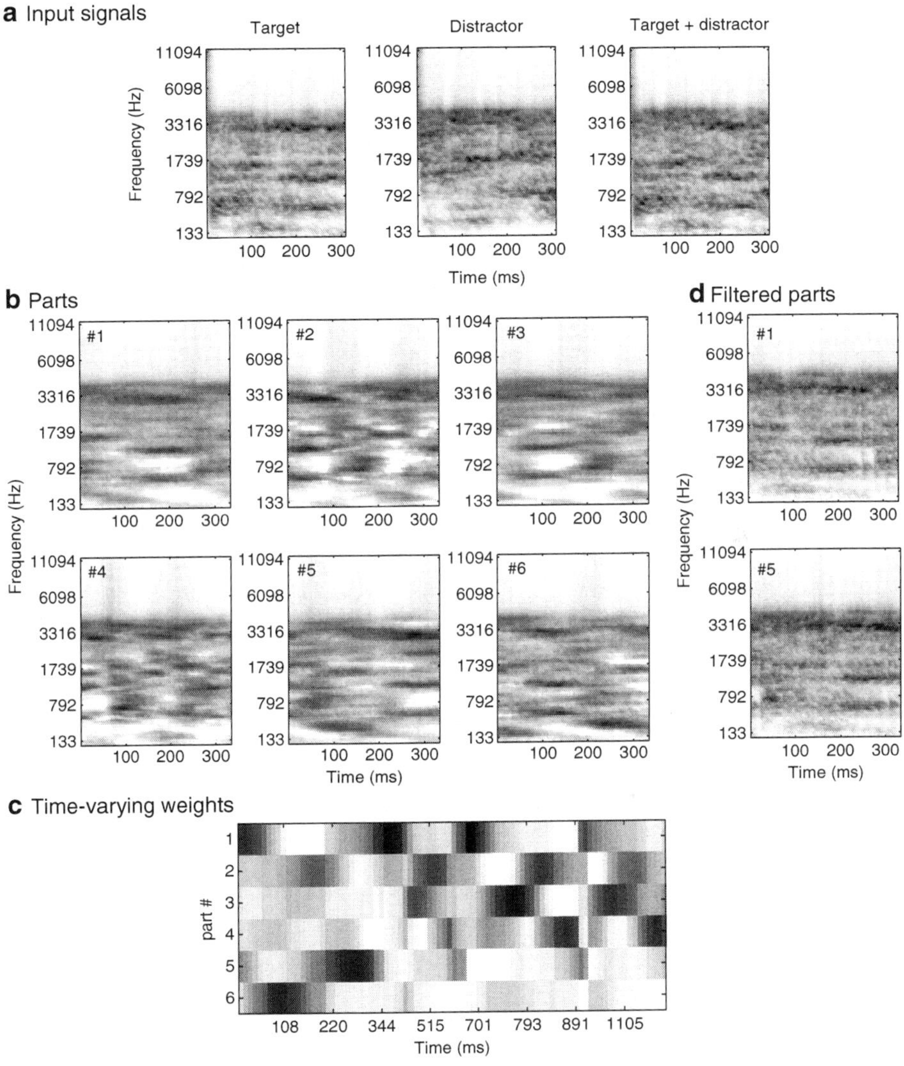

Fig. 57.4 Extraction of a repeating target sound from mixtures using nonnegative matrix factorization (NMF). (**a**) Cochleograms of a target sound (*left*), a distractor (*middle*), and the mixture of the two (*right*). (**b**) Six parts (bases) derived by NMF. (**c**) Time-varying weights for the six parts. (**d**) Two examples of filtered parts. Note the similarity between these filtered parts and the target

References

Agus TR, Thorpe SJ, Pressnitzer D (2010) Rapid formation of robust auditory memories: insights from noise. Neuron 66:610–618

Andreou L-V, Kashino M, Chait M (2011) The role of temporal regularity in auditory segregation. Hear Res 280:228–235

Fritz J, Elhilali M, Shamma S (2005) Active listening: task-dependent plasticity of spectrotemporal receptive fields in primary auditory cortex. Hear Res 206:159–176
Hupé J-M, Pressnitzer D (2012) The initial phase of auditory and visual scene analysis. Philos Trans R Soc Lond B Biol Sci 367:942–953
Kashino M, Kondo HM (2012) Functional brain networks underlying perceptual switching: auditory streaming and verbal transformations. Philos Trans R Soc Lond B Biol Sci 367:977–987
Kashino M, Okada M, Mizutani S, Davis P, Kondo HM (2007) The dynamics of auditory streaming: psychophysics, neuroimaging, and modeling. In: Kollmeier B, Klump G, Hohmann V, Langemann U, Mauermann M, Upperkamp S, Verhey J (eds) Hearing—from sensory processing to perception. Springer, Berlin, pp 275–283
Kondo HM, Kashino M (2009) Involvement of the thalamocortical loop in the spontaneous switching of percepts in auditory streaming. J Neurosci 29:12695–12701
Kondo HM, Pressnitzer D, Toshima I, Kashino M (2012) The effects of self-motion on auditory scene analysis. Proc Natl Acad Sci U S A 109:6775–6780
Lee DD, Seung HS (1999) Learning the parts of objects by non-negative matrix factorization. Nature 401:788–791
McDermott JH, Wrobleski D, Oxenham AJ (2011) Recovering sound sources from embedded repetition. Proc Natl Acad Sci U S A 108:1188–1193
Micheyl C, Tian B, Carlyon RP, Rauschecker JP (2005) Perceptual organization of tone sequences in the auditory cortex of awake macaques. Neuron 48:139–148
Shamma S, Elhilali M, Micheyl C (2011) Temporal coherence and attention in auditory scene analysis. Trends Neurosci 34:114–123
Slaney M (1988) Lyon's cochlear model. Apple technical report #13
Snyder JS, Carter OL, Lee SK, Hannon EE, Alain C (2008) Effects of context on auditory stream segregation. J Exp Psychol Hum Percept Perform 34:1007–1016
Ulanovsky N, Las L, Farkas D, Nelken I (2004) Multiple time scales of adaptation in auditory cortex neurons. J Neurosci 24:10440–10453
van Noorden LPAS (1975) Temporal coherence in the perception of tone sequences. Ph.D. Thesis, Eindhoven University of Technology
Winkler I, Denham S, Mill R, Bohn TM, Bendixen A (2012) Multistability in auditory stream segregation: a predictive coding view. Philos Trans R Soc Lond B Biol Sci 367:1001–1012

Chapter 58
A Naturalistic Approach to the Cocktail Party Problem

Ervin R. Hafter, Jing Xia, and Sridhar Kalluri

Abstract While studies of simple acoustic features have provided excellent bases for models of spatial hearing, we are seeking, here, to create a new paradigm for examination of shared attention and scene analysis in natural environments, where the listener is confronted with semantic information from multiple sources. In this new simulation of the cocktail party problem, a subject (S) is questioned, on-line, about information heard in multiple simultaneous stories spoken by different talkers. Questions based on brief passages in the stories are presented visually for manual response. To ensure that responses are based on semantic information rather than just keywords, the latter are replaced in the questions with synonyms. Pay is for performance, and S knows that while a majority of the questions come from a "primary talker," there is potential value in obtaining information from secondary sources. Results, to date, suggest that obtaining semantic information from separate stories is limited by two spatial factors, an exclusive filter that protects information from the attended talker and an inclusive filter that incorporates information from secondary talkers.

E.R. Hafter (✉)
Department of Psychology, University of California,
Berkeley, CA, USA
e-mail: hafter@berkeley.edu

J. Xia
Department of Psychology, University of California,
Berkeley, CA, USA

Starkey Hearing Research Center,
Berkeley, CA, USA

S. Kalluri
Starkey Hearing Research Center,
Berkeley, CA, USA

B.C.J. Moore et al. (eds.), *Basic Aspects of Hearing*,
Advances in Experimental Medicine and Biology,
DOI 10.1007/978-1-4614-1590-9_58,

1 Introduction

It is impressive to note how long the field of spatial hearing has been fascinated with our ability to single out speech from one talker in the presence of others, an issue beautifully elaborated in Colin Cherry's (1953) prescient discussion of the "cocktail party problem." What began as a way of talking about selective attention has grown into an important tool in auditory scene analysis (for reviews see Treisman 1969; Bronkhorst 2000).

Clearly, monaural cues such as level, pitch, inflection, and language support source segregation in the cocktail party, but much of the research has focused on binaural cues and sound localization. A common technique offered in Cherry (1953) utilizes the so-called Dichotic Listening (DL), where different steams of words are presented to the two ears via headphones. The instruction is to "shadow" the stimulus in the attended ear, that is, to repeat it as it goes along. Questions about features of the talkers or semantic content in both attended and unattended ears are generally saved to the end of a session. Typically, shadowing is accurate, but little is recalled about the unattended stimulus other than such acoustical features as fundamental frequency or prosody. However, some of the unattended words are held in short-term memory, as shown by querying the subject, S, immediately after a sudden cessation (Norman 1969). Further evidence of the processing of speech in the unattended ear comes from the finding that ~30 % of subjects noted the presence of their own name in the unattended ear (Moray 1969; Conway et al. 2001) and that there can be stammers in shadowing when a word in the unattended ear relates to a word in the attended ear (Lewis 1970). A major issue in DL is concerned with the act of shadowing, itself. The problem is that during a study intended to quantify the effects of shared attention between auditory tasks, S is doing another task, preparing for and executing speech production. Could attention to this motor task account for some of the difference in information derived from the shadowed and unshadowed ears? Nevertheless, DL has been important because it examines attention to streams of natural language and because it has pointed out the importance of acoustical segregation in shared attention.

A different issue with the DL paradigm is its relatively slow information rate. This is well addressed by a newer approach, the Coordinate Response Measure or CRM (Brungart 2001; Kidd et al. 2008). There, S listens to multiple, simultaneous talkers saying sentences that are identical in syntax and timing. A typical sentence is, "Ready, (call sign), go to (keyword-1), (keyword-2) now." In a common variation, the keyword lists contain a color and a number. The instruction is to identify the talker who says a given call sign and to repeat the two keywords spoken by that talker. An important advantage of the technique is that it allows study of informational masking through classification of errors based on intrusions of incorrect keywords spoken by unattended talkers (Ihlefeld and Shinn-Cunningham 2008). Another advantage is that by removing variance in syntax and shrinking the lists of possible responses, it creates a very high information rate, providing literally thousands of trials for testing of hypotheses. Like DL, CRM has highlighted the importance of acoustic differences such as the fundamental frequency of competing talkers and their locations in space. However, while the reduced variance afforded by closely matched stimuli increases statistical power, it may also be detrimental to understanding how listeners act in a real-world situation such as the cocktail party, where each talker tells a different story and the listener's task is to extract semantic meaning from the stories. In sharp contrast, the essential distinction in CRM is based on phonetic cues, that is, the sound of the keyword, rather than the gist it conveys. This is clearly illustrated by Woods and Kalluri (2011) who, using speech with CRM-like syntax, albeit with individual talkers staggered in time, report success accounting for identification of nonsense syllable keywords on the basis of audibility of the syllables.

In our new simulated cocktail party, S hears multiple talkers, each presenting a different stream of natural language (story). Reponses are answers to visually presented questions that are based on intrinsic meaning in selected passages, information requiring semantic rather than phonetic processing. Unlike DL, S does not speak (verbal shadowing), instead using a tactile device to answer questions. Attention is directed to the story of the "primary" talker by telling S that a majority of the questions are related to that story. S is paid for performance, so listening to the primary offers the highest payoff, but S is reminded that there is potential value in attending to secondary talkers, that is, eavesdropping. Here, we examine the utility of this paradigm in two experiments that look at spatial bandwidth for exclusion as well as inclusion of information from secondary talkers.

2 Methods

Stimuli are short stories taken from the internet, each lasting about 10 min. *N* simultaneous, but otherwise independent, stories are spoken by *N* different talkers. These are presented through separate loudspeakers placed at head height along a circle surrounding the S. At irregular moments during a session, phrases in stories are used to generate questions shown visually on a screen along with two possible answers. Information on which a question is based in local, in the sense that Ss cannot know the correct answer without attending to the story at the relevant moment. Most important is that the answers reflect semantic information in the relevant phrase rather than the phonetics of keywords. For example, the phrase in one story, "... though I was grateful to be there, I was more grateful that sleep was near...," is tested with the question "What was the narrator looking forward to?" and the two potential answers were "(1) going to bed" and "(2) exploring her new house." In the experiment, questions appear with a mean interval of 20 s.

Stories are read by professional radio announcers in anechoic space prior to sound editing. An inverse filter is used to smooth each story's time-varying sound pressure calculated through a running window. Next, the overall level is set to 70 dB SPL, before final adjustment in level in accord with equal-loudness judgments, by Ss comparing it to a "standard" story. Finally, 100 % performance in undistracted attention is ensured by rejecting questions that elicit an error when tested in the quiet.

A visual screen shows a cartoon of the spatial locations of *N* talkers in the experiment. One of these is colored to indicate the location of the "primary" talker. Questions appear below the cartoon ~1 s after the appropriate information appears in a story; S has 8 s to respond with a two-button box. Pay is for correct answers, regardless of source, but we assume that attention is focused on the "primary" story talker because S knows that it is the source for a majority of the questions. S must answer all questions and we point out that attending to a secondary source, if possible, could increase pay. All subjects go through a training procedure.

3 Experiment 1: Two Talkers

Stories from two female talkers were presented at ±7.5° or ±60° relative to the midline; one was labeled the primary. Ss were 28 young, native English-speaking subjects with normal hearing. In a control condition (A), 12 Ss knew that 100 % of the questions would come from the primary talker. In a shared-attention condition (B), the other 16 Ss knew that a majority of the questions (in actuality, 70 %) would come from the primary talker and the rest from the secondary. On half of the ~10-min sessions, the primary was on the right; on the other half, the primary was on the left.

3.1 Results and Discussion

Figure 58.1 shows performance plotted as a function of the angular distance between primary and secondary talkers. Solid and dashed lines represent answers to questions from primary and secondary talkers, respectively. High performance in the control condition (1A) indicates little interference by the mere presence of a secondary talker. The small effect of spatial separation is not significant ($t(11) = -1.219$, $p = 0.248$). With 30 % of the questions from the secondary (1B), performance on the primary fell, but this was accompanied by better than chance performance on questions from the secondary. A trade-off of this kind suggests division of attention between stories in accord with their perceived importance. Quite interesting is that performance on secondary stories was better when the talkers were closer together

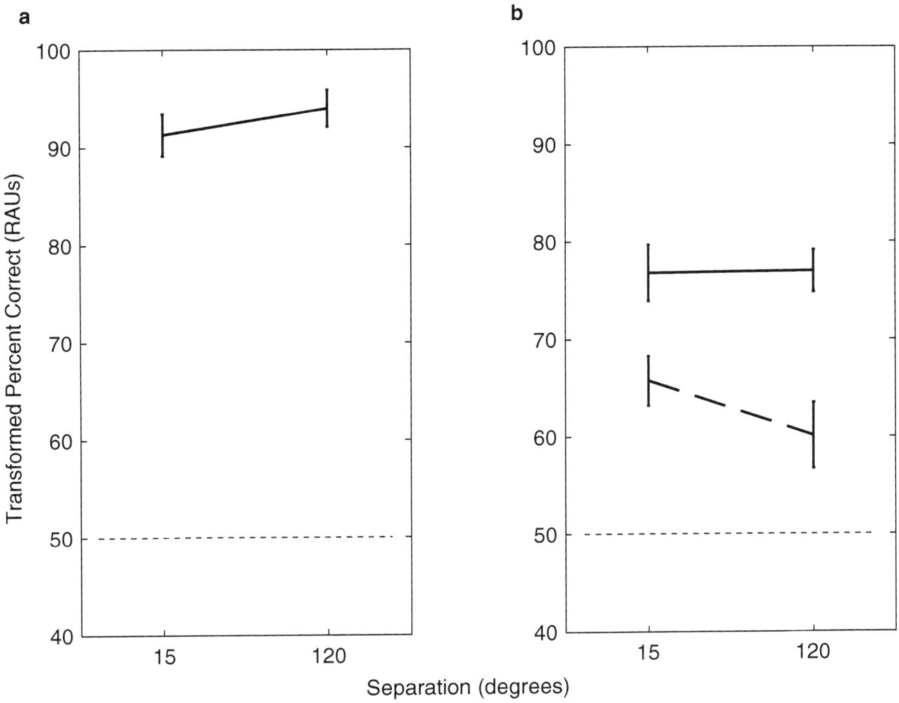

Fig. 58.1 Mean percent correct performance, transformed into rationalized arcsine units (RAU; Studebaker 1985) and plotted as a function of the angular separation between primary and secondary talkers. Chance performance is indicated by *dotted lines*. For Panel **a**, all questions were from the primary talker. For Panel **b**, 70 % of the questions were from a primary talker and 30 % from a secondary. *Solid lines* show performance based on questions from the primary talker. *Dashed lines* show performance based on questions from the secondary talker

($t(15)=2.139$, $p=0.045$), the opposite of what would be expected from a spatial release from masking.

4 Experiment 2: Three Talkers

Stories from three female talkers were presented at 0°, ±7.5°, and ±60° relative to the midline, with the one on the midline labeled "primary." Ss were a new set of 28 young, native English-speaking subjects with normal hearing. In the control condition (A), 12 Ss knew that 100 % of the questions would come from the primary talker. In the shared-attention condition (B), a different 16 Ss knew that a majority of the questions (actually, 60 %) would come from the primary talker, and the rest would be split evenly between secondary talkers (actually 20 and 20 %).

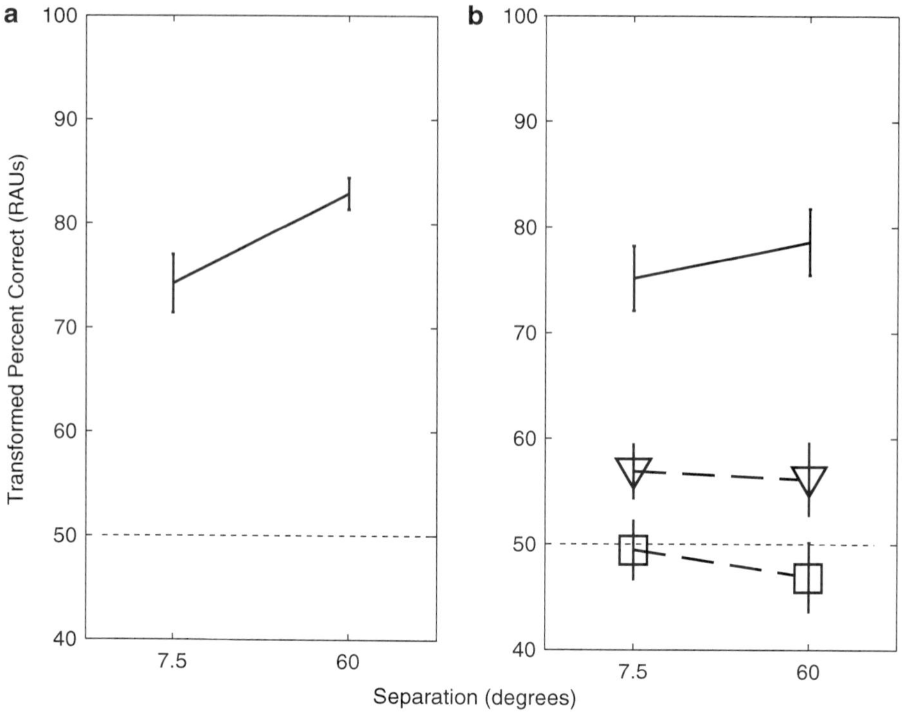

Fig. 58.2 Mean percent correct performance, transformed into rationalized arcsine units (RAU; Studebaker 1985) and plotted as a function of the angular separation between primary talker at 0°. Chance performance is indicated by *dotted lines*. For Panel **a**, all questions were from the primary talker. For Panel **b**, 60 % of the questions were from the primary talker and 20 % from each of the two secondary talkers. *Solid lines* show performance based on questions from a primary talker and *dashed lines* show performance based on the secondary talkers. Additionally, Panel **b** divides secondary performance for the two different secondary talkers; these are plotted by *triangles* and *squares*

4.1 Result and Discussion

Figure 58.2 shows performance plotted as a function of the angular distance from the primary and two secondary talkers. Solid and dashed lines represent answers to questions from primary and secondary talkers, respectively. With distraction from both sides (2A), performance in the control condition shows more interference than in Fig. 58.1a. However, there was a significant spatial release from masking, i.e., better performance with separations of 60° ($t(11) = -2.639$, $p = 0.027$). Perhaps the small but insignificant release seen in Fig. 58.1a reflects a ceiling effect that hid a release from masking. Comparison of Fig. 58.2a, b shows that primary performance was not further compromised in the shared-attention task, though the seeming spatial release for the primary talker in Fig. 58.2b is not significant ($t(15) = -1.157$, $p = 0.266$). Post-session comments from some Ss said that one of the secondary talkers had a particularly "high-pitched" and "animated" voice that made her seem to stand out. Results here for secondary talkers are thus parsed into two dashed lines, one (triangles) for the more distinctive talker and one (squares) for the others. For the squares, performance did not differ from chance, but for the triangles, performance was better than chance (one tailed $t(15)=2.870$, $p=0.0058$).

5 Summary and New Directions

In the SCP, speech flows rapidly and near-continuously as in natural discourse, and, unlike trial-based tasks, the speech is not interrupted by silent periods during which Ss respond. Also, cognitive demands are greater when Ss must maintain a more constant level of attention because they cannot anticipate which portion of the stories will be tested. Such factors are important if we hope to simulate the high-stress acoustic communications often encountered in real life. For future work, we are especially interested in how high stress exaggerates the detrimental effect of hearing loss or aging and whether technological interventions might help overcome the deficits.

Consistent with experiments using DL, our Ss showed strong limitations of semantic processing of simultaneous speech from multiple talkers. While subjects could derive semantic information from both of two talkers, when there were three, performance on one of the secondary talkers was no better than chance. Also consistent with studies using CRM, we found a larger spatial release from masking with two secondary talkers than with one.

What else may be concluded from this approach to the study of auditory attention in more natural environments? Although the paradigm is new and the data are somewhat preliminary, we will point to a few encouraging features. While there is usually a release from masking based on spatial separation, we found the opposite to be true for information from the secondary talker, as seen in Fig. 58.1b. In accord with Best et al. (2006), we propose another way of thinking about spatial bandwidths, one that assumes inclusion of information from a secondary talker when it falls into an attention band focused on the primary talker. Also interesting is that in our three-talker task, Ss were better able to derive information from one of the secondary sources when the (female) talker's voice was more readily distinguished from the primary source.

Results in Experiment 2 indicate that attention was shared between only two of the three talkers, but this limitation will be examined further with the introduction of more distinctiveness between voices. Further steps with this paradigm include direct comparisons between semantic and phonetic cues for attention, the speed with which attention can be switched between talkers, and ways in which hearing impairment and its treatment interact with listening in a real-world cocktail party. We are not surprised that processing of basic acoustic features such as fundamental frequency or location is present when listeners respond to the gist of a spoken message. Our hope is that the ability to examine the latter in our simulated cocktail party will offer new insights into top-down effects in auditory attention.

References

Best V, Gallun FJ, Ihlefeld A, Shinn-Cunningham BG (2006) The influence of spatial separation on divided listening. J Acoust Soc Am 120:1506–1516

Bronkhorst AW (2000) The cocktail party phenomenon: a review of research on speech intelligibility in multiple-talker conditions. Acustica 86:117–128

Brungart DS (2001) Informational and energetic masking effects in the perception of two simultaneous talkers. J Acoust Soc Am 109:1101–1109

Cherry E (1953) Some experiments on the recognition of speech, with one and with two ears. J Acoust Soc Am 25:975–979

Conway R, Cowan N, Bunting F (2001) The cocktail party phenomenon revisited: the importance of working memory capacity. Psychon Bull Rev 9:331–335

Ihlefeld A, Shinn-Cunningham BG (2008) Spatial release from energetic and informational masking in a selective speech identification task. J Acoust Soc Am 123:4369–4379

Kidd G Jr, Mason CR, Richards VM, Gallun FJ, Durlach N (2008) Informational masking. In: Yost WA (ed) Auditory perception of sound sources. Springer, New York, pp 143–189

Lewis JL (1970) Semantic processing of unattended messages using dichotic listening. J Exp Psychol 85:225–228

Moray N (1969) Attention in dichotic listening: affective cues and the influence of instructions. Q J Exp Psychol 11:56–60

Norman DA (1969) Memory while shadowing. Q J Exp Psychol 21:85–93

Studebaker GA (1985) A "rationalized" arcsine transform. J Speech Hear Res 28:455–462

Treisman AM (1969) Strategies and models of selective attention. Psychol Rev 76:282–299

Woods W, Kalluri S (2011) Cognitive and energetic factors in complex-scenario listening. In: First international conference on cognitive hearing science for communication. Linkoping, Sweden, pp 19–22

Chapter 59
Temporal Coherence and the Streaming of Complex Sounds

Shihab Shamma, Mounya Elhilali, Ling Ma, Christophe Micheyl, Andrew J. Oxenham, Daniel Pressnitzer, Pingbo Yin, and Yanbo Xu

Abstract Humans and other animals can attend to one of multiple sounds, and follow it selectively over time. The neural underpinnings of this perceptual feat remain mysterious. Some studies have concluded that sounds are heard as separate streams when they activate well-separated populations of central auditory neurons, and that this process is largely pre-attentive. Here, we propose instead that stream formation depends primarily on temporal coherence between responses that encode various features of a sound source. Furthermore, we postulate that only when attention is directed toward a particular feature (e.g., pitch or location) do all other temporally coherent features of that source (e.g., timbre and location) become bound together as a stream that is segregated from the incoherent features of other sources. Experimental

---

S. Shamma (✉) • P. Yin • Y. Xu
Department of Electrical and Computer Engineering,
Institute for Systems Research, University of Maryland,
College Park, MD 20742, USA
e-mail: sas@umd.edu

L. Ma
Bioengineering Program, University of Maryland, College Park, MD 20742, USA

Department of Electrical and Computer Engineering,
Institute for Systems Research, University of Maryland,
College Park, MD 20742, USA

M. Elhilali
Department of Electrical and Computer Engineering,
Johns Hopkins University, Baltimore, MD, USA

C. Micheyl • A.J. Oxenham
Department of Psychology, University of Minnesota, Minneapolis, MN, USA

D. Pressnitzer
Département d'études Cognitives, Equipe Audition,
Ecole Normale Superieure, Paris, France

Laboratoire de Psychologie de la Perception (UMR CNRS 8158),
Université Paris Descartes, Paris, France

B.C.J. Moore et al. (eds.), *Basic Aspects of Hearing*,
Advances in Experimental Medicine and Biology,
DOI 10.1007/978-1-4614-1590-9_59,

neurophysiological evidence in support of this hypothesis will be presented. The focus, however, will be on a computational realization of this idea and a discussion of the insights learned from simulations to disentangle complex sound sources such as speech and music. The model consists of a representational stage of early and cortical auditory processing that creates a multidimensional depiction of various sound attributes such as pitch, location, and spectral resolution. The following stage computes a coherence matrix that summarizes the pair-wise correlations between all channels making up the cortical representation. Finally, the perceived segregated streams are extracted by decomposing the coherence matrix into its uncorrelated components. Questions raised by the model are discussed, especially on the role of attention in streaming and the search for further neural correlates of streaming percepts.

1 Introduction

Listening in a complex acoustic environment fundamentally involves the ability to parse out and attend to one sound stream as the foreground source against the remaining background. In this view, streaming is an active listening process that engages attention and induces adaptive neural mechanisms that reshape the perceptual scene, presumably by enhancing responses to the target while suppressing responses to the background.

It is often conceptually useful to think of auditory streams as sequences of events or "tokens" that constitute the primitives of hearing, analogous to an alphabet. A token, such as a tone, a vowel, or a syllable, may have many concurrent perceptual attributes that arise very quickly through mechanical and hardwired neural mechanisms. Examples include a vowel's pitch, harmonic fusion, location, loudness, and the timbre of its spectral envelope. To segregate a sequence of tokens (be they phonemes or tones), it is necessary to satisfy a key condition – that the tokens be perceptually distinct from those associated with competing sequences, e.g., the pitches of two talkers or of two alternating tone sequences must be sufficiently different. This well-known principle of steaming has often been referred to as the "channeling hypothesis" implying that streams form when they activate distinct neuronal populations or processing channels (Bregman 1990; Hartmann and Johnson 1991). This requirement, however, is insufficient to explain stream formation, as we discuss next.

2 Feature Binding and Temporal Coherence

Forming a stream also requires *binding* of the parallel perceptual attributes of its tokens, to the exclusion of those belonging to competing streams. The simplest principle that explains how this phenomenon comes about is *temporal coherence* (Shamma et al. 2011). It asserts that any sequences of attributes that are temporally correlated will bind and form a stream segregated from uncorrelated tokens of perceptually different attributes. A simple example is the alternating two-tone sequences that stream apart when their pitches are sufficiently different (Bregman 1990). When the tones are made fully correlated (synchronous sequences), the streaming fails because the two pitch percepts bind together forming a repeating complex perceived as one stream (Elhilali et al. 2009).

We postulate that temporal coherence is the organizing principle necessary to make the correct perceptual assignments as to which tokens form a stream. More specifically, correlated tokens form a single stream regardless of the diversity of their associated percepts, e.g., whether they are simple synchronized tones of different pitches, or the far more complex voices of a choir of soprano and bass pitches all singing in unison. The importance of temporal coherence in streams is a natural consequence of the fact that environmental sources normally produce sounds with temporally coherent attributes. For instance, a speech signal typically fluctuates in amplitude at temporal rates of a few Hertz. Consequently, the salience of all instantaneous estimates of its attributes would fluctuate similarly, be it the salience of its pitch, its location, or its spectral envelope. This temporal pattern is unlikely to be correlated with that of another signal emanating from an independent source, and hence the lack of temporal coherence is the simplest direct cue to the segregation of the two signals. When multiple "physical sources" become correlated as in the example of the choir, or when an orchestra plays the same melody, the entire group is treated perceptually as one source (Shamma et al. 2011).

In this chapter, we briefly review a mathematical model of this idea (Elhilali et al. 2009; Ma 2011) and discuss its biological realization and results of physiological experiments to test its predictions. We also discuss some of the psychoacoustic implications of this model and relate it to earlier formulations of the streaming process based on the Kalman prediction (Elhilali and Shamma 2008).

3 The Temporal Coherence Model

The proposed computational scheme emphasizes two distinct stages in stream formation (Fig. 59.1): (1) extracting auditory features and representing them in a multidimensional space mimicking early cortical processing and (2) organizing the features into streams according to their temporal coherence. Many feature axes are potentially relevant including the tonotopic frequency axis, pitch, spectral scales (or bandwidths), location, and loudness. All these features are usually computed very rapidly (<50 ms). Tokens that evoke sufficiently distinct (nonoverlapping) features in a model of cortical responses are deemed perceptually distinguishable and hence potentially form distinct streams *if* they are temporally anti-correlated or uncorrelated over relatively long time periods (>100 ms), consistent with known dynamics of the cortex and stream buildup.

Figure 59.1 illustrates these processing stages. Inputs are first transformed into auditory spectrograms (Lyon and Shamma 1997) followed by a multiresolution analysis analogous to that thought to occur in the primary auditory cortex (Chi et al. 2006). For the purposes of this model, this transformation is implemented in two steps: (1) a multiscale (spectral) analysis that maps incoming spectrograms into multiscale (bandwidth) representations, followed by (2) temporal rate analysis in

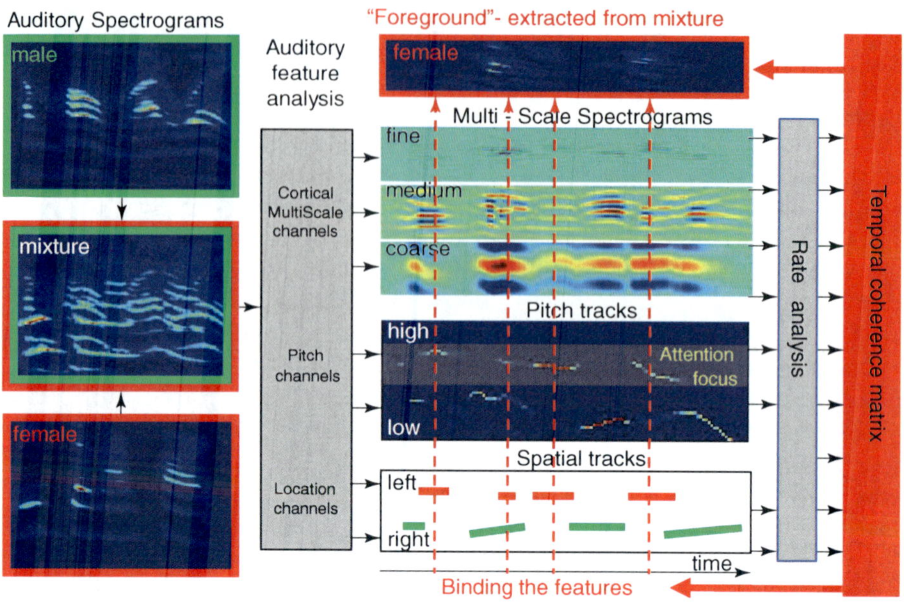

Fig. 59.1 Temporal coherence model. The mixture (sum of one male and one female sentences) is transformed into an auditory spectrogram. Various features are extracted from the spectrogram including a multiscale analysis that results in a repeated representation of the spectrogram at various resolutions; pitch values and salience are represented as a pitch-gram; location signals are extracted from the interaural differences. All responses are then analyzed by temporal modulation band-pass filters tuned in the range from 2 to 16 Hz. A pair-wise correlation matrix of all channels is then computed. When attention is applied to a particular feature (e.g., female pitch channels), all features correlated with this pitch track become bound with other correlated feature channels (indicated by the *dashed straight lines* running through the various representations) to segregate a foreground stream (female in this example) from the remaining background streams

which the temporal modulations of the (fine to coarse) multiscale spectrograms are analyzed by a filter bank tuned to rates from 2 to 16 Hz. In addition, other features such as pitch and location are estimated from the input spectrograms and the resulting tracks are later analyzed through the same rate analysis as for other channels, as illustrated in Fig. 59.1.

Subsequent to the feature and rate analysis, a pair-wise correlation matrix is computed among all scale-frequency-pitch-location channels, which is then used to group the channels into two sets representing the foreground and background streams. The responses are maximally correlated within each stream and least correlated across the two streams. One such factorization procedure is illustrated for the simple two-tone alternating (ALT) and synchronized (SYNC) sequences shown in Fig. 59.2. The correlation matrix cross-channel entries induced by these two sequences are quite different, being strongly positive (negative) for the SYNC (ALT) tones. A principal component analysis would then yield an eigenvector that can function as a "mask" to segregate the anti-correlated channels of the ALT stimulus, while grouping them together for the SYNC sequence, in agreement with their usual percept.

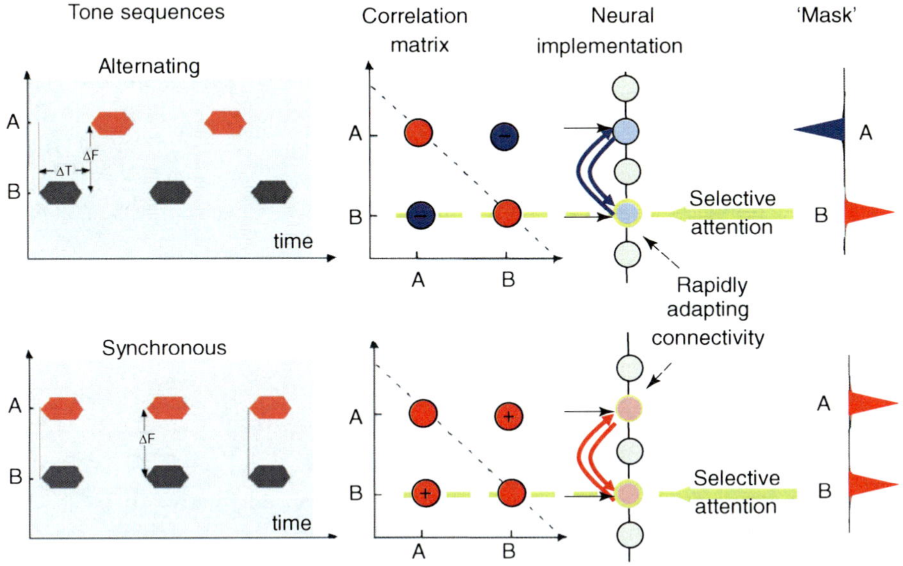

Fig. 59.2 Streaming of two-tone sequences. Alternating tone sequences are perceived as two streams when tones are far apart (large ΔF) and rates are relatively fast (small ΔT). Synchronous sequences are perceived as a single stream regardless of their frequency separation. The correlation matrices induced by these two sequences are different: pair-wise correlations between the two tones (A, B) are negative for the alternating sequence and positive for the synchronous tones. Neural implementation of this correlation computation can be accomplished by a layer of neurons that adapts rapidly to become mutually inhibited when responses are anti-correlated (alternating tones) and mutually excitatory when they are coherent (synchronous tones). When selective attention (*yellow arrow*) is directed to one tone (B in this example), the "row" of pair-wise correlations at B (along the *yellow dashed line*) can be used as a mask that indicates the channels that are correlated with the B stream. For the alternating sequence, tone A is *negatively* correlated with B, and hence, the mask is negative at A and eliminates this tone from the attended stream. In the synchronous case, the two tones are correlated, and hence, the mask groups both tones into the attended stream

4 Attention and Binding

It remains uncertain if the representation of streams in the brain requires attention or is simply modulated by it (Carlyon et al. 2001; Sussman et al. 2007). But it is intuitively clear that attending *selectively* to a specific feature such as the pitch of a voice (symbolized by the yellow-shaded pitch region in Fig. 59.1) results in binding the pitch with all other voice attributes in the foreground stream while relegating the rest of the concurrent sounds to the background. To explain how this process may occur, we consider the simpler two-tone stimulus in Fig. 59.2. When attention is directed to a particular channel (e.g., yellow arrow to tone B), the entries in the correlation matrix along the row of the selected channel can readily point to all the other channels that are highly correlated and hence may bind with it. Basically, this row is an approximation of the eigenvector of the correlation matrix and can be used as "mask" to assign the channels to the different streams (rightmost panel). Note that in such a model, the attentional focus is essential to bring out the stream, and without it the correlation matrix remains unused. This idea is implemented to segregate the two-talker mixture in Fig. 59.1. Specifically, the female speech could be readily extracted by simply focusing on the rows of the correlation matrix corresponding to the female pitch (shaded yellow in Fig. 59.1) and then using the correlation values as a mask to weight all correlated channels from the mixture.

5 Biological Realizations and Evidence for Temporal Coherence

The temporal coherence model suggests that streaming is a dynamic process in which responses of the attended stream become enhanced relative to the background. This requires computing a correlation matrix whose entries change rapidly according to the ongoing correlational structure of the stimulus. A simple biologically plausible neural implementation of these computations is depicted in Fig. 59.2, where an ordered array of feature channels (e.g., the tonotopic axis) project to a layer of neurons. Each pair of neurons is reciprocally connected with a sign and strength which is continuously updated to reflect the ongoing correlation between their inputs ("Hebb's rule"). If the inputs are anti-correlated, the connectivity is mutually inhibitory (top panels, Fig. 59.2); if highly correlated, it is mutually excitatory (bottom panels, Fig. 59.2).

When neuronal connections change, they effectively alter the response selectivity of the neurons or their receptive field properties. It has been shown that engagement in an auditory task with attention to the stimulus is essential for such rapid changes to occur (Fritz et al. 2007). Therefore, in the context of the coherence model, we postulate that the mutual connectivity would not adapt to reflect the correlation matrix in a passively listening animal. Once the animal attends to the stimuli, connectivity begins to form, partly influenced by the focus of the attention. Thus, if attention is *global*, then connectivity adapts to reflect the mutual correlations among all units. If attention, however, is directed to a particular neuron, then only the mutual connections to this neuron are adapted, thus gating the input of the neuronal layer by allowing through only those that are positively correlated to it while suppressing others.

6 Physiological Correlates of Streams in Behaving Ferrets

To explore these hypotheses, recordings were made in the auditory cortex of ferrets trained to attend globally to ALT or SYNC two-tone sequences and to detect a transition to a random cloud of tones by licking a waterspout for reward, as illustrated

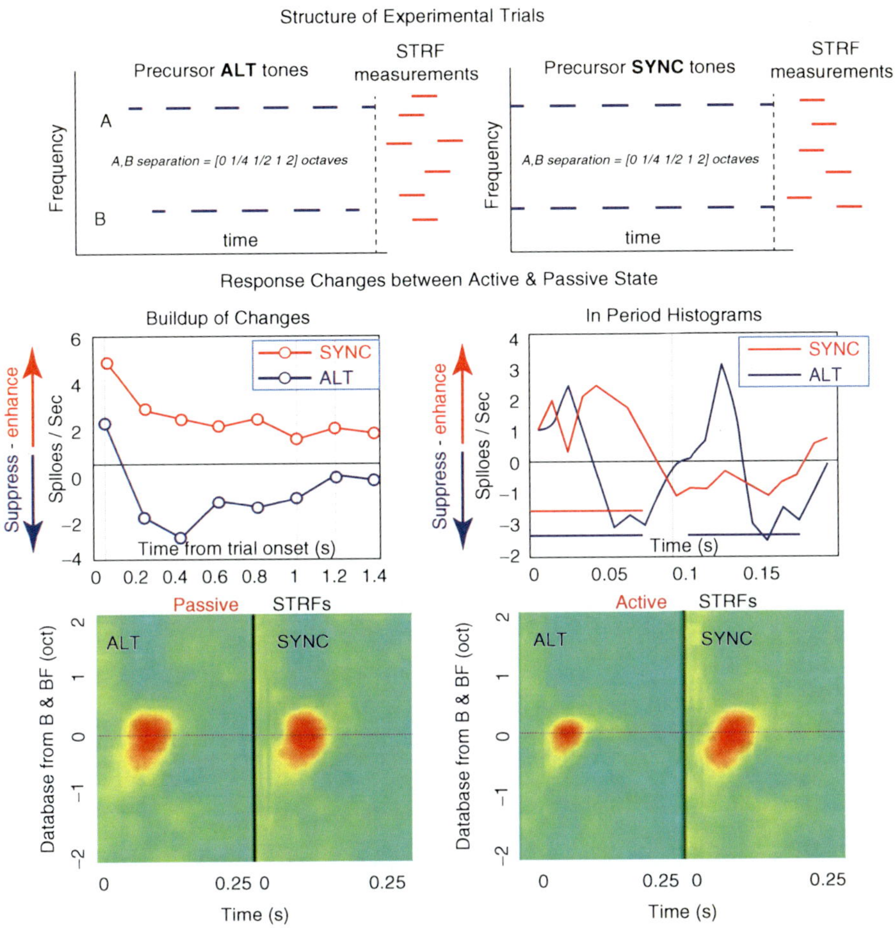

Fig. 59.3 Behavioral neurophysiology. (*Top Panels*) Structure of experimental trials. Ferrets listened to ALT or SYNC tone sequences presented for 1–3 s followed by a cloud of random tones (*red*) used to measure the STRF of the recorded neuron. (*Middle Panels*) Responses change when animals begin to listen attentively and globally to all tone sequences, i.e., not selectively to one tone. The responses become enhanced for the SYNC sequences (*red*) and attenuated for the ALT sequences (*blue*). Response changes (*left panel*) start immediately after onset of the trial but reach a plateau after three to four tone bursts (~0.5 s). Period histograms of responses to the tones (*red* and *blue bars* in *right panel*) reveal that SYNC tone responses (*red*) become significantly enhanced, while those of ALT tones become suppressed (*blue*). (*Bottom Panels*) STRFs measured at the end of tone sequences during the passive state show very little differences (*left panel*). During active attentive listening, STRFs become depressed after ALT compared to SYNC tone sequences (*right panel*)

in Fig. 59.3. The structure of the experimental trials is depicted in the top panels of Fig. 59.3. Responses were measured throughout the tone sequences to examine changes after trial onset as well as in the period histograms. Responses to the final random tone cloud were used to estimate the spectrotemporal receptive fields (STRFs) (deCharms et al. 1998). The type of sequence (ALT or SYNC) and its frequency combinations were randomly interleaved throughout a block of trials. Figure 59.3 (middle and bottom panels) displays results of recordings from 96 cells that were tuned at the frequency of the B tones, with A tone frequencies up to two octaves above and below that of the B tone.

The average responses to the tone sequences changed dramatically when the passive animal began to attend globally to the stimuli. In both SYNC and ALT conditions, average responses adapted rapidly to a steady state by about the third burst period (*left-middle panel*; Fig. 59.3). SYNC responses were significantly enhanced compared to their passive level, whereas ALT responses were suppressed. The changes in period histograms between the active and passive states for the SYNC and ALT stimuli are compared in Fig. 59.3 (*right-middle panel*). The SYNC response increases significantly during behavior; by contrast, the ALT response displays a strong but slightly delayed suppression soon after each tone's onset response.

Finally, the *bottom panels* contrast the STRFs measured after the end of the SYNC and ALT sequences during the passive and active states. When the animal was passive (Fig. 59.3: *left-bottom panel*), the average STRFs were similar. During behavior, however, there was a strong suppression of the STRFs following the ALT sequences. The average STRF was slightly enhanced after the SYNC sequence. These STRF changes persist but gradually weaken over the next few seconds.

7 Discussion

The physiological results are consistent with the postulates of the temporal coherence model. During SYNC sequences, responses become enhanced possibly reflecting mutually positive interactions. The opposite occurs during ALT sequences, where neurons decrease their overall responsiveness and compete as expected from mutually inhibitory interactions. Furthermore, we postulate that if attention had been directed to one of the ALT competing tones, it would have enhanced (to the perceptual foreground) the attended responses at the expense of the competing tone, consistent with previously published experimental results (Yin et al. 2006).

Finally, the temporal coherence model bears a close relationship to the Kalman predictive clustering-based algorithm described in Elhilali and Shamma (2008). This is because the principal eigenvector of the correlation matrix acts as a reduced feature "template" (or "mask" in Fig. 59.2) which combines and extracts the input feature vectors that match it. In the Kalman prediction model, the same matching operation is performed, but the "template" is computed by a classic on-line gradual clustering of the input patterns. Under certain conditions (e.g., input pattern normalization), the two types of algorithms are equivalent and yield similar clusters (Duda and Hart 1973).

References

Bregman A (1990) Auditory scene analysis: the perceptual organization of sound. MIT Press, Cambridge

Carlyon R, Cusack R, Foxton J, Robertson I (2001) Effects of attention and unilateral neglect on auditory stream segregation. J Exp Psychol Hum Percept Perform 27:115–127

Chi T, Ru P, Shamma S (2006) Multiresolution spectrotemporal analysis of complex sounds. J Acoust Soc Am 118:887–906

deCharms R, Blake D, Merzenich M (1998) Optimizing sound features for cortical neurons. Science 280:1439–1443

Duda R, Hart P (1973) Pattern classification and scene analysis. John Wiley and Sons, New York

Elhilali M, Shamma S (2008) A cocktail party with a cortical twist: how cortical mechanisms contribute to sound segregation. J Acoust Soc Am 124:3751–3771

Elhilali M, Ma L, Micheyl C, Oxenham A, Shamma S (2009) Temporal coherence in the perceptual organization and cortical representation of auditory scenes. Neuron 61:317–329

Fritz J, Shamma S, Elhiliali M (2007) Does attention play a role in dynamic receptive field adaptation to changing acoustic salience in A1? Hear Res 229:186–203

Hartmann W, Johnson D (1991) Stream segregation and peripheral channeling. Music Percept 9:155–184

Lyon R, Shamma S (1997) Computational strategies for pitch and timbre. In: Hawkins H, McMullen T, Popper A, Fay R (eds) Auditory computations. Springer, New York, pp 221–270

Ma L (2011) Auditory streaming: behavior, physiology, and modeling. PhD Thesis, Bioengineering Program, University of Maryland, College Park

Shamma S, Elhilali M, Micheyl C (2011) Temporal coherence and attention in auditory scene analysis. Trends Neurosci 34:114–123

Sussman E, Horvát J, Winkler I, Orr M (2007) The role of attention in the formation of auditory streams. Percept Psychophys 69:136–152

Yin P, Ma L, Elhilali M, Fritz J, Shamma S (2006) Primary auditory cortical responses while attending to different streams. In: Kollmeier B, Klump K, Hohmann V, Langemann U, Mauermann M, Uppenkamp S, Verhey J (eds) Hearing – from sensory processing to perception. Springer, New York, pp 257–266

Index

B.C.J. Moore et al. (eds.), *Basic Aspects of Hearing*, Advances in Experimental Medicine and Biology, DOI 10.1007/978-1-4614-1590-9, © Springer Science+Business Media New York 2013

M

N

O

P

R

S

T

V

W

Printed by Publishers' Graphics LLC
KSO130614.15.15.6